ADVANCED CATALYTIC MATERIALS - PHOTOCATALYSIS AND OTHER CURRENT TRENDS

Edited by **Luis Enrique Noreña**
and **Jin-An Wang**

Advanced Catalytic Materials - Photocatalysis and Other Current Trends

http://dx.doi.org/10.5772/60491
Edited by Luis Enrique Noreña and Jin-An Wang

Contributors

Jiatao Zhang, Muwei Ji, Jiajia Liu, Meng Xu, Hesun Zhu, Tifeng Jiao, Jie Hu, Qingrui Zhang, Qijin Chi, Arnab Halder, Minwei Zhang, Fuyi Chen, Adnan Qaseem, Nan Zhang, Agnieszka Wróblewska, Edyta Makuch, Ewelina Mójta, Boris Kharisov, H.V. Rasika Dias, Oxana V. Kharissova, Ubaldo Ortiz Mendez, Idalia Gomez De La Fuente, Alejandro Vazques Dimas, Yolanda Peña Mendez, Jacinto Sa, Jakub Szlachetko, Ying Zhou, Shan Yu, Yuhua Wang, Xinlong Ma, Hao Li, Bin Liu, Huihui Li, Shu Yin, Tsugio Sato, Fariba Dehghani, Bahareh Bahramian, Loic Stefan, Le Li, Minqiang Wang, Hiroaki Tada, A.Neren Ökte, Do Hyun Kim, Arvin Sangalang, Seunghwan Seok, Sandra Loera-Serna, Elba Ortiz

CBS, Edition **2017**

Published by InTech

Janeza Trdine 9, 51000 Rijeka, Croatia

Publishing Process Manager Iva Simcic

Technical Editor InTech DTP Team

Cover Designer InTech Design Team

Additional hard copies can be obtained from orders@intechopen.com

Advanced Catalytic Materials - Photocatalysis and Other Current Trends,
Edited by Luis Enrique Noreña and Jin-An Wang
p. cm.
ISBN 978-953-51-2244-9

INTECH

open science | open minds

Contents

Preface

Preface

Today's chemical industry processes worldwide largely depend on catalytic reactions and the desirable future evolution of this industry toward more selective products, more environmentally friendly products, more energy-efficient processes, a smaller use of hazardous reagents, and a better use of raw materials also largely involves the development of better catalysts and specifically, purposely designed catalytic materials. For instance, the world's petroleum industry (oil refining and petrochemical industries) almost completely rely on catalytic processes. The catalyst market is considered to amount for more than 20 billion USD in 2015 and currently growing at a rate of approximately 4% per year.

Most people know that a catalyst is a facilitator of a chemical reaction, as stated in secondary/medium level education text books; however, the careful study and the development of the new generation catalysts involve considerable or relatively large groups of specialists in universities, research centers, and industries, joining forces from different scientific and technical disciplines. In 1835, Berzelius described a catalyst as a substance capable of accelerating a chemical reaction that can be recovered at the end of the reaction. Catalysts can be divided into two broad general groups: homogeneous catalysts (those in the same physical phase as the reactants, mostly liquids or gases) and heterogeneous catalysts (mostly solids, thereafter also known as catalyst materials). Some of the first solid catalysts were metals o metal compounds, still in use nowadays, whereas many other different types of catalysts have arisen along the years. Platinum, palladium, gold, cobalt, nickel, copper, and iron are some important metal catalysts, either in bulk or supported. Metal oxides, such as alumina, titania, zirconia, magnesia, ceria, or iron oxide, that constitute a principal group of catalytic materials is also the case of microporous and mesoporous molecular sieves: silica, alumina, zeolite, MCM-41, SBA-15, aluminophosphates (ALPOs), silicoaluminophosphates (SAPOs), metal organic frameworks (MOFs), and activated carbon or pillared clays. To complete the picture, we should include a wide range of proteins and enzymes, forming the biocatalysts group.

This book has put together recent state-of-the-art topics on current trends in catalytic materials and consists of 16 chapters. To divide the book's 16 chapters into the sections has not been easy because many of the individual chapters do not deal only with one type of catalyst or with one type of application. Contributions come from around the world: Australia, China, Japan, Korea, Turkey, Denmark, Poland, Switzerland, Sweden, Italy, Spain, Mexico, and the United States. Part 1 is dedicated to powerful characterization techniques based on X-ray spectroscopy: resonant X-ray emission spectroscopy (RXES) and high-energy resolution off-resonant spectroscopy (HEROS). These characterization techniques provide new insights into the electronic structure of materials, molecular adsorption, and understanding catalytic reactions. Part 2 is dedicated to selected metal-based and structured porous catalysts (and their ample range of applications), such as nano-zero-valent-iron (nZVI), iron oxides, nanoferrites, zinc glutarate, metal-organic frameworks (MOFs), titanium-modified SBA-15, graphene oxide–hausmannite (Mn_3O_4) composites, and metal oxides, supported on silica. The largest section of the book is Part 3, comprising seven chapters on photocatalysis, from the reaction basics and a review of materials available to specific new materials and developments. Photocatalytic processes have received much attention because of the potential of using sun light (possibly the more available renewable energy) and new environmentally friendly "green" technologies. Part 4 is focused on materials for electrocatalytic applications (such as sensors and energy conversion and storage): silver–copper nanoalloys and graphene–metal oxide nanohybrid materials, covering the synthesis, structure, and properties of the materials. Final section, Part 5, reviews the extensive field of biotechnology applications of synthetic DNA catalysts.

Finally, the editors, we like to acknowledge the motivation and expertise provided by all the authors, some of them starting a promising career with bright ideas and some of them consolidated in the field with a remarkable background; constructing this book has been a several steps journey and they have maintained committed with the task. However, we also like to acknowledge InTech's editorial staff and their valuable support all along the way.

Dr. Luis Enrique Noreña
Department of Basic Sciences
Universidad Autonoma Metropolitana, Azcapotzalco,
Mexico

Dr. Jin-An Wang
ESIQIE, Instituto Politecnico Nacional,
Zacatenco,
Mexico

Section 1

Specialized Characterization

Chapter 1

X-Ray Spectroscopy — The Driving Force to Understand and Develop Catalysis

Jakub Szlachetko and Jacinto Sá

Additional information is available at the end of the chapter

http://dx.doi.org/10.5772/61940

Abstract

Catalysis is involved in about 90% of manmade chemicals. The development of novel or improved catalysts requires fundamental understanding of the commanding steps of a catalytic reaction. In simple terms, a catalytic transformation depends on the coupling between catalyst electronic structure and reagents' molecular orbitals. Herein, we report a spectroscopic technique capable of determining the electronic structure of metal containing catalysts under working conditions. The technique is called photon-in photon-out X-ray spectroscopy and can be employed to characterize materials, unveil substrate adsorption parameters, and follow changes in electronic structure during catalytic reactions.

Keywords: Photon-in photon-out X-ray spectroscopy, Electronic structure, Spectroscopy under working conditions

1. Introduction

Catalysis is a fascinating field located at the edge of a multitude of disciplines and fields. Spectroscopy dedicated to understanding catalytic phenomenon is among the most active research areas in field. The ability to visualize a catalytic transformation in real time has lured scientists due to the technical challenges and the potential for unprecedented understanding. The importance in accessing fundamental understanding of catalysis was masterfully highlighted in the 2007 report of the US Department of Energy, Basic Energy Sciences Workshop [1]:*[T]o realize the full potential of catalysis for energy applications, scientists must develop a profound understanding of catalytic transformations so that they can design and build effective catalysts with atom-by-atom precision and convert reactants to products with molecular precision. Moreover, they must build tools to make real-time, spatially resolved measurements of operating catalysts. Ultimately, scientists must use these tools to achieve a fundamental understanding of catalytic processes occurring in multiscale, multiphase environments.* However, this is applicable for any kind of catalytic system.

At its most fundamental level, catalysis is directly related to the electronic structure of the valence shells because they control chemical bond formation/rupture [2]. Several spectroscopies can provide information on valence shells but only X-rays do it directly. Hard X-rays are the epitome of spectroscopic probes because of their great penetration depth, and element specificity, which enables studies of catalytic states under working conditions [3]. X-ray photon-in photon-out core level spectroscopy is a powerful tool to understand catalytic reactions because it enables us to map the entire electronic structure of the catalyst under catalytic relevant conditions.

In this book chapter, we summarize our latest efforts in the use of synchrotron-based high-resolution X-ray spectroscopy to study a plethora of academic and industrial relevant catalytic systems. We will start by covering in a succinct manner the developments in the field that allow for the understanding of catalysis in situ/operando and time-resolved, keeping the high-energy resolution and element specificity. This section includes an overview on techniques, spectrometers, and experimental setups. After that we will describe the studies we carried out to understand catalysis and materials. Our work spans over three fundamental aspects of catalysis:

a. Materials characterization

b. Molecular adsorption

c. Study of catalytic reactions under real working conditions (*in situ*).

Vast majority of the studies were carried out on a dispersive von Hamos-type spectrometer [4]; however, when a higher peak-to-background signal level was necessary the experiments were carried out with a Johann-type spectrometer [5]. The studies were carried out both in homogeneous and heterogeneous catalytic systems. The latest includes also studies on photocatalysis.

The future of this spectroscopy concerns its application at the X-ray free electron lasers (XFELs). The advent of XFELs enabled scientist to achieve the goal of producing a movie of a catalytic transformation. XFEL light sources are in its beginning but their effect on the field of time-resolved X-ray science will be deep and comprehensive because reactions can be followed not only in situ but also in real time, this of course if the current limitations in respect to selective triggering and sample stability are overcome. Quoting Sá & Szlachetko: *in a cinematographic analogy, we have the camera (von Hamos Spectrometer), the film (HEROS), the set (catalytic reactor), the script (catalytic reaction) and the actors (catalyst and reagent molecules); what is missing is the director to shout 'action' and direct the scenes (pulse shaping), and that the actors do not fall ill (sample refreshment)* [6]. The insights gathered from the high-resolution X-ray spectroscopy provide deeper understanding of the systems and reactions, which can be used to improve or develop novel catalysts with enhanced properties.

2. Photon-in photon-out X-ray spectroscopy: principles and techniques

Unique properties of X-rays provide wide range of applications to study the catalytic materials, not only in steady-state environment but also more importantly at in situ and material's operating conditions. Ability to probe metal-site of catalytic material enclosed in gas or liquid environment is possible thanks to penetrating properties of X-rays. In the hard X-ray regime of few keV range, the attenuation length of X-rays is in the order of few thousands of micrometres for low Z materials (like carbon) to few micrometres in lead. Therefore, X-ray spectroscopy provides the depth of probe extensively larger as compared, for example, to electron-based techniques. The second important aspect of X-ray application is the ability to probe both, lowest unoccupied and highest occupied electronic states of an atom. By tuning the X-ray energy close to ionization threshold, the interaction of incidence photon with core-electron leads to electron excitation into unoccupied level above Fermi energy. The fine-tuning of incidence X-ray energy, nowadays easily possible at any synchrotron source, allows thus for precise scanning through the available excitation levels, and as a consequence provides electronic picture of above Fermi electronic states. The photon–electron interaction leading to the creation of the excited intermediate atomic state, decays radiatively to the final state by electron transition from higher electronic level into the created core-hole. This decay channel is accompanied by emission of an X-ray and may be used as probe of occupied electronic states. Therefore, the combination of X-ray absorption process and following X-ray emission event allow for obtaining the complete electronic picture, and hence chemical configuration, of absorbing atom.

In the following, we will discuss the latest developments and trends in X-ray spectroscopy as applied to study catalytic systems, ranging from steady-state measurements to real-time-resolved studies. The combination of X-ray absorption (XAS) and X-ray emission spectroscopies (XES) allowing for resonant X-ray emission spectroscopy (RXES) studies will be presented along with examples of technical developments allowing to extend RXES methodology into the time domain. Finally, the high-energy resolution off-resonant spectroscopy (HEROS) will be described in detail. The advantages and disadvantages of HEROS approach in in situ measurements will be illustrated further on with several practical examples.

2.1. Dispersive-type X-ray emission spectrometers

The high-energy resolution resonant and off-resonant X-ray emission spectroscopy relies on measurements of incidence and emission X-ray energies and intensities. In order to obtain meaningful X-ray emission data, the experimental resolution should be of the order of core-hole lifetime of probed atomic species. Typically, the core broadening is in the order of sub-eV up to few eV for K- and L-shells in the X-ray energy range of few keV [7]. The high-energy resolution of incidence X-rays is provided commonly at any synchrotron source by use of double-crystal monochromators. For X-ray detection, dedicated X-ray spectrometers are being developed, with different spectrometer geometries and arrangements depending on particular needs and goals of experiments. In the present section we focus solely on dispersive-type spectrometer because of its particular parameters and operating characteristics.

In comparison to other spectrometer solutions employed at synchrotron beamlines [5, 8] that use two-dimensional focusing, the dispersive-type spectrometer employs only one-dimensional focusing of X-rays. Dispersive-spectrometers are characterized by lower detection efficiency in comparison to, for example, spectrometers working in Johann geometry. On the other hand, the dispersive-spectrometer geometry allows for measurements of X-ray emission spectra in dispersive-mode, which enables detection of X-ray emission in a wide range of energies (few tens to few hundreds of eV) without any scanning elements during measurement. For the experiments requesting short acquisition times, the dispersive-spectrometer geometry may be thus regarded as optimal solution.

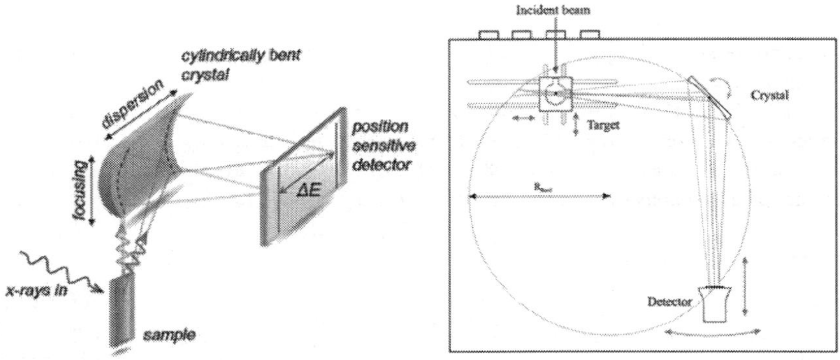

Figure 1. (Left) Schematic representation of von Hamos spectrometer geometry (Reprinted with permission from [4]. Copyright (2012), American Institute of Physics). (Right) Schematic view of the geometrical setup used in Johansson geometry (Reprinted with permission from [9]. Copyright (2012), American Institute of Physics)

There are two common geometries allowing for spectrometer setup in dispersive mode: Johannson-type [10] and von Hamos-type [11]. Schematic representation of von Hamos geometry is shown in Figure 1 (left) [4]. In such arrangement, the X-ray fluorescence from the sample is dispersed on cylindrically bent crystal. The dispersion axis and therefore energy range covered by the setup is limited by the length of the crystal/detector along dispersion axis. One-dimensional bending of the crystal aims at increasing the efficiency of the setup, as compared to flat crystal geometry, by focusing the diffracted X-rays onto the detector plane. The von Hamos setup provides good energy resolution being often below 1eV at relatively large Bragg angles. The Bragg angle domain is changed by linear displacement of the crystal and detector along dispersion axis, where the detector distance from the sample is always twice that of the crystal. Because of linear motions of the crystal/detector, the von Hamos spectrometer allows for flexible arrangements around the sample environment. Moreover, because of application of short curvature radiuses without loss on energy resolution, the spectrometer requires relatively small space for operation. Finally, the spectrometer geometry can be easily extended into multicrystal operation allowing for enhanced spectrometer efficiency or measurements of multiple X-ray emission lines [12].

In Johannson geometry, schematically shown in Figure 1 (right), the X-ray fluorescence from the sample is diffracted by cylindrically bent crystal; however, unlike in von Hamos geometry, the crystal curvature is positioned in dispersive axis. For Johannson geometry, the Bragg angle planes are not parallel to the crystal surface and therefore the energy broadening at lower Bragg angles is eliminated. Thanks to this, a wider Bragg angle range may be applied at high-energy resolution of the spectrometer. As a consequence, only few diffraction crystals are necessary to operate the spectrometer over broad energy range. By operating the Johannson spectrometer in the off-Rowland geometry the X-ray emission spectrum over certain band-width is measured, which is given by the detector length along dispersion axis. The drawback of off-Rowland setup is that only small part of the crystal contributes to the X-ray diffraction at one energy channel; however, this efficiency loss is somehow compensated by recording entire X-ray emission spectrum without scanning components.

Both spectrometers, von Hamos and Johannson types, yield absolute energy resolution significantly below the lifetime of characteristic emission lines, being crucial for detailed analysis of spectral features. The provided dispersive type of detection can be exploited to record time-resolved off-resonant, resonant and nonresonant X-ray emission studies. Because of ability of performing quick acquisitions of X-ray emission spectra, the spectrometers may be applied for in-situ spectroscopic studies of dynamic systems.

2.2. Kβ and valence-to-core RXES

The method of resonant emission X-ray scattering (RXES) is recognized as one of the most sensitive measurement techniques that allows the detection of very small changes in the electronic structure of studied element [13]. The method is based on X-ray scattering process in which the core electron is excited to unoccupied electronic states above Fermi level, and simultaneous detection of X-ray emission accompanied by atomic decay from intermediate to final state. By the monitoring of X-ray intensity and X-ray emission shape analysis, the detailed information on the electronic structure may be retrieved.

As an example, in Figure 2 we show the Kβ-RXES plane for TiO_2 anatase recorded around Ti K-absorption edge [18,21]. In the experiment, the incidence X-ray energy was tuned around 4968–4992eV range allowing probing by 1s electron the lowest unoccupied electronic states. Simultaneously, the Kβ and valence-to-core decay channels were measured by means of dispersive-type von Hamos spectrometer. The relatively large range of spectrometer crystal along dispersive axis, allowed to probe both Kβ and valence-to-core transition at once, so that only the incident beam energy had to be scanned to record the full RXES plane.

The measured RXES map exhibits several features depending on excitation and emission X-ray energy. The first X-ray emission signal appears at incidence X-ray energy of around 4968 eV. This weak pre-edge feature corresponds to the 1s-3d-like excitation and extends up to the excitation energy of 4974eV. The s→d types of electron excitations are quadrupole transitions that are characterized by relatively small excitation probability as compared to dipole excitations, which are commonly examined by RXES. However, as shown with the help of theoretical calculations, because of the strong d→p hybridization in TiO_2 compound, the 1s→3d excitation is characterized by a relatively high excitation probability. RXES measurements using transitions from higher electronic states (i.e., 3p→1s) proved to be extremely sensitive to determine the nature of the lowest unoccupied 3d states of Ti. With the excitation of an electron from a

1s shell to a 3d-localized state, excited electron interacts with the 3p electron, resulting in a shift of 3p→1s emission line by 1.8 eV to lower energies due to the electron–electron interaction effect. This effect can be seen in RXES plane in Figure J1 for lowest excitation energy of 4968eV. However, if the 3d excitation state is delocalized, the electron–electron interaction is negligible and shift of 3p→1s emission is thus not observed for pre-edge features at excitation energies above 4970eV. The Kβ and valence-to-core RXES measurements give unique opportunity to investigate simultaneously lowest unoccupied and highest occupied electronic states by X-ray absorption and X-ray emission spectral projections (see for details [18, 21]). We shall discuss shortly here the possibility of determining the absolute value of the energy band gap as well as possibility to employ RIXS measurements in cases where the commonly used optical methods are insufficient (i.e., in the case of "dark-samples" or samples containing so-called color-centers). The results obtained to date suggest that determination of the absolute band gap energy by Kβ-RIXS may be difficult due to interactions of the core excited electron with 3d valence electrons as well as because of the effects of core-hole shielding. Both processes can induce a small, but not negligible, change to the electron levels before Ti atom decays to its final state. However, the obtained Kβ-RIXS results based on comparison experimental RXES spectra of rutile and anatase TiO_2 structures suggest that the relative change in the band gap energy can be determined and examined to a high precision.

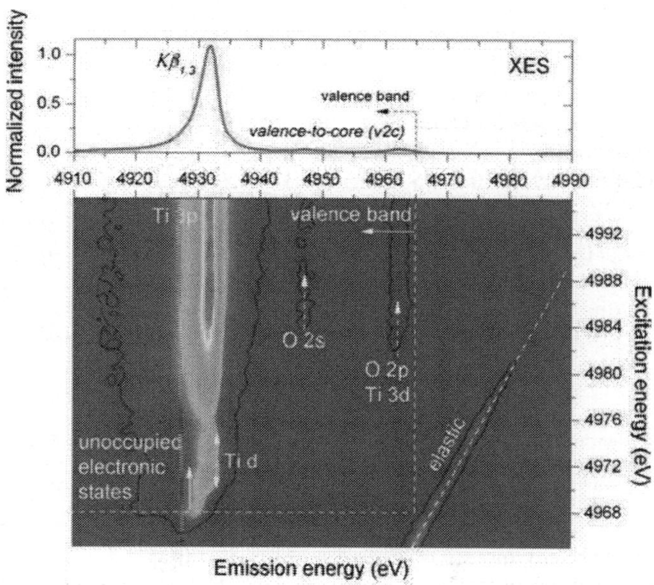

Figure 2. Kβ-RXES plane of TiO_2 anatase measured for Ti around K-absorption edge. The main detected spectroscopic features are labeled on the RXES plane. On top, the nonresonant Kβ and valence-to-core X-ray emission spectrum is shown, which was recorded for incidence X-ray energies above K-shell ionization threshold. (From [18] - Reproduced by permission of The Royal Society of Chemistry)

http://dx.doi.org/10.5772/61940

2.3. Time-resolved RXES

The use of X-ray spectroscopic methods for measuring a high time resolution data is limited by several technical factors. RXES measurements, employing two-dimensional focusing crystals, requires scanning of both incidence and emission X-ray energy. Therefore, the RXES method, despite the exceptional sensitivity of the measurement, could be applied only in experiments in which the sample was kept in a steady-state chemical equilibrium. The use of dispersive-type spectrometer for RXES spectroscopy significantly reduces the measurement time due to the dispersive type of detection of X-rays emitted from the sample. In such experimental configuration, the measurement time is determined only by the speed of the scanning incidence X-ray beam energy, which at present synchrotrons may be in the order of a several seconds [42, 43]. The acquisition time is an important aspect in the study of irreversible processes in which repeating the experiment at the same chemical conditions is very difficult or sometimes even impossible, for example, because of the small amount of available sample.

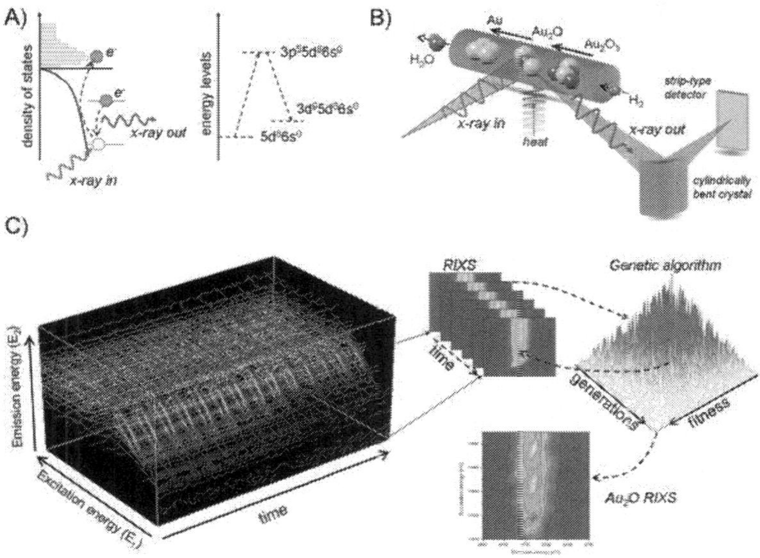

Figure 3. a) Scheme showing RXES process in gold. b) Experimental setup employed for in situ RXES spectroscopy. c) Series of RXES planes measured during the experiment (time-resolved RXES) together with schematic description of data analysis using genetic algorithm. (Reprinted with permission from [43] Copyright (2014) American Chemical Society)

As an example, we briefly discuss the aspect of time resolution in RIXS measurements by employing dispersive spectrometer to study the local electronic structure of Au in the temperature-programmed reduction (TPR) of Au_2O_3 gold oxide. During the measurements, the gold oxide Au_2O_3 was maintained under H_2 atmosphere and at the same time heated continuously at a rate of about 5°C per minute in the range from 20°C to 300°C. Schematic description of the experimental setup is shown in Figure 3, together with evolution of experimental time-resolved RXES spectra. RXES experiment, employing excitation from $2p_{3/2}$ to 5d and following decay transition from $3d_{3/2}$ to $2p_{3/2}$ final state, showed a change in electronic valence configuration of Au from $5d^8 6s^0$ to $5d^{10} 6s^1$, corresponding to gold oxide Au_2O_3 reduction to the metallic form Au^0 at temperatures above 150°C. However, as stressed in [43], thanks to obtained time resolution of RXES acquisition, the transitional forms of Au in the configuration of $5d^{10} 6s^0$ and corresponding to the formation of gold (I) oxide (i.e., Au_2O) were registered. The result was surprising, because the Au_2O compound is unstable and is quickly reducing to a form of Au^0. On the basis of RIXS measurements and by using analysis based on genetic algorithm, it was possible to determine the structure of 5d electronic states, as well as to determine the Au_2O crystallographic structure. The experimental results were examined with density functional theory (DFT) calculations, confirming the two-stage reduction process of Au_2O_3. The application of the RXES technique with high temporal resolution has been the key to obtain the described results. The time-resolved RXES enabled for the first time to investigate the structure of the unstable gold (I) oxide.

2.4. High energy resolution off-resonant spectroscopy

In many synchrotron studies of chemical processes the experiments are artificially slowed down in order to obtain good quality experimental data. The time limit, of the order of several seconds, drawn by the need of scanning the incident X-ray energy, was still a major constraint to study the electronic state of matter under working conditions as well as main restriction in ability to follow individual reaction steps. The idea for solving these technical problems was the application of off-resonant X-ray scattering, a process of photon–atom interaction for excitation energies being set below an absorption threshold [14]. The concept of using off-resonant X-ray scattering process dates back to 1982, when based on the Kramers-Heisenberg theory, J. Tulkki and T. Åberg developed simplified differential equations describing the interaction of photons with the atom [15]. In this theoretical study they found that for the incidence X-ray beam energy tuned below the absorption edge of an atom, the shape of the X-ray emission spectrum (XES) is proportional to the density of unoccupied electronic states, i.e., equivalent to X-ray absorption. However, the potential and possibilities of determining the electronic structure from the X-ray emission spectra at off-resonant conditions has not been extensively studied due to its very low cross section.

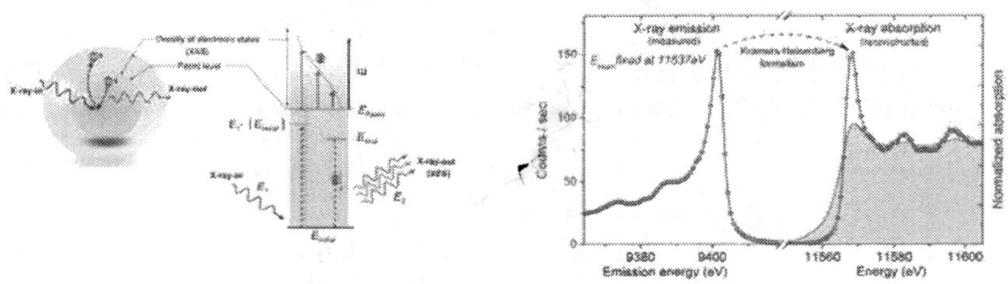

Figure 4. Schematic representation and energy level drawing for an off-resonant scattering process (left). On right we plot HEROS-XES for the $3d_{5/2}$-$2p_{3/2}$ transition of a Pt foil recorded at excitation energy of 11537eV. The reconstructed HEROS-XAS spectrum using Kramers-Heisenberg formalism is also plotted and compared to conventional total fluorescence yield XAS. (Adapted from [61] - Reproduced by permission of The Royal Society of Chemistry)

Correspondence of off-resonant scattering and X-ray absorption processes can be derived starting with simplified cross-section formulas describing photon–atom interaction in vicinity of ionization threshold. Within the Kramers-Heisenberg approach, the differential cross sections for resonant X-ray scattering can be expressed as follows [15] (Eq. 1):

$$\frac{d\sigma(\omega_1)}{d\omega_2} = 2\pi r_0^2 \int_{\omega_1}^{\omega_2} \frac{\dfrac{(\omega_i - \omega_f) g_{fi} (\omega_i - \omega) dg_i}{d\omega}}{(\omega_i + \omega - \omega_1)^2 + \dfrac{\Gamma_i^2}{4\hbar^2}} \frac{\dfrac{\Gamma_f}{2\hbar}}{(\omega_1 - \omega_f - \omega - \omega_2) + \dfrac{\Gamma_f^2}{4\hbar^2}} d\omega$$

where r_0^2 is the classical electron radius. The energies of the initial and final states are represented by $\hbar\omega_i$ and $\hbar\omega_f$, whereas $\hbar\omega_1$ and $\hbar\omega_2$ are the incoming and outgoing photon energies, respectively. The initial and final state broadenings are given by Γ_i and Γ_f. The g_{fi} stands for the oscillator strength of the X-ray transition from the final to initial vacancy state, and the $dg_i/d\omega$ represents the oscillator strength distribution for electron excitation.

The second term in Eq. 1 ensures the energy conservation given by $\omega=\omega_1-\omega_f-\omega_2$ and accounts for the final state broadening (Γ_f). By neglecting the final state width, the second term of Eq. 1 can be replaced by the Dirac delta function (Eq. 2):

$$\frac{d\sigma(\omega_1)}{d\omega_2} = 2\pi r_0^2 \int_{\omega_1}^{\omega_2} \frac{(\omega_i - \omega_f) g_{fi} (\omega_i - \omega) dg_i / d\omega}{(\omega_i + \omega - \omega_1)^2 + \Gamma_i^2 / 4\hbar^2} \delta(\omega_1 - \omega_f - \omega - \omega_2) d\omega$$

For the off-resonant excitations the oscillator strength distribution ($dg_i/d\omega$) is directly proportional to the X-ray absorption spectrum (XAS). Therefore, the shape of the X-ray emission spectrum (XES), which is proportional to the differential cross sections, can be described as follows (Eq. 3):

$$XES(E_2) = \int_{E_1}^{E_2} \frac{(E_i - E_f)(E_i - E) XAS(E)}{(E_i + E - E_1)^2 + \Gamma_i^2 / 4} \delta(E_1 - E_f - E - E_2) dE$$

the above equation the constants were omitted and frequencies replaced by E according to $\omega=E/\hbar$. By keeping E_1-E_f-E- E_2=0, Eq. 3 can be represented analytically for XAS(E) solution (Eq. 4):

$$XAS(E) = \frac{E_1}{E_2} \frac{(E_i - E_f - E_2)^2 + \Gamma_i^2 / 4}{(E_i - E_f)(E_i - E_f + E_1 - E_2)} XES(E_2)$$

This simplified formula provides the XAS(E) function that can be analytically solved for any measured XES(E_2). We should note that Eq. 4 is valid only at condition where $E_1 \ll E_i$, i.e., the measured off-resonant spectrum is free of any resonant features. Typically, the detuning of excitation energy should be as large as 5 x Γ_i. The final state broadening (Γ_f), the incoming beam energy distribution, and the resolution of the X-ray spectrometer are not considered in this equation. Therefore, the derived HEROS-XAS curves will be broadened by these three contributions.

By combining the aspects of off-resonant scattering, high energy and dispersion-type of detection to the measurements of unoccupied electronic states a high energy resolution off-resonant spectroscopy has been established. The proposed method does not require any scanning optical elements during the measurement, and therefore allows for X-ray spectroscopy measurements at timescales unattainable to XAS or RXES techniques [61, 64, 65, 16]. Figure 4 demonstrates the principle of HEROS. The Lα_1 X-ray emission ($3d_{5/2}$ - $2p_{3/2}$ transition) of a Pt foil was recorded at a fixed incident X-ray beam energy tuned below the L3 absorption edge. The off-resonant X-ray emission was recorded by means of von Hamos spectrometer that allowed to cover over 60eV energy range at fixed optical arrangement. The corresponding HEROS-XAS spectrum was reconstructed using the measured X-ray emission spectrum and Eqs. 1–4. The HEROS-XAS spectrum (open circles) is compared to the X-ray absorption spectrum (filled orange area) that was recorded by means of total fluorescence yield (TFY). We should stress here, that the derived HEROS-XAS spectrum exhibited more detailed information than the conventional XAS spectrum, due to the removal of the initial state broadening. The result implies that the HEROS-XES exhibits the same information as high-energy resolution XAS. Moreover, the HEROS experiment is performed at a fixed optical arrangement, meaning acquisition time resolution is simply controlled by experimental efficiency, not by the speed of scanning the incident energy axis. Finally, as recently demonstrated, the HEROS spectra are free of self-absorption effects [63]. Self-absorption belongs to one of the phenomena disturbing the absorption measurements [17] and leads to a modification of the shape of the measured absorption spectrum, visible in particular in the case of samples having high concentrations of the studied element. There are several methods, both computational and experimental, that allow minimizing the effects of self-absorption in the experimental spectra, but until now there was no spectroscopic method allowing direct measurement of the XAS spectra that does not contain the effect of self-absorption. No effect of self-absorption in the HEROS measurement is explained by a fixed experimental geometry (i.e., no scanning optical elements during measurements), allowing for the acquisition of spectra at fixed both incident beam and emission energies. Therefore, the HEROS method could be used not only for measurements requiring a high time resolution, but also in the experiments, where the exact knowledge of the electronic structure is essential for the chemical speciation or theoretical calculations.

3. Case studies

3.1. Materials' characterization

Photocatalytic properties are intimately related to the electronic structure of the employed semiconductors. They determine materials' optical properties and to a large extent surface reactivity. RXES measurements can determine the electronic structure of conduction band

http://dx.doi.org/10.5772/61940

(XAS) and valence band (v2c-XES). To demonstrate RXES abilities, we determined the electronic structure of N-doped TiO_2 [18]. TiO_2 is the most used photocatalyst, and N-doping is often used to change its band gap in order for the material to absorb in the visible range.

In the case of photocatalysts, and more specifically on N-doped TiO_2, the fundamental aspects for a high-performing material are: decent overlap between N and O orbitals (synergy effect), and dense and wide orbital levels in the conduction band (charge mobility). The first aspect defines how well-mixed the doped system is, a key parameter toward achieving the desired synergetic effect and visible light absorption. In the worst case scenario, the resultant material is composed of a solid solution of two different materials, namely, TiO_2 and TiN. The second aspect refers to charge mobility, which in the case of TiO_2 like any other semiconductor occurs via charge particles carriers trapping and detrapping mechanism [19]. Therefore, charge mobility depends on the number of available orbital levels (sub-bands) in the conduction and valence band and spacing between them.

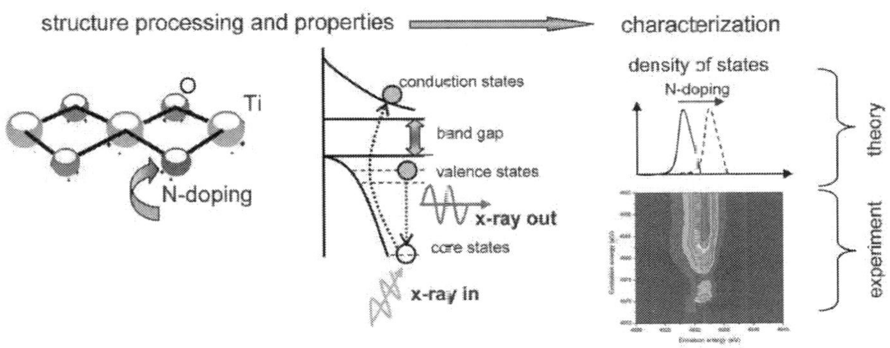

Figure 5. Schematic representation of combined theory and experimental approach for the design of improved doped semiconductors. (Reproduced from elsewhere [18] with permission)

By combining RXES measured at Ti K-edge and FEFF calculations [20], we were able to determine the electronic structure of conduction band (XAS) and valence band (valence-to-core X-ray emission spectroscopy (v2c-XES). Figure 5 represents schematically the procedure to attain information about semiconductor electronic structure.

Figure 6 shows the effect of substitutional N-doping of TiO_2. Starting with pristine TiO_2, the XAS (conduction band) is constituted primarily of Ti-d empty states. The two peaks in the spectrum relates to the octahedral crystal field separation of the d-orbitals. The lowest energy peak is associated to the t_{2g} orbitals and the highest to e_g orbitals. The valence band (v2c-XES) is composed of O-2p and occupied Ti-d states. The valence edge is composed entirely of oxygen orbitals. However, the states are well-hybridized, thus one cannot state that electrons excited from valence edge are from oxygen. Rather, one can state that they were located in the oxygen orbitals at the time of the excitation.

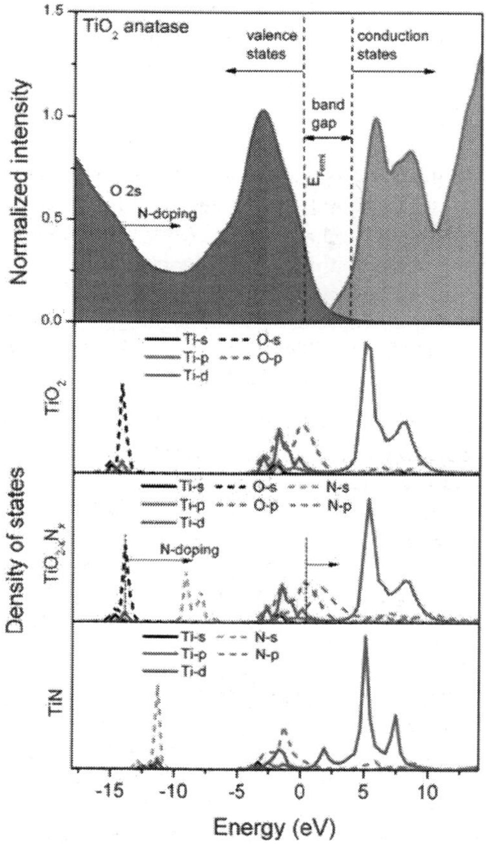

Figure 6. Electronic band structure of TiO_2 and TiN. (*Top*) Valence and conduction band electronic states extracted from measured RIXS plane. (*Below*) Calculated Ti, O, and N DOS for TiO_2, TiN, and $TiO_{2-x}N_x$, where x amounts to 2% N-dopant level. (Reproduced from elsewhere [18] with permission)

The addition of N led to a shift of the valence edge to higher energy, i.e., reduction of semiconductor band gap. The shift is due to the appearance of N-2p orbitals that are strongly hybridized with O-2p orbitals. However, too much N-doping results in the formation of TiN and consequent coloring in the sample, but according to the electronics of the materials, TiN could not work as a sensitizer. Plus, it is known that TiN has no catalytic activity and thus might decrease activity by blocking and/or reducing surface sites.

By confirming RXES measurements with FEFF calculations, we derived an elegant and cost-effective strategy for the rational design of novel materials used in the conversion of solar energy into chemical bonds. Together, one is able to map the material electronic structure with

http://dx.doi.org/10.5772/61940

high precision, requiring a very limited amount of sample (experimental work) or none (theory). The technique can be applied to other materials in all the forms they are normally found, namely nanopowders, and thin-films forms [21].

3.2. Molecular adsorption

Reagents' adsorption is the initial step of any catalytic reaction, and the interaction between reactants and active site is governed by the electronic structure of the catalyst, which determines adsorption strength and geometry [2]. Thus, modification of the active site electronic structure alters reactant adsorption parameters. It is therefore crucial to attain information about adsorption parameters, preferentially under relevant conditions. Surface science studies and theoretical calculations provided over the years a platform to acquire information about these parameters under model conditions, and idealized surfaces. However, this valuable information is difficult to correlate with reactions carried out in liquid phase and/or pressure.

Catalytic hydrogenations with molecular H_2 are unquestionably the workhorse of catalytic organic synthesis. The synthetic capacity of catalytic hydrogenation is superbly condensed in Rylander's following quotation [22]: *Catalytic hydrogenation is one of the most useful and versatile tools available to the organic chemist. The scope of the reaction is very broad; most functional groups can be made to undergo reduction, frequently in high yield, to any of several products. Multifunctional molecules can often be reduced selectively at any of several functions. A high degree of stereochemical control is possible with considerable predictability, and products free of contaminating reagents are obtained easily.* Most of the above comments apply to heterogeneous catalytic hydrogenations over supported Group VIII metals [23].

Alternatively to molecular hydrogen, hydrogen donors such as isopropanol or formic acid can be applied in transfer hydrogenations but they generate unwanted side products. Several important inventions have been accomplished in the last 150 years, such as the application of highly dispersed metals, e.g., nickel, in the hydrogenation of organic compounds [24], selective semi-hydrogenation of C≡C-bonds in the presence of Pd-Pb/$CaCO_3$ catalysts (Lindlar catalyst) [25], and more recently asymmetric hydrogenations [26], pioneered by Knowles and Noyori for which they received in 2001 the Nobel Prize in Chemistry. Catalytic hydrogenations can be carried out in a variety of ways, namely liquid or gas phase, and in batch-wise or continuous mode.

Hydrogenation reactions involving hydrogen dissociation on the metal surface have high reaction probability on many surface sites: top, bridge, and step sites. For this reason, hydrogenation reactions are usually structure-insensitive in respect to hydrogen activation. However, the reaction is considered structure-sensitive in respect to the hydrocarbon hydrogenation. Crespo-Quesada et al. [27] reported that semi-hydrogenation of alkynes occurred preferentially on terraces, whereas further hydrogenation to the alkane occurred at the edges. Similarly, Schmidt et al. [28] revealed that in ethyl pyruvate enantioselective hydrogenation the Pt(111)/Pt(100) ratio controlled reaction rate and enantiomeric excess.

Platinum group metals are highly active hydrogenation catalysts, operating at low temperatures and H_2 pressures. However, their scarcity and high cost intensified the research devoted in finding alternative catalysts deprived of noble metals. One should be made aware that fine chemical industries, such as the ones responsible for the production of vitamins and fragrance, afford products of medium added value. Accordingly, the use of noble metals and/or sophisticated ligands is only justified when the turnover number is higher than 1000 or even 10,000 depending on the cost of the target product [29]. Nonprecious metal catalysts, especially those based on nickel (e.g., Raney nickel [30], Urushibara nickel [31], CENTOPRIME [32] catalysts) have also been developed as economical alternatives, but they are often less active, requiring higher temperature and H_2 pressure to attain comparable performance.

The chemoselective hydrogenation of C=C and C=O-bonds is extensively used in the preparation of pharmaceuticals, fragrances, and vitamins precursors. Common substrates are unsaturated ketones, aldehydes, or esters, which have to be hydrogenated selectively either at the C=C or C=O bond depending on their end-use. Palladium and platinum are routinely used to hydrogenate C=C bonds [33]; however, the systems lose their effectiveness when the molecule contains several hydrogenable functionalities. Chemoselective hydrogenation of unsaturated aldehyde, ketones, and esters has been catalyzed over a plethora of 3d transition metal catalysts [34]. For example, Ni and Cu catalysts were found both active and selective in the hydrogenation of C=O instead of C=C [35].

As aforementioned, platinum is routinely used in the chemoselective hydrogenation of C=C and C=O-bonds. Recently, Manyar et al. [36] demonstrated that Pt supported on OMS-2 possesses different affinities in the chemoselective hydrogenation of C=C and C=O-bonds depending on the substrate used. 5 wt% Pt/OMS-2 (Pt average particle diameter 2 nm) supported on cryptomelane manganese oxide octahedral molecular sieve provides high selectivity for the hydrogenation of both ketoisophorone and cinnamaldehyde. However, in the case of ketoisophorone hydrogenation, 98% of the substrate was hydrogenated to (6R)-2,2,6-trimethylcyclohexane-1,4-dione (levodione) at 100% conversion, i.e., the catalyst selective reduced the C=C bond. Whereas in the case of cinnamaldehyde, 80% selectivity for the reduction of the C=O bond forming cinnamyl alcohol at 100% conversion was found using the same catalyst. The observed selectivity in the ketoisophorone hydrogenation contrasts with the expected Pt behavior, which is commonly preferentially to hydrogenate C=O versus C=C bonds. The expected selectivity was observed when the reaction was performed over 5 wt% Pt/Al_2O_3 under the same reaction conditions. A possible explanation for the differences is different adsorption geometry/strength of the substrates.

The change in the Pt electronic structure following the adsorption of an α, β-unsaturated aldehyde and ketone was followed by in situ HR-XAS in the liquid phase and the adsorption strength calculated with Density Functional Theory (DFT) [37]. Probably one of the most important outcomes of the experiment is that adsorption of a molecule on Pt surface led to an immediate change in XAS single, i.e., Pt electronic structure. The perceptual change relates to molecule coverage, which in this case is below 10% (Fig. 7a). All the molecules produced a Fermi level shift to higher energies, which suggests electronic density donation from Pt to the adsorbed molecule. Larger energy shifts equates to stronger interactions, and consequently higher adsorption energy. Thus, from the Fermi energy level shifts measured by Pt L_3-edge HR-XAS, one might interject that cinnamaldehyde has the strongest interaction with Pt followed by H_2 and then ketoisophorone (Fig. 7a).

http://dx.doi.org/10.5772/61940

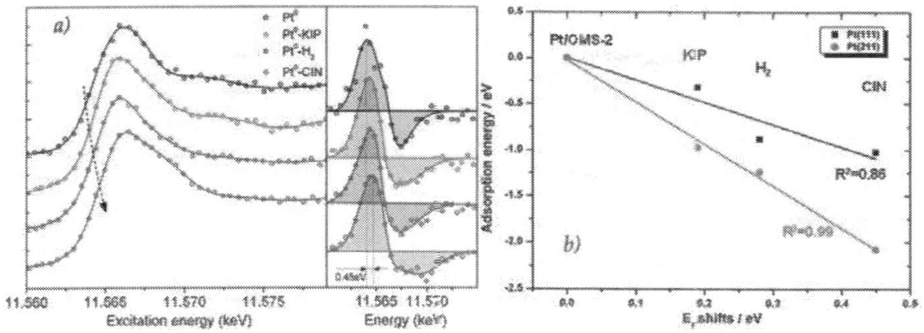

Figure 7. (a) Pt L_3-edge HR-XAS spectra and 1st derivative (right insert) of Pt/OMS-2 interacting with H_2, KIP (ketoisophorone), and CIN (cinnamaldehyde). The spectra were recorded in situ in 10 mol% methanol in water at 373 K. The lines are from a least squares fitting procedure. (b) Correlation between the experimentally measure shifts in the HR-XAS spectra (E_f) and the calculated adsorption energies via DFT for the Pt(111) (black) and Pt(211) (red) surfaces. (Reproduced from elsewhere [37] with permission)

The resulting shift in the Pt Fermi energy, measured at Pt L_3-edge, due to adsorption of molecules was found to be in good agreement with the molecule adsorption energy trends calculated by DFT, which provided valuable insight into the reaction selectivity (Fig. 7b). It should be mentioned that the experiments were carried out at 373 K and 10 bar pressure, i.e., catalyst working conditions. This was possible due to the development of a homemade cell comprised of a stainless steel autoclave reactor with window comprising of a polyether ether ketone (PEEK) insert [38]. The work confirmed that the combination of state-of-the-art spectroscopy (HR-XAS) and theoretical calculations is a powerful and versatile tool to reveal differences in adsorption behavior for reactants in the liquid phase under reaction conditions, with unprecedented resolution and sensitivity.

Carbon monoxide is often used as chemical probe in spectroscopy because it is very sensitive to the electronic structure of materials. Furthermore, CO is a key reagent in several reactions such as Fischer-Tropsch, PROX, and catalytic converter catalysis. RXES can accurately resolve the occupied and unoccupied d-density-of-states (DOS) of Pt after the adsorption of molecules, including CO [39], under reaction conditions. RXES was found capable of differentiating between the possible adsorption geometries of CO adsorbed on Pt, namely atop, bridged, and faced-bridging [40], making it the ideal tool to determine the changes in adsorption geometry caused by an external effect.

RXES measured at Pt L_3-edge was used to determine the changes in CO oxidation over platinum sites induced by a magnetic field [41]. CO molecules adsorbed on nonmagnetic Pt nanoparticles supported on a carbon capped Co magnetic nanocore on atop position in the absence of a magnetic field, as expected. Under magnetic field, part of the atop CO changed to the bridged position (Fig. 8), which is a nonreactive state. This result indicates that catalytic activity can be modified by the presence of a magnetic field even if the active site is not magnetic. This opens other potential applications such as molecular displacement on auto-

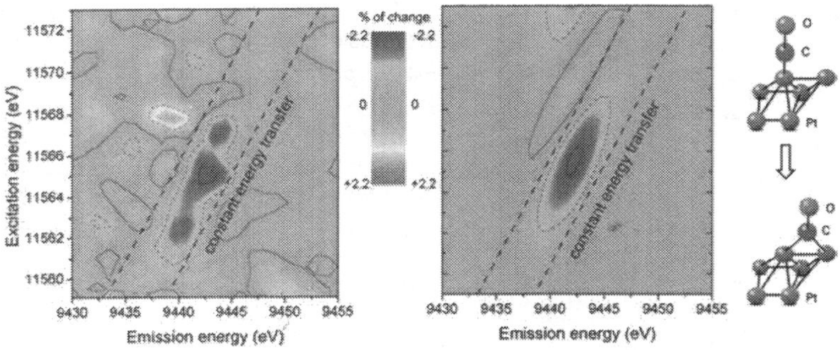

Figure 8. Pt L_3M_5 Δ-RXES due to the presence of a 50 mT magnetic field on Pt on Co with adsorbed CO (field OFF–field ON). (*Left*) Experimental map differences measured in situ. (*Right*) Calculated map differences. (Reproduced from elsewhere [41] with permission)

poisoning reactions, characterized by poisoning of reagents, products, and/or trace contaminates that cannot be removed or avoided. For example, hydrogen fuel cells are vulnerable to CO poisoning because CO binds strongly to platinum averting hydrogen adsorption, deterring commercial application. By changing adsorption from atop to bridge, CO adsorption strength decreases by ca. 0.1 eV, enabling its removal at lower temperature.

3.3. Study of catalytic reactions under real working conditions (in situ)

Determination of metal oxidation states under relevant working conditions is crucial to understanding catalytic behavior. Temperature-programmed reduction (TPR) is often used to determine the temperature at which the catalyst reduces; however, it fails to determine the oxidation state of the catalyst at this specific temperature. That information is normally attained when a spectroscopic measurement is coupled with the TPR. The reduction behavior of NiO [42] and Au_2O_3 [43] was determined using a novel approach, namely time-resolved RXES or 4D RXES. The experimental concept of such experiment is depicted in Figure 9.

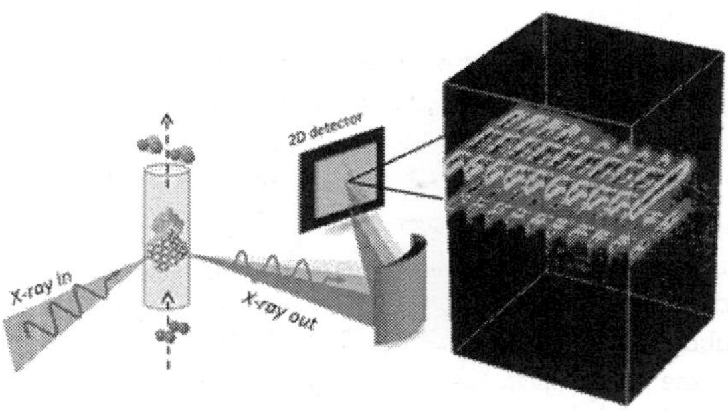

Figure 9. Experimental concept for time-resolved RXES measurements. (Reproduced from elsewhere [42] with permission)

Figure 10 shows the RXES map before and after reduction of nano-NiO. A shift in the main signal to lower incidence energies after reduction was observed, coherent with the formation of metallic nickel. From the RXES map one can excerpt information about unoccupied states (XAS analysis) and occupied states (XES analysis). In order to determine reaction pathway, the XAS absorption contribution of NiO and Ni metal was plotted as function of time. The transition from oxide to metallic state was found to be very fast, consisting of single-step reduction mechanism.

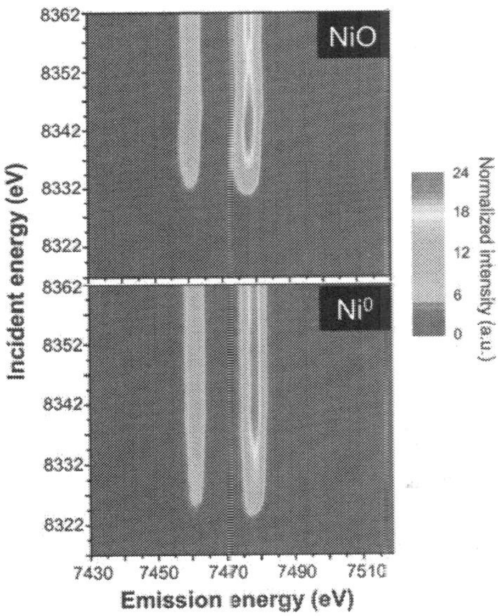

Figure 10. RXES map of the initial state (top) and the final state (bottom) of NiO nanoparticle reduction. (Reproduced from elsewhere [42] with permission)

Similar analysis can be performed with XES signals that carry the inherent benefit, when performed in a dispersive-type spectrometer, of high time resolution inaccessible to XAS technique. However, the chemical sensitivity of XES depends on where the analysis is performed. Since this is often unknown a priory, one needs to measure the full RXES map in order to determine the optimal incidence energy to perform XES analysis. The subsequent measurements can be performed only at this range, which decreases drastically the acquisition time, i.e., improved time resolution without losing chemical information. Based on RXES maps, emission analysis above the absorption edge, so-called non-resonant XES or simply XES, is less sensitive to follow the chemical state of atom as compared to resonant XES measured at incoming energies with high discriminating power, in this case at 8325.9 eV (NiO pre-edge) and 8331.9 eV (NiO inflection point) incident beam energy (Fig. 11). This illustrates the importance of knowing at which energy the XES data should be collected.

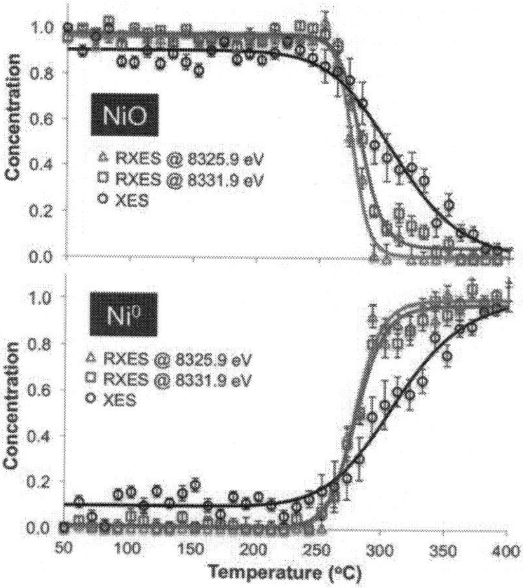

Figure 11. Temporal evolution of XES signals for NiO (top) and Ni^0 (bottom) at different incident energies. (Reproduced from elsewhere [42] with permission)

Gold catalysis gained significant popularity after Haruta's discovery that gold nanoparticles are extremely active [44]. Activity of gold is often associated to its metallic phase but some cases consider the involvement of gold in higher oxidations states [45]. The difficulty in detecting gold in oxidations states between 0 and +3 resides in the fact that these intermediate states are unstable and short-lived.

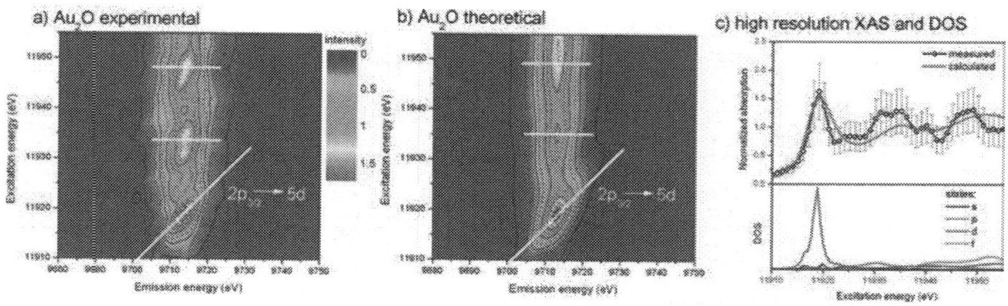

Figure 12. (a) Extracted RXES plane of Au_2O from the time-resolved RXES data set compare d to theoretical calculations (b). (c) Assessment between the extracted high-energy resolution X-ray absorption (HR-XAS) spectrum and the matching theoretical predictions. The influence of the DOS of the different orbitals to the X-ray absorption spectrum is plotted in the bottom panel. (Reproduced from elsewhere [46] with permission)

We performed Au_2O_3 TPR coupled with time-resolved RXES to see if one could detected some of these intermediate states. A short-lived Au_2O compound was detected for the first time under in situ conditions, permitting a better understanding of the reaction mechanism of Au_2O_3 reduction [46]. Based on time-resolved RXES data analysis combined with genetic algorithm methodology, we were able to determine the electronic and geometric structure of the unstable Au_2O transitional specie (Fig. 12). The data analysis revealed a larger value for the lattice constant of the intermediary Au_2O specie as compared to the theoretical predictions. DFT calculations revealed that such structure may indeed be formed, and the expanded lattice constant is justified by the termination of Au_2O on the Au_2O_3 structure. The temporal evolution of the species shows the characteristic behavior of a short-lived intermediate state, in this case Au_2O (Fig. 13).

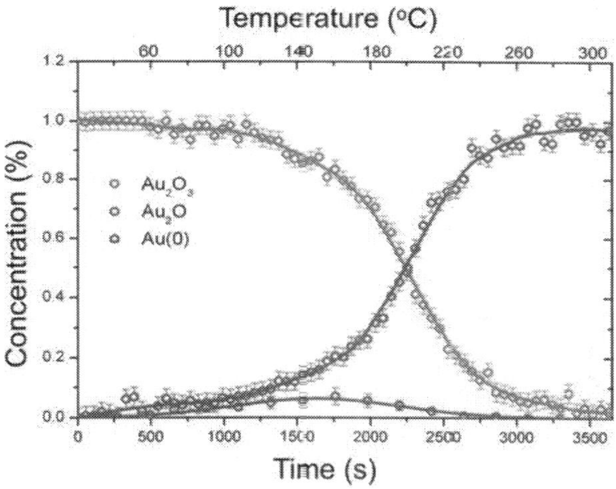

Figure 13. Concentration changes of Au_2O_3, Au_2O, and Au(0) during TPR of Au_2O_3. The data are plotted versus time (bottom scale) and temperature (top scale). (Reproduced from elsewhere [46] with permission)

Excessive amounts of oxidants, in particular H_2O_2, can injure proteins and lipids resulting in cell death. Oxidative stress has been connected to a multitude of pathophysiological conditions, such as Alzheimer's and Parkinson's diseases, aging, cancer, genetic damage, and tissue damage in cardiac ischemic/reperfusion injury. In nature, enzymes called catalases conduct the catalytic disproportionation of H_2O_2 to water and molecular oxygen.

Nitrogen-containing ligand such as imidazoles, is a class of single-site Mn catalase-like complexes. Imidazoles and their derivatives with strong π-donating ability were found to accelerate H_2O_2 disproportionation. The complexes display elevated activity in organic solvents but exceptionally small reactivity or even none in water under physiological pH values. The activity improved when a base was added as a promoter. We reported a novel manganese (II) complex with a pyridine substitute as a ligand (Fig. 14) [47]. The complex was stable and showed catalase-like activity in neutral aqueous solution.

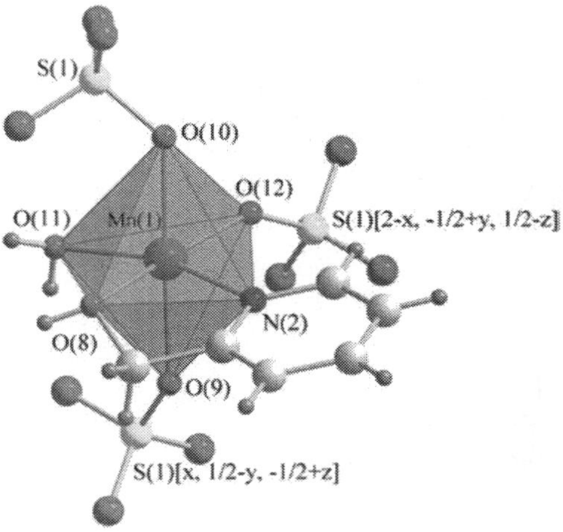

Figure 14. DIAMOND *diagram* showing the coordination environment of Mn(II) in $[Mn(2\text{-}CH_2OHpy)(SO_4)(H_2O)]_n$. (Reproduced from elsewhere [47] with permission)

The combination of RXES measured at Mn K-edge and theoretical calculations enables us to propose a reaction mechanism (Fig. 15) in which the manganese complex is firstly oxidized due to the loss of a bridge oxygen proton. This is followed by H_2O_2 coordination and proton abstraction by the sulfate group mediated by water. Release of molecular oxygen and catalyst regeneration entails the involvement of a second complex molecule. This affects catalytic reaction rate because they become diffusion-limited. The projected mechanism was the first effort in trying to understand single-site Mn complexes reactivity, and clearly more work needs to be done to establish and understand Mn-catalase mimics' reaction mechanism. Nonetheless, based on findings two novel Mn complexes were synthetized, displaying catalytic activities several orders of magnitude higher than the parent one [48].

The increase in the planet's human population put strong pressure on the food supply chain and energy requirements. Photocatalysis is at the forefront of technologies to produce solar fuels. The first significant breakthrough in converting light into chemical energy was published in 1972 by Fujishima and Honda, who reported the electrochemical photolysis of water assisted by a semiconductor under UV-A radiation [49]. However, despite all the scientific advances, it remains a significant challenge to construct a device capable of producing solar fuels, such as hydrogen, at a scale and cost able to compete with fossil fuels. Moreover, developments based on large band gap semiconductors' (>3 eV), such as TiO_2, are hampered by the fact that it requires UV-A irradiation to induce charge separation. Nevertheless, TiO_2 remains indisputably the best photocatalyst to date [50]. Several efforts are being made to improve sunlight

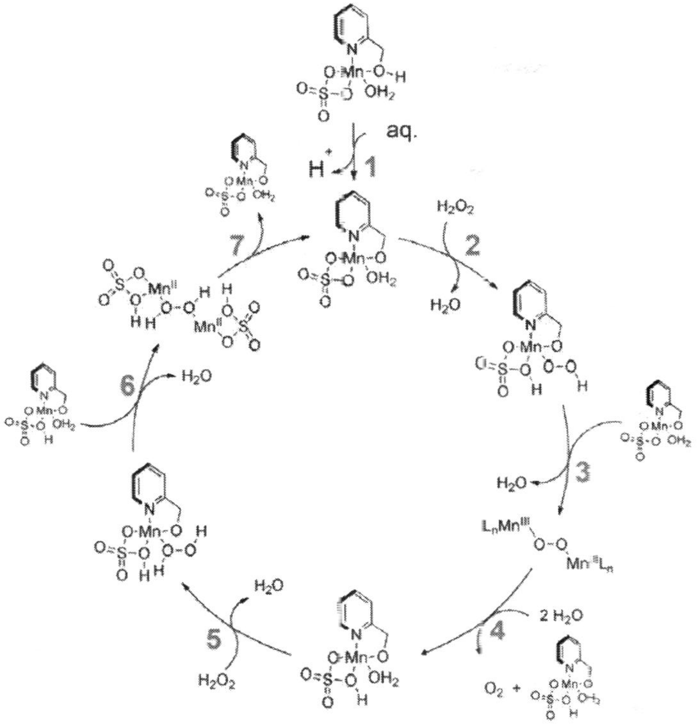

Figure 15. Proposed reaction mechanism for H_2O_2 disproportionation over $[Mn(2\text{-}CH_2OHpy)(SO_4)(H_2O)]_n$ in water. (Reproduced from elsewhere [47] with permission)

absorption, including band gap manipulation by doping with elements, such as N, C, and S [51]. However, these are often counterproductive because they decrease the breadth of reactions that can be photocatalyzed and overall performance.

A more promising strategy to circumvent TiO_2 deficiency in absorbing visible light is to use sensitizers capable of harvesting solar light and introducing hot electrons into the TiO_2 conduction band (CB). O'Regan and Grätzel [52] developed dye-sensitized solar cells (DSSCs), in which the dye component harvests sunlight and readily injects electrons ($\tau_{inj} < 1$ ps) into the TiO_2 CB. The success of the dye system depends on fast electron injection and slow back electron transfer. However, organic dyes have low stability, discrete absorption levels (narrow-band), and small optical cross sections, thus requiring a high dye coverage, which diminishes the space available for photocatalytic reactions. Besides, holes in oxidized dyes are unreactive, which hinders hydrogen evolution since a sustainable process demands for equal consumption of electrons and holes.

Metallic nanoparticles (NPs) are interesting sensitizer candidates because of their large optical cross sections related to the excitation of localized surface plasmons (LSP). Gold group metals exhibit plasmonic resonances in the visible region, and their absorption can be easily tuned by changing nanoparticles morphology (shape and size), enabling a good match with the solar spectrum. Moreover, their d^{10} configuration bestows them chemical stability. Recently, the excitation of Au and Ag LSP nanostructures was shown to improve charge transfer from sensitizer to semiconductor [53], increase the photocurrents under solar irradiation [54], and enhance photoinitiated catalytic oxidations [55]. The latest corroborates that the holes in these structures are reactive and can take part of photo-oxidations.

Hallett-Tapley et al. [56] defended that the plasmon excitation could drive reactions via thermal, electronic, and/or antenna processes. In the case of photocatalysis, the relevant process is the electronic one, in which electrons and holes are allegedly formed upon plasmon excitation [57]. In theory, plasmonic structures can be used directly in photocatalysts; however, the electron–hole pair is short-lived (few femto-second), making it problematic to drive chemical reactions. Interfacial reactions are relatively slow with kinetics in the milliseconds range. To increase charge separation lifetime, the charges can be confined to spatially separated sites where reactions take place. For example, charge separation lifetime can be dramatically improved by coupling LSP structures with a semiconductor (analogous to DSSCs). Mubeen et al. [58] prepared a proof-of-concept system conglomerating the findings on plasmonic nanostructures, water reduction, and oxidation catalysts to produce simultaneously H_2 and O_2 under visible light irradiation. This pioneering work demonstrated system's potential but highlighted the need for improvement.

Despite plasmonic nanostructures' potential, until recently there was no spectroscopic evidence that hot carriers are formed after photon absorption. Since this is essential for the utilization of these structures in photocatalysis, we decided to investigate hot carriers formation on plasmonic structures during illumination. The creation of electron–hole pairs due to LSP excitation was measured by high-resolution X-ray absorption spectroscopy at the Au L_3-edge, which determines Au 5d unoccupied electronic states. LSP excitation led to an upward shift of the ionization energy threshold by ca. 1.0 eV, and an increase of Au d-band hole population, consistent with hot electrons formation, and their promotion to high-energy states [59] (Fig. 16).

To evaluate if the hot electrons possess sufficient energy to be injected into TiO_2 CB, we carried out transient broadband mid-IR spectroscopy [19]. Free and trapped electrons in a semiconductor conduction band produce a broad mid-IR band. Upon excitation, a broad mid-IR band appeared, confirming the presence of electrons in the TiO_2 CB. The minimum in transmittance was observed at t=0 ps for both systems, which is the best possible overlap between pump and probe pulses, thus confirming fast injection into the TiO_2 and consequently increase in electron lifetime from few fs to 100s of ns. The result corroborates Furube et al. [60] speculated mechanism and more importantly, it indorses that these NPs can drive photocatalytic reactions under solar irradiation.

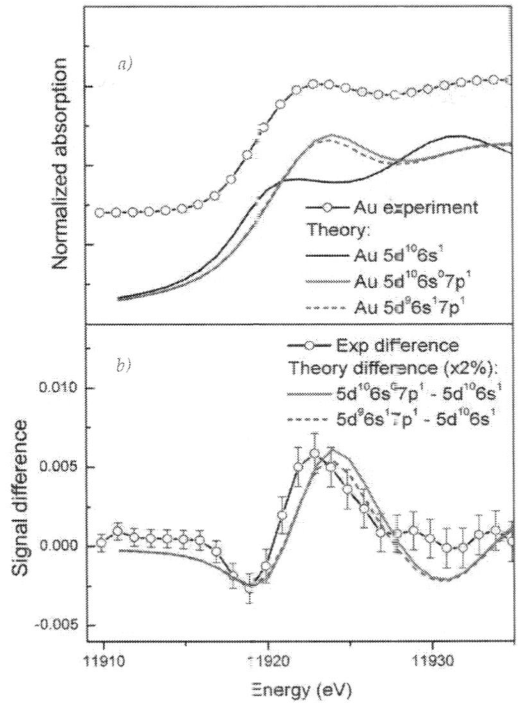

Figure 16. HR-XAS experiments portraying variations in the Au DOS induced by continuous wave laser excitation of plasmon resonance at 532 nm, with 100 mW power. (*a*) HR-XAS ground state spectrum of Au nanoparticles (open circles black trace), FDMNES calculated spectra of the ground state (black trace Au $5d^{10}6s^1$) and excited states (red trace $5d^{10}6s^07p^1$ and blue trace $5d^96s^17p^1$). (*b*) Difference spectra between the excited and ground state: experimental (open circles black trace), and calculated assuming 2% excitation Au $5d^{10}6s^17p^0 \rightarrow$ Au $5d^{10}6s^07p^1$ (red trace); Au $5d^{10}6s^17p^0 \rightarrow$ Au $5d^96s^07p^1$ (dashed blue trace). (Reproduced from elsewhere [59] with permission)

We hope that all the arguments mentioned so far demonstrate the capabilities of HR-XAS in understanding catalytic reactivity. However, HR-XAS suffers from an important limitation, namely its time resolution, which is at best of the order of tenths of seconds for ideal samples with high metal concentration. The limitation starts from the fact that HR-XAS measurements entail scanning of the incoming energy, which is restricted to the speed at which the monochromator can be moved. This makes it problematic to monitor a catalytic process in real time. Ideally, the experiments should be carried out in continuous mode in which data collection is carried out synchronously and uninterruptedly, enabling the identification of metastable regimes and intermediate species. HEROS can provide element-specific information about the unoccupied density of states [61], and due to the scanning free arrangement of von Hamos spectrometer, HEROS spectra can be recorded on a shot-to-shot basis with extraordinary time resolution (only depending on sample concentration and photon flux), while upholding spectral energy resolution [62]. Moreover, HEROS spectra are not perturbed by the self-absorption process [63]. This makes HEROS a commanding tool to identify and quantify the catalyst electronic changes during reaction and as it happens.

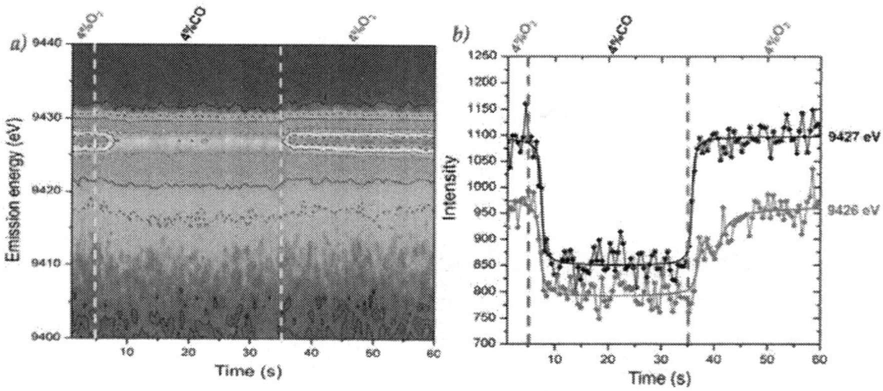

Figure 17. (a) Temporal evolution of the HEROS spectra during CO/O_2 switches at 300 °C on 1.3 wt% Pt/Al_2O_3. (b) Temporal evolution of the HEROS signals at 9426 and 9427 eV (whiteline region) during CO/O_2 switches. (Reproduced from elsewhere [64] with permission)

We performed an in situ time-resolved HEROS study with subsecond resolution providing insight into the oxidation and reduction steps of a Pt catalyst during CO oxidation with 500 ms resolution [64]. Figure 17 displays a gentle oxidation step, comprised of two distinguishing stages, namely dissociative adsorption of oxygen followed by partial oxidation of Pt subsurface. By comparing the experimental spectra with theoretical calculations, we found that the intermediate chemisorbed O on Pt is adsorbed on atop position, insinuating CO surface poisoning or surface reconstruction. This indicates HEROS ability to perform chemical speciation with subsecond time resolution, which opens exciting opportunities to follow catalysis in real time. Since the HEROS spectra are collected in a single shot, the time resolution can be further improved since it depends only on the number of incoming photons and element concentration, which makes it particularly suited for experiments at the XFELs. Recently, it was shown that experiments with 100 ms resolution could be achieved at the synchrotron [65].

4. Outlook and future trends in the field

Further developments in the field of X-ray spectroscopy applied to catalysis science are nowadays focused in multiple aspects. Most importantly, improvements in the experimental time resolution are necessary for deep insight knowledge of many chemical processes. For this, a range of complementary X-ray sources is necessary due to timescale changes in the chemical environment of the atomic species that extend over a wide time range from femtoseconds up to a few seconds. Therefore, a lot of expectation is associated with new, large-scale X-ray facilities being under development worldwide. A new synchrotron sources, like MAXIV in Sweden, are expected to deliver ultrabright X-ray pulses with pulse durations from tens of picoseconds down to sub-picoseconds times. Recently developed fourth-generation X-ray sources – X-ray free electrons lasers – pushed the time limit down to femto-second timescales. The two operating XFELs (LCSL, USA and SACLA, Japan) as well as those under construction (EuXFEL, Germany and SwissFEL, Switzerland) provide a unique opportunity to access timescales of charge-transfers and the following bond-breaking and bond-making mechanism. On the other hand, in order to fully explore capabilities of those new X-ray sources, further developments on X-ray spectroscopy techniques is necessary together with progress on sample delivery/treatments schemes for in situ investigations. Finally, further advancement and evolution in the field of X-ray spectrometers and detectors will be crucial in order to fully explore opportunities provided by the subpicoseconds and femtoseconds X-ray probes.

Author details

Jakub Szlachetko[1*] and Jacinto Sá[2]

*Address all correspondence to: jakub.szlachetko@psi.ch

1 Paul Scherrer Institute, Switzerland & Institute of Physics, Jan Kochanowski University of Kielce, Poland

2 Department of Chemistry, Uppsala University, Sweden & Institute of Physical Chemistry, Polish Academy of Sciences, Poland

References

[1] http://web.anl.gov/catalysis-science/publications/CAT_rpt.pdf

[2] a) B. Hammer, J. K. Nørskov, Nature 376 (1995) 238; b) B. Hammer, J. K. Nørskov, Adv. Catal. 45 (2000) 71; c) F. H. B. Lima,J. Zhang, M. H. Shao, K. Sasaki, M. B. Vukmirovic, E. A. Ticianelli, R. R. Adzic, J. Phys. Chem. C 111 (2007) 404; d) U. Heiz, E. L. Bullock, J. Mater. Chem. 14 (2004) 564; e) W. M. C. Sameera, F Maseras, WIREs Comput. Mol. Sci. 2 (2012) 375.

[3] a) M. Tada, S. Murata, T. Asakoka, K. Hiroshima, K. Okumura, H. Tanida, T. Uruga, H. Nakanishi, S.-i. Matsumoto, Y. Inada, M. Nomura, Y. Iwasawa, Angew. Chem. Int. Ed. 46 (2007) 4310; b) J. Zhang, K. Sasaki, E. Sutter, R. R. Adzic, Science 315 (2007) 220.

[4] J. Szlachetko, M. Nachtegaal, E. de Boni, O. Safonova, J. Sá, G. Smolentsev, M. Szlachetko, J. A. van Bokhoven, J.-C. Dousse, J. Hoszowska, Y. Kayser, P. Jagodzinski, A. Bergamaschi, B. Schmid, C. David, A. Lücke, Rev. Sci. Instrum. 83 (2012)103105.

[5] a) H. Johann, Z. Phys., 1931, 69, 185; b) E. Kleymenov, J. A. van Bokhoven, C. David, P. Glatzel, M. Janousch, R. Alonso-Mori, M. Studer, M. Willimann, A. Bergamaschi, B. Henrich, M. Nachtegaal, Rev. Sci. Instrum., 82 (2011) 065107.

[6] J. Sá, J. Szlachetko, Catal. Lett. 144 (2014) 197.

[7] J. L. Calmpbell, At. Data Nucl. Data Tables 77 (2001) 50.

[8] D. Sokaras et al., Rev. Sci. Instr. 84, (2013) 053102.

[9] M. Kavcic et al., Rev. Sci. Instr. 83 (2012) 033113.

[10] T. Johannson, Z. Phys. 71 (1932) 705.

[11] L. von Hamos, Naturwiss. 20 (1932) 705.

[12] R. Alonso-Mori et al., Rev. Sci. Instr. 83, (2012) 073114.

[13] For example, Special Issue on: Progress in Resonant Inelastic X-ray Scattering, J. Electron Spectroscopy Related Phenomena, 188, 1-182 (2013) and refs. therein.

[14] C. J. Sparks, Phys. Rev. Lett. 33 (1974) 262, H. Hayashi et al., Chem. Phys. Lett. 371 (2003) 125, H. Hayashi et al., Phys. Rev. B 68 (2003) 045122, J. Szlachetko et al., Phys. Rev. Lett. 97 (2006) 073001, J. Szlachetko et al., Phys. Rev. A 75 (2007) 022512.

[15] J. Tulkki, T. Aberg, J. Phys. B: At. Mol. Phys. 15 (1982) L435.

[16] M. Kavcic et al., Phys. Rev. B 87 (2013) 075106.

[17] S. Eisebitt et al., Phys. Rev. B 47 (1993) 14103.

[18] J. Szlachetko, J. Sá, Cryst. Eng. Comm. 15 (2013) 2583.

[19] a) L. J. Antila, F. G. Santomauro, L. Hammarström, D. L. A. Fernandes, J. Sá, Chem. Commun. 51 (2015) 10914; b) J. Sá, P. Friedli, R. Geiger, P. Lerch, M. H. Rittmann-Frank, C. J Milne, J. Szlachetko, F. G. Santomauro, J. A. van Bokhoven, M. Chergui, M. J. Rossi, H. Sigg, Analyst 138 (2013) 1966; c) P. Friedli, H. Sigg, J. Sá, Photochem. Photobiol. Sci. 13 (2014) 1393.

[20] a) A. L. Ankudinov, B. Ravel, J. J. Rehr, S. D. Conradson, Phys. Rev B 58 (1998) 7565; b) J. J. Rehr, R. C. Albers, Rev. Mod. Phys. 72 (2000) 621; c) A. L. Ankudinov, J. J. Rehr, J. Low, S. R. Bare, Phys. Rev. Lett. 86 (2001) 1642.

[21] a) J. Szlachetko, M. Pichler, D. Pergolosi, J. Sá, T. Lippert, RSC Adv. 4 (2014) 11420; b) J. Szlachetko, K. Michalow, M. Nachtegaal, J. Sá, J. Chem. Sci. 126 (2014) 511.

[22] P. N. Rylander, Catalytic Hydrogenation over Platinum Metals (1967) Academic Press, New York.

[23] R. L. Augustine, Heterogeneous Catalysis for the Synthetic Chemist, (1996), pp. 315–343, Marcel Dekker, New York.

[24] a) P. Sabatier, J. B. Sendrens, ComptesRendusHebdomadais des Seances de l'Academie des Science 128 (1899) 1173; b) P. Sabatier, J. B. Sendrens, ComptesRendusHebdomadais des Seances de l'Academie des Science 130 (1900) 1781.

[25] a) H. Lindlar, Helv. Chim. Acta 35 (1952) 446; b) H. Lindlar, R. Dubuis, Org. Synth. 45 (1966) 89.

[26] a) W. S. Knowles, Angew. Chem. Int. Ed. 41 (2002) 1998; b) R. Noyori, Angew. Chem. Int. Ed. 41 (2002) 2008.

[27] M. Crespo-Quesada, A. Yarulin, M. Jin, Y. Xia, L. Kiwi-Minsker, J. Am. Chem. Soc. 133 (2011) 12787.

[28] E. Schmidt, A. Vargas, T. Mallat, A. Baiker, J. Am. Chem. Soc. 131 (2009) 12358.

[29] C. Chapuis, D. Jacoby, Appl. Catal. A 221 (2001) 93.

[30] a) M. Raney, Method of producing finely-divided nickel (1927) U.S. patent 1,628,190; b) T.-K. Yang, D.-S. Lee, J. Haas,e-EROSEncyclopedia of Reagents for Organic Synthesis (2006) doi:10.1002/047084289X.rr001.pub2

[31] Y. Urushibara, Bull. Chem. Soc. Jpn 25 (1952) 280.

[32] http://catalysts.evonik.com/product/catalysts/en/catalyst-brands/products/activated-base-metal-catalysts/pages/default.aspx

[33] a) W. Bonrath, et al. Process for the preparation of saturated aliphatic ketones (2006) Worldwide Patent WO2006029737; b) T. Kawaguchi, et al. 2,3,5-Trimethylhydroquinone (1973) German Patent DE2250066.

[34] For example: a) B. H. Wu, H. Q. Huang, J. Yang, N. F. Zheng, G. Fu, Angew. Chem. Int. Ed. 51 (2012) 3440; b) M. S. Ide, B. Hao, M. Neurock, R. J. Davis, ACS Catal. 2 (2012) 671.

[35] a) A. Saadi, R. Merabti, Z. Rassoul, M.M. Bettahar, J. Mol. Catal. A 253 (2006) 79; b) F. Salman, C. Park, R. T. K. Baker, Catal. Today 53 (1999) 385; c) W. Lin, J. Catal. 303 (2013) 110; d) V. Gutiérrez, F. Nador, G. Radivoy, M. A. Volpe, Appl. Catal. A. 464-465 (2013) 109.

[36] H. G. Manyar, B. Yang, H. Daly, H. Moor, S. McMonagle, Y. Tao, G. D. Yadav, A. Goguet, P. Hu and C. Hardacre, ChemCatChem 5 (2013) 506.

[37] H. G. Manyar, R. Morgan, K. Morgan, B. Yang, P. Hu, J. Szlachetko, J. Sá, C. Hardacre, Catal. Sci. Technol. 3 (2013) 1497

[38] M. Makosch,C. Kartusch, J. SaÅL,R. B. Duarte, J. A. van Bokhoven, K. Kvashnina, P. Glatzel, J. Szlachetko, D. L. A. Fernandes, E. Kleymenov, M. Nachtegaal, K. Hungerbühler, Phys. Chem. Chem. Phys. 14 (2012) 2164.

[39] P. Glatzel, J. Singh, K. O. Kvashnina, J. A. van Bokhoven, J. Am. Chem. Soc. 132 (2010) 2555.

[40] O. V. Safonova, M. Tromp, J. A. van Bokhoven, F. M. F. de Groot, J. Evans, P. Glatzel, J. Phys. Chem. B 110 (2006) 16162.

[41] J. Sá, J. Szlachetko, M. Sikora, M. Kavčič, O. V. Safonova, Maarten Nachtegaal, Nanoscale 5 (2013) 8462.

[42] J. Sá, Y. Kayser, U. Hartfelder, C. J. Milne, D. L. A. Fernandes, J. Szlachetko, Phys. Chem. Chem. Phys. 16 (2014) 7692.

[43] J. Szlachetko, J. Sá, M. Nachtegaal, U. Hartfelder, J.-C. Dousse, J. Hoszowska, D. L. A. Fernandes, H. Shi, C. Stampfl, J. Phys. Chem. Lett. 5 (2014) 80.

[44] M. Haruta, T. Kobayachi, H. Sano and N. Yamada, Chem. Lett. (1987) 405.

[45] a) A. Goguet, M. Ace, Y. Saih, J. Sá, J. Kavanagh, C. Hardacre, Chem. Commun. (2009) 4889; b) J. Sá, A. Goguet, S. F. R. Taylor, R. Tiruvalam, C. J. Kiely, M. Nachtegaal, G. J. Hutchings, C. Hardacre, Angew. Chem. Int. Ed. 50 (2011) 8912.

[46] J. Szlachetko, J. Sá, M. Nachtegaal, U. Hartfelder, J.-C. Dousse, J. Hoszowska, D. L. A. Fernandes, H. Shi, C. Stampfl, J. Phys. Chem. Lett. 5 (2014) 80.

[47] M. Zienkiewicz,J. Szlachetko,C. Lothschütz, M. Hodorowicz, A. Jabłońska-Wawrzycka, J. Sá, B. Barszcz, Dalton Trans. 42 (2013) 7761.

[48] M. Zienkiewicz, A. Jabłońska-Wawrzycka, J. Szlachetko, Y. Kayser, K. Stadnicka, W. Sawka-Dobrowolska, J. Jezierska, B. Barszcz, J. Sá,Dalton Trans. 43 (2014) 8599.

[49] A. Fujishima, K. Honda, Nature238(1972) 37.

[50] K. Hashimoto, H. Irie, A. Fujishima, JpnJ. Appl. Phys. 44 (2005) 8269.

[51] R. Asahi, T. Morikawa, T. Ohwaki, K. Aoki, Y. Taga, Science 293 (2001) 269.

[52] a) B. O'Regan, M. Grätzel, Nature353 (1991) 737; b) M. Grätzel, Nature 414 (2001) 338.

[53] For example: a) K. R. Catchpole, A. Polman, Opt. Exp. 16 (2008) 21793; b) M. D. Brown, T. Suteewong, R. S. Kumar, V. D'Innocenzo, A. Petrozza, M. M. Lee, U. Wiesner, H. J. Snaith, Nano Lett. 11 (2011) 438; c) L. M. Peter, J. Phys. Chem. Lett. 2 (2011) 1861.

[54] a) Y. Nishijima, K. Ueno, Y. Yokota, K. Murakoshi, H. Misawa, J. Phys. Chem. Lett. 1 (2010) 2031; b) F. Wang, N. A. Melosh, Nano Lett. 11 (2011) 5426; c) Y. Tian, T. Tatsuma, J. Am. Chem. Soc. 127 (2005) 7632.

[55] a) P. Christopher, H. Xin, S. Linic, Nat. Chem. 3 (2011) 467; b) D. Tsukamoto, Y. Shiraishi, Y. Sugano, S. Ichikawa, S. Tanaka, T. Hirai, J. Am. Chem. Soc. 134 (2012) 6309; c) P. Christopher, H. Xin, A. Marimuthu, S. Linic,Nat. Mater. 11 (2012) 1044; d) S. Lin-

ic, P. Christopher, D. B. Ingram, Nat. Mater. 10 (2011) 911; e) Z. Zhang, J. E. Sader M. L. Roukes,Nano Lett. 13 (2013) 14.

[56] G. L. Hallett-Tapley, M. J. Silvero, M. González-Béjar, M. Grenier, J. C. Netto- Ferreira, J. C. Scaiano, J. Phys. Chem. C 115 (2011) 10784.

[57] M. L. Brongersma, N. J. Halas, P. Nordlander,Nat. Nanotechnol. 10 (2015) 25.

[58] S. Mubeen, J. lee, N. Singh, S. Krämer, G. D. Stucky. M. Moskovits, Nat.Nanotechnol. 8 (2013) 247.

[59] J. Sá, G.Tagliabue, P. Friedli, J. Szlachetko, M. H. Rittmann-Frank, F. G. Santomauro, C. J. Milne, H. Sigg, Energy Environ. Sci. 6 (2013) 3584.

[60] A. Furube,L. Du, K. Hara, R. Katoh, M. Tachiya,J. Am. Chem. Soc. 129 (2007) 14852.

[61] J. Szlachetko, M. Nachtegaal, J. Sá, J -C. Dousse, J. Hoszowska, E. Kleymenov, M. Janousch, O. V. Safonova, J. A. van Bokhoven, Chem. Commun. 48 (2012) 10898.

[62] J. Szlachetko, M. Nachetgaal, E. de Boni, O. Safonova, J. Sá, G. Smolentsev, M. Szlachetko, J. A.van Bokhoven, J.-C. Dousse, J. Hoszowska, Y. Kayser, P. Jagodzinski, A. Bergamaschi, B. Schmid, C. David, A. Lücke, Rev. Sci. Instrum. 83 (2012) 103105.

[63] W. Błachucki, J. Szlachetko, J. Hoszowska, J.-Cl. Dousse, Y. Kayser, M. Nachtegaal, J. Sá, Phys. Rev. Lett. 112 (2014) 173003.

[64] J. Szlachetko, D. Ferri, V. Marchionni, A. Kambolis, O. V. Safonova, C. J. Milne, O. Kröcher, M. Nachtegaal, J. Sá,J. Am. Chem. Soc. 135 (2013) 19071.

[65] J. Sá,Recycl. Catal.2 (2015) 23.

Section 2

New Perspectives in Metal-based Catalytic Materials and Structured Porous Materials

Chapter 2

Iron-based Nanomaterials in the Catalysis

Boris I. Kharisov, Oxana V. Kharissova, H.V. Rasika Dias, Ubaldo Ortiz Méndez, Idalia Gómez de la Fuente, Yolanda Peña and Alejandro Vázquez Dimas

Additional information is available at the end of the chapter

http://dx.doi.org/10.5772/61862

Abstract

Available data on catalytic applications of the iron-containing nanomaterials are reviewed. Main synthesis methods of nZVI, nano-sized iron oxides and hydroxides, core-shell and alloy structures, ferrites, iron-containing supported forms, and composites are described. Supported structures include those coated and on the basis of polymers or inert inorganic materials (*i.e.*, carbon, titania or silica). Description of catalytic processes includes the decomposition reactions (in particular photocatalytic processes), reactions of dehydrogenation, oxidation, alkylation, C–C coupling, among a series of other processes. Certain attention is paid to magnetic recovery of catalysts from reaction systems and their reuse up to several runs almost without loss of catalytic activity.

Keywords: Iron nanomaterials, ferrites, catalysis, bimetallic nanoparticles, core-shell nanoparticles

1. Introduction

1.1. General information on the iron-containing nanostructures

Nanomaterials on iron basis mainly include zero-valent iron (ZVI and nZVI (nano zero-valent iron) are nowadays classic terms), iron-based nanoalloys or core-shell nanoparticles, iron(II and III) oxides, and ferrites, among others. Metallic iron is normally covered with iron(II) and iron(III) oxides [1]. The iron oxides (iron oxide nanoparticles are also referred in several reports to as superparamagnetic iron-oxide nanoparticles (SPIONs) although SPIONS have inducible magnetic properties) [2, 3], belong to the most technologically important oxides of transition metals. The collective term "iron oxides" is also used for oxides, hydroxides, and oxy-hydroxides containing Fe(II) and/or Fe(III) cations and OH^- and/or O^{2-} anions. In total, sixteen pure iron oxide phases, *i.e.*, oxides, hydroxides or oxy-hydroxides, are currently known. These compounds are $Fe(OH)_3$, $Fe(OH)_2$, $Fe_5HO_8 \cdot 4H_2O$, Fe_3O_4, FeO, five polymorphs of FeOOH and four of Fe_2O_3. In these oxide compounds, which are generally low soluble and possess brilliant colors, the iron is present in the form of Fe(III). The extremely important advantages of nanostructured iron, in comparison with other nanomaterials, are its relatively low toxicity and capacity to be biodegradable. This metal, in addition, is non-expensive and commonly widespread material [4].

Particle diameters of nZVI are normally in the range from 10 to 100 nm [5], exhibiting a classic core-shell structure. Their core contains metallic iron phase, meanwhile the oxidation products of zero-valent iron form mixed valent [*i.e.*, Fe(II) and Fe(III)] oxide shell. If stabilizers in excess are present, these core-shell nanoparticles could be protected against further oxidation [6]. Among such stabilizers, a series of organic compounds can be used for nZVI functionalization to stabilize nZVI aqueous dispersions, inhibiting strongly their further agglomeration. Such compounds can be used to satisfy this purpose, for example, PEG, polyacrylic acid, 4-butane-diphosphonic acid, and methoxyethoxyethoxyacetic acid (MEEA) [7].

The magnetite (Fe_3O_4) and maghemite (γ-Fe_2O_3) are of a particular interest talking about iron oxides (SPIONs). The magnetite structure corresponds to an inverse spinel ferrite. The oxygen ions are the part of a close-packed cubic lattice, containing the iron ions between two different interstices, tetrahedral sites (A), and octahedral sites (B). In a chemical point of view, the magnetite/maghemite can be represented by the following formula: Fe^{3+} $[Fe^{2+}_{1-y}$ Fe^{3+}_{1-y} $Fe^{3+}_{1.67y}\square_{0.33y}]O_4$, where y=0 for pure magnetite and y=1 for pure maghemite (completely oxidized magnetite). From room temperature up to Curie temperature (T_c=860 K), the A sites are filled by Fe^{3+} ions and the B sites are filled by Fe^{3+} and Fe^{2+} ions in equal quantity. Although the lepidocrocite (γ-FeOOH) dehydration transforms into γ-Fe_2O_3, industrial fabrication of maghemite is based on a multistep process (1):

$$\begin{aligned}&(\alpha \text{ and/ or } \gamma)\text{ - FeOOH}(\text{oxidation}) \rightarrow \alpha \text{ - Fe}_2\text{O}_3(\text{reduction}) \rightarrow \\ &\rightarrow \text{Fe}_3\text{O}_4(\text{controlled oxidation}) \rightarrow \gamma \text{ - Fe}_2\text{O}_3\end{aligned} \tag{1}$$

In addition to the nZVI and SPIONs, a variety of composite inorganic iron-based nanomaterials have been discovered, in particular core-shell Fe(or Fe_xO_y)/Au or more complex trimetallic nanoparticles such as $(Fe_{60}Co_{49})_{core}/Au_{shell}$ [8]. These nanoparticles were classified [9] on the basis of on their complexity levels: 1) the nanostructures based of an iron-containing material with magnetic properties *different from iron oxide*; 2) the nanostructures with a non-spherical morphology (*e.g.* hollow structure); 3) the nanostructures with multi-material composition, *i.e.* each of them is constructed ≥2 more domains of joined together different inorganic materials.

2. Main synthesis methods

A number of currently used methods, shown below, are nowadays used for preparation of Fe-containing nanomaterials. At the same time, some well-known conventional *wet chemistry routes* have not been forgotten and are applied for nZVI synthesis [10] (for instance, by a borohydride reduction in laboratory scale) [11], Fe_2O_3 (sol-gel technique [12] or electrochemical deposition [13]), or Fe_3O_4 (urea- and NaOH-assisted hydrolysis of Fe^{3+} and Fe^{2+} salts and further *ultrasonic treatment* of $FeO(OH)/Fe(OH)_2$). Particle sizes and morphologies of the formed nanomaterials, synthesized by distinct methods, can vary depending on the synthesis conditions. For example, the Fe_3O_4 nanoparticles [14], obtained by *radio frequency nitrogen plasma* technique, represent a regular spherical form, meanwhile the nanoparticles synthesized by wet chemistry synthesis were seen as well shaped cubic form. In case of the plasma prepared iron oxides, their size distribution was wider resulting small (25–80 nm) and larger (>100 nm) particles.

2.1. nZVI and Fe-M nanoalloys and core-shell nanostructures

To produce nanopowders, the method of *electrical explosion of wire* was applied [15, 16], among other physical methods. Intermetallic phases can also be obtained by the *arc discharge* technique, for example Fe-Sn bimetallic nanoparticles [17]. Thus prepared nanoparticles have a core/shell structure consisting of a SnO_2 shell (5–10 nm in thickness) and a core containing polycrystalline intermetallic compounds. The intermetallic compounds $FeSn_2$ and Fe_3Sn_2 were shown to be generated; they coexist with the Sn phase as a single nanoparticle. *Microwave irradiation* (which is used to fabricate normally inorganic materials and composites, lesser organic- or organometallic-based materials), is an alternative method in comparison with classic heating of product precursors. Metal transformations in microwave field are extensively reviewed in a comprehensive recent book [18]. Elemental metal nanoparticles in various forms can also be fabricated using this method. For example, iron-based nanoparticles were obtained applying microwave–polyol route in ethyleneglycol at 100 and 150 °C with additives of polyvinyl pyrrolidone and dodecyl amine [19]. Also, pulsed excimer *laser radiation* (248 nm) was used to ablate a feedstock of permalloy (~2 μm, Ni 81%:Fe 19%) under both normal atmospheric conditions (in air) and in other gases, as well as under pressures [20]. α-Fe nanoparticles were also obtained with use of a modified *metal-membrane incorporation* method applying diffusing metal ions through a dialysis membrane [21] (the diffusion-time ≤15 min).

To get nZVI in laboratory conditions [22], the classic and usual synthesis technique is the reduction of Fe(II or III) salts using $NaBH_4$, $NaAlH_4$ or $LiAlH_4$ as reductants. Thus, nZVI was synthesized (reaction 2) [23] in ethanol medium by the method of reduction of FeX_n (X= Cl, OH, OR, CN, OCN, SCN) using sodium borohydride under atmospheric conditions [24].

$$2FeCl_3 + 6NaBH_4 + 18H_2O \rightarrow 2Fe^0 + 6NaCl + 6B(OH)_3 + 21H_2 \quad (2)$$

A patent [25] describes a route to metal nanoparticles by *thermal decomposition* of iron acetate $Fe(OOCCH_3)_2$, placed in a reaction vessel with a passivating solvent such as a glycol ether. Discussing the *pyrolysis* method, it should be noted the preparation of iron nanoparticles (embedded in a carbon matrix) from metal phthalocyanine as precursor [26] and carbon-encapsulated iron nanoparticles (size 5–20 nm) *via* a picric acid-*detonation-induced pyrolysis* of ferrocene as precursor; this route has such peculiarities as self-heating and extremely fast process [27]. Also, the *"greener" techniques* [28, 29, 30, 31, 32] have been applied for nanoparticle fabrication. Use of plant extracts and other natural products on polyphenole basis in these syntheses as reductants and capping agents at the same time for obtaining nZVI and several other Fe-containing nanoparticles is intriguing [33] as well. For instance, the *herbal tea extracts* were applied to reduce iron(III) chloride to elemental iron nanoparticles (50 nm) [34].

In addition, a variety of general physico-chemical methods have been applied for the production of as *Fe-containing bi- and polymetallic alloys* as core-shell nanostructures. For instance, high entropy Nd-Fe-Co-Ni-Mn alloy nanofilms were prepared [35] by *electrodeposition* at r.t. After preliminary preparation of alumina nanotemplates, Fe, Fe-Ni, and Fe-Pd nanowires were successfully electrodeposited within their porous structure. Also, the Fe-Pt nanocrystalline magnetic films (200 nm of thickness) with planar texture were obtained with use of *magnetron sputtering* and crystalline annealing in magnetic field [36].

2.2. Supported and coated iron nanoparticles

A number of publications are devoted to *carbon-supported* ZVI nanomaterials [37]. This type of protective carbon-cage encapsulation of iron nanoparticles can result hybrid core-shell nanomaterials with unique properties [38]. This way, carbon encapsulated iron core-shell nanoparticles (15–40 nm in size) were obtained *via* confined *arc plasma method* [39]. Resulting nanoparticles possessed a clear core-shell structure. The core (16 nm in diameter) of the particles corresponded to a BCC iron structure, and the shell (thickness 6–8 nm) was shown to be disorder carbon phase. A closely related *arc discharge technique* [40] is also frequently used for obtaining a variety of nanomaterials, in particular the iron containing ones. For instance, a simple, inexpensive and one-step arc discharge synthesis technique to prepare metal-containing carbon nanocapsules in aqueous solution is known [41]. It was established that iron nanoparticles can be *in situ* encapsulated in carbon shells when the arc discharge process was carried out in aqueous solutions of $FeSO_4$.

Combustion synthesis of iron oxide/iron-coated carbons such as cellulose fiber, anthracite, and activated carbon was reported a classic microwave oven with inverter technology [42]. The size of the iron oxide/iron nanoparticle-coated samples were determined to be in the range of 50–400 nm. It was found that iron oxide/iron nanoparticles exist in 4 main phases: γ-Fe_2O_3, α-Fe_2O_3, Fe_3O_4, and Fe, some of them had significant arsenic adsorption. In addition, carbon-coated iron nanoparticles with well-developed quasi-spherical shape were prepared with $Fe(NO_3)_3 \cdot 9H_2O$ and starch as carbon source [43, 44]. This is an efficient approach for the mass production of nanocage structures under mild conditions, which needs to be further explored for preparing various carbon coated metal nanomaterials. *Radiation methods* are being applied more widely last 10–15 years, in that case for preparation of Fe-carbon-supported nanostructures. This way, 2 types of amorphous carbon films (15 at.% iron containing film) were deposited onto Si substrates by a sputtering method and further exposed to an electron flow, where the energy and dose rate were much smaller compared to the electron beam in a TEM. In this case, graphitic structures were observed in amorphous matrix at temperatures up to 450 K. It was established that the graphitization progressed more intensively during the electron irradiation than in annealing at 773 K. This was attributed to thermal and catalytic effects, strongly related to grain growth of metal clusters.

2.3. Free and supported iron oxides and ferrites

Zeolites and closely related supporting materials represent an ideal basis for iron oxide nanocomposites. This way, the zeolite loading with nanoiron oxide by a simple chemical process was described [45]. Final crystallite sizes of the doped nanomaterials were in the range of 4–6 nm. It was shown that the zeolites become to have magnetic properties after being doped with nanoiron oxide. Mesoporous nanocomposites "iron oxide/silicate" Fe_2O_3-SBA-15 (SBA-15 is an abbreviation for hexagonally ordered mesoporous silica) with iron loadings of 1.2–35.8 wt.% were prepared hydrothermally [46]. It was revealed that these composites contain well-dispersed iron oxide nanoclusters in the walls of ordered mesoporous silica and high surface area. Certain number of composite nanomaterials based on Fe_3O_4 is known, for instance core/shell Fe_3O_4 coated gold nanoparticles (diameter 50–100 nm) [47]. Their possible formation mechanism was proposed as follows: pH-sensitive polymer owing to a shrunken or stretched structure of polyethyleneimine (PEI), led to the aggregation of the Fe_3O_4-gold seed nanoparticles, then gold reduces onto the surface of Fe_3O_4-gold seed nanoparticles. It was concluded that these core/shell multifunction nanomaterials will not only have external magnetic separation by the core of Fe_3O_4 but also detect the large biological molecules using the shell of gold. In addition, iron phthalocyanine prepolymer/Fe_3O_4 nano hybrid magnetic material [48] can be applied as high temperature-resistant polymer magnetic composite material. At last, ferrites having different sizes, from ultrasmall (2 nm) to 50 nm, can be fabricated by distinct techniques [49] mainly co-precipitation method (CPM), sometimes without using any capping agents/surfactants.

3. Catalytic applications

3.1. nZVI and supported Fe^0 nanocomposites

3.1.1. Catalyzed removal or decomposition of pollutants

nZVI has been extensively reported to resolve a series of environmental problems, related with destruction, adsorption, precipitation, reduction or oxidation of heavy metals, salt anions, hydrocarbons and halogenated organic pollutants (Fig. 1) [50], leading to their conversion to final non-hazardous products. In these processes, iron nanoparticles have shown high efficiency and practically no damage for the environment because of absence of toxicity. Thus, the efficiencies of *nitrate removal* from aqueous solution by single (TiO_2 and nZVI), and composite (nano-TiO_2–Fe^0 composite, NTFC) system under UV illumination were studied [51]. Among the three systems, both nZVI and NTFC can effectively remove nitrate. However, only NTFC can achieve satisfactory transformation of nitrate to N_2. Reactive materials for catalytic *degradation of chlorinated organic compounds* in water at ambient conditions have been prepared on the basis of silica-supported Pd-Fe nanoparticles [52]. Nanoscale Fe-Pd particles were synthesized inside porous silica supports using $(NH_4)_3[Fe(C_2O_4)_3]$ and $[Pd(NH_3)_4]Cl_2$ or Pd acetate as reaction precursors. The reduction of these supported precursors using hydrogen led to materials, revealed high activity in the processes of perchloroethene (PCE) degradation and 2-chlorobiphenyl (2-ClBP) dechlorination. It was established that highly dispersed amorphous Fe-Pd bimetallic nanoparticles on silica support possess superior catalytic activity against PCE dechlorination, comparing with the free-standing Fe-Pd nanoparticles. It was also established that the addition of vitamin B12 (it is known to be an effective electron mediator, having strong synergistic effects with nZVI for reductive dehalogenation reactions) can significantly enhance the reductive dechlorination of PCE by nZVI [53]. A remarkable reductive dechlorination of PCE ($0.25\pm0.01\ h^{-1}$) was observed in nZVI suspension (0.05 g/24 mL) with 0.5 mM vitamin B12 in 6 h, while no significant reductive dechlorination of PCE was observed in the nZVI suspension without vitamin B12. Similar composite material based on deposition of nZVI particles and cyanocobalamine (vitamin B12) on a diatomite matrix was also used for catalytic transformation of PCE and other organic contaminants in water [54]. The composite material rapidly degrades or transforms completely a large spectrum of water contaminants, including halogenated solvents like TCE, PCE, and *cis*-DCE, pesticides like alachlor, atrazine and bromacyl, and common ions like nitrate, within minutes to hours. In a related publication [55], iron nanoparticles were applied for remediation of PCB-contaminated soil, taking into account a maximization of PCB destruction in each treatment stage. The efficiency of PCB destruction during the first step treatment (mixing of soil and iron nanoparticles in water) can be improved by increasing the water temperature. The PCB destruction efficiency of minimum 95% can be achieved. In air at 300 °C, Fe_2O_3 is also a good catalyst for remediating PCB-contaminated soils. In addition, photo-Fenton like method using nZVI/UV/H_2O_2 was applied [56] for removing total *petroleum hydrocarbons* (TPH) and determining the optimal conditions using Taguchi method. The removal rate in optimal conditions was between 95% and 100%. The nZVI particles can be reused in a magnetic field. This process may enhance the rate of diesel degradation in polluted water and could be used as a pretreatment step for the biological removal of TPH from diesel fuel in the aqueous phase. Among other degradation applications, we note the use of nZVI/AC for catalytic wet peroxide oxidation of phenol [57]. The catalytic activity of phenol degradation was improved by application of nZVI/AC catalysts compared to that of Fe/AC. For the range 150–1,000 mg/L, the phenol conversion ≥90% can be achieved using these nanocatalysts during 15 min of the reaction in presence of the stoichiometric hydrogen peroxide for complete mineralization. At last, as an example of classic organic synthesis, we emphasize the conversion of synthesis gas to C2 through C4 olefins with up to 60% selectivity by carbon [58], using catalysts which comprise Fe promoted nanoparticles (5–30 nm in diameter) homogeneously dispersed on weakly interactive α-alumina or carbon nanofiber supports.

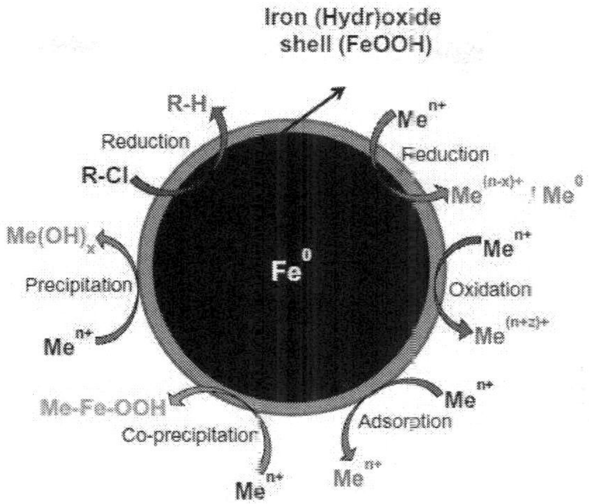

Figure 1. Core–shell structure of nZVI depicting various mechanisms for the removal of metals and chlorinated compounds. Adapted from Li et al. 2006 with permission.

Degradation of dyes with use of nZVI is carried out mainly on carbon-based nanocomposites. Thus, nZVI/activated carbon (nZVI/AC) was investigated as heterogeneous Fenton catalyst in 3D electrode system for methyl orange (MO) degradation [59]. The mineralization of MO was significantly improved by 20–35% compared to 2D AC system at the optimum conditions. A possible mechanism for decolorization and mineralization was proposed, which was attributed to the combination of adsorption, anodic oxidation, and Fenton oxidation in 3D nZVI/AC system. As an example of application of CNTs-based system, an ozone catalyst capable of working on acidic solution environments (labeled as CNTs-Fe^0), which was prepared by immobilizing nZVI onto the surface of MWCNTs and used for decomposition of methylene blue (MB) by formed hydroxyl radicals (HO·) [60]. At pH 3, the production of HO· was found to be considerably accelerated in the presence of CNTs-Fe^0 about 80 times in comparison with results using plain ozone. In the process of CNTs-Fe^0 catalytic ozonation, the CNTs support was analyzed to perform as effective "promoter" allowing the fast surface-mediated reactions, owing to the combination of its surface-active nature, conductivity, and chemical stability. All these and above technologies perfectly fit into advanced water treatment technologies, whose additional representative example is as follows. Thus, the *in situ* synthesis of air-stable nZVI embedded in cellulose fibers led to the assembly of highly reactive magnetic filter papers (membrane nanocomposites) [61]. This nZVI@FP nanocomposite (Fig. 2) showed high activity towards the removal of hexavalent chromium as well as an excellent catalytic ability to convert phenols into catechols, by simple filtration processes of the contaminated water solutions.

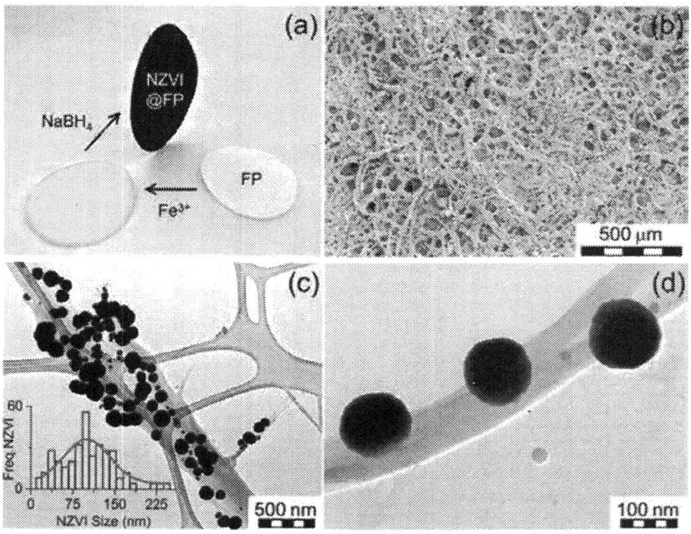

Figure 2. (a) Optical image showing the 2-step assembly of the magnetic nZVI@FP nanocomposite. (b) Scanning electron micrograph of nZVI@FP loaded with 5% in weight of nZVI. (c) TEM micrograph of NZVI entrapped over a cellulose fibre of FP. The inset shows the particle size distribution, as estimated from TEM (Freq.-nZVI *vs.* nZVI-size), together with the fitting analysis (red-line). (d) TEM micrograph of a magnified region of (c) revealing the intimate interaction between NZVIs and the fiber surface. FP = filter paper. Adapted from Datta et. al. 2014 with permission.

3.2. Fe–M nanoalloys, bimetallic NPs and core–shell nanostructures

Degradation of pollutants. Iron-based alloys, core-shell, and bimetallic nanoparticles, especially with noble metals, have been extensively applied in the catalysis and reviewed [62, 63, 64], so we will present in this section only their most representative recently reported examples. As well as nZVI described above, bimetallic Fe-containing nanoparticles (Fe with Pt, Ru, Rh, Ni, Co, Au, Cu, Ag) are used for the catalytic elimination of environmental pollutants. Reactions between them (pollutants and nanoparticles) can be mainly divided in 4 types [65]: a) catalytic replacement reactions for removal of heavy metals, b) hydrodehalogenation reactions (in case of halogenated hydrocarbons), c) azo and nitro hydrogenation reactions (for azo and nitro) compounds, and d) hydrodeoxygenation reactions (for oxyanions). In comparison with monometallic iron nanoparticles, the bimetallic iron NPs have considerable capacity to be separated and catalytic ability for degradation of non-biodegradable pollutants. Among them, Fe-Pt NPs are of an especial interest. Thus, Pt-Fe application as heterogeneous Fenton-like catalysts was reported for hydrogen peroxide decomposition and the decolorization of methylene blue [66]. FePt (and also Fe_3O_4, see more information below) NPs were prepared and tested as heterogeneous Fenton-like catalysts to evaluate and compare their efficiency toward the decolorization of MB dye in solution. Both FePt and Fe_3O_4 exhibited high activity toward the MB decolorization reaction though FePt exhibited a reaction rate that was 100 times faster for a 5 ppm catalyst concentration. Both FePt and Fe_3O_4 NPs are superparamagnetic and thus can be easily separated with a magnet and reused for subsequent catalytic cycles. The same objective was reached using core-shell NPs on the nZVI basis. Thus, a series of nanocomposites consisting of nZVI encapsulated in SiO_2 microspheres were applied for the degradation of organic dyes was investigated using MB as the model dye in the presence of H_2O_2 [67]. The degradation efficiency and apparent rate constant of the degradation reaction were significantly enhanced with increased nZVI encapsulated in SiO_2 microspheres, whereas the dosage of H_2O_2 remarkably promoted degradation rate without affecting degradation efficiency.

It should be also noted that Fe-containing *thin films* can be applied for dye degradation. Thus, 2D nano-TiO_2 and Fe-doped nano-TiO_2 thin films with large sizes were obtained [68] at low temperature in an aqueous system via molecular self-assembly approach. Degradation of methyl orange solution under action of UV and visible light radiation was applied for evaluation of the photocatalytic activity. The doped iron presence was shown to improve the TiO_2 photocatalytic activity. The degradation yields of methyl orange were 98.62% and 89.24%, respectively, under illumination by UV lamp and using visible light. As an example of another degradation process, we note that the longevity and reactivity of nZVI and palladized bimetallic particles (BNP) were evaluated in batch and column experiments for remediation of a trichloroethene (TCE)-contaminated plume within a clayey soil [69]. The particle behavior was found to be severely affected by clay sediments. Results of butch studies testified that TCE degradation in ORR clayey soil corresponds to a pseudo-first-order kinetic model with reaction rate constants (k) of 0.05–0.24 day^{-1} at varied iron-to-soil ratios. Despite of elevated reactivity in water phase, the BNP were less effective in the site-derived clay sediment resulting calculated TCE removal efficiencies of 98.7% and 19.59%, respectively.

Transformations of ketones using stabilized nZVI and bimetallic Fe NPs are known. Among these catalytic processes, we note that porous Pt-Fe bimetallic nanocrystals were found to effectively facilitate the manufacturing of 2-propanol from acetone [70]. It was suggested that the high reactivity is strictly related to the interface consist of bimetallic Pt-Fe alloy and the Fe_2O_{3-x}. As an example of reduction of substituted aromatic ketones to alcohols, iron nanoparticles (size 9 nm, 14 nm, and 17 nm), stabilized by polyethylene glycol (PEG), carboxymethyl cellulose (CMC), and poly N-vinyl pyrrolidone (PVP), were used as catalysts in the hydrogenation reaction of various substituted aromatic ketones to alcohols with $NaBH_4$ [71]. Fe-PEG NPs were found to act as better catalyst than Fe-CMC NPs and Fe-PVP NPs. The trend in the catalytic activity among metals falls in the line of decreasing size effect of the nanoparticles i.e., the order of the nanoparticle sizes increase as Fe-PEG < Fe-CMC < Fe-PVP. Also, effects of substituents in the aromatic ring of ketones revealed that +I substituents are better catalyzed than –I substituents. In addition, bimetallic ruthenium-iron nanoparticles constitute a magnetically recoverable heterogeneous catalyst for transfer hydrogenation with a pronounced selectivity for ketones over aldehydes and nitro groups [72]. The nanoparticles are recyclable up to five times without significant decrease in activity or leaching.

Other processes. An Fe group ternary nanoalloy FeCoNi (NA) catalyst (its synthesis see Fig. 3) enabled selective electrocatalysis towards CO_2-free power generation from highly deliverable ethylene glycol (EG) [73]. This FeCoNi nanoalloy catalyst exhibited the highest selectivities toward the formation of C_2 products and to oxalic acid, *i.e.*, 99% and 60%, respectively, at 0.4 V vs. the reversible hydrogen electrode (RHE), without CO_2 generation. The key feature was the formation of an atomically mixed FeCoNi alloy to enhance the synergetic effect of the Fe group elements on anti-self-oxidation and selective oxidation of EG to oxalic acid. Reduction processes also required polymetallic particles. Thus, trimetallic core/shell Pd/FePt NPs were applied in oxygen reduction reaction (ORR) catalysis [74]. The uniform FePt shell was formed by controlled nucleation of $Fe(CO)_5$ in the presence of a Pt salt and Pd NPs at designated reaction temperatures.

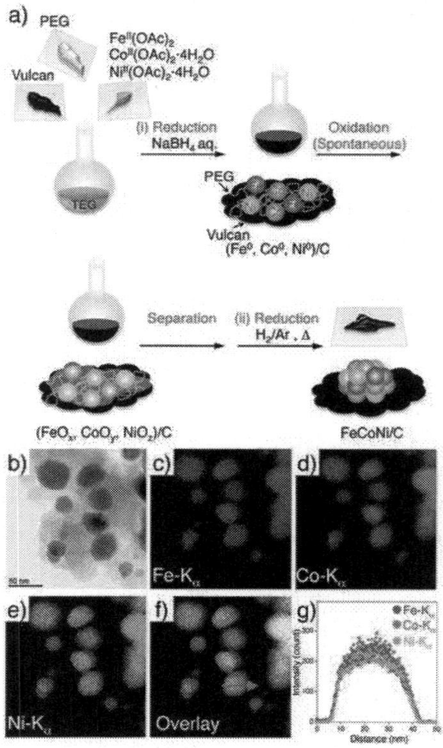

Figure 3. Synthetic scheme for the preparation of a FeCoNi nanoalloy catalyst supported on carbon (FeCoNi/C). Metallic Fe, Co and Ni form in the presence of polyethylene glycol (PEG) and a carbon support (vulcan) after the addition of an aqueous solution of $NaBH_4$. The metallic species are oxidised spontaneously, with production of an oxide mixture composed of Fe_3O_4, Co_3O_4, NiO, and so on is produced. FeCoNi/C was prepared by hydrogen reduction of the oxide mixture. Adapted from Matsumoto et al., 2014 with permission.

Iron-iron oxide core–shell nanoparticles were used as a catalyst for the *hydrogenation of olefins and alkynes* (reaction 3) under mild conditions in ethanol and in an aqueous medium [75]. The system is active in respect of a row of substrates and considerably selective for alkenes and alkynes over aromatic and carbonyl groups. The authors supposed that the presence of an oxide shell does not decrease its activity and provides a certain protection against oxidation by oxygen and water. In addition, highly active and well-defined AuPt nanoalloys, supported on the surface of ellipsoidal Fe@SiO_2 nanoparticles, were prepared by a method involving the loading of Pt NPs on the Fe_2O_3@SiO_2 nanocapsules *via* Sn^{2+} linkage and reduction, then *in situ* fabrication of Au nanoparticles by the galvanic replacement reaction between Au and Pt, and finally calcination and reduction to convert the nonmagnetic Fe_2O_3 to Fe core with high saturation magnetization [76]. The obtained Fe@SiO_2/AuPt samples exhibited a remarkably higher catalytic activity in comparison with the supported monometallic counterparts toward the *reduction of 4-nitrophenol to 4-aminophenol* by $NaBH_4$. The catalyst can be reused for several cycles with convenient magnetic separation.

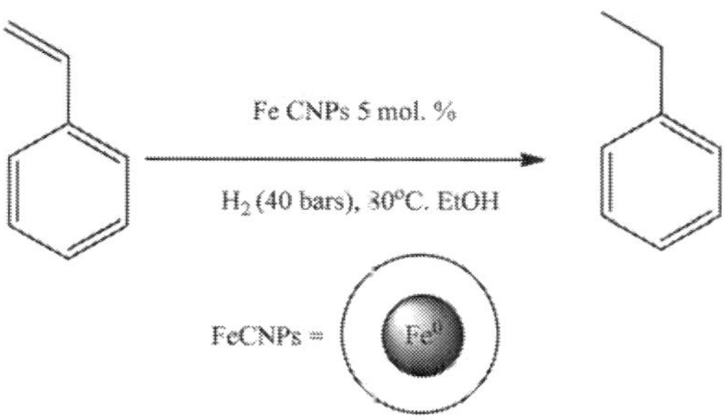

(3)

Hydrogenation of olefin catalyzed by Fe CSNPs.

3.3. Nano-Fe_2O_3 phases and their composites

Destruction of dangerous substances and pollutants. As well as Fe(0)-containing nanocomposites, oxidated iron forms are successfully used for degradation of several substances, both inorganic and organic. Thus, the decrease in temperature of *decomposition of ammonium perchlorate* in the presence of nano-ferric oxide was investigated [77] (see also section below on ferrites). It was shown that addition of nanometer-sized ferric oxide leads to a significant decrease in higher decomposition temperature of ammonium perchlorate. The catalytic activity of a colloidal catalyst (based on iron(III) oxides and obtained by hydrolysis followed by peptization of $FeCl_3 \cdot 6H_2O$ salt in water in the presence of 1% ethanol) in *decomposition of H_2O_2* was studied [78]. The obtained catalyst is mainly composed of α-Fe_2O_3 crystals with an admixture of other crystalline structures of iron oxides, as well as carbon-containing compounds. Its activity with respect to H_2O_2 decomposition varies nonlinearly and nonmonotonically and its particle size grows starting from 1–3 nm with an increase in the initial concentration of $FeCl_3 \cdot 6H_2O$ used to synthesize the catalyst. In addition, thermal decomposition of silver acetate at 200 °C in the presence of iron oxide microspheres in diphenyl ether led to the formation of iron oxide@Ag core–shell nanoparticles, exhibiting superparamagnetic behavior with a blocking temperature of about 42 K [79]. Their good catalytic activity and magnetic recovery was demonstrated by using two reactions, namely, *reduction of 4-nitrophenol* and *reduction of methylene blue* in aqueous solution. An especial case is the use of iron oxide / TiO_2 nanoparticles [80]. Thus, photocatalytic oxidation with TiO_2 nanoparticles (6–20 nm) was investigated as a promising water-treatment process [81, 82]. TiO_2 nanoparticles, after UV irradiation, are able to adsorb and degrade a huge variety of organic contaminants present in the environment. For example, in case of *organic arsenic species* (monomethylarsonic [MMA] and dimethylarsinic [DMA] acids), the strong affinity between the TiO_2 nanoparticles surface results a covalent bonding between MMA or DMA and the nanoparticle surface via formation of bidentate (AsMMA-Ti 3.32 Å) and monodentate (AsDMA-Ti 3.37 Å) inner-sphere complexes, respectively. Dopation of TiO_2 nanoparticles with Fe^{3+} ions at 0.1–0.5% may significantly increase the photocatalytic activity. The doped ions act as charge separators of the photoinduced electron–hole pair and enhanced interfacial charge transfers. Finally, an interesting application is known for *adsorption of elemental sulfur*. Thus, photoinduced sulfur desorption from the surfaces of Au Nps loaded on a series on metal oxides, in particular Fe_2O_3, was studied [83]. Elemental sulfur S_8 was selectively adsorbed on the Au Nps surfaces of Au/metal oxides in an atomic state. This phenomenon is applicable to the low temperature cleaning of sulfur-poisoned metal catalysts.

Organic synthesis. Free iron oxide(III) NPs and its nanocomposites have found numerous application in organic synthesis. Thus, hematite α-Fe_2O_3 NPs (diameters in the 7–18 nm range, synthesized by thermolysing a PVA-$Fe(OH)_3$ gel matrix at moderate temperatures) can effectively catalyze the *epoxidation of styrene* with *tert*-butyl hydroperoxide (TBHP) as the terminal oxidant [84]. Iron oxide nanoparticles supported on zirconia were tested in the gas-phase *conversion of cyclohexanol to cyclohexanone* in a fixed-bed flow type, Pyrex glass reactor, at 433–463 K [85]. Major detected products were cyclohexanone, cyclohexene and benzene, depending on the used catalyst. Experimental results showed that there was no leaching of metal, and that the catalyst was thus truly heterogeneous. In addition, core-shell Fe_2O_3/Pt nanoparticles with amorphous iron oxide cores exhibited superior catalytic activity with lower peak potential and enhanced CO_2 selectivity toward *methanol electrooxidation* in acidic medium [86]. This catalytic performance may be attributed to the uniform distribution of Pt particles on the amorphous Fe_2O_3 surface as well as interactions between the Pt particles and amorphous Fe_2O_3 cores. The catalytic activity of core-shell Fe_2O_3/Pt nanoparticles first increases and then decreases with decreasing Pt content. These nanomaterials also were found to have much higher structural stability and tolerance to the intermediates of methanol oxidation.

A magnetically separable core–shell iron oxide@nickel (IO@Ni) nanocatalyst, synthesized by reduction of Ni^{2+} ions in the presence of iron oxide (Fe^{2+}, Fe^{3+}) by a one-pot synthetic route using $NaBH_4$ as a reducing agent and starch as a capping agent, was found to have excellent activity for the *hydrogenation reactions* of aromatic nitro compounds under mild conditions using water as a green solvent (reaction 4). Excellent chemoselectivity and recyclability up to 30 cycles for the nitro group reduction was demonstrated. Nano propylsulphonated γ-Fe_2O_3 (NPS-γ-Fe_2O_3, reaction 5) was applied as a magnetically recyclable heterogeneous catalyst for the efficient one-pot synthesis of bis(pyrazolyl)methanes in water (reaction 6) [87]. The catalyst was easily isolated from the reaction mixture by a magnetic bar and reused at least five times without significant degradation in activity. Nanoporous α-Fe_2O_3 nanoparticles (about 100 nm in size and containing pores <10 nm) were synthesized *via* a hydrothermal method and applied in the catalytic *benzylation of benzene and benzyl chloride* (BC) in the fabrication of diphenylmethane (DPM) [88]. The BC conversion reached 100% in a reaction time of 3 min with 97.76% selectivity to DPM. The nanoporous α-Fe_2O_3 nanoparticles also have potential applications in other Friedel-Crafts alkylations, especially in large molecular reactions. Another example is related with chemical recycling of PET. Thus, easily recoverable superparamagnetic γ-Fe_2O_3 nanoparticles (10.5 nm in size and 147 $m^2 \cdot g^{-1}$ surface area, produced by calcining Fe_3O_4 nanoparticles prepared by the co-precipitation method) were used as a reusable catalyst for *PET glycolysis* [89]. At 300°C and a 0.05 catalyst/PET weight ratio, the maximum bis(2-hydroxyethlyl) terephthalate (BHET) monomer yielded reached more than 90% in 60 min. The catalyst was reused 10 times, giving almost the same BHET yield each time. In addition, heterogeneous photo-Fenton reaction which utilizes nanosized iron oxides as catalyst for maximizing the activity due to the enhanced physical or chemical properties brought about by the unique structures was described [90].

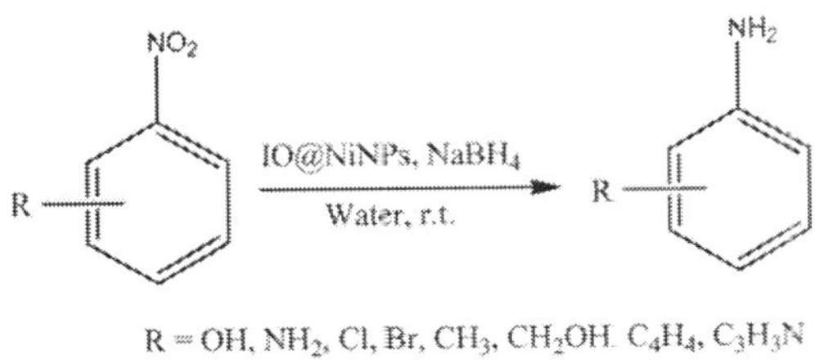

(4)

General scheme for the reduction of various nitroaromatics.

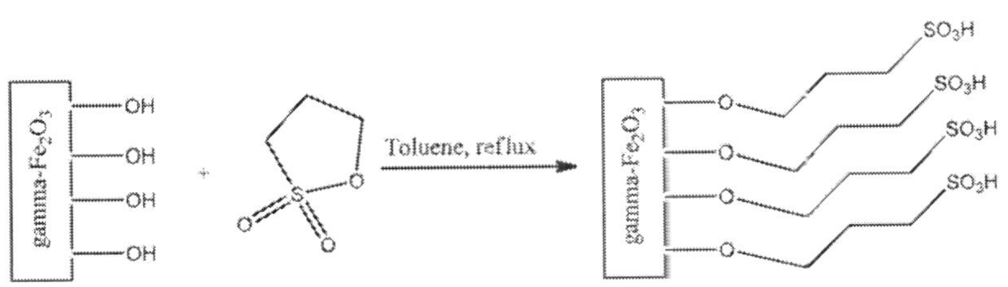

(5)

Synthesis of NPS-γ-Fe_2O_3.

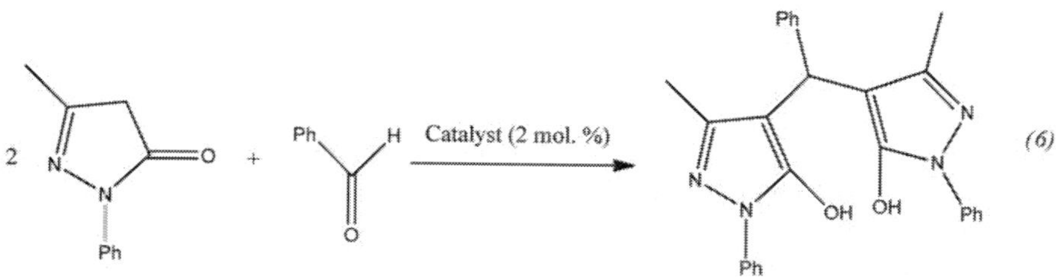

The reaction of benzaldehyde with 1-phenyl-3-methyl-5-pyrazolone.

Antibacterial activities. The doping effects of silver(I) and iron(III) on photocatalytic results using TiO_2 thin films were studied [91]. Ag and Fe doping and co-doping contents on nano-titania photocatalytic bactericidal films were obtained by sol-gel technique, thus uniting three classic active antibacterial species (Ag, Fe and TiO_2). The photocatalytic activity of TiO_2 films was confirmed by the sterilizing rate of the *E-coli* in each case. Applying fluorescent light irradiation, the optimal doping amounts of silver(I)/titania and iron(III)/titania were found to be 0.05% and 0.1%, respectively. In addition, a photocatalytic technique using visible light and carbon nanotubes and nano-sized Fe_2O_3 powder was used to *inhibit pathogenic bacterial growth* in water [92]. It was suggested that after careful design, this system can be used to disinfect drinking water, making it free of pathogenic bacteria.

3.4. Nano-Fe_3O_4 phases and their composites

Iron(II,III) oxide based nanostructures are slightly less explored in the organic synthesis. Thus, haemin-functionalized magnetic iron(II,III) oxide nanoparticles (Fe_3O_4/haemin) exhibited pronounced electrocatalytic activity towards *trichloroacetic acid* (TCA) like haemin itself (a linear detection range of 5–80 M and a detection limit of 0.3 M at 60 °C) [93]. This activity towards TCA was affected by detection temperature, which was controlled via electrically heated carbon paste electrodes. The maximal catalytic current was 5.8 times higher at 60 °C than at room temperature (25 °C). A process capable of synthesizing minor fractions of *aromatic hydrocarbons* (benzene, toluene, xylenes, and mesitylene) from CO_2 and H_2 at modest temperatures (T = 380 to 540°C) employing Fe/Fe_3O_4 nanoparticles as catalyst was designed [94]. The authors consider this technology as principally compatible with solar heat and hydrogen technology and having the potential to mitigate the impacts of global warming by making use of the existing distribution technology for gasoline. Also, *trisubstituted imidazoles* can be synthesized condensation reaction from 1,2-diketones, aromatic aldehydes, and ammonium acetate (reaction *7*) in high yield in the presence of sulphamic acid functionalized magnetic Fe_3O_4 nanoparticles (reaction *8*) as a solid acid catalyst under solvent-free classical heating conditions or using microwave irradiation [95].

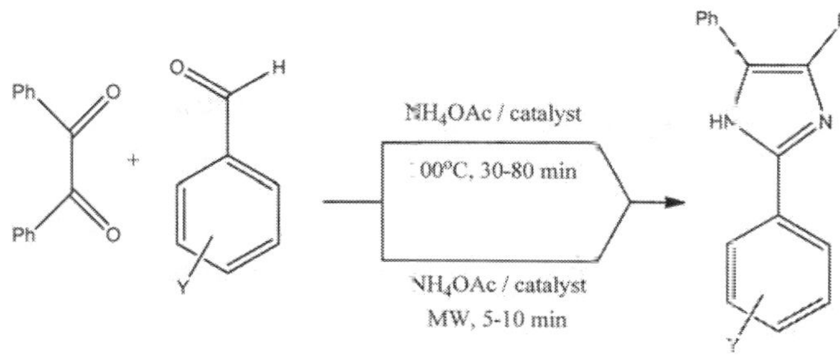

(7)

One-pot synthesis of 2,4,5-trisubstituted imidazoles catalyzed by sulphamic acid functionalized magnetic Fe_3O_4 nanoparticles under conventional heating conditions or using microwave irradiation.

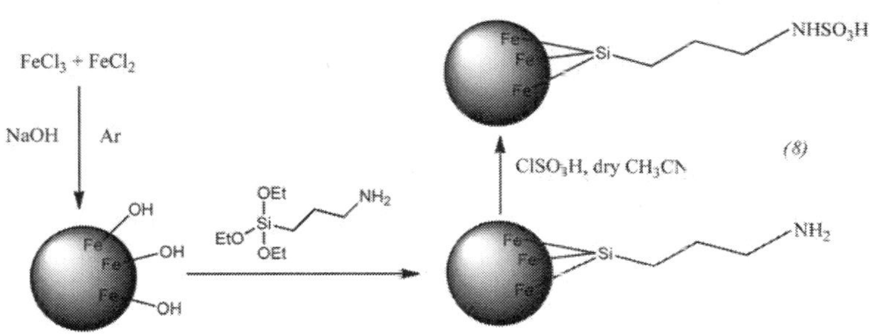

(8)

Preparation steps for fabricating sulphamic acid functionalized magnetic Fe_3O_4 nanoparticles.

Among other reactions, we note an efficient one-pot, three-component condensation reaction between 4-hydroxycoumarin, aryl glyoxals, and malononitrile catalyzed by Fe_3O_4 nanoparticles, which was carried out for the synthesis of several *dihydropyrano[c]chromenes* [96]. Also, an inexpensive and non-hazardous sulfuric acid functionalized magnetic Fe_3O_4 nanoparticles (efficiently catalyze one-pot multicomponent condensation of β-naphthol with aromatic and aliphatic aldehydes and amide derivatives (reaction 9) at 100 °C under solvent-free conditions to afford the corresponding *amidoalkyl naphthols* in excellent yields and in very short reaction times [97]. Silver(0) nanoparticles supported on silica-coated Fe_3O_4 (synthesis see reactions *10*) serve as an efficient and recyclable heterogeneous catalyst for oxidant-free *dehydrogenation of alcohols* to the corresponding carbonyl compounds [98] The catalyst can be easily recovered and reused for 8 reaction cycles without considerable loss of activity. At last, a nanocomposite of functionalized Ni(II) complex containing surface of pyridine, methoxysilanyl and amino groups with iron(II,III) oxide, Fe_3O_4@[-Ni(bpy)$_2$(py-tmos)] was found to be highly efficient green catalyst for the synthesis of a diverse range of 3,4-dihydropyrimidin-2(1H)-ones under solvent free conditions, and in addition it could be easily recovered by a simple magnetic separation and recycled at least 5 times without deterioration in catalytic activity [99].

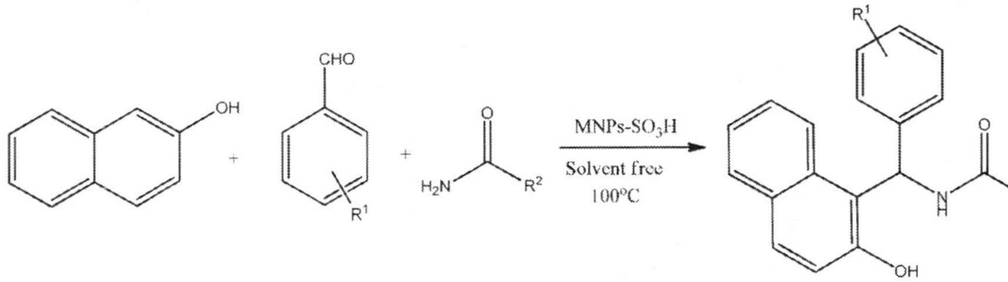

(9)

Synthesis of amidoalkyl naphthols

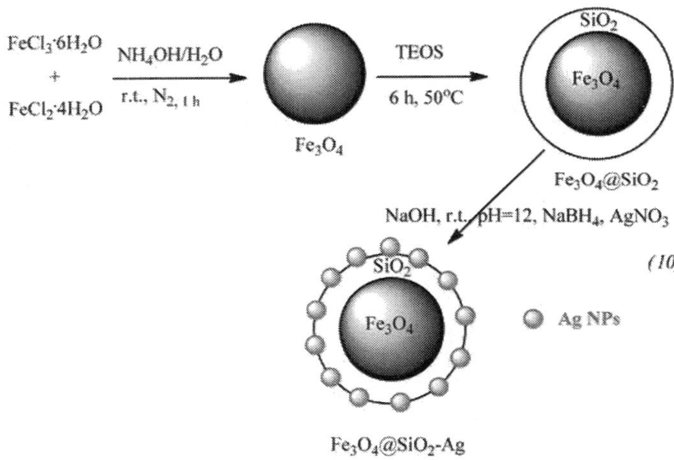

(10)

Preparation of magnetically recoverable heterogeneous nanocatalyst $Fe_3O_4@SiO_2$-Ag.

3.5. Ferrites

Cobalt ferrites having different sizes, from ultrasmall (2 nm) to 50 nm, can be fabricated by distinct techniques [100], mainly co-precipitation method (CPM), sometimes without using any capping agents/surfactants. Thus, the CPM was used to synthesize ultrasmall $CoFe_2O_4$ superparamagnetic nanoparticles (SPMNPs, 2–8 nm of an average size and high surface area of 140.9 m^2/g) without any surfactant [101]. Their catalytic activity was verified in the preparation of *arylidene barbituric acid derivatives* (reaction *11*) applying $CoFe_2O_4$ SPMNPs in aqueous ethanol as a reusable catalyst, which can be magnetically separated from the reaction system. Main advantages of this approach were found to be high yields, short reaction time and high turnover frequency, a clean reaction methodology, and chemoselectivity, among several others. More large-size $CoFe_2O_4$ magnetic nanoparticles (25 nm) were used as a catalyst for the

http://dx.doi.org/10.5772/61862

oxidation of various alkenes in the presence of *tert*-butylhydroperoxide (*t*-BuOOH) with almost quantitative yields (reaction *12*) [102]. It seemed that this heterogeneously catalysis system proceeds by coordination of *t*-BuOOH to the metal (Fe^{3+} cations) on the surface of the catalyst. The separation of the catalyst from the reaction media was easily achieved with the aid of an external magnet, and the catalyst can be reused several times with no loss of activity. In addition, combination of synthesis techniques can also be used for cobalt ferrite preparation. Thus, synthesis of spinel $CoFe_2O_4$ MNPs (average sizes 40–50 nm) was achieved by a combined sonochemical and co-precipitation technique in aqueous medium, also without any surfactant or organic capping agent [103]. These uncapped NPs were utilized directly for *aldol reaction* in ethanol (reaction 13). After the reaction was over, the nanoparticules were compartmented by using an external magnet.

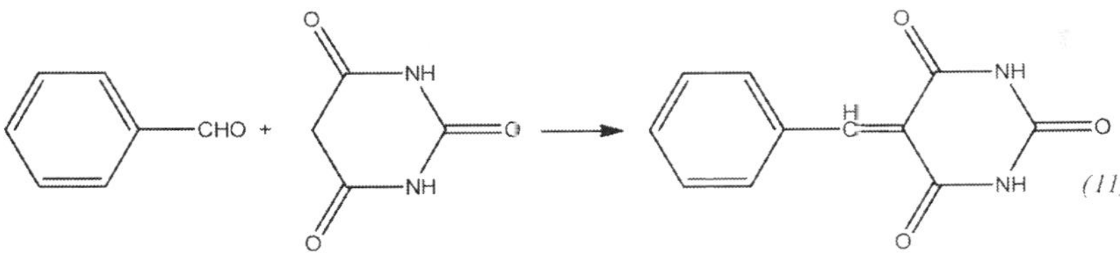

(11)

Optimization of reaction conditions using $CoFe_2O_4$ nanocatalyst. Yields 40–95%, best results in EtOH.

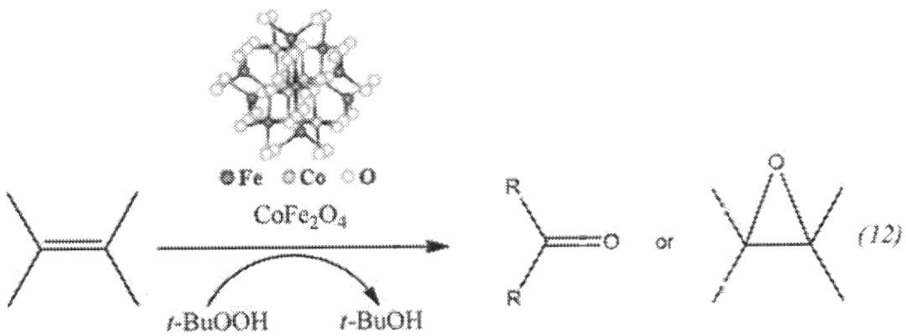

(12)

Oxidation of alkenes using $CoFe_2O_4$ catalyst.

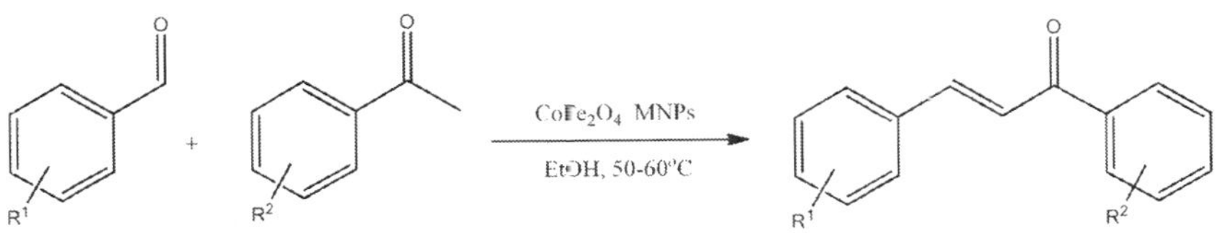

(13)

Figure 4. Aldol condensation reaction in presence of cobalt ferrite MNPs.

Nickel ferrites. Pure or doped nickel ferrites out of the nano-range size are common and frequently used in several catalytic processes. For instance, high reactivity of $NiFe_2O_4$ (111) surfaces (higher that in Fe_3O_4) is well-known; $NiFe_2O_4$ is an effective metal-doped ferrite catalyst in a typical industrial process such as the water-gas shift (WGS) reaction [104]. Similarly, $NiFe_2O_4$ was examined as catalyst in photocatalytic water oxidation using $[Ru(bpy)_3]^{2+}$ as a photosensitizer and $S_2O_8^{2-}$ as a sacrificial oxidant [105]. The catalytic activity of $NiFe_2O_4$ is comparable to that of a catalyst containing Ir, Ru, or Co in terms of O_2 yield and O_2 evolution rate under ambient reaction conditions. As an example of non-nano-sized doped ferrites, their catalysts (granules of ~1 mm diameter) of nickel, cobalt and copper, prepared by co-precipitation hydrothermal route and impregnated with palladium, cerium and lanthanum as promoters [106], were tested for carbon monoxide oxidation activities. The catalysis of $NiFe_2O_4$ nanoparticles on the *hydrogen storage performances* of magnesium hydride synthesized by high-energy ball milling was studied [107], showing that the initial dehydrogenation temperature of 7 mol.% $NiFe_2O_4$-doped MgH_2 is 191°C, which is 250°C lower than that of pristine MgH_2. The enhancement in the H_2 storage performances of MgH_2 by adding $NiFe_2O_4$ nanoparticles is primarily ascribed to intermetallic Fe_7Ni_3 and (Fe, Ni) phases during the desorption procedure, which act as the real catalyst in the 7 mol.% $NiFe_2O_4$-doped sample. As an example of another application, a magnetic acidic catalyst comprising Preyssler ($H_{14}[NaP_5W_{30}O_{110}]$) heteropoly acid supported on silica coated nickel ferrite nanoparticles ($NiFe_2O_4$@SiO_2) was investigated for the synthesis of *bis(dihydropyrimidinone)benzene and 3,4-dihydropyrimidin-2(1H)-ones derivatives* by the Biginelli reaction [108]. With the catalyst, the reactions occurred in less than 1 h with good to excellent yields.

Copper ferrites. Non-nanosized range copper ferrites have certain catalytic applications, such as, for example, for CO conversion to CO_2 [109]. In a difference of pure nickel ferrites, copper ferrite NPs are applied in organic catalysis in more uniform particle size (mainly about 20 nm). Thus, nano material on the basis of copper ferrite (20 nm) was applied as reusable heterogeneous initiator in the preparation of 1,4-dihydropyridines. The interaction of substituted *aromatic aldehydes, ethyl acetoacetate and ammonium acetate* (reaction *14*) was observed in presence of $CuFe_2O_4$ nano powders in ethanol at ambient conditions. The nano catalyst can be magnetically recovered and reused [110]. The same 20-nm size copper ferrite nano material also was reported as reusable heterogeneous initiator in the synthesis of β,γ-unsaturated ketones and allylation to acid chlorides in THF at r.t. without any additive/co-catalyst (reactions 15–16) [111]. The notable advantages are less expensive, heterogeneous reusable catalyst; mild reaction conditions, high yields of products, shorter reaction times, no isomerization during the reaction, and easy workup. In addition, 20-nm $CuFe_2O_4$ was applied as reusable heterogeneous initiator in the synthesis of *α-aminonitriles* by one-pot reaction of different aldehydes with amines and trimethylsilyl cyanides at r.t. in water as a solvent (reactions *17–18*) [112]. α-Aminonitriles are important in preparing a wide variety of amino acids, amides, diamines, and nitrogen containing heterocycles. In addition, a strategy for the synthesis of *benzoxazoles* from substituted N-(2-halophenyl)benzamides (reaction *19*) was developed [113], where inexpensive, readily available, air-stable, recyclable copper(II) ferrite serves as a nanocatalyst. Also, larger-size cubic copper ferrite $CuFe_2O_4$ nanopowders (24–51 nm in size) were synthesized *via* a hydrothermal route using industrial wastes (ferrous sulfate containing free sulfuric

http://dx.doi.org/10.5772/61862

acid ≈10%, 0.01% Zn^{2+} and 2% silica; copper waste 12.5% Cu, 8.7% Cl^- with minor Ni 0.001%) [114]. Study of photocatalytic degradation of the *methylene blue* (MB, $C_{16}H_{18}ClN_3S$) dye using copper ferrite powders showed a good catalytic efficiency (95.9%) at hydrothermal temperature 200 °C for hydrothermal time 24 h at pH = 12 due to high surface area (118.4 m²/g).

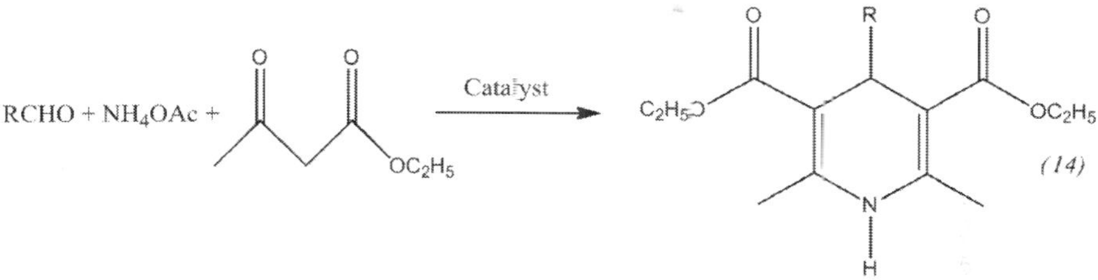

(14)

Catalyst: Copper ferrite (1 mol.%), R = a) Ph, b) 4-Me-*o*-C_6H_4, c) 4-ClC_6H_4, d) 4-NO_2-C_6H_4, e) 4-MeC_6H_4, (f) 3-NO_2-C_6H_4, (g) *n*-C_9H_9, (h) 2-NO_2-C_6H_4, (i) 2-furyl, (j) 2-Me-*o*-C_6H_4.

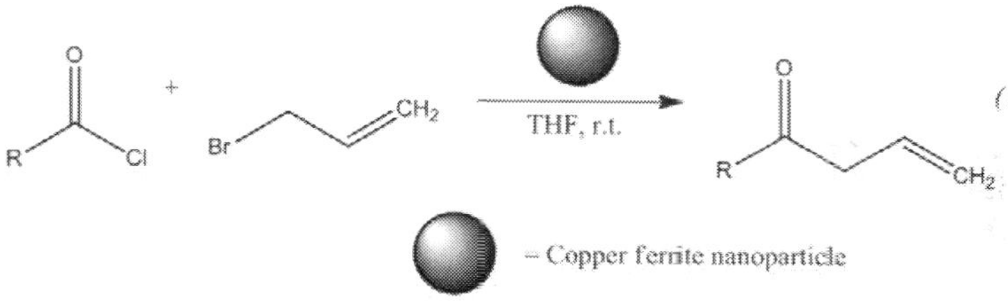

(15)

Synthesis of β,γ-unsaturated ketone using allyl bromide. R = (a) C_6H_5, (b) 2-ClC_6H_4, (c) 2-Br, 5-F, C_6H_3, (d) 2-Br, 5-F, C_6H_3, (e) furanyl, (f) 5-phenyl, 3-Methyl, 4-Isoxazolyl, (g) 5-(2,5-dichloro phenyl, 3-methyl, 4-isoxazolyl, (h) $(CH_3)_3C$-, (i) $C_{11}H_{23}$-, (j) $C_{15}H_{31}$-.

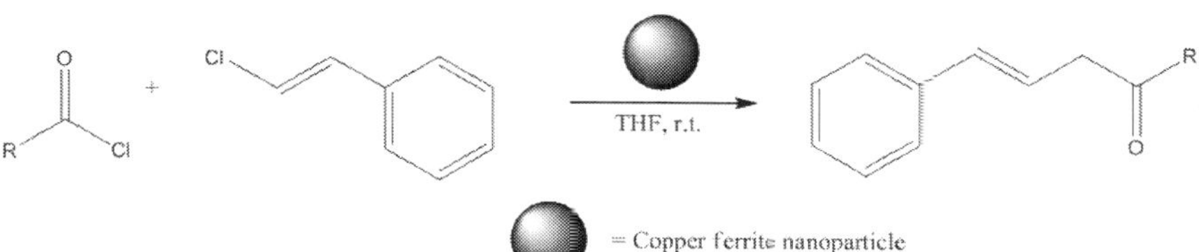

(16)

Synthesis of β,γ-unsaturated ketone using cinnamyl chlorides. Synthesis of β,γ-unsaturated ketone using cinnamyl chlorides. R = (a) C_6H_5, (b) 2-ClC_6H_4, (c) furanyl, (d)-$CH(CH_3)_2$.

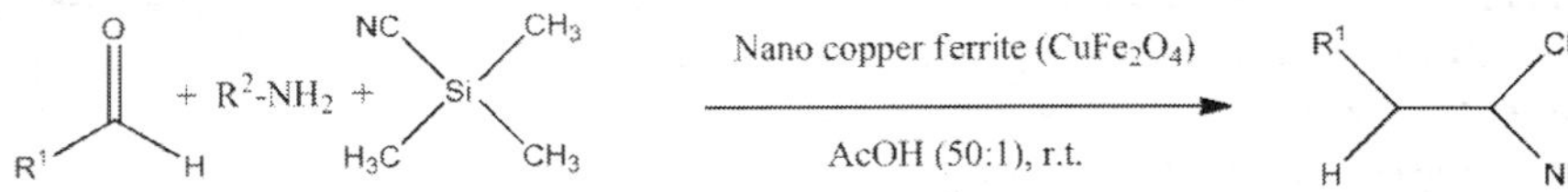

(17)

The synthesis of α-aminonitriles in the presence of nano $CuFe_2O_4$ in water as green solvent at r.t.

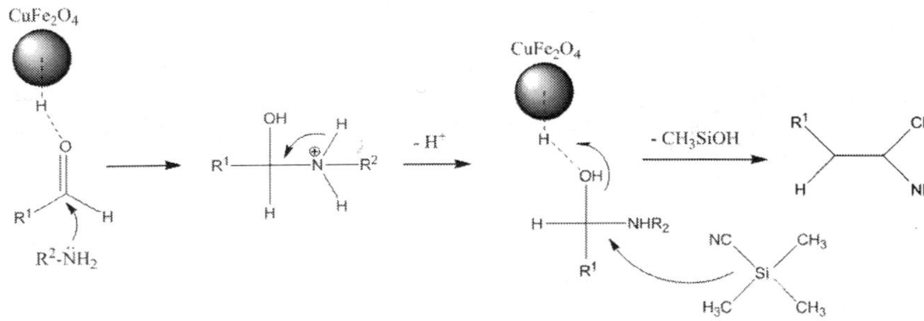

(18)

Suggested mechanism for the synthesis of α-aminonitriles derivatives in presence of acidic nano copper ferrite.

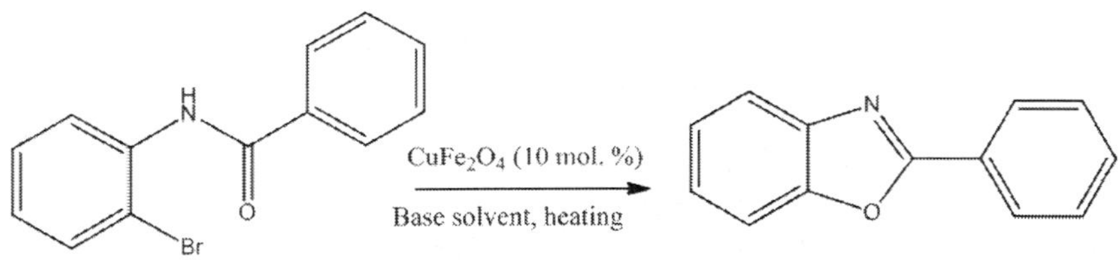

(19)

Catalyzed cyclization of N-(2-bromophenyl)benzamide to 2-phenyl-1,3-benzoxazole.

Zinc ferrite. Non-nano-sized zinc ferrites ($ZnFe_2O_4$) have been used in oxidative organic reactions. Thus, the catalytic behavior for *oxidative conversion of methane and oxidative coupling of methane* was investigated over pure and neodymium substituted zinc ferrites prepared by combustion method [115]. The catalytic activity proved to be strongly related to the oxide structure as well as to the specific defects created by substitution. The pure zinc ferrite ($ZnFe_2O_4$) and $ZnNd_2O_4$ exhibited high activity for coupling reaction whereas the neodymium substituted ferrites ($ZnFe_{1.75}Nd_{0.25}O_4$, $ZnFe_{1.5}Nd_{0.5}O_4$ and $ZnFeNdO_4$) was low active in this reaction. The order of the catalytic activities expressed as yields to C_2^+ were $ZnNd_2O_4$ > $ZnFe_2O_4$ > $ZnFe_{1.75}Nd_{0.25}O_4$ > $ZnFeNdO_4$ > $ZnFe_{1.5}Nd_{0.5}O_4$.

Analyzing pure zinc ferrite nanocatalysts, we note that mainly ultrasmall particles are currently applied in catalytic purposes. Thus, a nanosized highly ordered mesoporous zinc ferrite (ZF, 7–10 nm in size) was synthesized *via* co-precipitation method, further sulfated with ammonium sulfate solution to obtain sulfated ZF (SZF) and was used for the synthesis of nopol by Prins condensation of β-pinene and paraformaldehyde (reaction *20*) [116]. 70% *conversion of β-pinene* with 88% selectivity to nopol was observed; the spent catalyst was regenerated and reused successfully up to four cycles with slight loss in catalytic activity. The influence of various reaction parameters such as solvent, reaction temperature, effect of substrate stoichiometry and catalyst loading was investigated. In particular, very low conversion (9%) of β-pinene was observed in protic solvents such as methanol (solvent effect); in case of apolar–aprotic solvents such as hexane, ethyl acetate and toluene, increase in the β-pinene conversion as well as nopol selectivity was observed. In case of reaction temperature, as the temperature increased to 110 °C, β-pinene conversion increased to 72% with slight drop in selectivity for nopol (57%). 95 °C was identified as optimal reaction temperature for further studies. At molar ratio of β-pinene to paraformaldehyde 1:3 molar ratio, maximum conversion was observed. The β-pinene conversion increased with an increase the catalyst loading (0.12 to 0.16 g) without affecting the nopol selectivity.

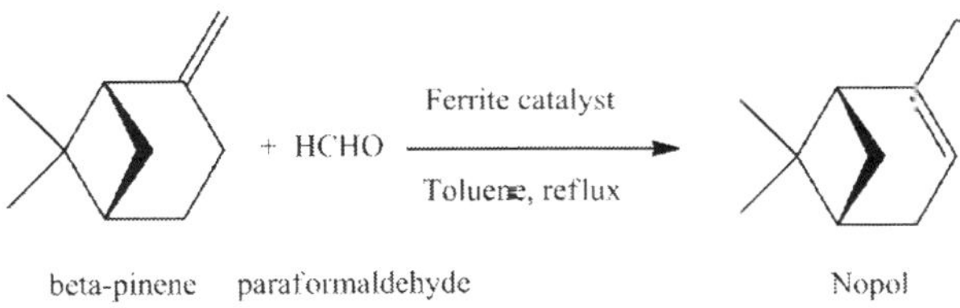

(20)

Prins condensation reaction of β-pinene and paraformaldehyde.

Other simple ferrites. The catalytic behavior of supported Au NPs for the process of *oxidation of benzyl alcohol* was elucidated in presence of gold nanoparticles [117]. Mg^{2+} ions, being present in the ferrite structure, led to an improvement of the catalytic activity of supported Au NPs to *ca.* 35% conversion, when an additional base was absent. Modifying the support with addition of MgO, the catalytic activity of supported Au nanoparticles was further improved to *ca.* 50% conversion; however, the catalyst was found to be deactivated in successive recycling tests. As well as nano-Fe_2O_3 mentioned above in the corresponding section, nano-$MnFe_2O_4$ particles (20–30 nm in size), synthesized by co-precipitation phase inversion method and low-temperature combustion method, using $MnCl_2$, $FeCl_3$, $Mn(NO_3)_2$, $Fe(NO_3)_3$, NaOH and $C_6H_8O_7$, were applied for thermal *decomposition of ammonium perchlorate* [118]. The catalytic mechanism was explained by the favorable electron transfer space provided by outer *d* orbit of transition metal ions and the high specific surface absorption effect of $MnFe_2O_4$ particles. Manganese ferrite nanoparticles were also applied in the synthesis of *spirooxindoles* (compounds **I-III**) *via* a one-pot and three-component reaction of isatins, malononitrile, and anilinolactones in the presence of a catalytic amount of $MnFe_2O_4$ NPs in PEG-400, as a nontoxic, green, and reusable solvent [119].

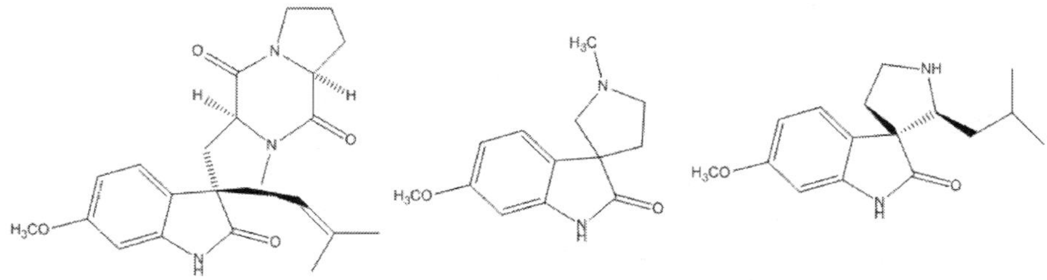

Selected spirooxindole natural products.

Mixed-metal or core-shell ferrites. Ferrites containing 2 metal ions, additionally to iron, are much more widespread in the nano-catalysis; their nanoparticle size can vary in a broad range, from ultrasmall particles (5–8 nm) up to 100 nm or more (in case of supported NPs). Both nano-sized and out-of-nano-sized mixed-metal ferrite NPs can be synthesized by a variety of methods, in particular classic sol-gel and co-precipitation methods or microwave heating ($MnZnFe_2O_4$ [120]).

***Cobalt**-based ferrite nanoparticles*. For cobalt-containing ferrite NPs, as well as for zinc ferrite above, one of important applications is the methanol decomposition to CO and hydrogen. Thus, $Cu_{1-x}Co_xFe_2O_4$ ($0<x<1$, 8–40 nm in size) was applied as a nanodimensional powder for this purpose [121]. The stabilization of the cubic structure with the substitution of copper ions by cobalt in mixed Cu-Co ferrites was observed. Cobalt containing ferrites exhibited higher and more stable catalytic activity and selectivity in *methanol decomposition* to CO and hydrogen in comparison with the $CuFe_2O_4$ one. Photocatalytic properties of the cobalt zinc ferrite $Co_{1-x}Zn_xFe_2O_4$ ($0<x<1$) nanoparticles (10.5–14.8 nm in size), prepared by a hydrothermal method, were studied on the example of *degradation of methyl blue* in aqueous solution [122]. It was elucidated that the oxidation-reduction potential of methyl blue aqueous solution in presence of the ferrite nano-particles at pH=7 under natural sunlight irradiation was negative and increased with increase in Zn content. The degradation rate of methyl blue also decreases as increase in Zn content in sunlight.

***Nickel**-based ferrite nanoparticles*. Similar to cobalt ferrites, several nickel-containing mixed or core-shell ferrites have been reported as nanocatalysts but in more narrow size range (18–50 nm). Thus, a magnetically separable catalyst consisting of ferric hydrogen sulfate (FHS) supported on silica-coated nickel ferrite nanoparticles (50 nm) was prepared [123]. This catalyst was shown to be an efficient heterogeneous catalyst for the *synthesis of 1,8-dioxodecahydroacridines* (reaction *21*) under solvent-free conditions. The catalyst can be recycled several times with no significant loss of catalytic activity.

Other mixed-metal ferrite nanoparticles. Ferrite nanoparticles, containing other metals and applied in the catalysis, are represented more chaotically in the available literature. Thus, the spinel ferrites $Cu_{1-x}Cd_x[Fe_{1-x}Al_xCr_{1-x}Mn_x]O_4$, where $0<x<1$, having unknown particle size, were

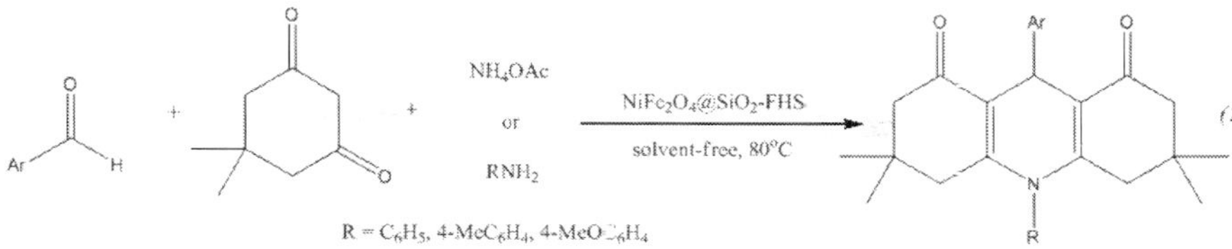

(21)

Synthesis of 1,8-dioxodecahydroacridines in the presence of $NiFe_2O_4@SiO_2$-FHS.

prepared by the coprecipitation technique [124]. Catalytic studies using *decomposition of* H_2O_2 as a model reaction between 303 and 343 K for 1–5 h using first order rate law suggested higher catalytic power for the composition $x = 0$ and then it decreases gradually. For the mixed spinel ferrite system $Mn_{1-x}Cu_xFe_2O_4$ ($x = 0$, 0.25, 0.5, 0.75, 1.0), the formation of phase pure spinels with FCC cubic structure with particle size ranging from 5.21 nm to 20 nm was observed at 333 K applying co-precipitation method with $MnCl_2$, $Fe(NO_3)_3 \cdot 9H_2O$ and $Cu(NO)_3 \cdot 3H_2O$ as precursors [125]. These ferrites were used as catalysts in the *alkylation of aniline*, showing a maximum conversion of 80.5% of aniline with selectivity of 98.6% towards N-methylaniline at 673 K, methanol/aniline molar ratio of 5:1 and weight hour space velocity of 0.2 h^{-1}. It was found that the yield is maximum for $CuFe_2O_4$. In addition, the catalytic performance of the ferrites was found to be proportional to surface area as well as acidity.

4. Conclusions

Iron-based nanoparticles, utilized in catalytic reactions described in this chapter, possess different sizes, from ultrasmall (2 nm) to 100 nm. They are obtained mainly by wet-chemical sol-gel or co-precipitation methods, sometimes combined with simple calcination at high temperatures, sonochemical technique, mechanical high-energy ball milling, or spark plasma sintering, among other methods. Microwave heating or hydrothermal route are also frequently used. Due to magnetic properties, these nano catalysts can be easily recovered from reaction systems and reused up to several runs almost without loss of catalytic activity.

Catalytic processes with application of iron-based nanocomposites are in a wide range. Notable attention is paid to methanol decomposition to CO and methane or to CO and hydrogen. Other catalyzed organic reactions consist of oxidation of various alkenes, aldol, alkylation and dehydrogenation reactions, synthesis of various organic compounds such as, for example, quinoxaline derivatives [126], β,γ-unsaturated ketones, arylidene barbituric acid derivatives, α-aminonitriles, nopol, 1,4-dihydropyridines, and 1,8-dioxodecahydro-acridines. Degradation/decomposition processes are also reported, for instance decomposition of H_2O_2 or photocatalytic degradation of methylene blue. Some of catalyzed reactions might have great practical applications, for instance transesterification of soybean oil to biodiesel. In addition, small iron-based particles could also be considered [127] as substituents of noble metals in a variety of catalytic transformations. Several iron nanomaterials could have biological applications, such as peroxidase-like catalytic activity of Fe_3O_4 ultrasmall NPs [128].

We note that the total number of nano-iron composites applications for catalytic purposes is still not high, so it could be a perfect research niche for further applications of these nanomaterials in a variety of organic processes.

Author details

Boris I. Kharisov[1], Oxana V. Kharissova[1*], H.V. Rasika Dias[2], Ubaldo Ortiz Méndez[1], Idalia Gómez de la Fuente[1], Yolanda Peña[1] and Alejandro Vázquez Dimas[1]

*Address all correspondence to: okhariss@mail.ru

1 Universidad Autónoma de Nuevo León, Monterrey, México

2 The University of Texas at Arlington, Arlington, TX, USA

References

[1] Li, X.Q.; Elliott, D.W.; Zhang, W. Zero-Valent Iron Nanoparticles for Abatement of Environmental Pollutants: materials and engineering aspects. Crit. Rev. Solid State Mater. Sci. 2006, 31, 111–122.

[2] Hofmann-Amtenbrink, M.; von Rechenberg, B.; Hofmann, H. Superparamagnetic nanoparticles for biomedical applications. In: Nanostructured Materials for Biomedical Applications, 2009, Transworld Research Network, Kerala, India.

[3] Mahmoudi, M.; Sahraian, M. A.; Shokrgozar, M. A.; Laurent, S. Superparamagnetic Iron Oxide Nanoparticles: Promises for Diagnosis and Treatment of Multiple Sclerosis. ACS Chem. Neurosci. 2011, 2, 118–140.

[4] Huber, D. L. Iron nanoparticles, in: Dekker encyclopedia of nanoscience and nanotechnology, Schwarz, J. A.; Contescu, C. I.; Putyera, K., Eds., 2008, vol. 3, pp. 1681–1687, CRC Press, Taylor and Francis Group, Boca Raton, FL.

[5] C.S. Rajan.Nanotechnology in Groundwater Remediation. International Journal of Environmental Science and Development, 2011, 2(3), 182-187.

[6] Cheong, S.; Ferguson, P.; Hermans, I.F.; Jameson, G.N.L.; Prabakar, S.; Herman, D.A.J.; Tilley, R.D. Synthesis and Stability of Highly Crystalline and Stable Iron/Iron Oxide Core/Shell Nanoparticles for Biomedical Applications. ChemPlusChem, 2012, 77, 135–140.

[7] Harm, U.; Schuster, J.; Mangold, K.-M. Modification of iron nanoparticles for ground water remediation. DECHEMA, Karl-Winnacker Institut, 2010, http://kwi.dechema.de/kwi_media/Downloads/ec/F564_Nanoeiser_Harm.pdf.

[8] Bai J, Wang J-P. High-magnetic-moment core-shell-type FeCo-Au/Ag nanoparticles. Appl. Phys. Lett., 2005, 87(15), 152502/1-152502/3.

[9] Figuerola, A.; Di Corato, R.; Manna, L.; Pellegrino, T. From iron oxide nanoparticles towards advanced iron-based inorganic materials designed for biomedical applications. Pharmaceutical Research, 2010, 62(2), 126–143.

[10] Qing Wei Ding, Tian Wei Qian, Hong Fang Liu, Xue Wang. Preparation of Zero-Valent Iron Nanoparticles and Study of Dispersion. 2011, Applied Mechanics and Materials, 55–57, 1748–1752.

[11] Jae-min Lee; Ji-hun Kim; Jin-wook Lee; Jae-hwan Kim; Ho-seok Lee; Yoon-seok Chang. Synthesis of Fe-nano Particles Obtained by Borohydride Reduction with Solvent. Paper A-068, in: Bruce M. Sass (Conference Chair), Remediation of Chlorinated and Recalcitrant Compounds—2008. Proceedings of the Sixth International Conference on Remediation of Chlorinated and Recalcitrant Compounds (Monterey, CA; May 2008). ISBN 1-57477-163-9, published by Battelle, Columbus, OH, www.battelle.org/chlorcon.

[12] Khodabakhshi, A.; Amin, M.M.; Mozaffari. M. Synthesis of magnetite nanoparticles and evaluation of its efficiency for arsenic removal from simulated industrial wastewater. Iran. J. Environ. Health. Sci. Eng., 2011, 8 (3), 189–200.

[13] Park, H.; Ayala, P.; Deshusses, M.A.; Mulchandani A.; Choi, H.; Myung, N.V. Electrodeposition of maghemite (γ-Fe_2O_3) nanoparticles. Chem. Eng. J., 2008, 139, 208–212.

[14] Grabis, J.; Heidemane, G.; RAŠMANE, D. Preparation of Fe_3O_4 and γ-Fe_2O_3 Nanoparticles by Liquid and Gas Phase Processes. Mat. Sci. (MEDŽIAGOTYRA). 2008, 14 (4), 292–295.

[15] Ballard, S.G. Apparatus and methods for the production of powders. US20056972115 (2005).

[16] Kim, W.-B., Park, J.-S., Suh, C.-Y., Kil, D.-S., Lee, J.-C 2007, US20070209477.

[17] Lei, J.P.; Dong, X.L.; Zhu, X.G. et al. Formation and characterization of intermetallic Fe–Sn nanoparticles synthesized by an arc discharge method. Intermetallics, 2007, 15(12), 1589–1594.

[18] Gupta, M.; Wong, E.; Leong, W. Microwaves and metals. Wiley-Interscience, 2007, 256 pp.

[19] Komarneni, S.; Katsuki, H.; Li, D.; Bhalla, A.S. Microwave–polyol process for metal nanophases. J. Phys.: Condens. Matter., 2004, 16, S1305-S1312.

[20] Becker, M.F.; Brock, J.R.; Cai, H. et al. Metal nanoparticles generated by laser ablation. Nanostruct. Mater. 1998, 10(5), 853–863.

[21] Rodrigues, A.R.; Soares, J.M.; Machado, F.L.A.; de Azevedo, W.M.; de Carvalho, D.D. Synthesis of α-Fe particles using a modified metalmembrane incorporation technique. J. Magnetism Magnetic. Mat., 2007, 310(2), Part 3: 2497–2499.

[22] Klaine, S.J.; Alvarez, P.J.J.; Batley, G.E.; Fernandes, T.F.; Handy, R.D.; Lyon, D.J.; Mahendra, S.; Mclaughlin, M.J.; Lead, J.R. Nanomaterials in the environment: behavior, fate, bioavailability, and effects. Environmental Toxicology and Chemistry, 2008, 27(9), 1825–1851.

[23] Yuvakkumar, R.; Elango, V.; Rajendran, V.; Kannan, N. Preparation and characterization of zero-valent iron nanoparticles. Digest Journal of Nanomaterials and Biostructures, 2011, 6(4), 1771–1776.

[24] Bönnemann, H.; Brijoux, W.; Brinkmann, R.; Dinjus, E.; Joussen, T.; Korall, B. Erzeugung von kolloiden Übergangsmetallen in organischer Phase und ihre Anwendung in der Katalyse. Angew. Chem., 1991, 103, 1344–1346; Angew. Chem. Int. Ed. Engl., 1991, 30, 1312.

[25] Harutyunyan, A.; Grigorian, L.; Tokune, T. Method for synthesis of metal nanoparticles. 2005, US20056974493.

[26] Klinke, C.; Kern, K. Iron nanoparticle formation in a metal-organic matrix: from ripening to gluttony. Nanotechnology, 2007, 18, 215601.

[27] Lu, Y.; Zhu, Z.; Liu, Z. Carbon-encapsulated Fe nanoparticles from detonation-induced pyrolysis of ferrocene. Carbon, 2005, 43(2), 369–374.

[28] Anastas, P. T.; Horvath, I. T. Green Chemistry for a Sustainable Future. Wiley; 1st edition, 2012, 350 pp.

[29] Chao-Jun Li, Paul T. Anastas. Handbook of Green Chemistry — Green Processes. Wiley-VCH; 3 Volume Set edition, 2012, 1326 pp.

[30] Ahluwalia, V. K. Green Chemistry: Environmentally Benign Reactions, Second Edition. CRC Press; 2nd edition, 2012, 326 pp.

[31] Patel, J. T.; Patel, O. B.; Raval, B. P. Green Chemistry: New Avenues in Chemical Research: Focus in Healthcare. LAP LAMBERT Academic Publishing, 2012, 60 pp.

[32] Luque, R. Green Chemistry. Nova Science Publishers, 2011.

[33] Hoag, G. E.; Collins, J. B.; Varma, R. S.; Nadagouda, M. Green synthesis of metal nanoparticles using plant extracts. PCT Int. Appl. 2009, WO 2009140694 A2 20091119.

[34] Kipkurgat Erick Tanui. Green Synthesis and Characterization of Iron Nanoparticles. http://chemistry.uonbi.ac.ke.

[35] Yao, C.; Ma, H.; Tong, Y. Electrochemical preparation and magnetic study of amorphous nanostructured Nd-Fe-Co-Ni-Mn high entropy alloy film. Yingyong Huaxue, 2011, 28(10), 1189–1194.

[36] Glebov, A.V.; Glebov, V.A.; Popova, O.I. Development of nanofilm Fe-Pt magnets for superdense-recording heads. Tsvetnye Metally, 2009, (12), 67–70.

[37] Wang, J.; Cheng, R. Preparation and application of activated carbon supported iron nanomaterial, with application to pentachlorophenol degradation. 2010, CN 101708457.

[38] Bystrzejewski, M. Synthesis of carbon-encapsulated iron nanoparticles via solid state reduction of iron oxide nanoparticles. Journal of Solid State Chemistry, 2011, 184(6), 1492–1498.

[39] Wei, Z.; Wang, X.; Yang, H. Preparation of carbon-encapsulated Fe core-shell nanostructures by confined arc plasma. Materials Science Forum, 2011, 688(Nano-Scale and Amorphous Materials), 245–249.

[40] Rao, C.N.R.; Thomas, P.J.; Kulkarni, G.U. Nanocrystals: Synthesis, Properties and Applications (Springer Series in Materials Science). Springer, 2007, 182.

[41] Bingshe, X.; Junjie, G.; Xiaomin, W.; Xuguang, L.; Hideki, I. Synthesis of carbon nanocapsules containing Fe, Ni or Co by arc discharge in aqueous solution. Carbon, 2006, 14(13), 2631–2634.

[42] Nadagouda, M.N.; Lytle, D.A. Microwave-assisted combustion synthesis of nano iron oxide/iron coated activated carbon, anthracite, cellulose fiber, and silica, with arsenic adsorption studies. Journal of Nanotechnology, 2011, 972486, 8 pp.

[43] Yuliang An; Jieshan Qiu. Synthesis of carbon encapsulated iron nanoparticles by carbonization of starch with iron as catalyst. 2007, The American Carbon Society. http://acs.omnibooksonline.com/data/papers/2004_C085.pdf

[44] An Yuilang; Wu Xiaojuan; Sui Zhiming; yuan Xia, Liu yanqiu. A novel method for synthesis of homogeneous carbon encapsulated Fe nanoparticles based on natural biopolymer. J. Rare Earths, 2007, 25, 452.

[45] Salah El-Din, T.A.; Elzatahry, A.A.; Aldhayan, D.M.; Al-Enizi, A.M.; Al-Deyab, S.S. Synthesis and Characterization of Magnetite Zeolite Nano Composite. Int. J. Electrochem. Sci.,2011, 6, 6177–6183.

[46] Liu, H.; Lan, G.; Yan, Y.; Tang, H.; Liu, H.; Li, Y. Direct hydrothermal synthesis of novel ordered magnetic mesoporous nanocomposites with high content of iron. Gongye Cuihua, 2011, 19(8), 11–15.

[47] Shi, Y.-f.; Zhou, X.; Zhong, L.-b.; Xu, W.-l.; Wang, Y., Zhang, Q.-q. Synthesis and formation mechanism of core-shell Fe_3C_4 coated gold nanomaterials. Dongnan Daxue Xuebao, Yixueban, 2011, 30(1), 6–10.

[48] Liu, X.; Xu, M.; Zhao, R.; Zhong, J. Iron phthalocyanine prepolymer/Fe_3O_4 nano hybrid magnetic material and its preparation method. Edited by: Casciano, D.A.; Sahu, S.C. Faming Zhuanli Shenqing, 2011, CN 102086304.

[49] Goodarz Naseri, M.; Saion, E.B.; Abbastabar Ahangar, H.; Halim Shaari, A.; Hashim, M. Simple Synthesis and Characterization of Cobalt Ferrite Nanoparticles by a Thermal Treatment Method. J. Nanomater. Volume 2010, Article ID 907686, 8 pp.

[50] Li, X.Q.; Elliot, D.W.; Zhang, W.X. Zero-valent iron nanoparticles for abatement ofenvironmental pollutants: materials and engineering aspects. Crit. Rev. Solid State, 2006, 31, 111–22.

[51] Pan, J.R.; Huang, C.; Hsieh, W.-P.; Wu, B.J. Reductive catalysis of novel TiO_2/Fe^0 composite under UV irradiation for nitrate removal from aqueous solution. Separation and Purification Technology, 2012, 84, 52–55.

[52] Kustov, L.M.; Al-Abed, S.R.; Virkutyte, J.; Kirichenko, O.A.; Shuvalova, E.V.; Kapustin, G.I.; Mishin, I.V.; Nissenbaum, V.D.; Tkachenko, O.P.; Finashina, E.D. Novel Fe-Pd/SiO_2 catalytic materials for degradation of chlorinated organic compounds in water. Pure Appl. Chem. 2014, 86(7), 1141–1158.

[53] Amir, A.; Lee, W.Enhanced reductive dechlorination of tetrachloroethene by nanosized zero valent iron with vitamin B12. Chem. Engin. J., 2011, 170, 492–497.

[54] Dror, I.; Jacov, O.M.; Cortis, A.; Berkowitz, B. Catalytic Transformation of Persistent Contaminants Using a New Composite Material Based on Nanosized Zero-Valent Iron. ACS Appl. Mater. Interfaces, 2012, 4, 3416–3423.

[55] Varanasi, P.; Fullana, A.; Sidhu, S. Remediation of PCB contaminated soils using iron nano-particles. Chemosphere, 2007, 66(6), 1031–1038.

[56] Dehghani, M., Shahsavani, E.; Farzadkia, M.; Reza Samaei, M. Optimizing photo-Fenton like process for the removal of diesel fuel from the aqueous phase. J. Environ. Health Sci. Eng., 2014, 12, 87.

[57] Messele, S.A.; Bengoa, C.; Stuber, F.; Fortuny, A.; Fabregat, A.; Font, J. Catalytic wet peroxide oxidation of phenol using nanoscale zero-valent iron supported on activated carbon. Desalination and Water Treatment, 2015, 55, 1–10. http://www.tandfonline.com/doi/full/10.1080/19443994.2014.1002011#preview

[58] Torres Galvis, H.M.; Bitter, J.H.; Khare, C.B.; Ruitenbeek, M.; Dugulan, A.I.; de Jong, K.P. Supported Iron Nanoparticles as Catalysts for Sustainable Production of Lower Olefins. Science, 2012, 335, 835.

[59] Zhang, C.; Zhou, L.; Yang, J.; Yu, X.; Jiang, Y.; Zhou, M.Nanoscale zero-valent iron/AC as heterogeneous Fenton catalysts in three-dimensional electrode system. Environ. Sci. Pollut. Res. 2014, 21, 8398–8405.

[60] Zhang, S.; Wang, D.; Zhang, X.; Fan, P. Zero-Valent Iron Immobilized on Multi-Walled Carbon Nanotubes for Heterogeneous Catalytic Ozonation of Methylene Blue as Model Compound. Clean – Soil, Air, Water, 2014, 42 (5), 609–616.

[61] Datta, K.K.R.; Petala, E.; Datta, K.J.; Perman, J.A.; Tucek, J.; Bartak, P., Otyepka, M.; Zoppellaro, G.; Zboril, R. NZVI modified magnetic filter paper with high redox and catalytic activities for advanced water treatment technologies. Chem. Commun., 2014, 50, 15673–15676.

[62] Varma, R.S. Nano-catalysts with magnetic core: sustainable options for greener synthesis. Sustainable Chemical Processes, 2014, 2, 11.

[63] Wei, S.; Wang, Q.; Zhu, J., Sun, L.; Line, H.; Guo, Z. Multifunctional composite core–shell nanoparticles. Nanoscale, 2011, 3, 4474–4502

[64] Govan, J.; Gun'ko, Yu.K. Recent Advances in the Application of Magnetic Nanoparticles as a Support for Homogeneous Catalysts. Nanomaterials, 2014, 4, 222–241.

[65] Liu, W.-J.; Qian, T.T.; Jiang, H. Bimetallic Fe nanoparticles: Recent advances in synthesis and application in catalytic elimination of environmental pollutants. Chemical Engineering Journal, 2014, 236, 448–463.

[66] Hsieh, S.; Lin, P.-Y. FePt nanoparticles as heterogeneous Fenton-like catalysts for hydrogen peroxide decomposition and the decolorization of methylene blue. J. Nanopart. Res. 2012, 14, 956.

[67] Mao, Z.; Wu, Q.; Wang, M.; Yang, Y.; Long, J.; Chen, X. Tunable synthesis of SiO_2-encapsulated zero-valent iron nanoparticles for degradation of organic dyes. Nanoscale Research Letters, 2014, 9, 501.

[68] Li, Q.-x.; Sun, Z.-y.; Wang, T. Edited by: Casciano, D.A.; Sahu, S.C. A study on preparation, characterization and photocatalytic activity of iron-doped nano-TiO_2 thin films based on self-assembled monolayers. Kuangwu Xuebao, 2011, 31(1), 102–107.

[69] Katsenovich, Y.P.; Miralles-Wilhelm, F.R. Evaluation of nanoscale zerovalent iron particles for trichloroethene degradation in clayey soils. Science of the Total Environment, 2009, 407, 4986–4993.

[70] Ji, Y.; Wu, Y.; Zhao, G.; Wang, D.; Liu, L.; He, W.; Li, Y. Porous bimetallic Pt-Fe nanocatalysts for highly efficient hydrogenation of acetone. Nano Research, 2015, 8(8), 2706-2713. http://link.springer.com/article/10.1007%2Fs12274–015-0777-z

[71] Parimala, L.; Santhanalakshmi, J. Studies on the Iron Nanoparticles Catalyzed Reduction of Substituted Aromatic Ketones to Alcohols. Journal of Nanoparticles. Volume 2014, Article ID 156868, 10 pp.

[72] Hudson, R.; Chazelle, V.; Bateman, M.; Roy, R.; Li, C.; Moores, A. Sustainable synthesis of magnetic ruthenium-coated iron nanoparticles and application in the catalytic transfer hydrogenation of ketones. ACS Sustainable Chem. Eng., 2015, 3 (5), 814–820.

[73] Matsumoto, T.; Sadakiyo, M.; Lee Ooi, M.; Kitano, S.; Yamamoto, T.; Matsumura, S.; Kato, K.; Takeguchi, T.; Yamauchi, M. CO_2-Free Power Generation on an Iron Group Nanoalloy Catalyst via Selective Oxidation of Ethylene Glycol to Oxalic Acid in Alkaline Media. Sci. Reports, 2014, 4, 5620, 6 pp.

[74] Mazumder, V.; Chi, M.; More, K.L.; Sun, S. Core/Shell Pd/FePt Nanoparticles as an Active and Durable Catalyst for the Oxygen Reduction Reaction. J. Am. Chem. Soc. 2010, 132, 7848–7849.

[75] Hudson, R.; Riviere, A.; Cirtiu, C.M.; Luska, K.L.; Moores, A. Iron-iron oxide core–shell nanoparticles are active and magnetically recyclable olefin and alkyne hydrogenation catalysts in protic and aqueous media. Chem. Commun., 2012, 48, 3360–3362.

[76] Zhang, Z.; Zhou, Y.; Zhang, Y.; Zhou, S.; Xiang, S.; Sheng, X.; Jiang, P. A highly reactive and magnetic recyclable catalytic system based on AuPt nanoalloys supported on ellipsoidal Fe@SiO_2. J. Mater. Chem. A, 2015, 3, 4642–4651.

[77] Mahinroosta, M. Catalytic effect of commercial nano-CuO and nano-Fe_2O_3 on thermal decomposition of ammonium perchlorate. Journal of Nanostructure in Chemistry, 2013, 3, 47.

[78] Lesin, V.I.; Pisarenko, L.M.; Kasaikina, O.T. Colloidal Catalysts Based on Iron (III) Oxides. 1. Decomposition of Hydrogen Peroxide. Colloid Journal, 2012, 74 (1), 85–90.

[79] Sharma, G.; Jeevanandam, P. A Facile Synthesis of Multifunctional Iron Oxide@Ag Core–Shell Nanoparticles and Their Catalytic Applications. Eur. J. Inorg. Chem., 2013, 6126–6136.

[80] Zaleska, A. Doped-TiO_2: A Review. Recent Patents on Engineering, 2008, 2, 157–164.

[81] Aufan, M. Nanoparticules d'oxydes mrtalliques: relations entre la reactivite de surface et des reponses biologiques. http://www.tel.archives-ouvertes.fr/docs/00/30/85/03/PDF/auffan_PhD.pdfSimilares.

[82] Jing, C.; Xiaoguang Meng; Suqin Liu; Salem Baidas; Ravi Patraju; Christos Christodoulatos; George P. Korfiatis. Surface complexation of organic arsenic on nanocrystalline titanium oxide. Journal of Colloid and Interface Science, 2005, 290(1), 14–21.

[83] Tada, H.; Soejima, T.; Ito, S.; Kobayashi, H. Photoinduced desorption of sulfur from gold nanoparticles loaded on metal surfaces. J. Amer. Chem. Soc. 2004, 126(49), 15952–15953

[84] Bepari, R.A.; Bharali, P.; Das, B.K. Controlled synthesis of α-and γ-Fe_2O_3 nanoparticles via thermolysis of PVA gels and studies on α-Fe_2O_3 catalyzed styrene epoxidation. Journal of Saudi Chemical Society, 2014, In press.

[85] Sadiq, M.; Zamin, G.; Razia; Ilyas, M. Synthesis and Characterization of Iron Oxide Nanoparticles Supported on Ziconia and Its Application in the Gas-Phase Oxidation of Cyclohexanol to Cyclohexanone. Modern Research in Catalysis, 2014, 3, 12–17

[86] Liu, Y.-T.; Yuan, Q.-B.; Duan, D.-H.; Zhang, Z.-L.; Hao, X.-G.; Wei, G.-Q.; Liu, S.-B. Electrochemical activity and stability of core-shell Fe_2O_3/Pt nanoparticles for methanol oxidation. Journal of Power Sources, 2013, 243, 622–629.

[87] Sobhani, S.; Pakdin-Parizi, Z.; Nasseri, R. Nano n-propylsulphonated γ-Fe_2O_3: A novel magnetically recyclable heterogeneous catalyst for the efficient synthesis of bis(pyrazolyl)methanes in water. J. Chem. Sci. 2013, 125 (5), 975–979.

[88] Cuong, N.D.; Hoa, N.D.; Hoac T.T.; Khieu, D.Q.; Quang, D.T.; Quang, V.V.; Hieu, N.V. Nanoporous hematite nanoparticles: Synthesis and applications for benzylation of benzene and aromatic compounds. J. Alloys Comp. 2014, 582, 83–87.

[89] Bartolome, L.; Imran, M.; Lee, K.G.; Sangalang, A.; Keun Ahnd, J.; Hyun Kim, D. Superparamagnetic γ-Fe_2O_3 nanoparticles as an easily recoverable catalyst for the chemical recycling of PET. Green Chem., 2014, 16, 279–286.

[90] Wang, C.; Liu, H.; Sun, Z. Heterogeneous Photo-Fenton Reaction Catalyzed by Nanosized Iron Oxides for Water Treatment. International Journal of Photoenergy, 2012, Article ID 801694, 10 pp.

[91] Han, J.; Zhang, S.-j.; Lu, Z.-g.; Wang, Y.-j. Doping Fe^{3+} to nano-Ag TiO_2 for photocatalytic performance improvement. Yingyong Guangxue, 2010, 31(5), 718–723.

[92] Sharon, M.; Pal, B.; Kamat, D.V. Photocatalytic killing of pathogenic bacterial cells using nanosize Fe_2O_3 and carbon nanotubes. Journal of Biomedical Nanotechnology, 2005, 1(3), 365–368.

[93] Yin, Z.-Z.; Li, Y.; Jiang, L.-P.; Kumar Rana, R.; Jun-Jie Zhu. Synthesis and electrocatalytic activity of haemin-functionalised iron(II,III) oxide nanoparticles. Analytica Chimica Acta, 2013, 781, 48–53.

[94] Wang, H.; Hodgson, J.; Shrestha, T.B.; Thapa, P.S.; Moore, D.; Wu, X.; Ikenberry, M.; Troyer, D.L.; Wang, D.; Hohn, K.L.; Bossmann, S.H. Carbon dioxide hydrogenation to aromatic hydrocarbons. Beilstein J. Nanotechnol. 2014, 5, 760–769.

[95] Safari, J.; Zarnegar, Z. Sulphamic acid-functionalized magnetic Fe_3O_4 nanoparticles as recyclable catalyst for synthesis of imidazoles under microwave Irradiation. J. Chem. Sci. 2013, 125 (4), 835–841.

[96] Khodabakhshi, S.; Karami, B. Mojtaba Baghernejad Iron(II,III) oxide nanoparticle-catalyzed selective synthesis of unknown dihydropyrano[c]chromenes under green conditions. Monatsh Chem., 2014, 145, 1839–1843.

[97] Safari, J.; Zarnegar, Z. A magnetic nanoparticle-supported sulfuric acid as a highly efficient and reusable catalyst for rapid synthesis of amidoalkyl naphthols. Journal of Molecular Catalysis A: Chemical, 2013, 379, 269–276.

[98] Bayat, A.; Shakourian-Fard, M.; Ehyaei, N.; Mahmoodi Hashemi, M. Silver nanoparticles supported on silica-coated ferrite as magnetic and reusable catalysts for oxidant-free alcohol dehydrogenation. RSC Adv., 2015, 5, 22503–22509.

[99] Girija, G.; Bhojya Naik, H.S.; Vinay Kumar, B.; Sudhamani, C.N.; Harish, K.N. Fe_3O_4 nanoparticle supported Ni(II) complexes: A magnetically recoverable catalyst for Biginelli reaction. Arabian Journal of Chemistry, 2014, Available from: http://www.sciencedirect.com/science/article/pii/S1878535214001701, In press.

[100] Goodarz Naseri, M.; Saion, E.B.; Abbastabar Ahangar, H.; Halim Shaari, A.; Hashim, M. Simple Synthesis and Characterization of Cobalt Ferrite Nanoparticles by a Thermal Treatment Method. J. Nanomater. Volume 2010, Article ID 907686, 8 pp.

[101] Kaur Rajput, J.; Kaur, G. $CoFe_2O_4$ nanoparticles: An efficient heterogeneous magnetically separable catalyst for "click" synthesis of arylidene barbituric acid derivatives at room temperature. Chinese Journal of Catalysis, 2013, 34, 1697–1704.

[102] Kooti, M.; Afshari, M. Magnetic cobalt ferrite nanoparticles as an efficient catalyst for oxidation of alkenes. Scientia Iranica F, 2012, 19 (6), 1991–1995.

[103] Kamal Senapati, K.; Phukan, P. Magnetically separable cobalt ferrite nanocatalyst for aldol condensations of aldehydes and ketones. Bulletin of the Catalysis Society of India, 2011, 9, 1–8.

[104] Kumar, P.V.; Short, M.P.; Yip, S.; Yildiz, B.; Grossman, J.C. High Surface Reactivity and Water Adsorption on $NiFe_2O_4$ (111) Surfaces. J. Phys. Chem. C., 2013, 117 (11), 5678–5683.

[105] Hong, D.; Yamada, Y.; Nagatomi, T.; Takai, Y.; Fukuzumi, S.Catalysis of Nickel Ferrite for Photocatalytic Water Oxidation Using $[Ru(bpy)_3]^{2+}$ and $S_2O_8^{2-}$. J. Am. Chem. Soc. 2012, 134, 19572–19575.

[106] Radhakrishnan Nair, T.D.; Aniz, C.U. Effect of Redox Nature of Impregnated Ferrite Catalysts on Their Carbon Monoxide Oxidation Activity. RRJMS, 2013, 1 (2), 45–52.

[107] Wan, Q.; Li, P.; Shan, J.; Zhai, F.; Li, Z.; Qu, Z. Superior Catalytic Effect of Nickel Ferrite Nanoparticles in Improving Hydrogen Storage Properties of MgH_2. J. Phys. Chem. C, 2015, 119, 2925–2934.

[108] Eshghi, H.; Javid, A.; Khojastehnezhad, A.; Moeinpour, F.; Bamoharram, F.F.; Bakavoli, M.; Mirzaei, M. Preyssler heteropolyacid supported on silica coated $NiFe_2O_4$ nanoparticles for the catalytic synthesis of bis(dihydropyrimidinone)benzene and 3,4-dihydropyrimidin-2(1H)-ones. Chinese Journal of Catalysis, 2015, 36, 299–307.

[109] Lou, J.-C.; Chang, C.-K. Catalytic Oxidation of CO Over a Catalyst Produced in the Ferrite Process. Env. Eng. Sci., 2006, 23 (6), 1024–1032.

[110] Kasi Viswanath, I.V.; Murthy, Y. L. N. One-Pot, Three-Component Synthesis of 1, 4-Dihydropyridines by Using Nano Crystalline Copper Ferrite. Chem. Sci. Trans., 2013, 2 (1), 227–233.

[111] Murthy, Y. L. N.; Diwakar, B.S.; Govindh, B.; Nagalakshmi, K.; Kasi Viswanath, I.V.; Singh, R. Nano copper ferrite: A reusable catalyst for the synthesis of β,γ-unsaturated ketones. J. Chem. Sci., 2012,124 (3), 639–645.

[112] Gharib, A.; Noroozi Pesyan, N.; Vojdani Fard, L.; Roshani, M Catalytic Synthesis of α-Aminonitriles Using Nano Copper Ferrite ($CuFe_2O_4$) under Green Conditions. Organic Chemistry International, Volume 2014, Article ID 169803, 8 pp.

[113] Yang, D.; Zhu, X.; Wei, W.; Jiang, M ; Zhang, N.; Ren, D.; You, J.; Wang, H. Magnetic Copper Ferrite Nanoparticles: An Inexpensive, Efficient, Recyclable Catalyst for the Synthesis of Substituted Benzoxazoles via Ullmann-Type Coupling under Ligand-Free Conditions. Synlett, 2014, 25, 729–735.

[114] Rashad, M.M.; Mohamed, R.M.; Ibrahim, M.A.; Ismail, L.F.M.; Abdel-Aal, E.A. Magnetic and catalytic properties of cubic copper ferrite nanopowders synthesized from secondary resources. Advanced Powder Technology, 2012, 23, 315–323.

[115] Papa, F.; Patron, L.; Carp, O.; Paraschiv, C.; Balint, I. Catalytic behavior of neodymium substituted zinc ferrites in oxidative coupling of methane. Rev. Roum. Chim. 2010, 55 (1), 33–38.

[116] Jadhav S.V.; Mohan Jinka, K.; Bajaj, H.C. Nanosized sulfated zinc ferrite as catalyst for the synthesis of nopol and other fine chemicals. Catalysis Today, 2012, 198, 98–105.

[117] de Moura, E.M.; Garcia, M.A.S.; Gonçalves, R.V.; Kiyohara, P.K.; Jardimc, R.F.; Rossi, L.M. Gold nanoparticles supported on magnesium ferrite and magnesium oxide for the selective oxidation of benzyl alcohol. RSC Adv., 2015, 5, 15035–15041.

[118] Han, A.; Liao, J.; Ye, M.; Yan, L.; Peng, X. Preparation of Nano-$MnFe_2O_4$ and Its Catalytic Performance of Thermal Decomposition of Ammonium Perchlorate. Chinese Journal of Chemical Engineering, 2011, 19(6), 1047–1051.

[119] Ghahremanzadeh, R.; Rashid, Z.; Zarnanic, A.-H.; Naeimi, H. Manganese ferrite nanoparticle catalyzed tandem and green synthesis of spirooxindoles. RSC Adv., 2014, 4, 43661–43670.

[120] Zhenyu, L.; Guangliang, X.; Yalin, Z. Microwave assisted low temperature synthesis of MnZn ferrite nanoparticles. Nanoscale Res. Lett. 2007, 2, 40–43.

[121] Velinov, N.; Koleva, K.; Tsoncheva, T.; Paneva, D.; Manova, E.; Kunev, B.; Mitov, I. Copper-cobalt ferrites as catalysts for methanol decomposition. 11th European Congress on Catalysis – EuropaCat-XI, Lyon, France, September 1st-6th, 2013.

[122] Wang, L.L.; He, H.Y. Surface Alkaline-Acidic Characteristics and Photo catalytic Properties of $Co_{1-x}Zn_xFe_2O_4$ Nano-particles Synthesized by Hydrothermal Method. J. Sci. Res. & Reports, 2014, 3 (2), 263–274; Article no. JSRR.2014.001.

[123] Khojastehnezhad, A.; Rahimizadeh, M.; Eshghi, H.; Moeinpour, F.; Bakavoli, M. Ferric hydrogen sulfate supported on silica-coated nickel ferrite nanoparticles as new and green magnetically separable catalyst for 1,8-dioxodecahydroacridine synthesis. Chinese J. Cat., 2014, 35, 376–382.

[124] Kumar B. Gupta, S.; Venkatachalam, A. Synthesis and characterization of spinel ferrites $Cu_{1-x}Cd_x[Fe_{1-x}Al_xCr_{1-x}Mn_x]O_4$. Rasayan J. Chem., 2010, 3 (4), 745–750.

[125] Dixit, R.; Gupta, P.; Saxena, R.; Dwivedi, R. Methylation of Aniline over Mn-Cu Ferrites Catalysts. Global Journal of Science Frontier Research Chemistry, 2013, 13 (7), 10 pp.

[126] Dandia, A.; Singh, R.; Joshi, J.; Maheshwari, S. Magnetically separable $CuFe_2O_4$ nanoparticles: an efficient catalyst for the synthesis of quinoxaline derivatives in tap-water under sonication. Eur. Chem. Bull., 2013, 2 (10), 825–829.

[127] Manova, E.; Estournès, C.; Paneva, D.; Reh, J.-L.; Tsoncheva, T.; Kunev, B.; Mitov, I. Mössbauer study of nanodimensional nickel ferrite – mechanochemical synthesis and catalytic properties. ICAME 2005 (Proceedings of the 28th International Conference on the Applications of the Mössbauer Effect (ICAME 2005), Montpellier, France, 4–9 September 2005, Volume I (Part I–II/V). Editors: Lippens, P.-E.; Jumas, J.-C.; Génin, J.-M. R.), 2007, 215–220.

[128] Wu, Y.; Song, M.; Xin, Z.; Zhang, X.; Zhang, Y.; Wang, C.; Li, S.; Gu, N. Ultra-small particles of iron oxide as peroxidase for immunohistochemical detection. Nanotech., 2011, 22, 225703, 8 pp.

Chapter 3

New Catalytic Systems for Fixation of Carbon Dioxide into Valuable Poly(Alkylene Carbonates)

Bahareh Bahramian and Fariba Dehghani

Additional information is available at the end of the chapter

http://dx.doi.org/10.5772/61969

Abstract

Fixation of carbon dioxide into valuable products is a promising method to mitigate the issues of global warming and decrease the consumption of fossil-fuel carbon sources. Poly alkylene carbonates (PACs) are environmentally friendly and low-cost polymers that are synthesized from copolymerization of carbon dioxide and epoxides. PACs are contemplated as an alternative to commercially available non-degradable polymers in the market for a broad range of applications. However, a burden for the synthesis of this group of polymers is the chemical activation of thermodynamically stable CO_2. It is, therefore, imperative to develop a catalyst with high efficiency to overcome this hurdle. In this chapter, we describe the development and recent advances in the catalytic systems that have been designed to activate CO_2 for copolymerization with epoxides. In particular, we will focus on the industrial trends presented in the patents for conversion of CO_2 into PACs.

Keywords: Carbon Dioxide Fixation, Epoxides, Polycarbonates, Catalysts

1. Introduction

Carbon dioxide is one of the main greenhouse gases that contributes to global warming [1]. One of the approaches to tackle this issue is to fix CO_2 as a raw material into valuable polymers [2]. The polymer industry is an emerging market that mostly relies on petrochemical products derived from fossil-fuel feedstocks. However, CO_2 is a cheap, renewable, non-toxic and the most abundant carbon source. The consumption of CO_2 as a reactant for the synthesis of carbon-based products is a key step to cut down the reliance on the fossil fuels.

Alternative copolymerization of CO_2 with epoxides, such as propylene oxide (PO) and cyclohexene oxide (CHO), results in the formation of biodegradable PACs. As the physical properties of PACs are comparable to conventional polymers, they can be used for a broad range of applications such as packaging, agricultural and biomedical industries [3]. One of the obstacles in the synthesis of PACs is the design of an efficient catalyst that can reduce the activation energy of thermodynamically stable CO_2 for such polymerization reaction [4]. In addition to activity, it is pivotal to design a selective catalyst that reduces the yield of by-products during the PAC copolymerization reactions [5]. Other factors that are contemplated in developing a catalyst are cost and toxicity.

Since 1696 that poly(propylene carbonate) (PPC) was synthesized, many efforts have been attempted to design catalysts [6]. These catalysts are mainly classified into two categories: (1) homogeneous and (2) heterogeneous. This classification is based on their solubility in the reaction media; therefore, the homogenous catalysts are those that are in the same phase as reactants and heterogeneous are those that are in another phase. Regardless of large number of research in this area, only a few of conventional metal-based heterogeneous catalysts have been used for the commercial PACs synthesis that possess acceptable activity and selectivity for large-scale production [3]. However, physical properties of this type of catalysts affect their selectivity and activity remarkably. For example, particle size, crystallinity, microstructure and morphology are key factors that show impact on the activity of a catalyst and the yield of the final product [7–9]. Many bench-scale studies have been conducted to promote the catalyst activity for the synthesis of PACs, and particularly poly(propylene carbonate) [3, 10]. In this chapter, we provide an insight about these types of catalysts. Prior to a discussion about catalyst, the main advantages of PACs are described. Then various catalysts that are available for their synthesis are introduced, followed by discussion about strategies that have been undertaken to promote the activity of these catalysts and finally, an overview of patents filed for the synthesis of PACs are briefly reviewed.

2. Copolymerization of CO_2 and epoxides

Direct copolymerization of CO_2 and epoxides such as ethylene oxide, propylene oxide, isobutylene oxide, cycloheptene oxide, cyclopentene oxide and cyclohexene oxide results in the formation of PACs, which is a typical example of fixation of CO_2 in polymers [11]. PACs have excellent physical properties, including low density, transparency, durability, colourability and processability, and also they exhibited decent electrical insulation properties [12]. Accordingly, these types of polymers have a broad range of applications in electronics, optical media and have been used for the preparation of sheets, automotive productions, medical devices and healthcare products [10]. Furthermore, they are commercially used as a binder, plasticizer and raw material for polyurethane synthesis [3]. However, the most important feature of PACs that has attracted attention in recent years, as a potent alternative for conventional polymers, is their biodegradability. PACs degrade into water and CO_2 when exposed to moisture and enzymes or non-sterilized soil such as landfills [3]. PACs can be considered as an alternative to non-degradable polymers to address the growing concerns about the disposal of plastics in municipal waste, shortage of allocated spaces for landfills and consumer pressure for sustainable products. The list of some companies that use PACs in commercial scale for different purposes is presented in Table 1. PPC is used in combination with acrylonitrile butadiene styrene (ABS) to make a scratch resistance plastic by Bayer MaterialScience [13]. Furthermore, poly(propylene carbonate) and poly(hydroxybutyrate) are blended with other polymeric materials or mixed with inorganic solids to form the alternative plastic to acrylonitrile butadiene styrene by Baden Aniline and Soda Factory (BASF) –a chemical company [14]. Cardia Bioplastics (CO_2 Starch Pty Ltd) has developed a method to manufacture commercially PPC/starch blends [15], and they currently manufacture huge amount of degradable plastic bags globally from this blend.

Name of Company	Country	Application	Website
Life Cycle Products	England	Plastic bag and waste management products	www.lifecycleproducts.co.uk
Nature Works Packaging	Australia	Shopping bag	www.natureworkspackaging.com
Drogaria Araujo	Brazil	Biohybrid™ bags	www.araujo.com.br
AZOmaterials	England	Organic binders for nanoparticles	www.azom.com
Novomer	USA	Polyurethane hot-melt adhesive	www.novomer.com
Mengxi High-tech Group	China	Medical dressings, biodegradable packaging	www.mengxigroup.com
Jiangsu Jinlong-CAS Chemical Co., Ltd	China	Biodegradable packaging	www.zkjlchem.com
TaiZhou BangFeng Plastic Co., Ltd	China	Disposable biodegradable dishes	www.cn-bf.net
Cardia Bioplastics™	Australia	Shopping bag	www.cardiabioplastics.com

Table 1. List of Some Companies That Use PACs in Commercial Scale

PACs can be used for biomedical applications as an alternative to current biodegradable polymers such as poly(lactic acid) (PLA). For example, enzymatic degradation of PPC leads to producing benign products that include only water and CO_2, which does not cause any inflammation in biological environment. However, PLA degrades by hydrolysis and generates lactic acid, reduces the pH in surrounding tissue leads to inflammation. In addition, studies by Zhong et al. in 2012 and Yang et al. in 2013 demonstrated the potential of using PPC for producing scaffolds for tissue engineering application [16–18].

PACs are superior polymers for the agricultural purposes. For example, they can be used for the production of plastic films for agricultural mulches as an alternative to non-degradable polyethylene. It is therefore eradicating the risk of accumulation of plastic residues in agricultural spots, the high cost of collecting mulches after the harvesting season and contamination with soil and dirt. PACs can also be used for designing biodegradable packaging products as they are transparent and have minimal permeability to oxygen and water [19].

It is pivotal to determine the presence of any impurities that may have an adverse effect on physical properties and also the toxicity of a product fabricated from a polymer for food and biomedical applications. For instance, in processing PACs, the type of epoxide, by-product, catalyst and any other impurities may have an impact on their properties [20]. In an ideal copolymerization, CO_2 and epoxide molecules form carbonate linkage. However, in reality, two molecules of epoxide may also bond and produce undesirable products. The presence of ether linkage has adverse effects on properties such as mechanical strength, thermal transition, thermal decomposition temperatures and molecular weight [21]. Therefore, the design of an active catalyst plays a critical role in promoting selectivity for the synthesis of favourable copolymer and reducing the rate of reaction of undesirable products.

3. PACs synthesis

The first discovery of PACs synthesis goes back to 1969 when Inoue et al. used a mixture of diethyl zinc ($ZnEt_2$) and water as a catalyst to conduct the copolymerization of CO_2 and propylene oxide and form PPC [22]. Shortly after that, the same group successfully used triethylaluminium as another organometallic catalyst for the synthesis of PPC [6]. However, the yield of the copolymerization reaction in the presence of these catalysts was low due to the formation of side products and presence of ether linkages on the polymer backbone. To address the issue of low catalyst activity, other hydrogen donor compounds rather than water have been used in combination with $ZnEt_2$ [23, 24]. It was found that donor compounds with two or three active hydrogen sites formed multi-site catalytic systems with higher activity and selectivity compared with mono-site components. Therefore, the reaction was conducted towards alternative copolymerization of CO_2 and PO, and the amount of undesirable cyclic propylene carbonate (cPC) reduced. Furthermore, a series of metal salts of acetic acid were used to catalyze the copolymerization of carbon dioxide and PO [25, 26]. Besides, a combination of zinc hydroxide with dicarboxylic acids was used to enhance the yield of desirable polymer product; among all, catalyst system derived from zinc, hydroxide and glutaric acid showed the superior activity [27].

One common mechanism proposed for the copolymerization of CO_2 and epoxides is coordination–insertion mechanism catalyzed via metal compounds with Lewis acid and Lewis base active sites [10, 28]. In the coordination step, the epoxide molecule is coordinated by the metallic centre of a catalyst (Lewis acid active site) and then attacked by nucleophile site (Lewis base site) and undergone ring opening to form metal-bound alkoxide [10]. The nucleophilic attack can take place by either the nucleophile active site on the metal catalyst (bifunctional homogenous catalysts) or a separate compound (binary catalysts) and resulted to activation of alkoxide [20]. CO_2 molecule then inserted into the metal-oxygen bond and initiated the reaction by forming metal carbonate. Up to this stage, all the steps were associated with the activity of the catalyst; however, the pathway of the reaction after this step relies on the

http://dx.doi.org/10.5772/61969

selectivity of the catalyst. In fact, selectivity is a function of the type of alkoxide. Commonly, the metal carbonate goes towards its ring closure and forms propylene carbonate or propagates by multiple coordination and insertion of CO_2 and produces polycarbonate chain [29]. If the second pathway is followed by the alternative coordination–insertion mechanism, the resulted polycarbonate has 100% carbonate linkage in its structure; however, some catalysts can also homopolymerize epoxides and form ether linkages on the backbone of polymer [10]. The schematic of suggested coordination–insertion mechanism for the copolymerization of epoxides and CO_2 is demonstrated in Figure 1.

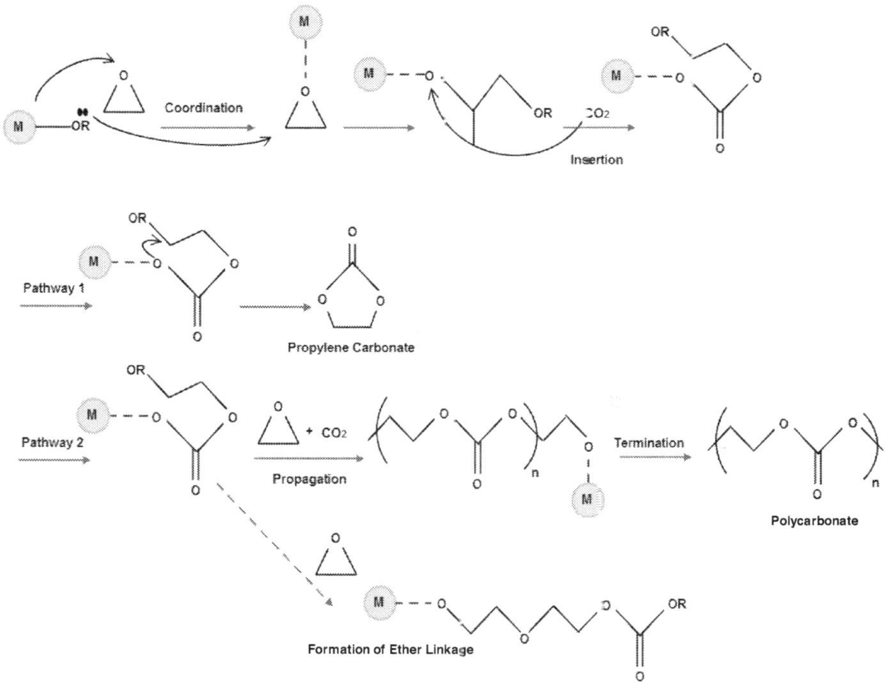

Figure 1. Coordination–insertion mechanism suggested for the copolymerization of epoxides with CO_2

4. Catalytic systems for CO_2/epoxide polymerization

Since 1696 when PPC was first synthesized, many efforts were carried out to develop both homogeneous and heterogeneous catalysts with high selectivity [30]. In the following section, various types of homogeneous and heterogeneous catalysts that have been used for the reaction between CO_2 and epoxides are described.

4.1. Homogenous catalysts

Triphenyl phosphate (TPP) compounds with a metal atom in the centre were the first group of homogeneous catalyst reported for copolymerization of CO_2 and epoxides [31, 32]. TPP compounds showed high activity; however, this catalysts were not very efficient, and the yield of PACs was very low even after more than one weak reaction. Nitrogen donors compounds such as N-methylimidazole and (4-dimethylamino)pyridine (DMAP) were added to TPP compounds to enhance the yield of reaction. However, this catalyst system could only be used for the copolymerization of COH and CO_2. For the other epoxides, only cyclic alkylene carbonates were produced [33, 34].

Chen et al. investigated the effect of metal type located in the centre of TPP compounds on the activation of propylene oxide. It was found that the type of metal influenced the acidity strength of catalysts and, therefore, the strength of epoxide metal bond and the yield of copolymer synthesis [35]. Among three types of metal used in this study, chromium-based catalyst exhibited higher selectivity compared to Al and Co towards the formation of PPC from the reaction between PO and CO_2.

In a metal complex catalyst, the metal acts as Lewis acid to attack the epoxide and after that the nucleophile part of the catalyst opens the epoxide ring. The nucleophile part of the catalyst can be either a ligand attached to the metal complex or a co-catalyst. In the former group, the nucleophile ligand can often be a salen, a porphyrin, a phthalocyanine or an amido macrocycle [30]. The nature of metal in the catalyst is critical in its activity and selectivity and also the quality of the resulted polymer. Indeed, the strength of epoxide metal bond can determine the pathway of reaction. When this bond is too weak, the reaction proceeds towards formation of ether linkages and backbiting while the high strength of the bond may lead to closure of the ring and formation of cyclic carbonates [36].

After TPP compounds, great number of metal complex catalytic systems such as phenoxide [37–39] and β-diiminate (BDI) [29, 40] were investigated in terms of their activity for copolymerization of CO_2 and epoxides that are discussed in this section.

4.1.1. Phenoxide complexes

Phenoxide compounds include an aryl ligand that plays the role of nucleophile site of a catalyst [41]. It was found that the activity of a catalyst is dramatically higher when the phenoxide compounds contain zinc atom as a metal centre [42]. It was found that this group of metal complexes are only able to activate and catalyze CHO to react with CO_2 and are not efficient in activation of other epoxides such as PO [37, 43, 44].

4.1.2. β-diiminate complexes

BDI is a family of complexes that include an aryl or acyl ligand with the molecular formula of L_nMOR (R: alkyl, acyl) [40]. BDIs have broad applications in inorganic as well as organic and polymer synthesis, especially in the polymerization of lactides [45]. In the case of copolymerization of epoxides and CO_2, BDIs have been used in a large number of studies. The results demonstrated that both the nature of metal centre and the aryl ligand have significant effect on the activity and selectivity of the catalyst [29, 40, 46]. For example, only zinc metal showed remarkable efficiency and selectivity towards the copolymerization of propylene oxide and carbon dioxide [47, 48].

4.1.3. Metal salen complexes

Salens are a class of organic compounds including 1,6-bis(2-hydroxyphenyl)-2,5-diaza-hexa-1,5-diene ligand that are broadly used in the synthetic chemistry [49]. Metal salen complexes were first used for the ring opening copolymerization of epoxides in 1995 [50]. This family of catalysts has several advantages over other complexes for the copolymerization of epoxides and CO_2. They are highly selective towards the synthesis of polycarbonates in the mild reaction condition (ambient temperature) [51]. Metal salen catalysts can be categorized according to the type of metal in the centre.

Chromium-based salen complexes ((salen)CrCl) were efficient in the copolymerization of epoxides and CO_2 [52]. Two years after the first discovery, Jacobsen et al. successfully synthesized PCHC with 100% carbonate linkage using chiral salen chromium chloride catalyst [53]. The chromium-based salen complexes, however, showed low activity in the copolymerization of CO_2 and PO due to the negligible differences between the activation energy of PO and PPC compared to cyclohexene carbonate (CHC) versus poly(cyclohexene carbonate) (PCHC) [54]. Therefore, the selectivity of the catalyst is a critical factor for the copolymerization of PO and CO_2. The addition of co-catalyst was found to be an efficient approach to tackle this issue. In 2003, Rieger et al. attempted to add co-catalyst 4-(N,N-dimethylamino)pyridine (DMAP) to chromium salen and were able to successfully synthesize PPC with 98% carbonate linkage and minimal PC side product [55]. They proposed that strong coordination of DMAP to chromium promoted the propagation of polymer chain and formation of carbonate linkages. The type of initiator was another factor that had an impact on the selectivity of chromium salen catalysts for the synthesis of PPC. It was found that changing the initiator from phosphines to azide in combination with Cr(salen)N_3 catalyst significantly stimulated the reaction pathway towards the formation of PPC rather than cyclic propylene carbonate [56].

The metal centre has a pivotal role in catalyst activity and selectivity of the metal salen complexes. The first cobalt salen complexes (Co(salen)AOc) was introduced in 2003 for the copolymerization of PO and CO_2, which resulted in producing PPC with 99% carbonate linkage [57]. In addition, it was found that similar to the chromium salen catalysts, the addition of co-catalyst to the cobalt salen complexes has an impact on their activity and selectivity. For instance, the addition of sub-stoichiometric amount of bis(triphenylphosphine)iminium (PPN) as co-catalyst to a cobalt salen increased the yield of PPC synthesis to 36% with 99% carbonate linkage [58]. However, any further effect on yield enhancement led to the formation of cyclic propylene compounds due to backbiting degradation of the polymer chain. To accomplish high activity of a catalyst without losing the selectivity, Nakano et al. designed a cobalt salen complex with piperidinyl and piperidinium arms [59]. This modification resulted in producing 99% PPC with 97% conversion yield. Basically in this reaction, the side arms played the role of in situ co-catalysts. This approach was a breakthrough in the copolymerization of CO_2 and epoxides as it reduced the catalyst loading significantly. Similarly, a cobalt complex containing two tertiary amine cations on pendant arms was designed and showed high activity and above 90% conversion yield for very low catalyst loadings such as 1:50,000 (catalyst/ PO) [60].

Many research activities have focused on the area of homogeneous catalysts for CO_2–epoxide copolymerization due to the design flexibility and high activity. However, none of them has been used in large scale due to complicated synthesis process and low selectivity for PACs copolymerization. On the other hand, heterogeneous catalysts are generally non-toxic and economically viable due to the simpler synthesis process and easier handling. In the next section, the heterogeneous catalysts that are designed for the synthesis of PACs are described.

4.2. Heterogeneous catalysts

4.2.1. Organometallic compounds

The organometallic compounds for the copolymerization of CO_2 and epoxides are designed from the combination of a hydrogen donor compound and a metal-based complex to activate CO_2 and open the ring of the epoxide. The commonly used hydrogen donors are water, primary amines, dihydric phenols, trihydric phenols, aromatic dicarboxylic acids and aromatic hydroxycarboxylic acids [61]. Several efficient organometallic catalysts with well-defined structures have been developed for the copolymerization of carbon dioxide and epoxide. However, the activity of the catalyst derived from zinc hydroxide and glutaric acid was superior compared to other compounds [27]. Ree et al. copolymerized PO and CO_2 using zinc glutarate (ZnGA) obtained from various zinc sources. As a result, zinc glutaric derived from zinc oxide and glutaric acid yielded the highest catalyst activity of 64 g /g of catalyst [62]. Discovery of the catalyst activity of zinc glutarate for copolymerization of CO_2 and epoxides was a breakthrough in the field especially after 1995 when Darensbourg et al. substituted organic solvents, as the reaction media, by supercritical fluid [63]. This makes the copolymerization process more environmentally friendly and economically viable.

4.2.2. Double-Metal Cyanide (DMC) complexes

DMCs are another group of heterogeneous catalyst that are efficiently used for the homopolymers of epoxides. The first DMCs that showed average activity for copolymerization of PO and CO_2 to produce PPC were $Zn_3[Fe(CN)_6]_2$ and $Zn_3[Co(CN)_6]_2$, a double-metal cyanide compound based on Zn and Fe or Co [64–66]. The activity of zinc–cobalt–DMC catalyst was comparable with zinc glutarate [3, 21]. However, it was found that the system suffered from low selectivity at low temperatures and poor activity at high temperatures [67]. In addition to these catalysts, Darensbourg et al. and Robertson et al. attempted to modify the catalyst structure by increasing the crystallinity of DMC-based catalyst [68, 69]. It was also found that the low molecular weight polyols could act as an initiator to activate DMCs and promote the copolymerization reaction [70].

http://dx.doi.org/10.5772/61969

4.2.3. *Rare earth metal catalysts*

In the very first attempts, rare earth metal complexes such as $Y(P_2O_4)_3$–$Al(i\text{-}Bu)_3$–glycerine were used for the copolymerization of PO and CO_2 and formation of high molecular weight PPC. However, the structure of the resulted copolymer consisted of up to 90% of ether linkages [71]. In the next discoveries, rare earth metal catalysts showed an increase in selectivity and reduction in synthesis time for copolymerization of CO_2 with epoxides [72–74]. Yttrium carboxylate in combination with $ZnEt_2$ significantly improved carbonate linkage percentage on the PPC and PCHC backbone [75]. A ternary catalytic system of rare earth complex, diethyl zinc and glycerine resulted in an extremely high molecular weight PPC [76]. Likewise, PO was successfully copolymerized with CO_2 and formed PPC by using $Y(CCl_3COO)_3$, $ZnEt_2$ and glycerine ternary system [76]. Rare earth metal catalytic systems, in general, can be used to adjust the microstructure of the polymer chain. For instance, head-to-tail linkage and molecular weight distribution of PPC was successfully adjusted by using yttrium carboxylate as a catalyst.

Comparing all heterogeneous catalytic systems created so far, conventional zinc glutarate is the only catalyst that has been used commercially for alternative copolymerization of CO_2 and PO. Zinc glutarate exhibits high activity and favourable selectivity [62, 77]. It is cheap, non-toxic and easy to synthesize and can be used for manufacturing relatively high molecular weight PPC with superior carbonate linkage percentage [5]. However, its activity is still one or two order of magnitude lower than the common catalysts used for synthesis of conventional polyolefins [78]. Therefore, enhancement of the catalytic activity of ZnGA was the topic of several studies [7–9].

4.2.4. *Catalyst activity of zinc glutarate*

Particle size, crystallinity, microstructure and morphology are the key factors that have impacts on activity and yield of the final product. These characteristics can be tuned by changing the source of zinc and glutarate, their particle size, purity, synthesis method and processing condition.

Ree et al. studied the effect of various zinc sources on the activity of ZnGA catalyst [62]. Their results showed that the highest activity of ZnGA was achieved when zinc was derived from zinc oxide. Zinc glutarate from various sources of glutarate was synthesized in another study carried out by Ree et al. [7]. Results of catalyst activity demonstrated that the yield of PPC was significant when zinc oxide and glutaric acid were used for the synthesis of ZnGA. The effect of zinc and glutarate source and synthesis media on the microstructure of ZnGA was also investigated in a study carried out by Kim et al. [8]. This study demonstrated that the highest activity was acquired when ZnGA was synthesized in toluene media due to achieving low surface area and perfect crystallinity.

Particle size and purity of reactants are other parameters that have effect on the activity of ZnGA [9]. It was shown that highly pure ZnO with large particle size resulted in ZnGA with the superior crystallinity and large particle size. In addition, in a bench-scale study, it was shown that when magnetic stirring was used rather than other methods of mixing, the degree of crystallinity of ZnGA was promoted and resulted in achieving higher yield and larger molecular weight PPC. In yet another study, Pluronic PE6400 was used as an amphiphilic template to increase the degree of crystallinity and enhance the surface area of ZnGA [79]. It is important to note that in this study a mixture of ethanol/water was used as a solvent for the synthesis of ZnGA, and they achieved 83 g PPC/g activity for ZnGA. The yield of PPC production was increased nearly 20% when this catalyst was used for the copolymerization reaction.

Reducing the particle size of crystalline particles of ZnGA is another strategy that has been carried out to increase the catalyst activity [9]. Ball milling method can be used to reduce the particle size; however, it is important to optimize the processing time as long-hour milling might reduce the degree of crystallinity. It was also attempted to increase the surface area to elevate the ZnGA catalyst activity. As an example, PPC with the yield of 115.2 g PPC/g catalyst was obtained when high surface montmorillonite (MMT) was used for ZnGA support [80].

Optimization of the reaction condition was the subject of several studies with the aim of increasing the yield of polymerization. High yield of 126 g PPC/g catalyst was reported for a zinc glutarate supported on a perfluorinated compound with PO/catalyst ratio of 200 ml/g under mechanical stirring [81]. An optimum PO/catalyst ratio of 312 ml/g resulted in PPC yield of 160 g PPC/g catalyst for ZnGA prepared using ultrasonic stirring method [9]. Duan et al. in 2010 investigated the effect of pressure on the yield of copolymerization, product composition, molecular structure and thermal stability of PPC [82]. The results showed a significant enhancement in catalyst selectivity towards the formation of PPC and reduction of by-products. Dehghani research group established a green method to synthesis ZnGA in super-critical CO_2 and eliminated usage of the organic solvent [2]. This study demonstrated that the crystallinity of catalyst produced in this solvent-free process was a function of processing parameters such as temperature and pressure. In addition, the yield of PPC copolymerization was significantly improved when catalyst processed by this technique was compared to ZnGA that was synthesized in toluene.

4.3. New advances in the copolymerization of CO_2 and epoxides

To enhance the catalytic activity and selectivity towards the copolymerization of CO_2 and epoxides, several measures have been taken such as bimetallic catalyst systems, non-metallic catalysts, ionic liquids (ILs), and supported catalysts.

4.3.1. Bimetallic catalyst systems

The idea behind the bimetallic pathway for the copolymerization of carbon dioxide and epoxides is to conduct the ring opening of epoxide and activation of carbon dioxide at the same time. In fact, one of the metal centres is in charge of the coordination and activation of the epoxide and the other metal centre associates with coordination and activation of CO_2 [28]. It was found that the bimetallic pathway involved a second-order dependence on the metal concentration, whereas the monometallic pathway associated with a first order kinetic [30]. For instance, a double zinc centre complex showed high activity and selectivity in the conversion of cyclohexene oxide and CO_2 to PCHC at very low CO_2 pressure [83]. Particularly, the bimetallic pathway showed remarkable improvement in the catalyst activity of the salen catalysts. In addition, a piperidinium end-capped (salen) Co(III) complex copolymerized PO and CO_2 without the generation of cyclic propylene carbonate [59].

4.3.2. *Non-metallic catalysts and ionic liquids*

An ionic liquid is a salt in which the ions are poorly coordinated. Accordingly, these solvents are liquid at the temperatures below 100°C, or even at room temperature. ILs are broadly used as reaction media, co-catalyst and catalyst for the chemical reactions. In the case of the reaction of epoxides and CO_2, several researchers have focused on the application of ILs. The mechanism of the reaction involved ring opening of the epoxide through a nucleophilic attack by the anion followed by insertion of carbon dioxide and ring closure [28]. Indeed, the absence of metallic centre as Lewis acid resulted in the selectivity of reaction towards the formation of cyclic carbonate compounds [30]. 1-*n*-butyl-3-methylimidazolium tetrafluoroborate successfully activated CO_2 and propylene oxide for the synthesis of propylene carbonate [84]. Under high-pressure condition, 100% conversion of propylene oxide was achieved by using 1-octyl-3-methylimidazolium tetrafluoroborate as catalyst [85]. However, the reaction was only selective towards the formation of cPC, and no polymer was achieved.

4.3.3. *Supported catalytic systems*

In these systems, the active catalyst should be anchored to the supporting material through a strong covalent bond in order to avoid possible leaching during the reaction. In fact, the immobilization of a catalyst on the solid support enables its rapid recovery and recycling. However, in some cases, immobilization shows adverse effect on catalyst activity due to limiting the diffusion of the reactants and intermediate compounds into and from the active sites of the catalyst [30]. Mesoporous materials with high surface area such as MCM-41 or SBA-15 have been used as supports to tackle this hurdle [23]. The zn β-diiminate catalyst was supported on controlled pore glass, and mesoporous SBA-15. This catalyst showed significant activity in the alternating copolymerization of CHO and CO_2. However, the resulted copolymer contains higher ether linkages than their corresponding non-supported catalysts due to the starvation of the active sites, particularly in the last stages of the reaction [86].

Recoverable aluminium salen catalysts bonded to poly(ethylene glycol bismethacrylate) and polystyrene exhibited comparable catalyst activity in the copolymerization of CO_2 and epoxides [87]. Chromium salen catalysts anchored on the aminopropylsilyl-modified silica in combination with an IL were used to catalyze the reaction of the styrene oxide and CO_2 [88]. The resulted catalysts exhibited 100% conversion yield; however, they were only selective towards the formation of cyclic carbonates.

Metal salen complexes were also immobilized on the surface of supporting material to facilitate their recovery after synthesis. For instance, zinc β-diiminate complex is immobilized on the surface of mesoporous silica (SBA-15), and this catalyst was used for the copolymerization of CHO and CO_2. However, ether linkage also occurred in the presence of this catalyst. In another study, chromium salen catalyst, immobilized on the surface of silicates, was used for the copolymerization of styrene oxide and CO_2 with nearly 50% activity and 80% selectivity [88]. In yet another example, aluminium salen catalyst, bonded on the surface of the polystyrene, was used for the copolymerization of CO_2 and styrene oxide [87]. In addition to the potential of being recovered, the supported aluminium salen catalyst exhibited similar activity to the unsupported aluminium salen compounds.

5. Industrial activities and commercial advances

A list of patents filed for the copolymerization of CO_2 and epoxide is shown in Table 2. Since 1969, most patents filed for the copolymerization of CO_2 and epoxides focused on developing heterogeneous catalysts, particularly zinc-based catalysts. The first catalyst, diethyl zinc, was developed by Inoue et al. in 1973 [89]. This patent that was sponsored by Nippon Oil Seal Industry Company demonstrated that the presence of co-catalyst significantly increased the yield of the copolymerization of ethylene oxide and CO_2 nearly threefold. The co-catalyst includes a hydrogen donor-based compound that was added to an organometallic compound, formed from metals that belonged to group II or group III [90].

A process for preparing soluble zinc catalysts by the reaction of zinc compounds with either a dicarboxylic acid anhydride or an alcohol or a monoester of a dicarboxylic acid was reported and filed in 1987 [91]. This catalyst system that was soluble in most of solvents was useful for the copolymerization of epoxides and carbon dioxide to form polycarbonates. In 1989 and 1990, two patents, both sponsored by Japanese companies such as Mitsui Petrochemical Industries, were filed for the development of several catalysts containing zinc oxide that reacted with an organic dicarboxylic acid such as glutaric acid or adipic acid [92, 93]. These patents introduced zinc glutarate from the source of zinc oxide and glutaric acid as the most efficient catalyst for the copolymerization of propylene oxide and carbon dioxide.

In 1998, dicarboxylic acids were used to successfully synthesize co-, ter-, and block-polymers from epoxides and carbon dioxide. This study that was sponsored by Buna Sow Leuna Olefinverbund GmbH consisted of two dicarboxylic acids, which could be aliphatic, aromatic or a mixture of the two, in combination with a divalent inorganic zinc salt [94]. In 2003 and 2004, new patents were filed to investigate catalyst efficiency of a range of zinc complexes synthesized from different dicarboxylic acids (chosen from pentane diol, hexane diol, 1,5-dibromopentane, 1,5-dichloropentane, 1,6-dichlorohexane, 1,6-dibromohexane, glutaronitrile, adiponitrile, glutarimide, adipamide, glutaralaldehyde and adipaldehyde) precursor and zinc precursor (chosen from zinc acetate dihydrate, zinc hydroxide, zinc nitrate hexahydrate, zinc perchlorate hexahydrate, zinc oxide and zinc sulphate) [95, 96]. Among them, the complex synthesized of zinc perchlorate hexahydrate and glutaronitrile showed the highest yield of 67% over a period of 30-h reaction for copolymerization of PO and carbon dioxide.

As mentioned before, zinc carboxylates, particularly zinc glutarate, are the only group of catalysts that could be effective in the commercial and large-scale synthesis of polycarbonates. However, the activity of this group of catalysts is still remarkably lower than the conventional catalysts used for the synthesis of polyolefins Consequently, many research and development activities are still focusing on novel catalysts for the synthesis of polycarbonates. For this purpose, most patents in the last decade focused on designing bimetallic or metal cyanide compounds to increase both yield and selectivity for the synthesis of polycarbonates [97]. For example, highly selective and active cobalt containing cyanide catalysts for producing poly(alkylene carbonates) from alkylene oxide and carbon dioxide was documented in 2005 [98]. However, the final product has still 10% cyclic carbonate by-product. In 2009, bimetallic complex of Zn(H), Co(II), Mn(II), Mg(II), Fe(II), Cr(III)-X or Fe(III)-X was reported for copolymerization of carbon dioxide and cyclohexene oxide or propylene oxide [99]. Following these developments in 2012, a catalyst was invented from the reaction of one double-metal cyanide compound, one organic complexing agent and one primary alcohol having 6–24 carbon atoms (sponsored by Henkel Ag & Co.). The compound was effective in catalyzing both homopolymerizations of epoxides or copolymerization of epoxides with carbon dioxide [100]. In the same year, a process for the synthesis of a polycarbonate from carbon dioxide and epoxides using a bimetallic catalyst system and a chain transfer agent was reported [101]. This invention that was sponsored by Imperial Innovations Limited exhibited that bimetallic complexes were successful in improvement of both activity and selectivity of PACs synthesis and accomplished above 90% yield of copolymerization with 99% carbonate linkage.

Patent Number	Catalyst	Epoxide	Activity (g/g catalyst)	Reference
US 3585168	Diethyl zinc and water [a]	Propylene oxide Ethylene oxide Styrene oxide Isobutylene oxide		[89]
US 4783445	Soluble zinc-based catalysts as a result of the reaction of zinc compounds (zinc oxide and zinc acetate) and dicarboxylic acids (maleic anhydride and monomethyle adipate) [a]	Ethylene oxide Propylene oxide Cyclohexene oxide		[91]
US 4943677	Zinc polycarboxylate catalyst (from zinc oxide and glutaric acid) in present of a hydrogen donor compound such as phenol and benzoic acid (acidity range 3-20 pKa) [a]	Propylene oxide		[102]
US 4981948	zinc containing solid catalyst by contacting zinc oxide with an organic dicarboxylic acid (oxalic acid, malonic acid, succinic acid, glutaric acid, adipic acid, 1,5-pentane dicarboxylic acid, 1,6-hexane dicarboxylic acid, 1,8-octane	Propylene oxide	60.7 (max)	[92]

Patent Number	Catalyst	Epoxide	Activity (g/g catalyst)	Reference
	dicarboxylic acid and 1,10-decane dicarboxylic acid) and aromatic dicarboxylic acids (phthalic acid, isophthalic acid, terephthalic acid, 1,2-naphthalene dicarboxylic acid, 1,3-naphthalene dicarboxylic acid, 1,4-naphthalene dicarboxylic acid, 1,5-naphthalene dicarboxylic acid, 1,6-naphthalene dicarboxylic acid, 1,7-naphthalene dicarboxylic acid, 1,8-naphthalene dicarboxylic acid, 2,3-naphthalene dicarboxylic acid, 2,5-naphthalene dicarboxylic acid, 2,6-naphthalene dicarboxylic acid and 2,7-naphthalene dicarboxylic acid) [a]			
US 5026676	Zinc carboxylate catalyst by reaction of zinc compounds (zinc oxide, $Zn(OH)_2$, $Zn(CO_3)_2$ and $Zn(OAc)_2$) with a dicarboxylic acid (glutaric acid or adipic acid) [a]	Propylene oxide	26 (max)	[93]
WO 2003029325	Zinc glutarate and zinc-cobalt based double metal cyanide catalyst [a]	Propylene oxide Ethylene oxide		[97]
WO 2004000912	Catalysts from the oxidizing of a dicarboxylic acid precursor (chosen from pentanediol, hexanediol, 1,5-dibromopentane, 1,5-dichloropentane, 1,6-dichlorohexane, 1,6-dibromohexane, glutaronitrile, adiponitrile, glutarimide, adipaimide, glutaraldehyde and adipaldehyde) and a zinc precursor (chosen from zinc acetate dihydrate, zinc hydroxide, zinc nitrate hexahydrate, zinc perchlorate hexahydrate, zinc oxide and zinc sulphate) [a]	Propylene oxide		[95]
US 20060089252	Cobalt salen catalyst [b]	Propylene oxide		[98]
WO 2009130470	Bimetallic complex of Zn(II), Co(II), Mn(II), Mg(II), Fe(II), Cr(III)-X or Fe(III)-X [b]	Cyclohexene oxide Propylene oxide		[99]
WO 2012136658	Double-metal cyanide compound [b]	Propylene oxide		[100]

Patent Number	Catalyst	Epoxide	Activity (g/g catalyst)	Reference
WO 2013034750	Bimetallic catalyst with metal centre from (Mg(II), Zn(II), Ca(II), Ge(II), Co(II), Mn(II), Ti(II), Fe(II), Cr(II), Cr(III)-X, Co(III)-X, Mn(III)-X, Fe(III)-X, A1(III)-X, Ti(III)-X, V(III)-X, Ge(IV)-$(X)_2$ or Ti(IV)-$(X)_2$) [b]	Cyclohexene oxide		[101]

[a] Heterogeneous catalysts, [b] Homogeneous catalysts

Table 2. List of the Filed Patents in the Field of CO_2 and Epoxide Copolymerization

6. Conclusions

This review demonstrated the significant advances that have been accomplished for developing active catalysts for the copolymerization of carbon dioxide and epoxides in the last four decades. In fact, a broad range of catalysts have been designed and examined with high activity and selectivity for the synthesis of biodegradable poly(alkylene carbonates). In addition, the mechanistic aspects of CO_2 cycloaddition to epoxides have been recognized and reported. Regardless of the structure of the catalyst and some slight differences, the overall mechanism proposed for the copolymerization of CO_2 and epoxides is coordination–insertion mechanism catalyzed via metal compounds with Lewis acid and Lewis base active sites.

Homogeneous catalysts for CO_2–epoxide copolymerization were proposed such as phenoxide, β-diiminate and metal salen complexes. The two former groups only showed activity for copolymerization of CHO and CO_2. In addition, salen catalysts exhibited activity and selectivity for a broader range of epoxides copolymerized by CO_2. However, none of these homogenous catalysts was used in commercial scale due to complicated synthesis process and low selectivity. Among several heterogeneous catalysts designed, organometallic complexes, particularly zinc glutarate, are the most effective catalysts because of both high activity and selectivity, low cost, non-toxicity and simple manufacturing process. In the last decade, attempts have been undertaken to design bimetallic catalysts to enhance activity towards copolymerization of CO_2 with epoxides. However, these studies are still in their infancy and have not yet used in the commercial scale.

In summary, the biodegradable polycarbonates are promising polymers that are produced from bonding CO_2 with epoxides that are partially renewable with huge market value for production of broad range of products such as food packaging materials, healthcare devices and agricultural mulches. Developing an efficient catalyst that reduces the cost and minimizes the catalyst consumption and by-product is still attractive as an alternative to non-degradable polymers in many applications to remedy the global issue of plastic wastes in landfills.

Author details

Bahareh Bahramian* and Fariba Dehghani

*Address all correspondence to: fariba.dehghani@sydney.edu.au

School of Chemical & Biomolecular Engineering, University of Sydney, Sydney, Australia

References

[1] Song C: Global challenges and strategies for control, conversion and utilization of CO_2 for sustainable development involving energy, catalysis, adsorption and chemical processing. Catalysis Today. 2006; 115: pp. 2–32 DOI: http://dx.doi.org/10.1016/j.cattod.2006.02.029

[2] Zhong X, Dehghani F: Solvent free synthesis of organometallic catalysts for the copolymerization of carbon dioxide and propylene oxide. Applied Catalysis B: Environmental. 2010; 98: pp. 101–111

[3] Qin Y, Wang X: Carbon dioxide-based copolymers: Environmental benefits of PPC, an industrially viable catalyst. Biotechnology Journal. 2010; 5: pp. 1164–1180 DOI: 10.1002/biot.201000134

[4] Sakakura T, Choi J-C, Yasuda H: Transformation of carbon dioxide. Chemical Reviews. 2007; 107: pp. 2365–2387 DOI: 10.1021/cr068357u

[5] Luinstra G A: Poly(propylene carbonate), old copolymers of propylene oxide and carbon dioxide with new interests: Catalysis and material properties. Polymer Reviews. 2008; 48: pp. 192–219

[6] Inoue S, Koinuma H, Tsuruta T: Copolymerization of carbon dioxide and epoxide with organometallic compounds. Die Makromolekulare Chemie. 1969; 130: pp. 210–220 DOI: 10.1002/macp.1969.021300112

[7] Ree M, Bae J Y, Jung J H, Shin T J, Hwang Y T, Chang T: Copolymerization of carbon dioxide and propylene oxide using various zinc glutarate derivatives as catalysts. Polymer Engineering and Science. 2000; 40: pp. 1542–1552

[8] Kim J S, Ree M, Shin T J, Han O H, Cho S J, Hwang Y T, Bae J Y, Lee J M, Ryoo R, Kim H: X-ray absorption and NMR spectroscopic investigations of zinc glutarates prepared from various zinc sources and their catalytic activities in the copolymerization of carbon dioxide and propylene oxide. Journal of Catalysis. 2003; 218: pp. 209–219

[9] Meng Y Z, Du L C, Tiong S C, Zhu Q, Hay A S: Effects of the structure and morphology of zinc glutarate on the fixation of carbon dioxide into polymer. Journal of Polymer Science, Part A: Polymer Chemistry. 2002; 40: pp. 3579–3591

[10] Kember M R, Buchard A, Williams C K: Catalysts for CO_2/epoxide copolymerization. Chemical Communications. 2011; 47: pp. 141–163

[11] Wang S J, Du L C, Zhao X S, Meng Y Z, Tjong S C: Synthesis and characterization of alternating copolymer from carbon dioxide and propylene oxide. Journal of Applied Polymer Science. 2002; 85: pp. 2327–2334

[12] Thorat S D, Phillips P J, Semenov V, Gakh A: Physical properties of aliphatic polycarbonates made from CO_2 and epoxides Journal of Applied Polymer Science. 2003; 89: pp. 1163–1176 DOI: 10.1002/app.12355

[13] http://www.materialscience.bayer.com/en: Access Date 16/06/2015

[14] http://www.basf.com.au/apex/ANZ/en/: Access Date 16/06/2015

[15] Changping C, Scheirs J, Polymer/thermoplastic starch compositions, 2011, Google Patents

[16] Zhong X, Lu Z, Valtchev P, Wei H, Zreiqat H, Dehghani F: Surface modification of poly(propylene carbonate) by aminolysis and layer-by-layer assembly for enhanced cytocompatibility. Colloids and Surfaces B: Biointerfaces. 2012; 93: pp. 75–84

[17] Zhong X, Dehghani F: Fabrication of biomimetic poly(propylene carbonate) scaffolds by using carbon dioxide as a solvent, monomer and foaming agent. Green Chemistry. 2012; 14: pp. 2523–2533

[18] Yang G, Su J, Gao J, Hu X, Geng C, Fu Q: Fabrication of well-controlled porous foams of graphene oxide modified poly(propylene-carbonate) using supercritical carbon dioxide and its potential tissue engineering applications. The Journal of Supercritical Fluids. 2013; 73: pp. 1–9 DOI: http://dx.doi.org/10.1016/j.supflu.2012.11.004

[19] Seo J, Jeon G, Jang E S, Bahadar Khan S, Han H: Preparation and properties of poly(propylene carbonate) and nanosized ZnO composite films for packaging applications. Journal of Applied Polymer Science. 2011; 122 pp. 1101–1108

[20] Taherimehr M, Pescarmona P P: Green polycarbonates prepared by the copolymerization of CO_2 with epoxides. Journal of Applied Polymer Science. 2014; 131: pp. 41141/1–41141/17 DOI: 10.1002/app.41141

[21] Zhang X H, Wei R J, Sun X K, Zhang J F, Du B Y, Fan Z Q, Qi G R: Selective copolymerization of carbon dioxide with propylene oxide catalyzed by a nanolamellar double metal cyanide complex catalyst at low polymerization temperatures. Polymer. 2011; 52: pp. 5494–5502

[22] Inoue S, Koinuma H, Tsuruta T: Copolymerization of carbon dioxide and epoxide. Journal of Polymer Science Part B: Polymer Letters. 1969; 7: pp. 287–292 DOI: 10.1002/pol.1969.110070408

[23] Kuran W, Pasynkiewicz S, Skupińska J: On the mechanism of the carbon dioxide/propylene oxide alternating copolymerization in the presence of organozinc cata-

lysts. Die Makromolekulare Chemie. 1977; 178: pp. 2149–2158 DOI: 10.1002/macp.1977.021780802

[24] Kuran W, Pasynkiewicz S, Skupińska J, Rokicki A: Alternating copolymerization of carbon dioxide and propylene oxide in the presence of organometallic catalysts. Die Makromolekulare Chemie. 1976; 177: pp. 11–20 DOI: 10.1002/macp.1976.021770102

[25] Soga K, Uenishi K, Hosoda S, Ikeda S: Copolymerization of carbon dioxide and propylene oxide with new catalysts. Die Makromolekulare Chemie. 1977; 178: pp. 893–897 DOI: 10.1002/macp.1977.021780325

[26] Soga K, Uenishi K, Ikeda S: Homopolymerization of propylene oxide and copolymerization of propylene oxide and carbon dioxide with metal salts of acetic acid. Journal of Polymer Science: Polymer Chemistry Edition. 1979; 17: pp. 415–423 DOI: 10.1002/pol.1979.170170211

[27] Soga K, Imai E, Hattori I: Alternating copolymerization of CO_2 and propylene oxide with the catalysts prepared from $Zn(OH)_2$ and various dicarboxylic acids. Polymer Journal. 1981; 13: pp. 407–410 DOI: 10.1295/polymj.13.407

[28] Pescarmona P P, Taherimehr M: Challenges in the catalytic synthesis of cyclic and polymeric carbonates from epoxides and CO_2. Catalysis Science & Technology. 2012; 2: pp. 2169–2187 DOI: 10.1039/c2cy20365k

[29] Moore D R, Cheng M, Lobkovsky E B, Coates G W: Mechanism of the alternating copolymerization of epoxides and CO_2 using β-diiminate zinc catalysts: Evidence for a bimetallic epoxide enchainment. Journal of the American Chemical Society. 2003; 125: pp. 11911–11924 DOI: 10.1021/ja030085e

[30] Laugel G, Rocha C C, Massiani P, Onfroy T, Launay F: Homogeneous and heterogeneous catalysis for the synthesis of cyclic and polymeric carbonates from CO_2 and epoxides: A mechanistic overview. Advanced Chemistry Letters. 2013; 1: pp. 195–214 DOI: 10.1166/acl.2013.1036

[31] Aida T, Ishikawa M, Inoue S: Alternating copolymerization of carbon dioxide and epoxide catalyzed by the aluminum porphyrin-quaternary organic salt or triphenylphosphine system. Synthesis of polycarbonate with well-controlled molecular weight. Macromolecules. 1986; 19: pp. 8–13 DOI: 10.1021/ma00155a002

[32] Kruper W J, Dellar D D: Catalytic Formation of cyclic carbonates from epoxides and CO_2 with chromium metalloporphyrinates. The Journal of Organic Chemistry. 1995; 60: pp. 725–727 DOI: 10.1021/jo00108a042

[33] Aida T, Inoue S: Catalytic reaction on both sides of a metalloporphyrin plane. Alternating copolymerization of phthalic anhydride and epoxypropane with an aluminum porphyrin-quaternary salt system. Journal of the American Chemical Society. 1985; 107: pp. 1358–1364 DOI: 10.1021/ja00291a041

[34] Kojima F, Aida T, Inoue S: Fixation and activation of carbon dioxide on aluminum porphyrin. Catalytic formation of a carbamic ester from carbon dioxide, amine, and

epoxide. Journal of the American Chemical Society. 1986; 108: pp. 391–395 DOI: 10.1021/ja00263a008

[35] Chen P, Chisholm M H, Gallucci J C, Zhang X, Zhou Z: Binding of propylene oxide to porphyrin– and salen–M(III) cations, where M = Al, Ga, Cr, and Co. Inorganic Chemistry. 2005; 44: pp. 2588–2595 DOI: 10.1021/ic048597x

[36] Srivastava R, Bennur T H, Srinivas D: Factors affecting activation and utilization of carbon dioxide in cyclic carbonates synthesis over Cu and Mn peraza macrocyclic complexes. Journal of Molecular Catalysis A: Chemical. 2005; 226: pp. 199–205 DOI: http://dx.doi.org/10.1016/j.molcata.2004.10.034

[37] Darensbourg D J, Niezgoda S A, Holtcamp M W, Draper J D,Reibenspies J H: Syntheses, structures, and binding constants of cyclic ether and thioether adducts of soluble cadmium(II) carboxylates. Intermediates in the homopolymerization of oxiranes and thiiranes and in carbon dioxide coupling processes. Inorganic Chemistry. 1997; 36: pp. 2426–2432 DOI: 10.1021/ic9701120

[38] Darensbourg D J, Wildeson J R, Yarbrough J C, Reibenspies J H: Bis 2,6-difluorophenoxide dimeric complexes of zinc and cadmium and their phosphine adducts: Lessons learned relative to carbon dioxide/cyclohexene oxide alternating copolymerization processes catalyzed by zinc phenoxides. Journal of the American Chemical Society. 2000; 122: pp. 12487–12496 DOI: 10.1021/ja002855h

[39] Darensbourg D J, Zimmer M S, Rainey P, Larkins D L: Solution and solid-state structures of phosphine adducts of monomeric zinc bisphenoxide complexes. Importance of These Derivatives in CO_2/Epoxide Copolymerization Processes. Inorganic Chemistry. 2000; 39: pp. 1578–1585 DOI: 10.1021/ic990594a

[40] Cheng M, Moore D R, Reczek J J, Chamberlain B M, Lobkovsky E B, Coates G W: Single-site β-diiminate zinc catalysts for the alternating copolymerization of CO_2 and epoxides: Catalyst synthesis and unprecedented polymerization activity. Journal of the American Chemical Society. 2001; 123: pp. 8738–8749 DOI: 10.1021/ja003850n

[41] Geerts R L, Huffman J C, Caulton K G: Soluble zinc bis(aryloxides). Inorganic Chemistry. 1986; 25: pp. 1803–1805 DOI: 10.1021/ic00231a018

[42] Darensbourg D J, Niezgoda S A, Draper J D, Reibenspies J H: Mechanistic aspects of the copolymerization of CO_2 and epoxides by soluble zinc bis(phenoxide) catalysts as revealed by their cadmium analogues. Journal of the American Chemical Society. 1998; 120: pp. 4690–4698 DOI: 10.1021/ja9801487

[43] Darensbourg D J, Holtcamp M W: Catalytic activity of zinc(II) phenoxides which possess readily accessible coordination sites. Copolymerization and terpolymerization of epoxides and carbon dioxide. Macromolecules. 1995; 28: pp. 7577–7579 DOI: 10.1021/ma00126a043

[44] Darensbourg D J, Holtcamp M W, Struck G E, Zimmer M S, Niezgoda S A, Rainey P, Robertson J B, Draper J D, Reibenspies J H: Catalytic activity of a series of Zn(II)

phenoxides for the copolymerization of epoxides and carbon dioxide. Journal of the American Chemical Society. 1999; 121: pp. 107–116 DOI: 10.1021/ja9826284

[45] Chamberlain B M, Cheng M, Moore D R, Ovitt T M, Lobkovsky E B, Coates G W: Polymerization of lactide with zinc and magnesium β-diiminate complexes: Stereocontrol and mechanism. Journal of the American Chemical Society. 2001; 123: pp. 3229–3238 DOI: 10.1021/ja003851f

[46] Byrne C M, Allen S D, Lobkovsky E B, Coates G W: Alternating copolymerization of limonene oxide and carbon dioxide. Journal of the American Chemical Society. 2004; 126: pp. 11404–11405 DOI: 10.1021/ja0472580

[47] Allen S D, Moore D R, Lobkovsky E B, Coates G W: High-activity, single-site catalysts for the alternating copolymerization of CO_2 and propylene oxide. Journal of the American Chemical Society. 2002; 124: p. 14284-14285 DOI: 10.1021/ja028071g

[48] Chisholm M H, Navarro-Llobet D, Zhou Z: Poly(propylene carbonate). 1. More about poly(propylene carbonate) formed from the copolymerization of propylene oxide and carbon dioxide employing a zinc glutarate catalyst. Macromolecules. 2002; 35: pp. 6494–6504

[49] Venkataramanan N S, Kuppuraj G, Rajagopal S: Metal–salen complexes as efficient catalysts for the oxygenation of heteroatom containing organic compounds—synthetic and mechanistic aspects. Coordination Chemistry Reviews. 2005; 249: pp. 1249–1268 DOI: http://dx.doi.org/10.1016/j.ccr.2005.01.023

[50] Martinez L E, Leighton J L, Carsten D H, Jacobsen E N: Highly enantioselective ring opening of epoxides catalyzed by (salen)Cr(III) complexes. Journal of the American Chemical Society. 1995; 117: pp. 5897–5898 DOI: 10.1021/ja00126a048

[51] Klaus S, Lehenmeier M W, Anderson C E, Rieger B: Recent advances in CO_2/epoxide copolymerization—New strategies and cooperative mechanisms. Coordination Chemistry Reviews. 2011; 255: pp. 1460–1479 DOI: http://dx.doi.org/10.1016/j.ccr.2010.12.002

[52] Jacobsen E N: Asymmetric catalysis of epoxide ring-opening reactions. Accounts of Chemical Research. 2000; 33: pp. 421–431 DOI: 10.1021/ar960061v

[53] Darensbourg D J, Yarbrough J C: Mechanistic aspects of the copolymerization reaction of carbon dioxide and epoxides, using a chiral salen chromium chloride catalyst. Journal of the American Chemical Society. 2002; 124: pp. 6335–6342 DOI: 10.1021/ja012714v

[54] Darensbourg D J, Yarbrough J C, Ortiz C, Fang C C: Comparative kinetic studies of the copolymerization of cyclohexene oxide and propylene oxide with carbon dioxide in the presence of chromium salen derivatives. In situ FTIR measurements of copolymer vs cyclic carbonate production. Journal of the American Chemical Society. 2003; 125: pp. 7586–7591 DOI: 10.1021/ja034863e

[55] Eberhardt R, Allmendinger M, Rieger B: DMAP/Cr(III) Catalyst ratio: The decisive factor for poly(propylene carbonate) formation in the coupling of CO_2 and propylene oxide. Macromolecular Rapid Communications. 2003; 24: pp. 194–196 DOI: 10.1002/marc.200390022

[56] Darensbourg D J, Mackiewicz R M, Phelps A L, Billodeaux D R: Copolymerization of CO_2 and epoxides catalyzed by metal salen complexes. Accounts of Chemical Research. 2004; 37: pp. 836–844 DOI: 10.1021/ar030240u

[57] Qin Z, Thomas C M, Lee S, Coates G W: Cobalt-based complexes for the copolymerization of propylene oxide and CO_2: Active and selective catalysts for polycarbonate synthesis. Angewandte Chemie International Edition. 2003; 42: pp. 5484–5487 DOI: 10.1002/anie.200352605

[58] Cohen C T, Chu T, Coates G W: Cobalt catalysts for the alternating copolymerization of propylene oxide and carbon dioxide: Combining high activity and selectivity. Journal of the American Chemical Society. 2005; 127: pp. 10869–10878 DOI: 10.1021/ja0517441

[59] Nakano K, Kamada T, Nozaki K: Selective formation of polycarbonate over cyclic carbonate: Copolymerization of epoxides with carbon dioxide catalyzed by a cobalt(III) complex with a piperidinium end-capping arm. Angewandte Chemie International Edition. 2006; 45: pp. 7274–7277 DOI: 10.1002/anie.200603132

[60] Noh E K, Na S J, S S, Kim S-W, Lee B Y: Two components in a molecule: Highly efficient and thermally robust catalytic system for CO_2/epoxide copolymerization. Journal of the American Chemical Society. 2007; 129: pp. 8082–8083 DOI: 10.1021/ja071290n

[61] Sugimoto H, Inoue S: Copolymerization of carbon dioxide and epoxide. Journal of Polymer Science Part A: Polymer Chemistry. 2004; 42: pp. 5561–5573 DOI: 10.1002/pola.20319

[62] Ree M, Bae J Y, Jung J H, Shin T J: A new copolymerization process leading to poly(propylene carbonate) with a highly enhanced yield from carbon dioxide and propylene oxide. Journal of Polymer Science Part A: Polymer Chemistry. 1999; 37: pp. 1863–1876 DOI: 10.1002/(sici)1099-0518(19990615)37:12<1863::aid-pola16>3.0.co;2-k

[63] Darensbourg D J, Stafford N W, Katsurao T: Supercritical carbon dioxide as solvent for the copolymerization of carbon dioxide and propylene oxide using a heterogeneous zinc carboxylate catalyst. Journal of Molecular Catalysis A: Chemical. 1995; 104: pp. L1–L4 DOI: http://dx.doi.org/10.1016/1381-1169(95)00142-5

[64] Kruper W J, Swart D J, Carbon dioxide oxirane copolymers prepared using double metal cyanide complexes, 1985, Google Patents

[65] Chen L-B: Activation and copolymerization of CO_2 by macromolecule-metal complexes. Makromolekulare Chemie. Macromolecular Symposia. 1992; 59: pp. 75–82 DOI: 10.1002/masy.19920590108

[66] Kim I, Yi M J, Byun S H, Park D W, Kim B U, Ha C S: Biodegradable polycarbonate synthesis by copolymerization of carbon dioxide with epoxides using a heterogeneous zinc complex. Macromolecular Symposia. 2005; 224: pp. 181–192 DOI: 10.1002/masy.200550616

[67] Kim I, Yi M J, Lee K J, Park D-W, Kim B U, Ha C-S: Aliphatic polycarbonate synthesis by copolymerization of carbon dioxide with epoxides over double metal cyanide catalysts prepared by using ZnX2 (X = F, Cl, Br, I). Catalysis Today. 2006; 111: pp. 292–296 DOI: http://dx.doi.org/10.1016/j.cattod.2005.10.039

[68] Darensbourg D J, Adams M J, Yarbrough J C: Toward the design of double metal cyanides for the copolymerization of CO_2 and epoxides. Inorganic Chemistry. 2001; 40: pp. 6543–6544 DOI: 10.1021/ic0155941

[69] Robertson N J, Qin Z, Dallinger G C, Lobkovsky E B, Lee S, Coates G W: Two-dimensional double metal cyanide complexes: highly active catalysts for the homopolymerization of propylene oxide and copolymerization of propylene oxide and carbon dioxide. Dalton Transactions. 2006; 45: pp. 5390–5395

[70] Chen S, Hua Z, Fang Z, Qi G: Copolymerization of carbon dioxide and propylene oxide with highly effective zinc hexacyanocobaltate(III)-based coordination catalyst. Polymer. 2004; 45: pp. 6519–6524 DOI: http://dx.doi.org/10.1016/j.polymer.2004.07.044

[71] Chen X, Shen Z, Zhang Y: New catalytic systems for the fixation of carbon dioxide. 1. Copolymerization of carbon dioxide and propylene oxide with new rare-earth catalysts-RE(P204)3-Al(i-Bu)3-R(OH)n. Macromolecules. 1991; 24: pp. 5305–5308 DOI: 10.1021/ma00019a014

[72] Shen Z, Chen X, Zhang Y: New catalytic systems for the fixation of carbon dioxide, 2. Synthesis of high molecular weight epichlorohydrin/carbon dioxide copolymer with rare earth phosphonates/triisobutyl-aluminium systems. Macromolecular Chemistry and Physics. 1994; 195: pp. 2003–2011 DOI: 10.1002/macp.1994.021950610

[73] Guo J-T, Wang X-Y, Xu Y-S, Sun J-W: Copolymerizations of carbon dioxide and epoxides in the presence of rare earth coordinate catalyst. Journal of Applied Polymer Science. 2003; 87: pp. 2356–2359 DOI: 10.1002/app.11923

[74] Liu B, Zhao X, Wang X, Wang F: Copolymerization of carbon dioxide and propylene oxide with neodymium trichloroacetate-based coordination catalyst. Polymer. 2003; 44: pp. 1803–1808 DOI: http://dx.doi.org/10.1016/S0032-3861(03)00034-X

[75] Quan Z, Wang X, Zhao X, Wang F: Copolymerization of CO_2 and propylene oxide under rare earth ternary catalyst: design of ligand in yttrium complex. Polymer. 2003; 44: pp. 5605–5610 DOI: http://dx.doi.org/10.1016/S0032-3861(03)00561-5

[76] Tao Y, Wang X, Zhao X, Li J, Wang F: Double propagation based on diepoxide, a facile route to high molecular weight poly(propylene carbonate). Polymer. 2006; 47: pp. 7368–7373 DOI: http://dx.doi.org/10.1016/j.polymer.2006.08.035

[77] Ree M, Hwang Y, Kim J S, Kim H, Kim G: New findings in the catalytic activity of zinc glutarate and its application in the chemical fixation of CO_2 into polycarbonates and their derivatives. Catalysis Today. 2006; 115: pp. 134–145

[78] Pokasermsong P, Praserthdam P: Comparison of activity of Ziegler-Natta catalysts prepared by recrystallization and chemical reaction methods towards polymerization of ethylene. Engineering Journal. 2009; 13: pp. 57–64

[79] Kim J S, Kim H, Yoon J, Heo K, Ree M: Synthesis of zinc glutarates with various morphologies using an amphiphilic template and their catalytic activities in the copolymerization of carbon dioxide and propylene oxide. Journal of Polymer Science, Part A: Polymer Chemistry. 2005; 43: pp. 4079–4088

[80] Wang J T, Zhu Q, Lu X L, Meng Y Z: ZnGA-MMT catalyzed the copolymerization of carbon dioxide with propylene oxide. European Polymer Journal. 2005; 41: pp. 1108–1114

[81] Zhu Q, Meng Y Z, Tjong S C, Zhao X S, Chen Y L: Thermally stable and high molecular weight poly(propylene carbonate)s from carbon dioxide and propylene oxide. Polymer International. 2002; 51: pp. 1079–1085 DOI: 10.1002/pi.847

[82] Jintang D, Jiajun W, Lianfang F, Long W, Xueping G: Pressure dependence of the CO_2/propylene oxide copolymerization catalyzed by zinc glutarate. Journal of Applied Polymer Science. 2010; 118: pp. 365–371

[83] Lee B Y, Kwon H Y, Lee S Y, Na S J, Han S-i, Yun H, Lee H, Park Y-W: Bimetallic anilido-aldimine zinc complexes for epoxide/CO_2 copolymerization. Journal of the American Chemical Society. 2005; 127: pp. 3031–3037 DOI: 10.1021/ja0435135

[84] Peng J, Deng Y: Cycloaddition of carbon dioxide to propylene oxide catalyzed by ionic liquids. New Journal of Chemistry. 2001; 25: pp. 639–641 DOI: 10.1039/b008923k

[85] Kawanami H, Sasaki A, Matsui K, Ikushima Y: A rapid and effective synthesis of propylene carbonate using a supercritical CO_2-ionic liquid system. Chemical Communications. 2003; 7: pp. 896–897 DOI: 10.1039/b212823c

[86] Yu K, Jones C W: Silica-immobilized zinc β-diiminate catalysts for the copolymerization of epoxides and carbon dioxide. Organometallics. 2003; 22: pp. 2571–2580 DOI: 10.1021/om030209w

[87] Alvaro M, Baleizao C, Carbonell E, El Ghoul M, García H, Gigante B: Polymer-bound aluminium salen complex as reusable catalysts for CO_2 insertion into epoxides. Tetrahedron. 2005; 61: pp. 12131–12139 DOI: http://dx.doi.org/10.1016/j.tet.2005.07.114

[88] Alvaro M, Baleizao C, Das D, Carbonell E, García H: CO_2 fixation using recoverable chromium salen catalysts: use of ionic liquids as cosolvent or high-surface-area silicates as supports. Journal of Catalysis. 2004; 228: pp. 254–258 DOI: http://dx.doi.org/10.1016/j.jcat.2004.08.022

[89] Shohei I, Hideomi K, Teiji T: Process for producing copolymer of epoxide and carbon dioxide, 1971, Google Patents

[90] Shohei I, Masaki K, Nobuyuki M, Tadamichi T, Masanori Y, Catalyst for copolymerizing epoxy compounds with carbon dioxide, 1975, Google Patents

[91] Sun H, Soluble epoxide/carbon dioxide copolymerization catalysts, 1988, Google Patents

[92] Kawachi H, Minami S, Armor J N, Rokicki A, Stein B K, Zinc-containing solid catalyst, process of preparing same and process for preparing polyalkylene carbonate, 1991, Google Patents

[93] Motika S A, Pickering T L, Rokicki A, Stein B K, Catalyst for the copolymerization of epoxides with CO_2, 1991, Google Patents

[94] Kogut S, Rom C, Schimmel K H, Wagenknecht E, Catalyst system for producing polyalkylene carbonates, 1999, Google Patents

[95] Jong-Sung K, Kie-Soo K, Seung-Jae M, Moon-Hor R, Method of preparing catalyst for polymerization of aliphatic polycarbonate and method of polymerizing aliphatic polycarbonate using same, 2003, Google Patents

[96] Döring M P D, Kröger M D, New substituted N-phenyl-3-phenylaminoacrylimidate zinc complexes useful as catalysts, especially for polycarbonate production by reaction of epoxides with carbon dioxide, 2005, Google Patents

[97] Bohres E, Heinemann J, Luinstra G, Verfahren zur herstellung aliphatischer polycarbonate, 2003, Google Patents

[98] Coates G, Qin Z, Cohen C, Polycarbonates made using highly selective catalysts, 2006, Google Patents

[99] Williams C K, Kember M, Knight P, Bimetallic catalytic complexes for the copolymerization of carbon dioxide and an epoxide, 2009, Google Patents

[100] Klein J, Kragl U, Kunze C, Marquardt M, Paetzold E, Zander L, Katalysator zur epoxidpolymerization und zur copolymerization von epoxid mit kohlendioxid, 2012, Google Patents

[101] Williams C, Kember M, Buchard A, Jutz F, Method of synthesising polycarbonates in the presence of a bimetallic catalyst and a chain transfer agent, 2013, Google Patents

[102] Rokicki A, Making poly(alkylene carbonates) of controlled molecular weight, 1990, Google Patents

Chapter 4

Catalytic Applications of Metal-Organic Frameworks

Sandra Loera-Serna and Elba Ortiz

Additional information is available at the end of the chapter

http://dx.doi.org/10.5772/61865

Abstract

In recent years, metal organic frameworks (MOF) have received considerable interest due to their physicochemical properties, such as structures' flexibility, high surface area, tunable pore size, and topologies, among others, which have lead to promising applications, particularly in the area of catalysis. In this chapter, we present the most important results of research conducted with MOF in catalytic applications; mainly in the design of its structure, synthesis, characterization, and possible limitations.

Keywords: Metal-organic frameworks, Heterogeneous catalysis, Porous materials

1. Introduction

MOFs and the related researches have become more and more important not only in chemistry but also in general science and technology. MOFs are a class of porous materials composed of metal-containing nodes connected by organic linkers through strong chemical bonds. The union of these two building units produces different coordination modes, depending on the symmetry of the linker and the coordination number of the metal center. The flexibility or rigidity of the added linker can allow the articulation of the clusters into a highly crystalline three-dimensional framework, which can exhibit higher surface area and pore volume than most porous zeolites [1]. Depending on the architecture of the obtained MOFs, they can be synthesized with high purity and also, they can be engineered to have a high skeletal density but constructed from relatively light elements. Therefore, most of the important related work is aimed at designing compounds possessing very large pores and high surface areas in order to load these materials with atoms, molecules, or even biomolecules. Due to these loading possibilities, wide applications of MOFs have emerged in different fields, such as in catalysis [2–4], guest adsorption (molecular recognition) [5], drug delivery [6, 7], gas storage [8–13], optical applications [14–16], composites [17], water treatment [18, 19], and sensor technologies [20], among others [21–26].

Some materials as metals in solution (transition metal complexes or metal salts) have been used in catalysis with excellent results. These materials are able to catalyze a variety of organic reactions, in many cases, reaching high yields and regenerating the material after the reaction. However, in many cases the metals are hardly recovered and/or decompose during the reaction due to the conditions. To achieve control these limitations, researchers have developed methods using porous materials as carriers, to achieve well-isolated, uniform single sites that don't interact between them, preventing the decomposition [2, 27]. Active sites on MOFs are located at the metal nodes on the crystalline structure; when the reaction occurs, the framework protects their active sites and increases the efficiency and resistance of catalyst [28].

Given the variety of metallic nodes and organic linkers, it is possible to control the synthesis of MOFs to design them with modular properties, functionalized with specific sites or specific assets to catalyze organic reactions. In this chapter, we present the main results of research with MOFs in the field of catalysis, with special focus on design, relationship between structure and activity, formation of active sites and limitations of these materials.

2. Design of MOFs

2.1. Crystal engineering of MOF

The term of metal organic framework was introduced by Yaghi in 1995 [29, 30], however, such structures were known until 1964 when Bailar first reported them [31]. The resurgence of the structures has been accompanied by the application of these materials in various areas, including: catalysis [2–4], guest adsorption (molecular recognition) [5], drug delivery [6, 7], gas storage [8–13], optical applications [14–16], composites [17], water treatment [18, 19] and sensor technologies [20], among others [21–26].

The structural characteristics of the MOFs are mainly determined by the nature of the metal center and the organic linker, yet, during the synthesis of these materials, solvents and/or counterions are typically used [32] and they also play an important role. The counterions change the environment of the metal ion and may generate overlaps with the structure resulting in weak interactions with the MOF. Meanwhile, solvent molecules with the MOF generally crystallize during synthesis, modifying the pore size.

Generally, the transition metal ions used can generate a wide range of structures. The properties of these metals, including the oxidation state and coordination number (typically varies from 2 to 7), produce a linear, trigonal, square planar, tetrahedral, trigonal pyramidal, trigonalbipyramidal, octahedral, and pentagonal bipyramidal geometries as well as some other distorted forms [32]. The lanthanoidions, whose coordination number varies between 7 and 10, have polyhedral geometries and can generate MOFs with particular topologies [33].

In the formation of MOFs, the organic linkers must meet certain requirements to form coordination bonds, mainly being multidentate having at least two donor atoms (N-, O- or S-) and being neutral or anionic. The structure of MOF is also affected by the shape, length, and functional groups present in the organic linker. The linkers commonly used in the MOFs synthesis are piperazine [34], 4,4′-bipyridine [34–37] (neutral ligands), and polycarboxylates (anionic ligands). Polycarboxylates may be di- [38–43], tri- [38, 40–43], tetra- [44, 45], or hexacarboxylates [46, 47].

The binding of a linker to the metal center may generate a one-dimensional (1D), two-dimensional (2D) or three-dimensional (3D) arrangement, which depends on the metal center (Figure 1) [48]. In a 1D network, two ligand molecules are coordinated to the metal center to generate a chain, while in a 2D network, three or four molecules of the organic linker are coordinated to generate a plane, and it grows in two dimensions. In a 3D MOF, the metal center, with high coordination number, joins three more linker molecules, along the three spatial dimensions, generating a three-dimensional structure with pores and cavities defined.

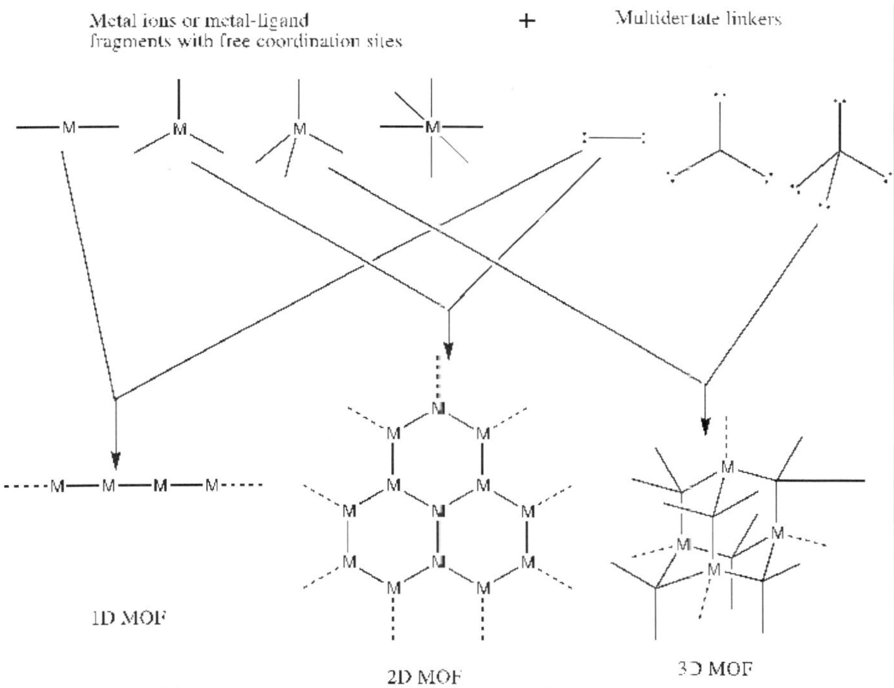

Figure 1. Basic building units of one-, two-, and three-dimensional MOFs [48].

Figure 2 shows examples of MOF with different dimensionalities. The helix (1D) is constituted by distorted tetrahedrons mercury (II), formed by the union of two nitrogen atoms (from two different linkers) and two terminal bromine atoms [49].

The 2D structures with grid shape are generally synthesized with a molar ratio between the ligand and the metal center of 1:2. An example of such structures is shown in Figure 3, the MOF is constituted by cobalt metal centers and ligands *N*-(3-pyridyl) nicotinamide [50]. The metal ions are coordinated with four molecules of ligand, which result in a two-dimensional flat-shaped structure.

1D	2D	3D
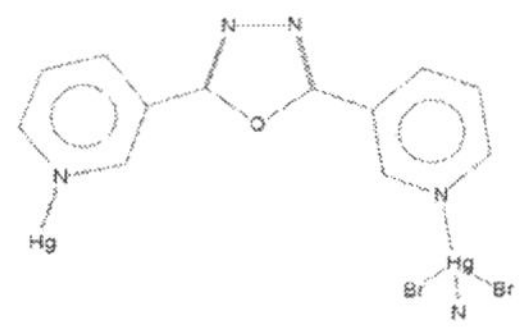	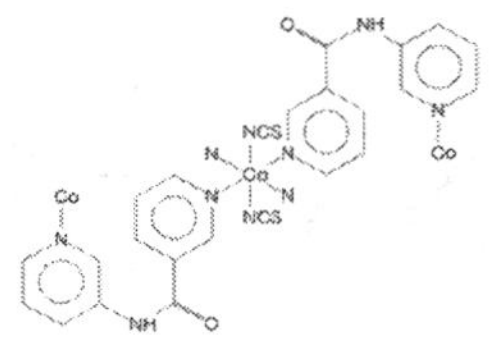	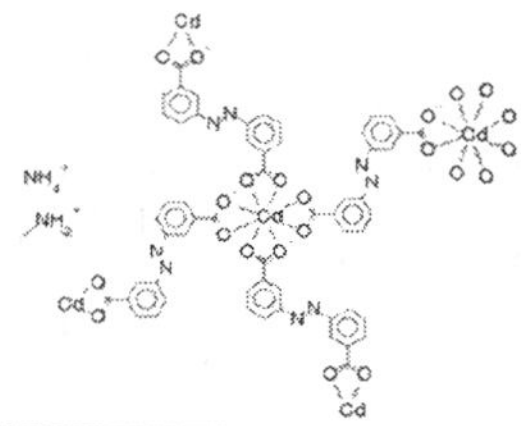
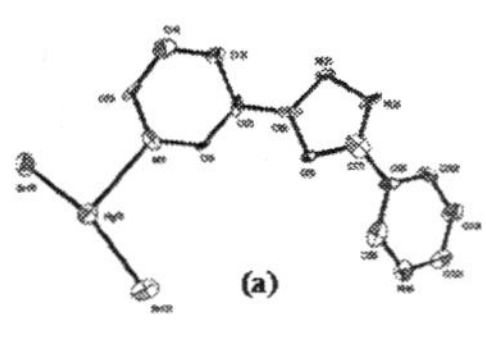	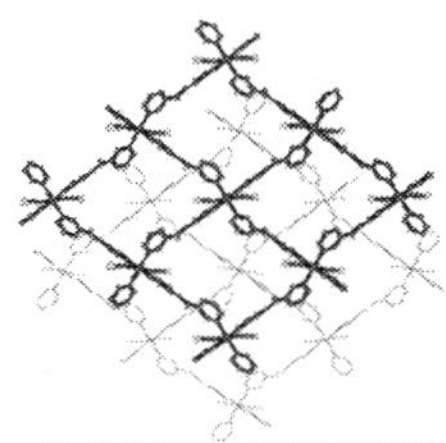	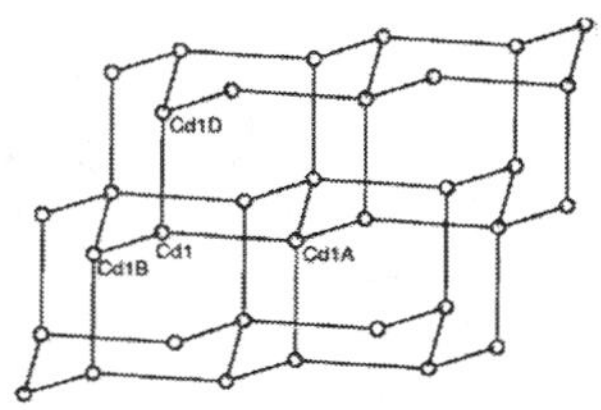
$HgBr_2$(2,5-bis(3-piridil)-1,3,4-oxadiazole) [49]	Co$(SCN)_2$(3-pna)$_2$]}$_\infty$ (SCN = tiocianate), (3-pna = N-(3-pyridil) nicotinamide) [50]	[Cd(3,3′-AZDB)$_2$]$(H_2NMe)_2$ (NH_4)(3,3′-AZDB = 3,3′-azodibenzoate) [51]

Figure 2. Examples of MOF structures 1D, 2D, and 3D.

The three-dimensional MOFs are formed by the interaction of one-dimensional chains in all three directions. Connectivity of the construction nodes depends on the metal center, and the formed structures are usually tetrahedral or octahedral. An example of such structures, wherein the metal is cadmium center and has an octahedral coordination, is given in Figure 3. The bidentate linker forms connections, where the four terminals of each linker involves oxygen atoms. The three-dimensional growth of the framework generates cavities; generally occupied by solvent molecules [51].

2.2. Synthesis of MOF

The physicochemical characteristics of MOFs can be modulated and it is clear that all these properties can be modified in the material from the synthesis process. The solvothermal synthesis is the most common way of obtaining MOFs. However, other recently studied methods of synthesis, which may cause significant changes in the MOF's properties, include (i) mechanochemical, (ii) electrochemistry, (iii) assisted synthesis (by ultrasound or micro-wave), and (iv) subcritical water.

2.2.1. *Solvothermal synthesis*

The solvothermal synthesis comprises the reaction of the metal salt and the organic ligand in the presence of organic solvents or solvent mixtures typically involving formamide [52–55], alcohols [56, 57], or pyrrolidones [58]. The reaction temperature is usually considerably less than 523 K, but depends on the boiling point of the solvent used, for example, in the case of methanol, the synthesis temperature is 338 K. Upon heating, the blends in sealed containers such as teflon or glass pressure tubes produced compounds incorporating solvent molecules in the pores of the material. These organic solvents are often toxic, carcinogenic, and/or dangerous to the environment. For example, dimethylformamide (DMF), one of the most commonly used solvents in the synthesis of MOFs is contaminant, mutagenic, and toxic [59]. Additionally, DMF decomposes when heated at high temperatures for long periods and therefore cannot be reused [60].

The concern about the use of these organic solvents has increased due to their negative impact on the environment [61]. Therefore, the replacement of polluting solvents in the synthesis of MOFs with solvents that do not pose a risk to the environment has been established as a primary objective, committing to the principles of Green Chemistry [62]. When the solvent is water, the main problem encountered during synthesis is the low or nonsolubility of the reactants and organic ligands, which generally prevents the formation of coordination polymer.

Other important parameters in the solvothermal synthesis are: temperature, concentration of reactants (which can be varied over a wide range) and the pH of the solution.

One of the advantages of this synthesis are the yields ranging from 60% to 95%; however, removing solvent molecules occluded in the pores of the MOF is not a simple process, the solids must be washed several times, which can take several days and considerably reduce performance.

2.2.2. *Mechanochemical synthesis*

The mechanochemical synthesis is named due to the chemical reaction that occurs as a consequence of mechanical energy supplied to the system [63]. In the mechanical grinding process, several phenomena occur, such as

a. Fragmentation of grains

b. Generation of new surfaces

c. Formation of dislocations and defects in the crystal structure

d. Polymorphic phase transformation material, and

e. Chemical reactions (decomposition, ion exchange, oxidation-reduction, complexation, etc.).

In the synthesis of MOFs, the metal salt and the organic linker are ground together in the absence of solvents. In 2002, Belcher et al. [64] reported the synthesis of a 1D copper coordination polymer, using mechanochemical synthesis (Figure 3).

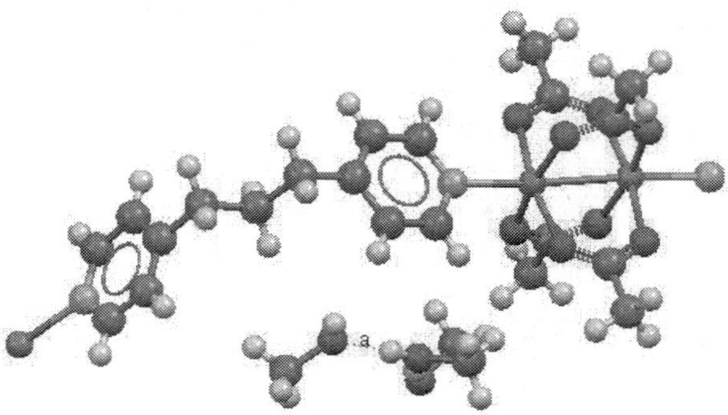

Figure 3. Basic unit of construction of coordination polymer $[Cu(O_2C\text{-}Me)_2]_2(\mu\text{-dpp})$ dpp = 2,3-bis(2-pyridyl)pyrazine. Gray, red, white, violet, and orange spheres correspond to atoms of C, H, N, O and Cu, respectively.

In other recent studies [65], MOFs were synthesized using 12 metal salts and 5 organic linkers to obtain 60 different solids. As a result, 38 microcrystalline MOFs were identified using X-ray diffraction techniques. Their structure patterns are found on the CSD database (Cambridge Structure Database), including microporous MOFs $[Cu(INA)_2]n$ (INA = isonicotinate) and $Cu_3(BTC)_2$.

2.2.3. Electrochemical synthesis

Electrochemical synthesis of MOFs was reported by BASF in 2005 [66], in order to eliminate the use of anions such as nitrate, perchlorate, and chloride, which act as counterions or as impurities in the network. In this synthesis method, the metal salts are replaced by metal ions produced from the anodic dissolution in the reaction medium. The dissolution also contained organic linkers; in cathodes, metal deposition occurred. In particular, for the synthesis of $Cu_3(BTC)_2$, copper metal bars that function as electrodes (anode and cathode) were employed in the electrochemical cell with organic linker (BTC = benzene-1,3,5-tricaboxylic), dissolved in methanol [67], with an applied voltage between 12 and 19 V and a current of 1.3 A for 150 min. The result was the oxidation of the copper bar acting as the anode to form Cu^{2+}, which reacts with the organic linker. Furthermore, in cathode, water reduction took place to produce hydrogen. At the end of the reaction, a greenish-blue precipitate was formed, which was filtered and dried to obtain $Cu_3(BTC)_2$. Using these synthesis pathways, materials can be produced with high purity and ease of being industrially scalable.

Other studies on the electrochemical synthesis of MOFs are presented in Table 1.

2.2.4. Microwave or ultrasound-assisted synthesis

Synthesis assisted by ultrasound or microwave is an alternative to the solvothermal synthesis. In microwave-assisted synthesis, the reaction mixture is subjected to nonionizing radiation,

MOF	Type of synthesis	Reference
Zn and Cu-carboxylates	Systematic study of Zn, Cu, Co and Mg as anode 1,2,3-H_3BTC, 1,2,5-H_3BTC, H_2BCD y H_2BCD-$(OH)_2$	[66]
Zn-Imidazolates	Synthesis of $Zn(MeIm)_2$ and $Zn(BIm)_2$	[68]
$Cu_3(BTC)_2$	Synthesis and y growth of galvanic displacement layered	[67]

H_3BTC = benzene-1,3,5-tricarboxilic acid; H_2BCD = terephtalic acid; MeIm = 2-methyl-1H-imidazole; BIm = benzimidazole.

Table 1. MOFs synthesized electrochemically.

which does not change the electronic structure of the material. The energy supplied to the material by electromagnetic waves through interactions of molecular type offers a number of advantages, such as a uniform controlled heating as well as a great speed with which energy is generated [69]. The characteristic frequency of this radiation is between 300 MHz and 300 GHz (wavelengths between 0.01 and 1 m). This synthetic method has been applied to organic molecules [70] and inorganic materials [71].

Generally, the microwave synthesis is carried out in minutes and offers a better method to control the morphology of the material and the selectivity of the phases. For example, MOF-5 (Figure 4) was synthesized using microwave at 368 K for 9 min, with a yield of 27% [72], while using the solvothermal synthesis, a yield of 60% was achieved after 7 days.

Table 2 shows the conditions of microwave-assisted synthesis of MOFs.

MOF	Time (min)	Temperature (K)
IRMOF1, IRMOF2, IRMOF3	0.42	
$(EMIm)_2[Ni_3(TMA)_2(OAc)_2](EMIm)_2[Co_3(TMA)_2(OAc)_2]$	50	473
$[Ni_{20}(C_5H_6O_4)_{20}(H_2O)_8]\cdot 40H_2O$	1	423–493
MIL(Cr)-101	1–40	483
$[Cu_2(pyz)_2(SO_4)(H_2O)_2]_n$	20, 360	453
$[Cu_2(oba)_2(DMF)_2]\cdot 5.25(DMF)$	1–150	433
$[Mn_3(BTC)_2(H_2O)]_6$	10	393
$[Cd(H_3IDC)(bbi)_{0.5}]$	20	
MOF-5	9–60	368–408
$[Cu_3(BTC)_2(H_2O)_3]\ [Cu_2(OH)(BTC)(H_2O)]\cdot 2nH_2O$	60	413
$[Cu(H_2BTC)_2(H_2O)_2]\cdot 3H_2O$	10	443
$[Zn_2(NDC)_2(DPNI)]$	60	393
$[Co_3(NDC)_3(DMF)_4]$ y $[Mn_3(NDC)_3(DMF)_4]$	30	383

EMIm = 1-ethyl-3-methylimidazolium; TMA = trimesate; pyz = pyrazine; oba = 4,4′-oxydibenzoic acid; BTC = benzene-1,3,5-tricarboxilic; H_3IDC = 4,5-imidazoledicarboxylic acid; bbi = 1,1′-(1,4-butanediyl)bis(imidazole); NDC = 2,7-naphthalene dicarboxylate; DPNI = *N,N* di(4-pyridyl)-1,4,8-naphthalenetetracarboxydiimide.

Table 2. Conditions of microwave-assisted synthesis of MOFs [73].

Ultrasound-assisted synthesis is another route for obtaining materials, where you can get MOFs with small crystal size in a short reaction time. In this synthesis, the reaction mixture is subjected to ultrasound (part of the spectrum of the sound whose frequency is approximately 19 kHz) to generate high temperatures (above 5000 K) and pressures at specific locations within the mixture. Such increases in temperature and pressure are due to the phenomenon of "cavitation", which involves the creation, expansion, and destruction of small bubbles that appear when the reaction mixture is treated with ultrasound [74]. In this case, acoustic radiation mechanical energy is converted into thermal energy. Among the MOFs synthesized by this method are MOF-5, MOF-177, Cu_3BTC_2, Zn-2,2'bipiridina-5,5'dicaboxilato, $Zn_3(BTC)_2$ $12H_2O$ [Zn (1,4-bencendicarboxilato) $(H_2O)]_n$ [75].

2.2.5. *Synthesis of MOFs using near supercritical water conditions*

Motivated by the resolution of the problem that exists with the use of solvents (1.3.1), Schröder and Poliakoff [76] developed a new methodology for the synthesis of MOFs, building for its acronym high-temperature water (HTW). Due to these properties, the HTW has been studied as a means of organic reactions [77, 78], destruction of contaminants [79] and formation of nanoparticles [80, 81]. Water properties change dramatically as it approaches its critical point (647 K, 220 bar) [82]. For example, the dielectric constant decreases to values of typical nonpolar solvents, and therefore, organic compounds, such as organic ligands of MOFs, can be dissolved.

Water can potentially be reused after the reaction has been completed and, if necessary, ion exchange may be employed in order to remove any traces of unreacted organic ligand and metal ions. HTW presents some technical difficulties due to the high pressures and accelerated corrosion of the reactors. However, Schröder and Poliakoff [76], in 2012, first reported the possibility of using HTW (573 K) as solvent for the synthesis of a MOF with high performance. The new MOF, $\{[Zn_2(L)]\,(H_2O)_3\}_\infty$; (L = 1,2,4,5-tetrakis(4-carboxyphenyl)-benzene) (Figure 4), is synthesized using only water as reaction medium at 573 K and 80 bar.

Figure 4. View of the crystal structure of the structure $\{[Zn_2(L)]\,(H_2O)_3\}_\infty$. Green, red, black and gray represent Zn atoms, O, C, and H, respectively [76].

2.3. Characterization and evaluation methods of MOFs

The different methods of synthesis of MOFs can generate homogeneous solids that allow carrying out processes of heterogeneous catalysis. Once the reaction finished, it is desirable that the physicochemical characteristics of material prevail. There are different characterization techniques for determining the homogeneity of the material, structural characteristics, and stability of the MOF. Analytical methods that are useful and applicable are listed below. However, others characterizations may exist which are also useful in the evaluation of MOFs, such as heterogeneous catalysis.

Powder X-ray diffraction (XRD) is used in determining the crystallographic MOFs by comparing the diffractogram of MOF before and after the catalytic process. In certain processes, the stability of the structure is also determined. Additionally, it is possible to determine the purity of the catalyst and some crystallographic parameters as red parameter, size of lattice, and crystal size.

Fourier transform infrared spectroscopy (FTIR) provides information about functional groups present in the network of the MOF. It is possible to make a comparison to determine the changes once the network has carried out the catalytic reaction.

Nuclear magnetic resonance (NMR) is a widely used technique in the characterization of products, by-products, and intermediates of the catalysed reaction. It is possible to determine the chemical environment inside the catalyst using probe molecules.

Nitrogen physisorption. The texture parameters such as surface area, pore volume, and average pore size are determined by this technique. The shape of the isotherm provides information about the homogeneity of the solid.

Ultraviolet-visible diffuse reflectance spectrum provides information about the environment metal coordination before and after carrying out a catalysed reaction.

Thermogravimetric analysis (TGA) is useful to determine the thermal stability of the MOF. In some processes, it is necessary to conduct a heat treatment prior to the catalysed reaction and treatment parameters are determined by TGA. It is possible to obtain a model which is highly suitable for the process reaction desired.

Scanning electron microscopy (SEM) and transmission electron microscopy (TEM) are able to show the morphology, defects, grain boundaries, mixtures of crystalline phases, and grain size, among others. In some solids, even the crystallinity and porosity can be determined by these techniques.

Energy-dispersive X-ray (EDX) is used to determine the elemental materials analysis. The MOF is studied before and after the catalytic process for identifying the presence of new elements and their percentage.

Gas absorption analyser can be used to analyze the adsorption capacity for MOF for a particular gas or vapour.

Gas/liquid chromatography-mass spectrometry (GC/LC-MS) is a powerful technique to analyse the catalytic reaction and determine the amount and type of products.

Raman spectroscopy is widely used in the characterization of noncrystalline or low-crystalline catalysts. Comparing the spectra before and after the reaction provides information about the incorporation of new components into the MOF network.

Temperature-programmed reduction (TPR) is used in determining redox reaction parameters. The catalytic activity in redox conditions can be determined by this technique.

3. MOF's structure using catalytic reaction

The active sites of MOFs can be designed depending on the type of catalytic process. The Rosseinsky group reported the methanolysis of rac-propylene oxide and expected to yield 2-methoxy-1-propanol and 1-methoxy-2-propanol reaction. They used the postsynthesis modification of a porous homochiral Ni(L-asp)bipy 0.5, 1 (L-asp = L-aspartate, bipy = 4,4-dipyridyl), leading to a functional Brønsted acidic material. These compounds are amino acids (L- or D-aspartate) together with dipyridyls as struts. The coordination chemistry is such that the amine group of the aspartate cannot be protonated by added HCl, but one of the aspartate carboxylates can. Thus, the framework-incorporated amino acid can exist in a form that is not accessible for the free amino acid. While the nickel-based compounds are marginally porous, on account of tiny channel dimensions, the copper versions are clearly porous [83].

The results showed that the carboxylic acids behave as Brønsted acidic catalysts, facilitating (in the copper cases) the ring-opening methanolysis of a small, cavity-accessible epoxide at up to 65% yield. These researchers pointed out that the superior homogeneous catalysts existed, but emphasized that the catalyst formed here is unique to the MOF environment, thus representing an interesting proof of concept [84].

Lewis acid solids are commonly used in selective oxidation. An example of this type of catalysts is trinuclear networks containing Cu^{2+}, which have shown a high activity and selectivity for the peroxidative oxidation process of cyclohexane into the corresponding alcohols and ketones ($MeCN/H_2O/HNO_3$ media) [85]. The structure of such MOFs is composed of the secondary building unit of $\{Cu_3(\mu_3\text{-}OH)(\mu\text{-pyrazole})\}$ with tetracoordinate metal centers in axial positions of easy access.

Other structures with these types of catalytic sites on the $Cu_3(BTC)_2$ coordinated network are made of copper links. It is feasible to prepare this MOF with modulated amounts of physisorbed (molecules placed into the channels) or chemisorbed (molecules occupying CuX coordination sites) water molecules with high surface area straight from the reaction vessel without any postsynthetic steps [8]. Different reaction models have been tested in this MOF including: citronellal cyclization [86], benzaldehydecyanosilylation [87], rearrangement of ethylene acetal of 2-bromopropiophenone, isomerization of alpha-pinene oxide [86], among others [28].

Another example of MOF with high concentration of Lewis acidic sites is $Mn_3[(Mn_4Cl)_3(BTT)_8(CH_3OH)_{10}]_2$; H_3BTT = 1,3,5-benzene-tristetrazol-5-yl. Mn^{2+} ions that are exposed on the surface of the framework might serve as potent Lewis acids, and catalyze the cyanosilylation of aromatic aldehydes and ketones, as well as the more demanding Mukaiyama-aldol reaction. Moreover, in each case, a pronounced size-selectivity effect consistent with the pore dimensions is observed [88].

Different types of MOF have been used in catalytic process as base catalysis, Brønsted acid catalysis, Lewis acid catalysis, C–C bond formation and polymerization, enantio selective catalysis, and catalysis by organometallic complex supported on MOFs, among others. Table 3 summarizes the MOF structures used in some catalytic processes reported so far. The most common ions catalysis are: Ag^{+}, Al^{3+}, Bi^{3+}, Ce^{4+}, Cr^{3+}, Co^{2+}, Cu^{2+}, Fe^{3+}, Mn^{2+}, Mg^{2+}, Pd^{2+}, Sc^{3+}, V^{4+}, Zn^{2+}, and Zr^{4+}.

MOF	Catalysed reaction	Reference
$Al_2(bdc)_3$(MIL-53(Al))	Reduction of carbon–carbon multiple bonds	[89]
$[Ag_3(tpha)_2]$ $3BF_4$ (Ag-tpha)	1,3-Dipolar cycloaddition	[90]
Bi(btb)	Hydroxymethylation of 2-methylfuran	[91]
Ce-mdip1	Cyanosilylation of aldehydes	[92]
$Cd(4\text{-btapa})_2(NO_3)_2$	Knoevenagel condensation	[93]
$[Cd(bpy)2](NO_3)_2$	Cyanosilylation of aldehyde	[94]
Cd3Cl6L13	Alkylation of aldehyde	[95]
Co(BPB)	Oxidation of olefin	[96]
Co2(dhbdc)(H_2O) (Co-MOF-74)	Cycloaddition of CO2 and epoxides	[97]
$[Co(DMA)_6]_3[(Co_4Cl)_3{}^-(btt)_8(H_2O)_{12}]_2$(Co-btt)	Ring opening of epoxides; oxidation of hydrocarbons	[98]
$Cr_3X(H_2O)_2O(bdc)_3$ X = F, OH, (MIL101(Cr))	Knoevenagel condensation; Heck coupling; Cyanosilylation of aldehydes Oxidation of hydrocarbons Oxidation of sulfides Cycloaddition of CO_2 and epoxides	[99–103]
$[PW_{11}TiO_{40}]^{5-}$@$[Cr_3F(H_2O)_2O(bdc)_3]$, and $[PW_{11}CoO_{39}]^{5-}$ @$[Cr_3F(H_2O)_2O(bdc)_3]$	Oxidation of olefin	[104]
$Cu(2\text{-pymo})_2]$ and $[Co(PhIM)_2$	Aerobic oxidation of olefin	[105]
Cu(bpy)($H_2O)_2(BF_4)_2$(bpy)	Ring-opening of epoxide	[106]
Cu(D-asp)$bpe_{0.5}$and Cu(L-asp)$bpe_{0.5}$]	Methanolysis of epoxide	[83]
$Cu(L2)_2(H_2O)_2$, $Cu(L3)_2(H_2O)(Py)_2$, $Cu(L3)_3(H_2O)Cl$ and Co(sal)($H_2O)(Py)_3$	Epoxidation of olefin	[107]
$Cu(SO_4)$(pbbm) and $(Cu(Ac)_2(pbbm))CH_3OH$	Oxidative self-coupling	[108]

MOF	Catalysed reaction	Reference
$Cu_3(btc)_2$	Isomerization; cyclization; rearrangement Oxidation of polyphenol Cyanosilylation of aldehyde	[86, 87, 109–113]
$Cu_2(papa)_2Cl_2$	Biginelli reaction; 1,2-additionof a,b-unsaturated ketones	[115]
$Cu_3(pdtc)(pvba)_2(H_2O)_3$	Henry reaction	[115]
$Cu(2\text{-}pymo)_2$	Click reaction Three-component couplings of amines, aldehydes and alkynes	[116, 117]
Cu(tcba)(DMA)	Epoxidation of olefins	[118]
$Cu_2(bpdc)_2(bpy)$	Cross-dehydrogenative coupling reaction	[119]
$Cu_2I_2(bttp4)$	Three-component coupling of azides, alkynes, and amines	[120, 121]
Cu-MOF-$SiF_6$and Cu-MOF-NO_3	Oxidation of benzylic compounds	[122]
CuPhos-Br and CuPhos-Cl and CuPhos-PF_6	Ketalization reaction	[123]
$Fe_3F(H_2O)_2O(btc)_2$ (MIL-100(Fe))	Friedel–Crafts benzylation Oxidation of hydrocarbons Ring-opening of epoxides Claisen Schmidt condensation Oxidation of thiophenol to diphenyldisulfide Isomerization of a-pinene oxide	[124–129
In(OH)(hippb)	Acetalization of aldehyde	[130]
$In_2(OH)_3(bdc)_{1.5}$	Reduction of nitroaromatic; oxidation of sulfide	[131]
$Mg_3(pdc1)(OH)_3(H_2O)_2$	Aldol condensation reactions	[132]
$Mg(pdc2)(H_2O)$	Aldol condensation reactions	[133]
Mn(porphyrin)@[$In_{48}(HImDC)_{96}$]	Oxidation of alkane	[134]
Ln(OH)(1,5-NDS) H_2O	Epoxidation of olefin	[135]
$(Mn(TpCPP)Mn_{1.5})(C_3H_7NO)_5\ C_3H_7NO$	Epoxidation of olefin; oxidation of alkane	[136]
$[Mn_3((Mn_4Cl)_3BTT_8(CH_3OH)_{10})]_2$	Cyanosilylation of aldehyde; Mukaiyama-aldol	[88]
$Mn_2(pvia)_2(H_2O)_2$	Alcohol oxidation	[137]
$(Na_{20}(Ni_8L4_{12})(H_2O)_{28})(H_2O)_{13}(CH_3OH)_2$	Oxidation to CO_2	[138]
$Sc_3(OH)(H_2O)_2O(btc)_2$ (MIL-100(Sc))	Intermolecular carbonyl ene reaction; Michael addition reaction; ketimine and aldimine formation	[139]

http://dx.doi.org/10.5772/61865

MOF	Catalysed reaction	Reference
Pd(2-pymo)$_2$	Oxidation of alcohol; Suzuki–Miyaura coupling; hydrogenation of olefin	[140–142]
Tb[V_6O_{13}{$(OCH_2)_3C(NH_2CH_2C_6H_4$-4- CO_2)}{$(OCH_2)_3C$-($NHCH_2C_6H_4$-4-CO_2)}$_2$]$^{4-}$	Oxidation of sulfide	[143]
$V^{IV}O$(bdc) (MIL-47(V))	Oxidation of hydrocarbons	[144, 145]
Zn_2(bpdc)$_2$L5	Epoxidation of olefins	[146]
$Zn_2(Py_2(PhF_5)_2$PorZn)(TCPB)	Intermolecular transfer of acyl	[147]
[$Zn_3(m_3$-O)($O_2CR)_6(H_2O)_3]^{n+}$	Transesterification	[148]
Zn_4O(bdc)$_3$] and [Zn_4O(ndc)$_3$	Friedel–Crafts alkylation	[149]
[(Zn_4O)(bdc-NH_2)$_3$]Vsal$_{0.4}$	Oxidation of olefin	[150]
Zn_2(bdc)(L-lac)(dmf)	Oxidation of thioethers	[151]
Zn_3(chirbtb-1)$_2$	Mukaiyama aldol reaction	[152]
Zn_3(chirbtb-2)$_2$	Mukaiyama aldol reaction	[152]
Zn(Meim)$_2$(ZIF-8)	Cycloaddition of CO_2 and epoxides	[153]
$Zr_6O_4(OH)_4$(bdc)$_6$ (UiO-66)	Aldol condensation reactions Cyclization of citronellal	[154–156]]

Ac = acetyl; bdc = 1,4-benzenedicarboxylate; BPB = 1,4-bis(4′-pyrazolyl)benzene; bpdb = 1,4-bis(3,5-dimethyl-1H-pyrazol-4-yl)benzene; bpdc = biphenyldicarboxylate; bpe = trans-1,2-bis(4-pyridyl)ethylene); bpy = 4,4′-bipyridine; btc = benzene-1,3,5-tricarboxylate; btapa = 1,3,5-benzene tricarboxylic acid tris[*N*-(4-pyridyl)amide]; btb = 5 -(4-carboxyphenyl)-[1,1′:3′,1″-terphenyl]-4,4′-dicarboxylate; btt = 1,3,5-benzenetris(tetrazol-5-yl); bttp4 = benzene-1,3,5-triyl triisonicotinate; ChirBTB-1 = 5′-(4-carboxy-3-((S)-4-isopropyl-2-oxooxazolidin-3-yl)phenyl)-3-((S)-4-isopropyl-2-oxooxazolidin-3-yl)-3′-(3-isopropyl-5-oxooxazolidin-4-yl)-[1,1′:3′,1″-terphenyl]-4,4″-dicarboxylate;ChirBTB-2 = 3,3″-bis((S)-4-benzyl-2-oxooxazolidin-3-yl)-5′-(3-(3-benzyl-5-oxooxazolidin-4-yl)-4-carboxyphenyl)-[1,1′:3′,1″-terphenyl]-4,4″-icarboxylic; D-asp = D-aspartate; bdc = benzene-1,4-dicarboxylate; dhbdc = 2,5-dihydroxyisophthalic; ImDC = 4,5-imidazole dicarboxylate; ippb = 4,4′-(hexafluoroisopropyl-idene)bis(benzoate); L1 = (R)-6,6′-dichloro-2,2-dihydroxy-1,1′-binaphthyl-4,4′-bipyridine; L2 = (4-formylphenoxy)acetic acid; L3 = 2-[2-[[(2-aminoethyl)imino]methyl]phenoxy]acetic acid; L4 = 4,5-imidazoledicarboxylic acid; L5 = (*R*,*R*)-(-)-1,2-cyclohexanediamino-*N*,*N*-bis(3-tert-butyl-5-(4-pyridyl)salicylidene)Mn^{III}Cl; L-lac = L-lactic acid; mdip = 5,50 -methylenediisophthalic; meim = 2-methyl-1H-imidazole;nds = naphthalenedisulfonic acid; papa = (S)-3-hydroxy-2-((pyridin-4-ylmethyl)amino)propanoic;pbbm = 1,1′-(1,5-pentanediyl)bis(1H-benzimidazole); pdc-1 = pyrazole-3,5-dicarboxylate; pdc-2= pyridine-2,5-dicarboxylate; PhIM = phenyl imidazolate; ptdc = pyridine-2,3,5,6-tetracarboxylic; pymo = 2-hydroxypyrimidinolate; $Py_2(PhF_5)_2$Por = 5,15-dipyridyl-1′,2′-bis(pentafluorophenyl)porphyrin; pvia= (E)-5-(2-(pyridin-4yl)vinyl)isophthalic; pvba = (E)-4-(2-(pyridin-4-yl)vinyl)benzoic; sal = salicylidene moiety; tcba = 4,′4″,′4‴-nitrilotris([1,10-biphenyl]-4-carboxylic); TCPB = 1,2,4 5-tetrakis(4-carboxyphenyl)benzene; TpCPP = tetra-(*p*-carboxyphenyl)porphyrin; tpha = tris(4-((*E*)-1-(2-(pyridin-2-yl)hydrazono)ethyl)phenyl)amine.

Table 3. MOF structure used for catalytic reaction.

4. Limitation of MOF structures

MOFs are excellent candidates for certain catalytic processes because: (1) they can be designed on a rational basis according to specific requirements; (2) their well-defined structure allows the assessment of structure–activity relationships; (3) the uniform catalytic sites; and (4) the intrinsic nature of their pores.

However, the synthesis of MOF requires a series of steps to allow an activation network free of solvent molecules and to expose the active sites, which can often require coordinating solvents that may be toxic, carcinogenic, and/or dangerous to the environment [59]. In some processes, the structure collapses, and activation prevents further use in catalysis.

Furthermore, when washed with solvent, they typically require energy, which increases the time synthesis process and drastically affects the efficiency. For example, with MOF-5 synthesized using solvothermal processes, a yield of 60% was achieved after 7 days [72].

The use of coordinating solvents during the synthesis of MOF such as dimethylformamide (DMF) or diethylformamide (DEF) may interfere with the availability of molecules to interact with the active sites. DMF and DEF decompose when heated at high temperatures for long periods and therefore cannot be reused [60]. The study of local defects is also crucial since catalytic processes can be favoured with the appearance of the same or conversely the process is catalysed not by excess thereof.

The MOF's purity can be affected by the formation of other crystalline compounds or the presence of reagents in the network. However, characterization of MOFs' purity and homogeneity can seldom be found in scientific papers about catalysis.

The thermal and chemical stability of MOFs is also a limitation for use in some catalytic process. The zirconium MOF reported by Hafizovic Cavka et al. [157], which has a thermal resistance above 500°C, resistance to most chemicals, and they remain crystalline even after exposure to 10 tons/cm^2 of external pressure, whereas other MOFs have a lower thermal and chemical stability.

Author details

Sandra Loera-Serna* and Elba Ortiz

*Address all correspondence to: sls@azc.uam.mx

Universidad Autonoma Metropolitana, Azcapotzalco, Distrito Federal, Mexico

References

[1] Janiak C, Vieth J K, MOFs, MILs and more: concepts, properties and applications for porous coordination netkorks (PCNs). New J. Chem. 2010;34:2366–2388. DOI: 10.1039/C0NJ00275E

[2] Farrusseng D, Aguado S, Pinel C, Metal-organic frameworks: opportunities for catalysis. Angew. Chem. Int. Ed. 2009;48:7502–7513.DOI: 10.1002/anie.200806063

[3] Lui J, Chen L, Cui H, Zhang J, Zhang L, Su C, Application of metal-organic frameworks in heterogeneous supramolecular catalysis. Chem. Soc. Rev. 2014;43:6011–6061. DOI: 10.1039/c4cs00094c

[4] Gascon J, Corma A, Kaptejin F, Llabrés F X, Metal organic framework catalysis; Quo vadis? ACS Catal. 2014;4:361–378. DOI: 10.1021/cs400959k

[5] Chen L, Xiang S C, Qian G D, Metal–organic frameworks with functional pores for recognition of small molecules. Acc. Chem. Res., 2010;43:1115–1124. DOI: 10.1021/ar100023y

[6] Horcajada P, Chalati T, Serre C, Gillet B, Sebrie C, Baati T, Eubank J F, Heurtaux D, Clayette P, Kreuz C, Chang J S, Hwang Y K, Marsaud V, Bories P N, Cynober L, Gil S, Férey G, Couvreur P, Gref R, Porous metal–organic-framework nanoscalecarriers as a potential platform for drug delivery and imaging. Nature Mater. 2010;9:172–178. DOI: 10.1038/NMAT2608

[7] Sun C Y, Qin C, Wang C G, Su Z M, Wang S, Wang X L, Yang G S, Shao K Z, Lan Y Q, Wang E B, Chiral nanoporous metal-organic frameworks with high porosity as materials for drug delivery. Advanced Mater. 2011;23:5629–5632. DOI: 0.1002/adma.201102538

[8] Loera-Serna S, López L, Flores J, López-Simeon R, Beltrán H I, An alkaline one-pot metathesis reaction to give a $[Cu_3(BTC)_2]$ MOF at r.t., with free Cu coordination sites and enhanced hydrogen uptake properties. RSC Advances. 2013;3:10962–10972. DOI: 10.1039/c3ra40726h

[9] Rowsell L C, Spencer E C, Eckert J; Howard J A K, Yaghi O M, Gas adsorption sites in a large-pore metal-organic framework. Science, 2005;309:1350–1354. DOI: 10.1126/science.1113247

[10] Llewellyn P L, Maurin G, Devic T, Loera-Serna S, Rosenbach N, Serre C, Bourrelly S, Horcajada P, Filinchuk Y, Férey G, Prediction of the conditions for breathing of metal organic framework materials using a combination of X-ray powder diffraction, microcalorimetry, and molecular simulation. J. Am. Chem. Soc. 2008;130:12808–12814. DOI: 10.1021/ja803899q

[11] Llewellyn P L, Horcajada P, Maurin G. Devic T, Rosenbach N, Bourrelly S, Serre C, Loera-Serna S, Vincent D, Filinchuk Y, Férey G, Complex adsorption of short linear alkanes in flexible metal-organic-frameworks MIL-53(Fe). J. Am. Chem. Soc. 2009;131:13002–13008. DOI: 10.1021/ja902740r

[12] Trung T K, Trens P, Tanchoux N, Bourrelly S, Llewellyn P L, Loera-Serna S, Serre C, Loiseau T, Fajula F, Férey G, Hydrocarbon adsorption in the flexible metal organic

frameworks MIL-53(Al, Cr). J. Am. Chem. Soc. 2008;130:16926–16932.DOI: 10.1021/ja8039579

[13] Li J R, Kuppler R J, Zhou H C, Selective gas adsorption and separation in metal-organic frameworks. Chem. Soc. Rev. 2009;38:1477–1504. DOI: 10.1039/B802426J

[14] Hu Z, Deibert B J, Li J, Luminiscent metal-organic frameworks for chemical sensing and explosive detection. Chem Soc. Rev. 2014;43:5815–5840. DOI: 10.1039/c4cs00010b

[15] Gandara F, Snejko N, Andres A D, Fernandez J R, Gomez-Sal J C, Gutierrez-Puebla E, Monge A, Stable organic radical stacked by in situ coordination to rare earth cations in MOF materials, RSC Advances, 2012;2:949–955. DOI: 10.1039/C1RA00447F

[16] Yan D, Tang Y, Lin H, Wang D, Tunable two-color luminescence and host–guest energy transfer of fluorescent chromophores encapsulated in metal-organic frameworks. Scientific Reports. 2014;4:4337:1–7.DOI: 10.1038/srep04337

[17] Zhu Q L, Xu Q, Metal-organic frameworks composites. Chem. Soc. Rev. 2014;43:5468–6176. DOI: 10.1039/c3cs60472a

[18] AdeyemoA A, Adeoye I O, Bello O S, Metal organic frameworks as adsorbents for dye adsorption: overview, prospects and future challenges. Toxicological Env. Chem. 2012;94:1846–1863. DOI: 10.1080/02772248.2012.744023

[19] Hu J, Dai W, Yan X, Comparison study on the adsorption performance of methylene blue and congo red on Cu-BTC. Desalination Water Treat. 2014:1–9. DOI: 10.1080/19443994.2014.988654

[20] Yang W, Feng J, Song S, Zhang H, Microwave-assisted modular fabrication of nanoscale luminescent metal-organic framework for molecular sensing. Chem. Phys. Chem. 2012;13:2734–2738.DOI: 10.1002/cphc.201200265

[21] Barea E, Montoro C, Navarro J A R, Toxic gas removal—metal–organic frameworks for the capture and degradation of toxic gases and vapours. Chem. Soc. Rev. 2014;43:5419–5430. DOI: 10.1039/c3cs60475f

[22] Zhang C F, Qiu L G, Ke F, Zhu Y J, Yuan Y P, Xua G S, Jianga X, A novel magnetic recyclable photocatalyst based on a core–shell metal–organic framework Fe_3O_4@MIL-100(Fe) for the decolorization of methylene blue dye. J. Mater. Chem. A. 2013;1:14329–14334. DOI: 10.1039/c3ta13030d

[23] Bourrelly S, Moulin B, Rivera A, Maurin G, Devautour-Vinot S, Serre C, Devic T, Horcajada P, Vimont A, Clet G, Daturi M, Lavalley J C, Denoyel R, Llewellyn P, Férey G, Loera-Serna S, Explanation of the adsorption of polar vapors in the highly flexible metal organic framework MIL-53(Cr). J. Am. Chem. Soc. 2010;132:9488–9498. DOI: 10.1021/ja1023282

[24] Loera-Serna S, Oliver-Tolentino M A, López-Núñez M L, Santana-Cruz A, Guzmán-Vargas A, Cabrera-Sierra R, Beltrán H I, Flores J, Electrochemical behavior of

[$Cu_3(BTC)_2$] metal–organic framework: The effect of the method of synthesis. J. Alloys Comp. 2012;540:113–120. DOI: 10.1016/j.jallcom.2012.06.030

[25] Van de Voorde B, Bueken B, Denayer J, De Vos D, Adsorptive separation on metal–organic frameworks in the liquid phase. Chem. Soc. Rev. 2014;43:5766–5788. DOI: 10.1039/c4cs00006d

[26] Falcaro P, Ricco R, Doherty C M, Liang K, Hill A J, Styles M J, MOF positioning technology and device fabrication. Chem. Soc. Rev. 2014;43:5513–5560. DOI: 10.1039/c4cs00089g

[27] Bianchini C, Cole-Hamilton D J, van Leeuwe P W N M, Heterogenized homogeneous catalysts for fine chemicals production. Catalysis by Metal Complex 2010. Ed. Springer. DOI: 10.1007/978-90-481-3696-4

[28] Luz I, Llabrés F X, Corma A, Bridging homogneous and heterogeneous catalysis with MOFs: "Clik" reactions with Cu-MOF catalysis. J. Cat. 2010;276:134–140. DOI: 10.1016/j.cat.2010.09.010

[29] Yaghi O M, Li G, Li H, Selective binding and removal of guests in a microporous metal–organic framework. Nature, 1995;378:703–706. DOI: 10.1038/378703a0

[30] Yaghi O M, Li H, Hydrothermal synthesis of a metal-organic framework containing large rectangular channels. J. Am. Chem. Soc., 1995;117:10401–10402. DOI: 10.1021/ja00146a033

[31] Archer R D, Preparative inorganic reactions. Volume 1. J. Am. Chem. Soc. 1965;87:1151–1152. DOI: 10.1021/ja01083a051

[32] Seo J, Sakamoto H, Matsuda R, Kitagawa S, Chemistry of porous coordination polymers having multimodal nanospace and their multimodal functionality J. Nanosci. Nanotechnol. 2010;10:3–20. DOI: http://dx.doi.org/10.1166/jnn.2010.1494

[33] Blake A J, Champness N R, Hubberstey P, Li W S, Withersby M A, Schröder M, Inorganic crystal engineering using self-assembly of tailored building-blocks. Coord. Chem. Rev. 1999;183:117–138. DOI: 10.1016/S0010-8545(98)00173-8

[34] Lu J, Paliwala T, Lim S C, Yu C, Niu T Y, Jacobson A J, Coordination polymers of $Co(NCS)_2$ with pyrazine and 4,4′-bipyridine: Syntheses and structures. Inorg. Chem. 1997;36:923–929.DOI: 10.1021/ic961158g

[35] Zaworotko M J, Superstructural diversity in two dimensions: crystal engineering of laminated solids. Chem. Commun. 2001;1:1–9.DOI: 10.1039/B007127G

[36] Kitagawa S, Kondo M, Functional micropore chemistry of crystalline metal complex-assembled compounds. Bull. Chem. Soc. Jpn. 1998;71:1739–1753. DOI: 10.1246/bcsj.71.1739

[37] Moulton B, Zaworotko M J, From molecules to crystal engineering: Supramolecular isomerism and polymorphism in network solids. Chem. Rev. 2001;101:1629–1658.DOI: 10.1021/cr9900432

[38] Eddaoudi M, Moler D B, Li H L, Chen B L, Reineke T M, O'Keeffe M, Yaghi O M, Modular chemistry: secondary building units as a basis for the design of highly porous and robust metal-organic carboxylate frameworks. Acc. Chem. Res. 2001;34:319–330. DOI: 10.1021/ar000034b

[39] Burrows A D, Harrington R W, Mahon M F, Price C E, The influence of hydrogen bonding on the structure of zinc co-ordination polymers. J. Chem. Soc. Dalton Trans. 2000;21:3845–3854. DOI: 10.1039/B003210G

[40] Choi H J, Lee T S, Suh M P, Selective binding of open frameworks assembled from nickel(II) macrocyclic complexes with organic and inorganic guests. J. Incl. Phenom. Macrocycl. Chem. 2001;41:155–162. DOI: 10.1023/A:1014436406651

[41] Gutschke S O H, Price D J, Powell A K, Wood P T, Hydrothermal synthesis, structure, and magnetism of $[Co_2(OH)\{1,2,3\text{-}(O_2C)_3C_6H_3\}(H_2O)]\cdot H_2O$ and $[Co_2(OH)\{1,2,3\text{-}(O_2C)_3C_6H_3\}]$: magnetic Δ-chains with mixed cobalt geometries. Angew. Chem. Int. Edit. 2001;40:1920–1923. DOI: 10.1002/1521-3773(20010518)40:10<1920::AID-ANIE1920>3.0.CO;2-2

[42] Prior T J, Rosseinsky M J, Crystal engineering of a 3-D coordination polymer from 2-D building blocks. Chem. Commun. 2001;5:495–496.DOI: 10.1039/B009455M

[43] Yaghi O M, Li H L, Groy T L, Construction of porous solids from hydrogen-bonded metal complexes of 1,3,5-benzenetricarboxylic acid. J. Am. Chem. Soc. 1996;118:9096–9101. DOI: 10.1021/ja960746q

[44] Lin X, Jia J H, Zhao X B, Thomas K M, Blake A J, Walker G S, Champness N R, Hubberstey P, Schröder M, High H_2 adsorption by coordination-framework materials. Angew. Chem.-Int. Edit. 2006;45:7358–7364. DOI: 10.1002/anie.200601991

[45] Yang S, Lin X, Blake A J, Thomas K M, Hubberstey P, Champness N R, Schröder M, Enhancement of H_2 adsorption in Li^+-exchanged co-ordination framework materials. Chem. Commun. 2008;46:6108–6110. DOI: 10.1039/B814155J

[46] Yan Y, Lin X, Yang S, Blake A J, Dailly A, Champness N R, Hubberstey P, Schröder M, Exceptionally high H_2 storage by a metal–organic polyhedral framework. Chem. Commun. 2009;9:1025–1027. DOI: 10.1039/B900013E

[47] Yan Y, Telepeni I, Yang S, Lin X, Kockelmann W, Dailly A, Blake A J, Lewis W, Walker G S, Allan D R, Barnett S A, Champness N R, Schröder M, Metal–organic polyhedral frameworks: high H_2 adsorption capacities and neutron powder diffraction studies. J. Am. Chem. Soc. 2010;132:4092–4094. DOI: 10.1021/ja1001407

[48] James S L, Metal-organic frameworks. Chem. Soc. Rev. 2003;32:276–288. DOI: 10.1039/B200393G

[49] Cheng J Y, Dong Y B, Huang R Q, Smith M D, Synthesis and characterization of new coordination polymers generated from oxadiazole-containing ligands and IIB metal ions. Inorg. Chim. Acta, 2005;358:891-898. DOI: 10.1016/j.ica.2004.10.034

[50] Uemura K, Kitagawa S, Kondo M, Fukui K, Kitaura R, Chang H C, Mizutani T, Novel flexible frameworks of porous cobalt(II) coordination polymers that show selective guest adsorption based on the switching of hydrogen-bond pairs of amide groups. Chem. Eur. J. 2002;8:3586–3600. DOI: 10.1002/1521-3765(20020816)8:16<3586::AID-CHEM3586>3.0.CO;2-K

[51] Chen Z F, Xiong R G, Abrahams B F, You X Z, Che C M, An unprecedented six-fold anion-type chiral diamondoid-like eight-coordinate Cd(II) coordination polymer with a second-order nonlinear optical effect. J. Chem. Soc. Dalton Trans. 2001;17:2453–2454. DOI: 10.1039/B105130J

[52] Kim J, Chen B, Reineke T M, Li H, Eddaoudi M, Moler D B, O'Keeffe M, Yaghi O M, Assembly of metal–organic frameworks from large organic and inorganic secondary building units: new examples and simplifying principles for complex structure. J. Am. Chem. Soc. 2001;123:8239–8247. DOI: 10.1021/ja010825o

[53] Farha O K, Malliakas C D, Kanatzidis M G, Hupp J T, Control over catenation in metal–organic frameworks via rational design of the organic building Block.J. Am. Chem. Soc. 2010;132:950–952. DOI: 10.1021/ja909519e

[54] Sun D, Ma S, Ke Y, Collins D J, Zhou H C, An interweaving MOF with high hydrogen uptake. J. Am. Chem. Soc. 2006;128:3896–3897. DOI: 10.1021/ja058777l

[55] Li J R, Yakovenko A A, Lu W, Timmons D J, Zhuang W, Yuan D, Zhou H C, Ligand bridging-angle-driven assembly of molecular architectures based on quadruply bonded Mo–Mo dimers. J. Am. Chem. Soc. 2010;132:17599–17610. DOI: 10.1021/ja1080794

[56] Hartmann M, Kunz S, Himsl D, Tangermann O, Adsorptive separation of isobutene and isobutane on $Cu_3(BTC)_2$.Langmuir. 2008;24:8634–8642. DOI: 10.1021/la8008656

[57] Guo H, Zhu G, Hewitt I J, Qiu S, "Twin copper source" growth of metal–organic framework membrane: $Cu_3(BTC)_2$ with high permeability and selectivity for recycling H_2. J. Am. Chem. Soc. 2009;131:1646–1647. DOI: 10.1021/ja8074874

[58] Son W J, Kim J, Kim J, Ahn S, Sonochemical synthesis of MOF-5. Chem. Commun. 2008; 6336–6338. DOI: 10.1039/B814740J

[59] Pohanish R P, Sittig's (2008) Handbook of Toxic and Hazardous Chemicals and Carcinogens, 5th ed., William Andrew Publishing. ISBN: 978-1-4377-7869-4

[60] Muzart J, *N,N*-dimethylformamide: much more than a solvent. Tetrahedron. 2009;65:8313–8323. DOI: 10.1016/j.tet.2009.06.091

[61] Tateyama Y, Ohki Y, Suzuki Y, OuchiA, The crystal and molecular structure of diaquadihydroxotetrakis(*m*-nitrobenzenesulfonato)discandium(III) in a linear polymeric

form, $[Sc_2(OH)_2(O_2NC_6H_4SO_3)_4(H_2O)_2]_n$. Bull. Chem. Soc. Jap. 1998;61:2214–2216. DOI: 10.1246/bcsj.61.2214

[62] Anastas P T, Warner J C, Green Chemistry: Theory and Practice, Oxford University Press, 1998. ISBN: 9780198506980

[63] Fernández-Bertran J F, Mechanochemistry: an overview. Pure Appl. Chem. 1999;71:581–586. DOI: 10.1351/pac199971040581

[64] Belcher W J, Longstaff C A, Neckenig M R, Steed J W, Channel-containing 1D coordination polymers based on a linear dimetallic spacer. Chem. Comm. 2002;15:1602–1603. DOI: 10.1039/B202652J

[65] Pichon A, James S L, Conditions—insights and trends. Cryst. Chem. Comm. 2008;10:1839–1847. DOI: 10.1039/B810857A

[66] Mueller U, Puetter H, Hesse M, Wessel H, Method for electrochemical production of a crystalline porous metal organic skeleton material. Patent; 2005. WO 2005/049892 A1.

[67] Mueller U, Schubert M, Teich F, Puetter H, Shierle-Arndt K, Pastré J, Metal–organic frameworks—prospective industrial applications. J. Mater. Chem. 2006;16:626–636. DOI: 10.1039/B511962F

[68] Richter I, Schubert M, Muller U, Porous metal organic framework based on pyrroles and pyridinones. Patent;2007. WO 2007/131955 A1.

[69] Hill J, Marchant R, Applied. Modelling microwave heating. Mathematical Modern. 1996;20:3–15. DOI: 10.1016/0307-904X(95)00107-U

[70] Larhed M, Moberg C, HallbergA, Microwave-accelerated homogeneous catalysis in organic chemistry. Acc. Chem. Res. 2002;35:717–727. DOI: 10.1021/ar010074v

[71] Jhung S H, Yoon J W, Hwang J S, Cheetham A K, Chang J S, Facile synthesis of nanoporous nickel phosphates without organic templates under microwave irradiation. Chem. Mater. 2005;17:4455–4460. DOI: 10.1021/cm047708n

[72] Choi J Y, Kim J, Jhung S H, Kim H K, Chang J S, Chae H K, Microwave synthesis of a porous metal-organic framework, zinc terephthalate MOF-5. Bull. Korean Chem. Soc. 2006;27:1523–1524. DOI: 10.5012/bkcs.2006.27.10.1523

[73] Klinowski J, Almeida Paz F A, Silva P Rocha J, Microwave-assisted synthesis of metal–organic frameworks. Dalton Trans. 2011;40:321–330. DOI: 10.1039/C0DT00708K

[74] Suslick K S, Hammerton D A, Cline R E, Sonochemical hot spot. J. Am. Chem. Soc. 1986;108:5641–5642. DOI: 10.1021/ja00278a055

[75] Meek S T, Greathouse J A, Allendorf M D, Metal-organic frameworks: A rapidly growing class of versatile nanoporous materials. Adv. Mater. 2011;23:249–267.DOI: 10.1002/adma.201002854

[76] Ibarra I A, Bayliss P A, Pérez E, Yang S, Blake A J, Nowell H, Allan D R, Poliakoff M, Schröder M, Near-critical water, a cleaner solvent for the synthesis of a metal–organic framework. Green Chem. 2012;14:117–122. DOI: 10.1039/C1GC15726D

[77] Savage P E, Organic chemical reactions in supercritical water. Chem. Rev. 1999;99:603–621. DOI: 10.1021/cr9700989

[78] Fraga-Dubreuil J, Poliakoff M, Organic reactions in high-temperature and supercritical water. Pure Appl. Chem. 2006;78:1971–1982. DOI: 10.1351/pac200678111971

[79] Brunner G, Near and supercritical water. Part II: Oxidative processes. J. Supercrit. Fluids. 2009;47:382–390. DOI: 10.1016/j.supflu.2008.09.001

[80] Aymonier C, Loppinet-Serani A, Reveron H, Garrabos Y, Cansell F, Review of supercritical fluids in inorganic materials science. J. Supercrit. Fluids, 2006;38:242–251. DOI: 10.1016/j.supflu.2006.03.019

[81] Cabañas A, Poliakoff M, The continuous hydrothermal synthesis of nano-particulate ferrites in near critical and supercritical water. J. Mater. Chem. 2001;11:1408–1416. DOI: 10.1039/B009428P

[82] Weingartner H, Franck E U, Supercritical water as a solvent. Angew. Chem. Int. Ed. 2005;44:2672–2692. DOI: 10.1002/anie.200462468

[83] Ingleson M J, Perez Barrio J, Bacsa J, Dickinson C, Park H, Rosseinsky M J, Generation of a solid Brønsted acid site in a chiral framework. Chem. Commun. 2008;11:1287–1289. DOI: 10.1039/b718443c

[84] Lee J Y, Farha O K, Roberts J, Scheidt K A, Nguyen S T, Hupp J T, Metal–organic framework materials as catalysts. Chem. Soc. Rev. 2009;38:1450–1459. DOI: 10.1039/b807080f

[85] Nicola C D, Karabach Y Y, Kirollov A M, Monar M, Pandolfo L, Pettinari C, Pombeiro A J L, Supramolecular assemblies of trinuclear triangular copper(II) secondary building units through hydrogen bonds. Generation of different metal–organic frameworks, valuable catalysts for peroxidative oxidation of alkanes. Inorg. Chem. 2007;46:221–230. DOI: 10.1021/ic061595n

[86] Alaerts L, Seguin E, Poelman H, Thibault-Starzyk F, Jacobs P A, De Vos D E, Probing the Lewis acidity and catalytic activity of the metal–organic framework [$Cu_3(BTC)_2$] (BTC = benzene-1,3,5-tricarboxylate). Chem. Eur. J. 2006;12:7353–7363. DOI: 10.1002/chem.200600220

[87] Schlichte K, Kratzke T, Kaskel S, Improved synthesis, thermal stability and catalytic properties of the metal-organic framework compound $Cu_3(BTC)_2$. Micro. Meso. Mater. 2004;73:81–88. DOI: 10.1016/j.micromeso.2003.12.027

[88] Horike S, Dinca M, Tamaki K, Long J R, Size-selective Lewis acid catalysis in a microporous metal-organic framework with exposed Mn^{2+} coordination sites. J. Am. Chem. Soc. 2008;130:5854–5855. DOI: 0.1021/ja800669j

[89] Dhakshinamoorthy A, Alvaro M, García H, Metal-organic frameworks (MOFs) as heterogeneous catalysts for the chemoselective reduction of carbon-carbon multiple bonds with hydrazine. Adv. Synth. Catal. 2009;351:2271–2276. DOI: 10.1002/adsc.200900362

[90] Jing X, He C, Dong D, Yang L, Duan C, Angew. Homochiral crystallization of metal–organic silver frameworks: Asymmetric [3+2] cycloaddition of an azomethineylide. Chem., Int. Ed. 2012;51:10127–10131. DOI: 10.1002/anie.201204530

[91] Feyand M, Mugnaioli E, Vermoortele F, Bueken B, Dieterich J M, Reimer T, Kolb U, de Vos D, Stock N, Automated diffraction tomography for the structure elucidation of twinned, sub-micrometer crystals of a highly porous, catalytically active bismuth metal–organic framework. Angew. Chem., Int. Ed. 2012;51:10373–10376. DOI: 10.1002/anie.201204963

[92] Dang D, Wu P, He C, Xie Z, Duan C, Homochiral metal–organic frameworks for heterogeneous asymmetric catalysis, J. Am. Chem. Soc. 2010;132:14321–14323. DOI: 10.1021/ja101208s

[93] Hasegawa S, Horike S, Matsuda R, Furukawa S, Mochizuki K, Kinoshita Y, Kitagawa S, Three-dimensional porous coordination polymer functionalized with amide groups based on tridentate ligand: Selective sorption and catalysis. J. Am. Chem. Soc. 2007;129:2607–2614. DOI: 10.1021/ja067374y

[94] Fujita M, Kwon Y J, Washizu S, Ogura K, Preparation, clathration ability, and catalysis of a two-dimensional square network material composed of cadmium(II) and 4,4′-bipyridine. J. Am. Chem. Soc. 1994;116:1151–1152. DOI: 10.1021/ja00082a055

[95] Wu C D, Hu A, Zhang L, Lin W, A homochiral porous metal–organic framework for highly enantioselective heterogeneous asymmetric catalysis. J. Am. Chem. Soc. 2005;127:8940–8941. DOI: 10.1021/ja052431t

[96] Tonigold M, Lu Y, Mavrandonakis A, Puls A, Staudt R, Mollmer J, Sauer J, Volkmer D, Pyrazolate-based cobalt(II)-containing metal–organic frameworks in heterogeneous catalytic oxidation reactions: elucidating the role of entatic states for biomimetic oxidation processes. Chem. Eur. J. 2011;17:8671–8695. DOI: 10.1002/chem.201003173

[97] Cho H-Y, Yang D A, Kim J, Jeong S Y, Ahn W S, CO_2 adsorption and catalytic application of Co-MOF-74 synthesized by microwave heating. Catal. Today. 2012;185:35–40. DOI: 10.1016/j.cattod.2011.08.019

[98] Biswas S, Maes M, Dhakshinamoorthy A, Feyand M, De Vos D E, Garcia H, Stock N, Fuel purification, Lewis acid and aerobic oxidation catalysis performed by a microporous Co-BTT (BTT^{3-} = 1,3,5-benzenetristetrazolate) framework having coordina-

tively unsaturated sites. J. Mater. Chem. 2012;22:10200–10209. DOI: 10.1039/C2JM15592C

[99] Henschel A, Gedrich K, Kraehnert R, Kaskel S, Catalytic properties of MIL-101. Chem. Commun. 2008;4192–4194. DOI: 10.1039/B718371B

[100] Kim J, Bhattacharjee S, Jeong K E, Jeong S Y, Ahn W S, Selective oxidation of tetralin over a chromium terephthalate metal organic framework, MIL-101. Chem. Commun. 2009;3904–3906. DOI: 10.1039/B902699A

[101] Hwang Y K, Hong D Y, Chang J S, Seo H, Yoon M, Kim J, Jhung S H, Serre C, Férey G, Selective sulfoxidation of aryl sulfides by coordinatively unsaturated metal centers in chromium carboxylate MIL-101. Appl. Catal., A, 2009;358:249–253. DOI: 10.1016/j.apcata.2009.02.018

[102] Maksimchuk N V, Kovalenko K A, Fedin V P, Kholdeeva O A, Heterogeneous selective oxidation of alkenes to α,β-unsaturated ketones over coordination polymer MIL-101. Adv. Synth. Catal. 2010;352:2943–2948. DOI: 10.1002/adsc.201000516

[103] Zalomaeva O V, Chibiryaev A M, Kovalenko K A, Kholdeeva O A, Balzhinimaev B S, Fedin V P, Cyclic carbonates synthesis from epoxides and CO_2 over metal–organic framework Cr-MIL-101. J. Catal. 2013;298:179–185. DOI: 10.1016/j.jcat.2012.11.029

[104] Maksimchuk N V, Timofeeva M N, Melgunov M S, Shmakov A N, Chesalov A Y, Dybtsev D N, Fedin V P, Kholdeeva O A, Heterogeneous selective oxidation catalysts based on coordination polymer MIL-101 and transition metal-substituted polyoxometalates. J. Catal. 2008;257:315–323. DOI: 10.1016/j.jcat.2008.05.014

[105] Llabrés i Xamena F X, Casanova O, Galiasso Tailleur R, Garcia H, Corma A, Metal organic frameworks (MOFs) as catalysts: A combination of Cu^{2+} and Co^{2+} MOFs as an efficient catalyst for tetralin oxidation. J. Catal. 2008;255:220–227. DOI: 10.1016/j.jcat.2008.02.011

[106] Jiang D, Mallat T, Krumeich F, Baiker A, Copper-based metal-organic framework for the facile ring-opening of epoxides. J. Catal. 2008;257:390–395. DOI: 10.1016/j.jcat.2008.05.021

[107] Pramanik A, Abbina S, Das G, Molecular, supramolecular structure and catalytic activity of transition metal complexes of phenoxy acetic acid derivatives. Polyhedron, 2007;26:5225–5234. DOI: 10.1016/j.poly.2007.07.033

[108] Xiao B, Hou H, Fan Y, Catalytic applications of CuII-containing MOFs based on N-heterocyclic ligand in the oxidative coupling of 2,6-dimethylphenol. J. Organomet. Chem. 2007;692:2014–2020. DOI: 10.1016/j.jorganchem.2007.01.010

[109] Chui S S Y, Lo S M F, Charmant J P H, Orpen A G, Williams I D, A chemically functionalizable nanoporous material $[Cu_3(TMA)2(H2O)3]_N$ Science. 1999;283:1148–1150. DOI: 10.1126/science.283.5405.1148

[110] Corma A, Iglesias M, Llabrés i Xamena F X, Sanchez F, Cu and Au metal–organic frameworks bridge the gap between homogeneous and heterogeneous catalysts for alkene cyclopropanation reactions. Chem. Eur. J. 2010;16:9789–9795. DOI: 10.1002/chem.201000278

[111] Pérez-Mayoral E, Cejka J, [$Cu_3(BTC)_2$]: A metal–organic framework catalyst for the Friedländer reaction. Chem. Cat. Chem. 2011;3:157–159. DOI: 10.1002/cctc.201000201

[112] Pérez-Mayoral E, Musilova Z, Gil B, Marszalek B, Polozij M, Nachtigall P, Cejka J, Synthesis of quinolines via Friedländer reaction catalyzed by CuBTC metal–organic-framework. Dalton Trans. 2012;41:4036–4044. DOI: 10.1039/C2DT11978A

[113] Opanasenko M, Shamzhy M, Cejka J, Solid acid catalysts for coumarin synthesis by the Pechmann reaction: MOFs versus zeolites. Chem. Cat. Chem. 2013;5:1024–1031. DOI: 10.1002/cctc.201200232

[114] Wang M, Xie M H, Wu C D, Wang Y G, From one to three: a serine derivate manipulated homochiral metal-organic framework. Chem. Commun. 2009, 2396–2398. DOI: 10.1039/B823323C

[115] Shi L X, Wu, C D, A nanoporous metal–organic framework with accessible Cu2+ sites for the catalytic Henry reaction. Chem. Commun. 2011;47:2928–2930. DOI: 10.1039/C0CC05074A

[116] Luz I, Llabrés i Xamena F X, Corma A, Bridging homogeneous and heterogeneous catalysis with MOFs: "Click" reactions with Cu-MOF catalysts. J. Catal. 2010;276:134–140. DOI: 10.1016/j.jcat.2010.09.010

[117] Luz I, Llabrés i Xamena F X, Corma A, Bridging homogeneous and heterogeneous catalysis with MOFs: Cu-MOFs as solid catalysts for three-component coupling and cyclization reactions for the synthesis of propargylamines, indoles and imidazopyridines. J. Catal. 2012;285:285–291. DOI: 10.1016/j.jcat.2011.10.001

[118] Shi D, Ren Y, Jiang H, Cai B, Lu J, Synthesis, structures, and properties of two three-dimensional metal–organic frameworks, based on concurrent ligand extension. Inorg. Chem. 2012;51:6498–6506. DOI: 10.1021/ic202624e

[119] Phan N T S, Vu P H L, Nguyen T T, Expanding applications of copper-based metal–organic frameworks in catalysis: oxidative C–O coupling by direct C–H activation of ethers over $Cu_2(BPDC)_2(BPY)$ as an efficient heterogeneous catalyst. J. Catal. 2013;306:38–46. DOI: 10.1016/j.jcat.2013.06.006

[120] Yang T, Cui H, Zhang C, Zhang L, Su C Y, Porous metal–organic framework catalyzing the three-component coupling of sulfonyl azide, alkyne, and amine. Inorg. Chem. 2013;52:9053–9059. DOI: 10.1021/ic4012229

[121] Yang T, Cui H, Zhang C, Zhang L, Su C Y, From homogeneous to heterogeneous catalysis of the three-component coupling of oxysulfonyl azides, alkynes, and amines. Chem. Cat. Chem. 2013;5:3131–3138. DOI: 10.1002/cctc.201300241

[122] Wang S, Li L, Zhang J, Yuan X, Su C Y, Anion-tuned sorption and catalytic properties of a soft metal–organic solid with polycatenated frameworks. J. Mater. Chem. 2011;21:7098–7104. DOI: 10.1039/C1JM10394F

[123] Tan X, Li L, Zhang J, Han X, Jiang L, Li F, Su C Y, Three-dimensional phosphine metal–organic frameworks assembled from Cu(I) and pyridyl diphosphine. Chem. Mater. 2012;24:480–485. DOI: 10.1021/cm202608f

[124] Horcajada P, Surble S, Serre C, Hong D Y, Seo Y K, Chang J S, Greneche J M, Margiolaki I, Férey G, Synthesis and catalytic properties of MIL-100(Fe), an iron(III) carboxylate with large pores. Chem. Commun 2007;2820–2822. DOI: 10.1039/B704325B

[125] Dhakshinamoorthy A, Alvaro M, Garcia H, Metal organic frameworks as efficient heterogeneous catalysts for the oxidation of benzylic compounds with t-butylhydroperoxide. J. Catal. 2009;267:1–4. DOI: 10.1016/j.jcat.2009.08.001

[126] Dhakshinamoorthy, A, Alvaro M, Garcia H, Claisen–Schmidt condensation catalyzed by metal-organic frameworks. Adv. Synth. Catal. 2010;352:711–717. DOI: 10.1002/adsc.200900747

[127] Dhakshinamoorthy A, Alvaro M, Garcia H, Aerobic oxidation of thiols to disulfides using iron metal–organic frameworks as solid redox catalysts. Chem. Commun. 2010;46:6476–6478. DOI: 10.1039/C0CC02210A

[128] Dhakshinamoorthy A, Alvaro M, Garcia H, Metal–organic frameworks as efficient heterogeneous catalysts for the regioselective ring opening of epoxides. Chem. – Eur. J. 2010;16:8530–8536. DOI: 10.1002/chem.201000588

[129] Dhakshinamoorthy A, Alvaro M, Chevreau H, Horcajada P, Devic T, Serre C, Garcia H, Iron(III) metal–organic frameworks as solid Lewis acids for the isomerization of α-pinene oxide. Catal. Sci. Technol. 2012:2:324–330. DOI: 10.1039/C2CY00376G

[130] Gándara F, Gomez-Lor B, Gutiérrez-Puebla E, Iglesias M, Monge M A, Proserpio D M, Snejko N, An indium layered MOF as recyclable Lewis acid catalyst. Chem. Mater. 2008;20:72–76. DOI: 10.1021/cm071079a

[131] Gomez-Lor B, Gutierrez-Puebla E, Iglesias M, Monge M A, Ruiz-Valero C, Snejko N, $In_2(OH)_3(BDC)_{1.5}$ (BDC = 1,4-Benzendicarboxylate): an In(III) supramolecular 3D framework with catalytic activity. Inorg. Chem. 2002;41:2429–2432. DOI: 10.1021/ic0111482

[132] Sen R, Saha D, Koner S, Controlled construction of metal-organic frameworks: hydrothermal synthesis, X-ray structure, and heterogeneous catalytic study. Chem. Eur. J. 2012;18:5979–5986. DOI: 10.1002/chem.201102953

[133] Saha D, Sen R, Maity T, Koner S, Porous magnesium carboxylate framework: synthesis, X-ray crystal structure, gas adsorption property and heterogeneous catalytic aldol condensation reaction. Dalton Trans. 2012;41:7399–7408. DOI: 10.1039/c2dt00057a

[134] Alkordi M H, Liu Y, Larsen R W, Eubank J F, Eddaoudi M, Zeolite-like metal-organic frameworks (ZMOFs) as platforms for applications: On metalloporphyrin-based catalysts. J. Am. Chem. Soc. 2008;130:12639–12641. DOI: 10.1021/ja804703w

[135] Gándara F, García-Cortés A, Cascales C, Gómez-Lor B, Gutiérrez-Puebla E, Iglesias M, Monge A, Snejko N, Rare earth arenedisulfonate metal–organic frameworks: An approach toward polyhedral diversity and variety of functional compounds. Inorg. Chem. 2007;46:3475–3484. DOI: 10.1021/ic0617689

[136] Suslick K S, Bhyrappa P, Chou J H, Kosal M E, Nakagaki S, Smithenry D W, Wilson S R, Microporous porphyrin solids. Acc. Chem. Res. 2005;38:283–291. DOI: 10.1021/ar040173j

[137] Xie M H, Yang X L, Wu C D, From 2D to 3D: A single-crystal-to-single-crystal photochemical framework transformation and phenylmethanol oxidation catalytic activity. Chem. Eur. J. 2011;17:11424–11427. DOI: 10.1002/chem.201101321

[138] Zou R Q, Sakurai H Xu Q, Preparation, adsorption properties, and catalytic activity of 3D porous metal–organic frameworks composed of cubic building blocks and alkali-metal ions. Angew. Chem., Int. Ed. 2006;45:2542–2546. DOI: 10.1002/anie.200503923

[139] Mitchell L, Gonzalez-Santiago B, Mowat J P S, Gunn M E, Williamson P, Acerbi N, Clarke M L, Wright P A, Remarkable Lewis acid catalytic performance of the scandium trimesate metal organic framework MIL-100(Sc) for C–C and C[double bond, length as m-dash]N bond-forming reactions. Catal. Sci. Technol. 2013;3:606–617. DOI: 10.1039/C2CY20577G

[140] Llabrés i Xamena F, Abad A, Corma A, Garcia H, MOFs as catalysts: activity, reusability and shape-selectivity of a Pd-containing MOF. J. Catal. 2007;250:294–298. DOI: 10.1016/j.jcat.2007.06.004

[141] Opelt S, Krug V, Sonntag J, Hunger M Klemm E, Investigations on stability and reusability of $[Pd(2\text{-pymo})_2]_n$ as hydrogenation catalyst. Micropor. Mesopor. Mater. 2012;147:327–333. DOI: 10.1016/j.micromeso.2011.07.003

[142] Schuster S, Klemm E, Bauer M, Chem. Eur. J. 2012;18:15831–15837.

[143] Han J W, Hill C L, A coordination network that catalyzes O_2-based oxidations. J. Am. Chem. Soc. 2007;129:15094–15095. DOI: 10.1021/ja069319v

[144] Leus K, Muylaert I, Vandichel M, Marin G B, Waroquier M, Van Speybroeck V, Van der Voort P, The remarkable catalytic activity of the saturated metal organic framework V-MIL-47 in the cyclohexene oxidation. Chem. Commun. 2010;46:5085–5087. DOI: 10.1039/C0CC01506G

[145] Leus K, Vandichel M, Liu Y Y, Muylaert I, Musschoot J, Pyl S, Vrielinck H, Callens F, Marin G Detavernier B C, Wiper P V, Khimyak Y Z, Waroquier M, Van Speybroeck V, Van Der Voort P, The coordinatively saturated vanadium MIL-47 as a low leach-

ing heterogeneous catalyst in the oxidation of cyclohexene. J. Catal. 2012;285:196–207. DOI: 10.1016/j.jcat.2011.09.014

[146] Cho S H, Ma B, Nguyen S T, Hupp J T Albrecht-Schmitt T E, A metal–organic framework material that functions as an enantioselective catalyst for olefin epoxidation. Chem. Commun. 2006;2563–2565. DOI: 10.1039/B600408C

[147] Shultz A M, Farha O K, Hupp J T, Nguyen S T, A catalytically active, permanently microporous MOF with metalloporphyrin struts. J. Am. Chem. Soc. 2009, 131, DOI: 10.1021/ja900203f.

[148] Seo J S, Whang D, Lee H, Jun S I, Oh J, Jeon Y J, Kim K, A homochiral metal–organic porous material for enantioselective separation and catalysis. Nature. 2000;404:982–986. DOI: 10.1038/35010088

[149] Ravon U, Domine M E, Gaudillere C, Desmartin-Chomel A, Farrusseng D, MOFs as acid catalysts with shape selectivity properties. New J. Chem. 2008;32:937–940. DOI: 10.1039/B803953B

[150] Ingleson M J, Barrio J P, Guilbaud J B, Khimyak Y Z, Rosseinsky M J, Framework functionalisation triggers metal complex binding. Chem. Commun. 2008;2680–2682. DOI: 10.1039/B718367D

[151] Dybtsev D N, Nuzhdin A L, Chun H, Bryliakov K P, Talsi E P, Fedin V P, Kim K, A homochiral metal–organic material with permanent porosity, enantioselective sorption properties, and catalytic activity. Angew. Chem. Int. Ed. 2006;45:916–920. DOI: 10.1002/anie.200503023

[152] Gedrich K, Heitbaum M, Notzon A, Senkovska I, Frohlich R, Getzschmann J, Mueller U, Glorius F, Kaskel S, A family of chiral metal–organic frameworks. Chem. Eur. J. 2011;17:2099–2106. DOI: 10.1002/chem.201002568

[153] Miralda C M, Macias E E, Zhu M, Ratnasamy P, Carreon M A, Zeolitic imidazole framework-8 catalysts in the conversion of CO_2 to chloropropene carbonate. ACS Catal. 2012;2:180–183. DOI: 10.1021/cs200638h

[154] Vermoortele F, Ameloot R, Vimont A, Serre C, De Vos D, An amino-modified Zr-terephthalate metal–organic framework as an acid–base catalyst for cross-aldol condensation. Chem. Commun. 2011;47:1521–1523. DOI: 10.1039/C0CC03038D

[155] Vermoortele F, Vandichel M, Van de Voorde B, Ameloot R, Waroquier M, Van Speybroeck V, De Vos D E, Electronic effects of linker substitution on Lewis acid catalysis with metal–organic frameworks. Angew. Chem. Int. Ed. 2012;51:4887–4890. DOI: 10.1002/anie.201108565

[156] Vermoortele F, Bueken B, Le Bars G, Van de Voorde B, Vandichel M, Houthoofd K, Vimont A, Daturi M, Waroquier M, Van Speybroeck V, Kirschhock C, De Vos D E, Synthesis modulation as a tool to increase the catalytic activity of metal–organic

frameworks: the unique case of UiO-66(Zr). J. Am. Chem. Soc. 2013;135:11465–11468. DOI: 10.1021/ja405078u

[157] Hafizovic Cavka J, Jakobsen S, Olsbye U, Guillou N, Lamberti C, Bordiga S, Petter Lillerud K, A new zirconium inorganic building brick forming metal organic frameworks with exceptional stability. J. Am. Chem. Soc. 2008;13850–13851. DOI: 10.1021/ja8057953

Chapter 5

Studies on Obtaining Diglycidyl Ether from Allyl-Glycidyl Ether over the Mesoporous Ti-SBA-15 Catalyst

Agnieszka Wróblewska, Edyta Makuch and Ewelina Mójta

Additional information is available at the end of the chapter

http://dx.doi.org/10.5772/61881

Abstract

The work presents the studies on the epoxidation of allyl-glycidyl ether (AGE) to diglycidyl ether (DGE) over the mesoporous Ti-SBA-15 catalyst and with 60 wt% hydrogen peroxide. The influence of the following parameters was studied: the temperature 0–100°C, the molar ratio of AGE/H_2O_2 = 0.03:1 – 4:1, the content of Ti-SBA-15 catalyst 0.0–0.5 wt%, and the reaction time 15–240 min. The studies showed that it is possible to obtain DGE with the selectivity of 100 mol% (for reaction time below 60 min) but at low conversion of AGE – about 4 mol%. The prolongation of the reaction time decreases the selectivity of DGE because the following competitive reactions take place: (1) hydration of the epoxide ring in AGE and 3A12PD formation, (2) collapsing of the ethers by hydrolysis of the ether groups, and (3) the epoxidation and the hydration of the products of collapsing and obtaining glycerol. The explanation of the very high ineffective decomposition of hydrogen peroxide and possible ways of increasing its efficiency of conversion are also presented.

Keywords: Epoxidation, allyl-glycidyl ether, diglycidyl ether, hydrogen peroxide, Ti-SBA-15

1. Introduction

1.1. Epoxidation of allylic compounds, especially diallyl ether and allyl-glycidyl ether, and applications of allyl-glycidyl ether epoxidation product

Research on the epoxidation of allylic compounds: propylene [1–5], allyl chloride [6–12], allyl alcohol [13–17], methallyl alcohol, crotyl alcohol, and 1-butene-3-ol [18] with hydrogen peroxide over the titanium silicate catalysts have been carried out for over 10 years. There are not many reports concerning the epoxidation of allyl-glycidyl ether (AGE) or diallyl ether

INTECH

open science | open minds

(DAE) [19] with hydrogen peroxide over these catalysts. It can be caused by the fact that the epoxidation of AGE and DAE is very complicated, as in addition to the epoxidized ether derivatives, by-products formed by the decomposition of these ethers and their epoxidized derivatives, as well as products of further transformations of these decomposed products, i.e. glycidol and glycerol, are received. Moreover, during epoxidation, the epoxy rings may be opened and diols can be formed. However, due to numerous applications of diglycidyl ether – DGE (production of linear, branched, and cyclic oligoglycerols used in the production of surfactants, preparation of anti-arrhythmia agents, production of components of other pharmaceuticals and medicines, for example, cryptands, preparation of lubricating oil additives, and synthesis of aminoethers), the epoxidation of AGE with hydrogen peroxide over Ti-SBA-15 is very interesting and worth further examinations. The simultaneous utilization of Ti-SBA-15 and a low-cost, environment friendly oxidant - hydrogen peroxide in the epoxidation of AGE, makes this method of DGE production a modern and environment friendly as the only product of hydrogen peroxide transformation is water and Ti-SBA-15 can be easily recovered from reaction mixtures, recycled into the process, and regenerated if it loses its activity [20]. An additional advantage is this process can be carried out in an aqueous medium also.

2. The titanium silicate materials used for diallyl ether and allyl-glycidyl ether epoxidation [19]

Until now, only the TS-1 and Ti-MWW titanium silicate materials have been used in the epoxidation of DAE and AGE – with hydrogen peroxide. In the studies performed by Wu et al. [19], AGE was the semi-product which was formed during DAE epoxidation and it underwent among others further transformation to DGE. In the above mentioned research, at the first stage, the epoxidation of DAE with hydrogen peroxide was performed under vigorous stirring in a 20 ml glass flask, connected to a condenser. In the typical run, the appropriate amounts of DAE, solvent (acetonitrile, acetone, water, methanol, ethanol, dioxane), and the catalyst were mixed in the flask and heated to the desired temperature under the agitation. Next, aqueous hydrogen peroxide (30 wt%) was added to the mixture to start the reaction. The reaction was carried out at the temperature of 60°C for 0.5 h in case of Ti-MWW and for 1.5 h in case of TS-1. Both Ti-MWW and TS-1 materials showed different solvent effect. During the examinations over Ti-MWW catalyst, the highest conversion of DAE was obtained for acetonitrile (about 40 mol%) and acetone (about 39 mol%) as the solvents. The selectivity of AGE was the highest for acetonitrile, methanol, ethanol, and dioxane and amounted to about 71–79 mol%. The selectivity of DGE, which was formed as the product of AGE transformation (epoxidation of the next double bond), was the highest for water as the solvent (about 40 mol %). High values of the selectivity of this product also allow to obtain such solvents as: acetonitrile (about 29 mol%) and acetone (about 25 mol%). The efficiency of hydrogen peroxide conversion was the highest for examinations in acetonitrile (about 95 mol%), acetone (about 92 mol%), and dioxane (about 94 mol%). In methanol and in water, the efficiency of hydrogen peroxide conversion amounted to about 82–87 mol%. The lowest value of this function of the process was obtained for ethanol (61 mol%). The total conversion of hydrogen peroxide was the highest in acetonitrile and acetone (about 99–100 mol%). These studies showed that the Ti-MWW catalyst favoured aprotic solvents such as acetonitrile or acetone. A little worse results were obtained for water as the solvent.

During the studies over the TS-1 catalyst, the highest conversion of DAE was obtained in acetone, methanol, and ethanol as the solvents. The selectivity of AGE was usually high and amounted to about 61–89 mol%. This function was the highest for dioxane. The highest values of DGE selectivity were obtained for acetone and methanol: about 22 and 21mol%, respectively. Efficiency of hydrogen peroxide conversion was the highest for acetone and amounted to 70 mol%. The studies over the TS-1 material showed that methanol and acetone were the most effective solvents in this process. The comparison of the results of DAE epoxidation present that Ti-MWW material was more efficient than TS-1 material in catalytic activity, epoxide selectivity, and hydrogen peroxide conversion when choosing acetonitrile or acetone as the solvent [19].

At the second stage, in which only Ti-MWW material was used, also the influence of temperature from 0 to 67°C was tested in the epoxidation of DAE over the Ti-SBA-15 material. The studies were performed in acetone and at the same conditions as in the first stage of these studies. The examinations showed that the selectivities of AGE and DGE changed of about 10 mol% during the rising of temperature (the selectivity of AGE from about 80 to about 70 mol %, and the selectivity of DGE from about 20 to 30 mol%). Conversion of DAE reached 40 mol % above the temperature of 60°C. The efficiency of hydrogen peroxide conversion slightly decreased with increasing temperature as a result of the thermal decomposition of hydrogen peroxide, but the value of this function was higher than 95 mol% [19].

At the third stage, the influence of the Ti-MWW amount and reaction time were studied. The examinations were performed in acetonitrile and at the temperature of 60°C. The studies showed that the more the catalyst was used, the higher the catalytic activity of Ti-MWW material. The DAE conversion raised rapidly within 30 min and then gradually with the time. The decrease in the reaction rate for longer time was mainly caused by the pore blocking by heavy organic species [19].

The comparison of the results obtained for Ti-MWW and TS-1 catalysts show that Ti-MWW material exhibited more benefits in catalytic activity using less harsh reaction conditions [19].

3. The Ti-SBA-15 – titanium silicate mesoporous material, its synthesis and characteristic

One of the newer titanium silicate catalysts is Ti-SBA-15. It is a mesoporous material, which is much more durable than Ti-MCM-41 catalyst [20, 21]. Greater durability of this catalyst is likely due to its construction – a honeycomb structure, wherein unlike Ti-MCM-41, in Ti-SBA-15 the main cylindrical channels are connected together by transverse channels which introduces additional porosity and, at the same time, strengthens the structure. Furthermore, Ti-SBA-15 is characterized by thicker pore walls, and its synthesis is carried out in the presence of a biodegradable template – Pluronic P123, as opposed to Ti-MCM-41 whose synthesis is carried out in the presence of an ammonium compound (hexadecyltrimethylammonium bromide) and is uncomfortable to the environment due to formation of amines during calcination of this catalyst [22–31]. Our studies on the epoxidation of allylic alcohols over the Ti-SBA-15 catalyst showed that this catalyst is very active in this process and these compounds can be effectively converted to epoxides [18]. This is the main reason why this catalyst has been chosen to conduct the epoxidation of AGE.

Ti-SBA-15 was obtained by the method of Berube et al. [30] and the following raw materials were used in its synthesis: Pluronic P123 (Aldrich, MW = 5800) as structure-directing agent, tetraethylortosilicalite (TEOS 98%, Aldrich) as a silicon source, and tetraisopropyl orthotitanate (TiPOT >98%, Merck) as a titanium precursor. The characterization of Ti-SBA-15 was performed with the following instrumental methods: XRD (X'Pert PRO Philips diffractometer, Co Kα radiation), IR (Shimadzu FTIR-8100 spectrometer, KBr pellet technique), UV-vis (SPECORD M40 type V-530 with the attachment for solid materials measurements), X-ray microanalysis (Oxford X-ray analyzer ISIS 300), SEM (JOEL JSM-6100 instrument), and porosimetric analysis (porosimetry analyzer ASAP Micromeritics).

The XRD pattern of the obtained Ti-SBA-15 catalyst was similar to the literature data [28, 32–33]. For the SBA-15 material are typical: one characteristic very intensive diffraction peak at 2 Theta angle of 1.01° and two weak peaks at 2 Theta angle of 1.75° and 2.05°, corresponding to the (100), (110), and (200) Bragg reflections. These reflections confirm the 2-d hexagonal symmetry (P6mm) of the SBA-15. The IR spectrum of the obtained Ti-SBA-15 was also similar to the literature data [23, 34–35]. To the main bands characterized, this kind of materials belong to: the band at 800 cm^{-1} which is assigned to symmetric stretching vibrations of Si-O-Si in SiO4 units the same as the band at 1, 000–1, 300 cm^{-1}, the band at 1625-1650 cm^{-1} which is assigned to adsorbed water molecules, the band at 3, 000–3, 700 cm^{-1} which is characteristic for surface Si-OH groups and adsorbed water molecules, and the band 957 cm^{-1} which is associated with Ti=O or Si-O-Ti vibrations. The UV-VIS spectrum of the obtained Ti-SBA-15 catalyst showed an intense absorption band at 211 nm, associated with ligand-to-metal charge transfer from oxygen to Ti of an isolated tetrahedral Ti species. This band is direct evidence for titanium atoms incorporated into the framework of the silica [36]. In this spectrum also a shoulder with a maximum around of 290 nm was present. This band is connected with the presence of Ti atoms in fivefold and sixfold coordination. This fivefold and sixfold coordination is generated through hydration by one or two water molecules of the tetrahedral titanium ion in the first coordination sphere [36]. Moreover, in the spectra was visible a weak band at 340 nm, which is assigned to the presence of anatase in the sample. The obtained UV-VIS spectrum is similar to the literature data [32, 36].

An X-ray microanalysis showed that the amount of Ti in the sample after calcination was 2.9 wt%. According to the porosimetric analysis, the specific surface area of the obtained catalyst amounted to 620 m^2/g, the size of the pores achieved was 5.0 nm, and the pore volume was 0.6 cm^3/g. The SEM analyses showed that the Ti-SBA-15 catalyst consisted of large and long, branched, pipe-like structures with diameter of about 4 μm. These structures consisted of smaller particles with diameter of about 0.3–0.6 μm and length about 1–2 μm. This morphology is typical for structures such as SBA-15 [28, 37].

4. The studies on the epoxidation of AGE over the Ti-SBA-15 catalyst and the comparison of the results obtained with the previous results obtained for Ti-MWW and TS-1 catalysts

In the epoxidation of AGE, the following raw materials were used: AGE (99%, Aldrich), hydrogen peroxide (60 wt% water solution, Chempur), and Ti-SBA-15 catalyst. The initial technological parameters were as follows: the molar ratio of AGE/H_2O_2 = 1:1, catalyst content 3 wt%, the reaction time 2 h, and intensity of stirring 500 rpm.

The process of AGE epoxidation was carried out in glass vials with the capacity of 12 cm^3 equipped with a rubber septum and a capillary. The raw materials were introduced into the vials at the ambient temperature and in the following order: hydrogen peroxide, catalyst, and AGE. Then the vials were closed with the rubber equipped with the capillary, located in a shaker holder and immersed in a water bath having the appropriate temperature. In order to calculate the mass balance, the unreacted hydrogen peroxide was determined by iodometric method and the remaining products and the unreacted AGE were analyzed by the GC method. The analyses were performed on the FOCUS apparatus with a flame-ionization detector fitted with Quadrex capillary columns filled with methyl-phenyl-siloxanes. After calculating the mass balance, the main functions describing the process were determined: the selectivity of transformation to DGE in relation to AGE consumed and also selectivities of the by-products in relation to AGE consumed, the conversion of AGE, and the selectivity of transformation to organic compounds in relation to hydrogen peroxide consumed (effective conversion of H_2O_2).

The studies on the epoxidation of AGE to DGE was carried out by one-variable method, changing the values of the following parameters: temperature 0–100°C, molar ratio of AGE/H_2O_2 0.03:1–4:1, content of the Ti-SBA-15 catalyst (0–5 wt%), and reaction time 15–240 min. The main results of the studies on the influence of temperature on the course of AGE epoxidation were presented in Figure 1.

During the studies on the influence of temperature, only three products were obtained: DGE, 3-allyloxy-1, 2-propanediol (3A12PD), and glycerol. The selectivity of transformation of AGE to the product of epoxidation of the unsaturated bond – DGE, increases during the increase of the temperature from 0 mol% (the temperature of 0°C) to 38 mol% (the temperature of 20°C) and then decreases to 9 mol% (the temperature of 100°C). Figure 1 shows that DGE is not the main product of this process, because for all investigated temperatures the main product of the process is 3A12PD (with exception at the temperature of 0°C at which the reaction does not proceed). The selectivity of this products changes from about 53–54 mol% (the temperatures 10–20°C) to 80–90 mol% (the temperatures 40–100°C).. The selectivity of glycerol amounts to 13 mol% for the temperature of 10°C, and then decreases to about 2 mol% (for the temperatures of 40–100°C). For the description of the process of AGE epoxidation, the reactions presented in Figure 2 can be proposed.

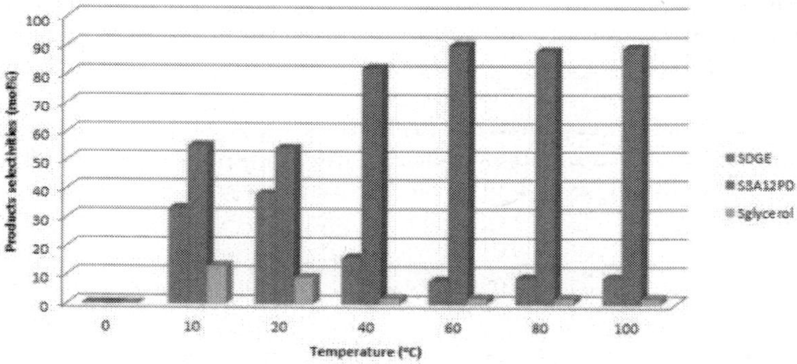

Figure 1. The influence of temperature on the selectivities of the products of AGE epoxidation process: S_{DGE} – the selectivity of DGE, S_{3A12PD} – the selectivity of 3-allyloxy-1, 2-propanediol, and $S_{glycerol}$ – the selectivity of glycerol (the molar ratio of AGE/H_2O_2 = 1:1, the content of the catalyst 3 wt%, and the reaction time 120 min).

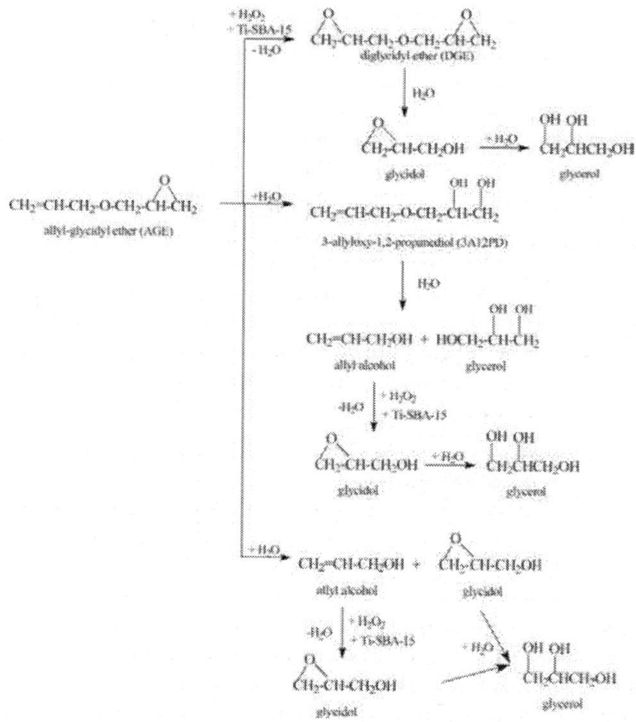

Figure 2. The main reactions of the process of AGE epoxidation.

Figure 2 shows that the process can proceed in the three directions: (1) the epoxidation of allylic group in AGE and formation of DGE, (2) the hydration of the epoxide ring in AGE and formation of 3A12PD, and (3) the formation of glycerol. Moreover, the two first directions are the main directions of the process in low temperatures. The glycerol formation is a very complicated process and this product can be formed as a product of hydrolysis of the ether groups of AGE, DGE, 3A12PD, and, simultaneously, as a result of the secondary reactions of the products of the hydrolysis of these ethers (epoxidation of allylic group in allyl alcohol to glycidol and next hydration of the epoxide ring in glycidol to glycerol). Allyl alcohol and glycidol were not detected in post-reaction mixtures. It shows that these compounds were very reactive at the investigated conditions and right away underwent secondary reactions. The tendency towards the formation of 3A12PD rises during increasing the temperature of the performing process. On the other hand, the selectivities of the DGE and glycerol decrease. It shows that at higher temperatures the epoxidation of AGE to DGE is hindered and the hydrolysis of the ether groups in AGE, DAE, and 3A12PD is stepped or it proceeds very slowly. The main reaction is hydration of the epoxide ring in AGE and formation of 3A12PD. The formation of 3A12PD as the main product in this process can be explained taking into account the acidic character of the mesoporous Ti-SBA-15 material. This character is mainly connected with: (1) the silanol groups located on the surface of this catalyst [19, 38], (2) the species of Ti present on the surface of the catalyst – tetrahedral Ti(IV) active sites, titanium-containing species in the form of dimmers or very small oligomers [38–41], and (3) the active species of titanium with hydrogen peroxide which are formed during the oxidation process, for example, five-membered active complexes – titanium hydroperoxo species with the specific structure, which are formed from tetrahedral Ti(IV) active sites, protic solvent (for example, methanol or water), and hydrogen peroxide and are present on the surface of the catalyst [42].

Among others, the formation of 3A12PD can also be under influence of the active species of titanium with hydrogen peroxide which are formed during the oxidation process and some of them can have the acidic character. A few structures have been proposed until now for explanations of the structures of these active species. Among these structures are: the peroxide structure, the hydroperoxide structure, and superoxide structures (radical species) [40, 42–49]. Indeed, mainly the hydroperoxide structure was described as the structure responsible for the epoxidation of olefinic compounds [45, 47]. It exists in equilibrium with the peroxide structure, which is a dominant structure in the water solution because it is stabilized by water molecules [45]. In the medium in which epoxidation takes place, the excess of olefins causes that the peroxide structure is converted to hydroperoxide structure [45]. Hydroperoxide structures in the presence of protic solvent create the five-membered active complexes – titanium hydroperoxo species which are composed from tetrahedral Ti(IV) active sites, protic solvent (in case of these studies, from water), and hydrogen peroxide[40, 42, 50, 51].

Bhaumik et al. [42] described that under the influence of the titanium hydroperoxo species undergoes acid-catalyzed cleavage of the oxirane rings in the epoxide compounds; this reaction has considerable SN1 character and the nucleophilic attack is easy to occur at the more crowded carbon atom that can best accommodate the positive charge. Taking into account that this data can be propose the possible way of 3A12PD formation from the AGE presented in Figure 3.

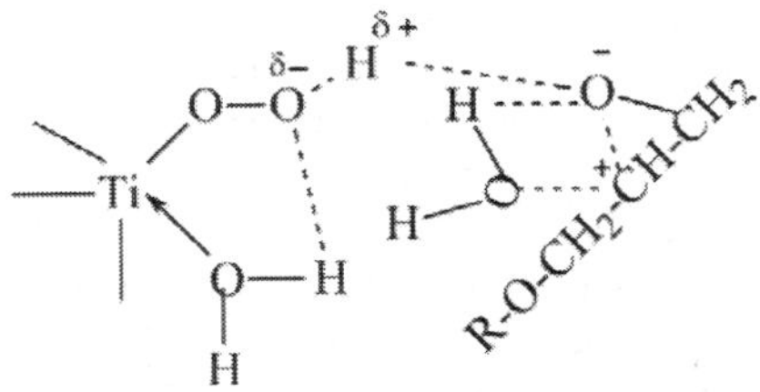

Figure 3. The possible way of 3A12PD formation from the AGE, where: R = CH_2=CH-CH_2-.

Also the hydration of the ether groups in AGE, DGE, and 3A12PD can be explained taking into account the acidic character of various species which are present in the mesoporous Ti-SBA-15 material (silanol groups, species of Ti and titanium hydroperoxo species).

During the studies on the influence of temperature, the conversion of AGE was very low and it changes from 0 mol% (the temperature of 0°C) to about 4–5 mol% for the highest temperatures. The changes of the effective conversion of H_2O_2 are very similar. Very low values of this function show that at the studied conditions the catalyst was very active in the ineffective decomposition of hydrogen peroxide to water and oxygen, which takes place at the active centres of Ti in the catalyst even at very low temperatures (the total conversion of hydrogen peroxide changed from 83–90 mol%). The ineffective decomposition of hydrogen peroxide over titanium silicate catalysts has been described in a great number of works [52–57], and this phenomenon is typical for these catalysts, for example, it was shown in the literature that titanium hydroperoxo species can decompose hydrogen peroxide molecules via formation of Ti-O* radical and hydroperoxo radical (HOO*) [52]. As the results presented in this work showed, the Ti-SBA-15 catalyst was very active in the decomposition of hydrogen peroxide. The epoxidation of olefinic bonds undergoes slower than the ineffective decomposition of hydrogen peroxide at the five-membered active species. Probably, very small reactivity of the AGE is connected with the steric hindrances connected with the structure of this ether when the molecules of AGE are close to the active species of Ti with hydrogen peroxide. These steric hindrances cause that the decomposition of AGE molecules also takes place. The increased, ineffective decomposition of hydrogen peroxide can be also caused by the presence of TiO_2 domains in the structure of the Ti-SBA-15 catalyst. Taking into account the UV-VIS spectrum of the Ti-SBA-15 catalyst used in this work, the broad absorption peak at the 211 cm^{-1} and the shoulder at the 290 cm^{-1}, which is not only connected with the presence of Ti atoms in fivefold and sixfold coordination [36] but also can be assigned to the oligomerized titanium-oxygen species – formation of Ti-O-Ti bonds by clusterization of octahedrally coordinated titanium ions [46, 58] or to octahedral titanium species in the form of highly dispersed TiO_2 particles with the particle size smaller than 5 nm [48]), and the results of the X-Ray microanalysis (amount of Ti 2.9 wt%), it can be assumed that the Ti-SBA-15 catalyst contains the titanium-oxygen species in the form of dimmers or very small oligomers (TiO_2 domains, Ti aggregates) [53, 55].

A lot of works present the strategies to increase the oxidant efficiency. The hydrogen peroxide decomposition is strongly dependent on the pH of the reaction mixtures and on the surface concentration of the hydroxyl groups of the catalytic material [52]. The improving of the efficiency of the hydrogen peroxide conversion can be done by: (1) addition of additives such as for example: CH_3COOH, $KHSO_4$, KH_2PO_4, KHF_2, Na_2SO_4, $NaHCO_3$, K_2CO_3, K_3PO_4, K_2HPO_4, or KH_2PO_4 [52]; (2) slow addition of hydrogen peroxide [52, 55, 57]; (3) choosing of the appropriate solvent for the epoxidation process – the most beneficial are methanol, acetonitrile, and acetone [44, 54, 55] or co-solvent, for example, sulfolane [59]; (4) increasing of the acidity of the catalyst by the addition of metal oxide, for example, of TiO_2, and utilization of the appropriate temperature of the calcination [56]; and (5) the surface hydrophobization of mesoporous titanium silicates [46]. We would like to test in our future work some of the ways of improving the efficiency of hydrogen peroxide conversion: choosing the appropriate solvent, additives, and slow addition of hydrogen peroxide.

Taking into account the results of the studies on the influence of temperature on the course of AGE epoxidation, the temperature of 20°C was taken as the most beneficial for the next studies.

The main results of the studies on the influence of the molar ratio of AGE/H_2O_2 on the course of AGE epoxidation were presented in Figure 4. The studies were performed at the range of molar ratios of AGE/H_2O_2 0.03:1 – 4:1. The other parameters were as follows: the temperature of 20°C, the content of the catalyst 3 wt%, and the reaction time 120 min.

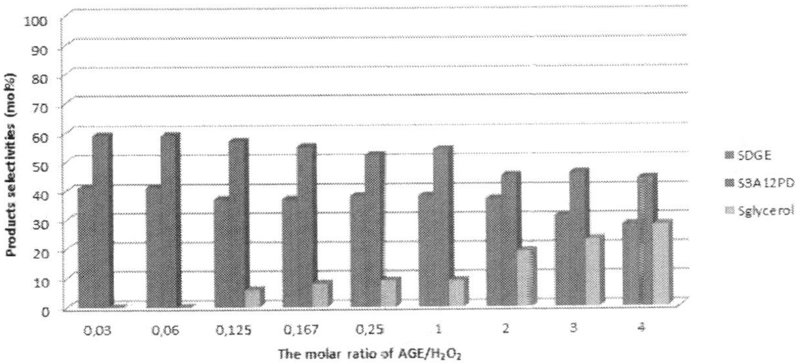

Figure 4. The influence of molar ratio of AGE/H_2O_2 on the selectivities of the products of AGE epoxidation process: S_{DGE} – the selectivity of DGE, S_{3A12PD} – the selectivity of 3-allyloxy-1, 2-propanediol, and $S_{glycerol}$ – the selectivity of glycerol (the temperature 20°C, the content of the catalyst 3 wt%, and the reaction time 120 min).

The studies show that the conversion of AGE was the highest for the lowest molar ratio of reactants and amounted of 11 mol% and next it decreased to 1 mol% for the molar ratio of AGE/H_2O_2 = 4:1. The effective conversion of H_2O_2 had very low values independent of the studied molar ratios, even for the molar ratios of AGE/H_2O_2 > 1. Figure 4 shows that independent of the molar ratio of reactants the main product of the process was 3A12PD, but its selectivity decreased during increasing the molar ratio of reactants. It shows that the excess of hydrogen peroxide or AGE molecules in the reaction mixture does not cause that the epoxidation of AGE is intensified and the hindering of the hydration of epoxide ring in AGE is not observed. The ethers molecules are unstable in reaction medium and underwent decomposition by the hydration of the ether groups. Simultaneously, the results obtained show that the surface of the catalyst independent of the molar ratio of reactants was very active in the reactions of hydration of epoxide rings and ether groups. Also the formed allyl alcohol and glycidol undergo secondary reactions (epoxidation and hydration of the epoxide ring) very easily. On the basis of the results obtained, the molar ratio of AGE/H_2O_2 = 0.03 was taken as the most beneficial for the next stages of the studies.

The main results of the studies on the influence of the Ti-SBA-15 catalyst content on the course of AGE epoxidation were presented in Figure 5. These studies were performed for the following parameters: the temperature 20°C, the molar ratio AGE/H_2O_2 = 0.03, and the reaction time of 120 min.

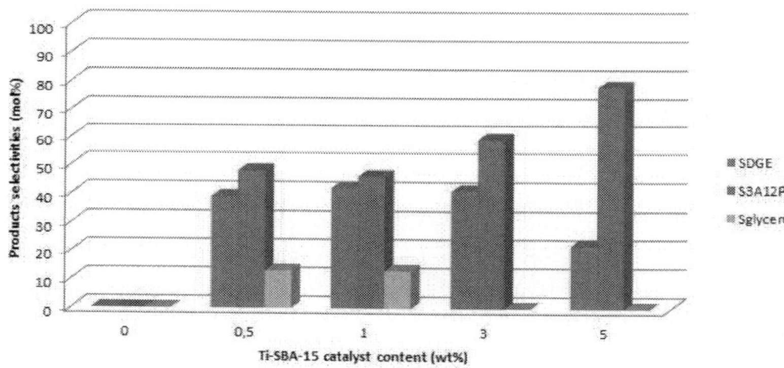

Figure 5. The influence of the Ti-SBA-15 catalyst content on the selectivities of the products of AGE epoxidation process: S_{DGE} – the selectivity of DGE, S_{3A12PD} – the selectivity of 3-allyloxy-1, 2-propanediol, and $S_{glycerol}$ – the selectivity of glycerol (the temperature 20°C, the molar ratio of AGE/H_2O_2 = 0.03, and the reaction time 120 min).

The studies show that for the amount of the catalyst of 0 wt% no one reaction proceeded. The conversion of AGE raised from 0 mol% (for 0 wt% of Ti-SBA-15) to 11 mol% (for 3 wt% of Ti-SBA-15) and next did not change. The effective conversion of H_2O_2 was very low and amounted to about 1 mol%, independent of the studied Ti-SBA-15 content.

Figure 5 shows that during the rising of the content of the catalyst the selectivity of 3A12PD increased from 0 mol% (the Ti-SBA-15 content 0 wt%) to 78 mol% (the Ti-SBA-15 content 5 wt %). It presents that with the increase of the Ti-SBA-15 content, the hydration of the epoxide ring in AGE was intensified, but the hydrolysis at the ether group of AGE, DGE, and 3A12PD was hindered. Only for the Ti-SBA-15 content of 0.5 and 1 wt%, glycerol was present in the post-reaction mixtures. On the other hand, at higher catalyst content, the phenomenon of ineffective decomposition of hydrogen peroxide at the active centers of Ti in the catalyst was intensified and the water molecules obtained during ineffective decomposition of hydrogen peroxide could probably take part in hydration of the epoxide ring in AGE. The content of the catalyst amounting to 3 wt% was chosen as the most beneficial for the last stage of the studies.

The main results of the studies on the influence of the reaction time on the course of AGE epoxidation were presented in Figure 6. These studies were performed for the following parameters: the temperature 20°C, the molar ratio AGE/H_2O_2 = 0.03, and the Ti-SBA-15 catalyst content 3 wt%.

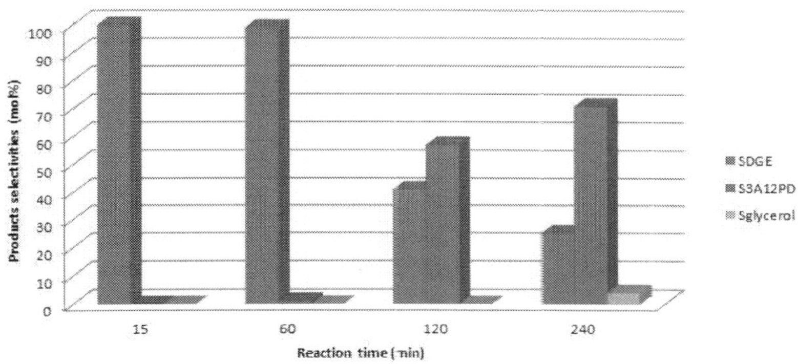

Figure 6. The influence of the reaction time on the selectivities of the products of AGE epoxidation process: S_{DGE} – the selectivity of DGE, S_{3A12PD} – the selectivity of 3-allyloxy-1, 2-propanediol, and $S_{glycerol}$ – the selectivity of glycerol (the temperature 20°C, the molar ratio of AGE/H_2O_2 = 0.03, and the Ti-SBA-15 catalyst content 3 wt%).

The results show that with the prolongation of the reaction time from 15 min to 240 min, the selectivity of DGE decreased from 100 mol% to 25 mol%. Only for the reaction time of 15 min and 50 min, it was possible to obtain only DGE as the product in the post-reaction mixtures. The conversion of AGE increased in the range of the studied reaction time from 3 mol% to 18 mol%, but the effective conversion of H_2O_2 was very low and amounted to about 1 mol%. Figure 6 presents that for the reaction time of 120 min the second product of this process was established – 3A12PD. Glycerol – the third product of this process appeared in the post-reaction mixture for the reaction time of 240 min. It shows that it is possible to obtain only one product in the post-reaction mixture – DGE (very desirable) only for short reaction time – 15 min. and 60 min. At this reaction time, it is only possible to stop the hydration of the epoxide ring and formation of 3A12PD.

The comparison of our results obtained for the Ti-SBA-15 material with the results obtained previously by Wu et al. for the Ti-MWW and TS-1 materials [19] shows that the main difference between the epoxidation of AGE over Ti-SBA-15 and the epoxidation of DAE and AGE over Ti-MWW and TS-1 is formation of the considerable amounts of 3-allyloxy-1, 2-propanediol over Ti-SBA-15 and low efficiency of hydrogen peroxide conversion for this catalyst. It is probably connected with the pore size of the Ti-SBA-15 mesoporous material and the structure of the surface of this catalyst, especially with the presence of various species of Ti and hydroxyl groups.

5. Conclusions and future possibilities

The results presented in these studies show that the best conditions established for the epoxidation of AGE to DGE on the Ti-SBA-15 catalyst in water solution are as follows: the temperature of 20°C, the molar ratio of AGE/H_2O_2 = 0.03:1, the content of the catalyst 3 wt%, and the reaction time of 60 min. At these mild and relatively safe conditions, the selectivity of DGE amounts to 100 mol%; the conversion of AGE and the selectivity of transformation to organic compounds in relation to hydrogen peroxide consumed amount to 4 mol% and 1 mol %, respectively. These studies also show that this process is very complicated because of the secondary reactions which proceed in reaction medium – hydration of epoxide ring in AGE (formation of 3A12PD), hydrolysis at the epoxide group in AGE, DGE, 3A12PD, epoxidation of the formed allyl alcohol, and hydration of the epoxide ring in glycidol and formation of glycerol. However, it is possible to choose such a way of carrying up this process in which only one product – DGE (the most desirable) – is obtained. The process of obtaining DGE is performed at very mild conditions, thus the danger of the explosive decomposition of hydrogen peroxide or compounds with epoxide group is very little, mainly taking into account the very low temperature of this process which amounts 20°C. Hydrogen peroxide used in this process is a relative cheap oxidizing agent and moreover, the methods of production of hydrogen peroxide are still developed and modernized. The method of AGE epoxidation with hydrogen peroxide is also ecologically friendly because only one product of its transformation is water. The presented studies showed that for this process hydrogen peroxide should be used in excess in relation to AGE. Mainly, it is connected with the phenomenon of ineffective decomposition of hydrogen peroxide (not explosive decomposition) at the active centers of Ti in the structure of the catalyst. This phenomenon causes that only a little amount of hydrogen peroxide takes part in epoxidation process and utilization of the excess of hydrogen peroxide in relation to AGE improves effective utilization of hydrogen peroxide molecules in the process of epoxidation. On the other hand, hydrogen peroxide undergoes very easy ineffective decomposition, thus even at the excess of hydrogen peroxide its concentration in reaction mixtures is low. The main cause of this situation is high content of the Ti-SBA-15 catalyst in reaction mixture (3 wt%). There are some possible ways of improving the hydrogen peroxide conversion efficiency by, for example, (1) addition of additives (CH_3COOH, $KHSO_4$, KH_2PO_4, KHF_2, Na_2SO_4, $NaHCO_3$, K_2CO_3, K_3PO_4, K_2HPO_4, or KH_2PO_4); (2) the changing of the way of hydrogen peroxide addition; (3) choosing the appropriate solvent (methanol, acetonitrile or acetone), or co-solvent (sulfolane); (4) increasing the acidity of the catalyst by the addition of metal oxide, for example, of TiO_2, and utilization of the appropriate temperature of the calcination; and 5) the surface hydrophobization of mesoporous titanium silicates. We would like to test in our future works some of the ways of improving the efficiency of hydrogen peroxide conversion: choosing of the appropriate solvent, additives, and slow addition of hydrogen peroxide. A large advantage of the presented process is also performing the process of epoxidation in water solution without any other solvents. Moreover, water was not additionally introduced in the reactor, only with the solution of the oxidizing agent it was formed during the process. Water is now regarded as a very ecological solvent for organic processes.

Author details

Agnieszka Wróblewska*, Edyta Makuch and Ewelina Mójta

*Address all correspondence to: Agnieszka.Wroblewska@zut.edu.pl

Institute of Organic Chemical Technology, West Pomeranian University of Technology Szczecin, Szczecin, Poland

References

[1] Chen LY, Chuan GK, Jaenicke S. Propylene epoxidation with hydrogen peroxide catalyzed by molecular sieves containing framework titanium. J Molecul Catal A: Chem 1998;132:281–92. DOI: 10.1016/S1381-1169(97)00276-8

[2] Li G, Wang X, Yan H, Liu Y, Liu X. Epoxidation of propylene using supported titanium silicalite catalysts. Appl Catal A: Gen 2002;236:1–7. DOI: 10.1016/S0926-860X(02)00288-0

[3] Wang R, Guo X, Wang X, Hao J. Propylene epoxidation over silver supported on titanium silicalite zeolite. Catal Lett 2003;90:57–63. DOI: 10.1023/A:1025816226760

[4] Huang J, Takei T, Akita T, Ohashi H, Haruta M. Gold clusters supported on alkaline treated TS-1 for highly efficient propene epoxidation with O_2 and H_2. Appl Catal B: Environ 2010;95:430–8. DOI: 10.1016/j.apcatb.2010.01.023

[5] Lee W-S, Akatay C, Stach EA, Ribeiro FH, Delgass WN. Gas-phase epoxidation of propylene in the presence of H_2 and O_2 over small gold ensembles in uncalcined TS-1. J Catal 2014;313:104–12. DOI: 10.1016/j.jcat.2014.02.013

[6] Gao H, Lu G, Sou J, Li S. Epoxidation of allyl chloride with hydrogen peroxide catalyzed by titanium silicalite. Appl Catal A: Gen 1996;138:27–38. DOI: 10.1016/0926-860X(95)00276-6

[7] Wang Q, Mi Z, Wang Y, Wang L. Epoxidation of allyl chloride with molecular oxygen and 2-ethyl-antrahydroquinone catalyzed by TS-1. J Molecul Catal A: Chem. 2005;229:71–5. DOI: 10.1016/j.molcata.2004.11.006

[8] Kumar Pandey R, Kumar R. Eco-friendly synthesis of epichlorohydrin catalyzed by titanium silicate (TS-1) molecular sieve and hydrogen peroxide. Catal Commun. 2007;8:379–82. DOI: 10.1016/j.catcom.2006.07.007

[9] Wang L, Liu Y, Xie W, Zhang H, Wu H, Jing Y, He M, Wu P. Highly efficient and selective production of epichlorohydrin through epoxidation of allyl chloride with hydrogen peroxide over Ti-MWW catalyst. J Catal 2007;246:205–14. DOI: 10.1016/j.jcat.2006.12.003

[10] Wang L, Zhou Y, Mi Z. Epoxidation of allyl chloride and hydrogen peroxide over titanium silicalite film on SiO_2 pellet support. J Chem Technol Biotechnol 2007;82:414–20. DOI: 10.1002/jctb.1689

[11] Danov SM, Sulimov AV, Sulimova AV. Solvent effect on epoxidation of allyl chloride with hydrogen peroxide on titanium-containing silicalite. Russ J Appl Chem 2008;81(11):1963–6. DOI: 10.1134/S1070427208110189

[12] Xia H, Zhou Ch-H, Tong D, Chen J, Yu W, Liu S. Preparation and catalysis in epoxidation of allyl chloride of zeolitic titanosilicate-1/smectitic clay minerals. Appl Clay Sci 2011;53:279–87. DOI: 10.1016/j.clay.2011.04.014

[13] Hutching GJ, Lee DF, Minihan AR. Epoxidation of allyl alcohol to glycidol using titanium silicalite TS-1, effect of the method of preparation. Catal Lett 1995;33:369–85.

[14] Wu P, Tatsumi T. A novel titanosilicate with MWW structure: III. Highly efficient and selective production of glycidol through epoxidation of allyl alcohol with H_2O_2. J Catal 2003;214:317–26. DOI: 10.1016/S0021-9517(02)00170-7

[15] Wróblewska A, Milchert E. Epoxidation of allyl alcohol with hydrogen peroxide over titanium silicalite TS-2 catalyst. J Chem Technol Biotechnol 2007;82:681–6. DOI: 10.1002/jctb.1739

[16] Wróblewska A, Fajdek A, Milchert E, Grzmil B. The Ti-MWW catalyst - its characteristic and catalytic properties in the epoxidation of allyl alcohol by hydrogen peroxide. Pol J Chem Technol 2010;12(1):29–34. DOI: 10.2478/v10026-010-0006-1

[17] Wróblewska A, Fajdek A. Epoxidation of allyl alcohol to glycidol over the microporous TS-1 catalyst. J Hazard Mater 2010;179:258–65. DOI: 10.1016/j.jhazmat.2010.02.088

[18] Wróblewska A, Makuch E. The utilization of Ti-SBA-15 catalyst in the epoxidation of allylic alcohols. React Kinet Mech Catal 2012;105:451–68. DOI: 10.1007/s11144-011-0405-1

[19] Wu P, Liu Y, He W, Tatsumi T. A novel titanosilicate with MWW structure: Catalytic properties in selective epoxidation of diallyl ether with hydrogen peroxide. J Catal 2004;228:183–91. DOI: 10.1016/j.jcat.2004.09.001

[20] Wróblewska A, Makuch E. Regeneration of the Ti-SBA-15 catalyst used in the process of allyl alcohol epoxidation with hydrogen peroxide. J Adv Oxidation Technol 2014;17(1):44–52.

[21] Wróblewska A, Makuch E. Studies on the deactivation of Ti-MCM-41 catalyst in the process of allyl alcohol epoxidation. Pol J Chem Technol 2013;15(4):1115. DOI: 10.2478/pjct-2013-0078

[22] Srivastava R, Srinivas D, Ratnasamy P. CO_2 activation and synthesis of cyclic carbonates and alkyl/aryl carbamates over adenine-modified Ti-SBA-15 solid catalysts. J Catal 2005;233:1–5. DOI: 10.1016/j.jcat.2005.03.023

[23] Wu P, Tatsumi T, Komatsu T, Yashima T. Postsynthesis, characterization, and catalytic properties in alkene epoxidation of hydrothermally stable mesoporous Ti-SBA-15. Chem Mater 2002;14:1657–64. DOI: 10.1021/cm01091v

[24] Berube F, Klietz F, Kaliaguine S. Surface properties and epoxidation catalytic activity of Ti-SBA15 prepared by direct synthesis. J Mater Sci 2009;44:6727–35. DOI: 10.10007/s10853-009-3566-9

[25] Wu S, Han Y, Zou Y-C, Song J-W, Zhao L, Di Y, Liu S-Z, Xiao F-S. Synthesis of heteroatom substituted SBA-15 by the "pH-adjusted" method. Chem Mater 2004;16:486–92. DOI: 10.1021/cm0343857

[26] Mandache I, Parvulescu VI, Popescu A, Parvulescu I, Banciu MD, Amoros P, Beltran D, Trong On D, Kaliaguine S. Epoxidation of dibenzocycloalkenes on Ti-Ge-MCM-41 and Ti-SBA-15 catalysts. Microporous Mesoporous Mater 2005;81:115–24. DOI: 10.1016/j.micromeso.2005.01.024

[27] Morey MS, O'Brien S, Schwarz S, Stucky GD. Hydrothermal and postsynthesis surface modification of cubic MCM-48 and ultralarge pore SBA-15 mesoporous silica with titanium. Chem Mater 2000;12:898–911. DOI: 10.1021/cm9901663

[28] Wittmann G, Demeester K, Dombi A, Dewulf J, Van Langenhove H. Preparation, structural characterization and photocatalytic activity of mesoporous Ti-silicates. Appl Catal B: Environ 2005;61:47–57. DOI: 10.1016/j.apcatb.2005.04.010

[29] Berube F, Nohair B, Kleitz F, Kaliaguine S. Controlled postgrafting of titanium chelates for improved synthesis of Ti-SBA-15 epoxidation catalysts. Chem Mater 2010;22:1988–2000. DOI: 10.1021/cm9030667

[30] Berube F, Kleitz F, Kaliaguine S. A comprehensive study of titanium-substituted SBA-15 mesoporous materials prepared by direct synthesis. J Phys Chem C 2008;112:14403–11. DOI: 10.1021/jp803853m

[31] Kruk M, Jaroniec M, Hyun Ch, Ryoo R. Characterization of the porous structure of SBA-15. Chem Mater 2000;12:1961–8. DOI: 10.1021/cm000164e

[32] Sacaliuc E, Beale AM, Weckhuysen BM, Nijhuis TA. Propene epoxidation over Au/Ti-SBA-15 catalysts. J Catal 2007;248:235–48. DOI: 10.1016/j.jcat.2007.03.014

[33] Feng J, An D, Chen B, Li Y, Ding K, Xie Y, Liu Z. Post-synthesis of Ti-SBA-15 in supercritical CO_2-ethanol solution. Clean 2009;37(7):527–33. DOI: 10.1002/clen.200900030

[34] Chen L-F, Zhu P-J, Qiao M-H, Shen W, Xu L-H, Fan K-N. Preparation of Cu/SBA-15 catalysts by different methods for the hydrogenolysis of dimethyl maleate to 1, 4-butanediol. Appl Catal A: Gen 2009;356:129–36. DOI: 10.1016/j.apcata.2008.12.029

[35] Nava R, Ortega RA, Alonso G, Ornelas C, Pawelec B, Fierro JLG. CoMo/Ti-SBA-15 catalysts for dibenzothiophene desulfurization. Catal Today 2007;127:70–84. DOI: 10.1016/j.cattod.2007.02.034

[36] Vinu A, Srinivasu P, Miyahara M, Ariga K. Preparation and catalytic performances of ultralarge-pore TiSBA-15 mesoporous molecular sieve with very high Ti content. J Phys Chem 2006;110:801–6. DOI: 10.1021/jp055851d

[37] Lopez-Munoz M-J, Van Grieken R, Aguado J, Marugan J. Role of the support on the activity of silica-supported TiO2 photocatalysts: Structure of the TiO2/SBA-15 photocatalysts. Catal Today 2005;101:307–14. DOI: 10.1016/j.cattod.2005.03.017

[38] Marchese L, Gianott E, Dellarocca V, Maschmeyer T, Rey F, Coluccia S, Thomas JM. Structure-functionality relationships of grafted Ti-MCM41 silicas. Spectroscopic and catalytic studies. Phys Chem Chem Phys 1999;1:585–92. DOI: 10.1039?A808225A

[39] Gianotti E, Bisin Ch, Marchese L, Guidotti M, Ravasin N, Psaro R, Coluccia S. Ti(IV) catalytic centers grafted of different siliceous materials: spectroscopic and catalytic study. J Phys Chem C 2007;11:5083–9. DOI: 10.1021/jp067506+

[40] Xie HI, Fan YX, Zhou Ch, Du ZX, Min EZ, Ge ZH, Li XN. A review on heterogeneous solid catalysts and related catalytic mechanisms for epoxidation of olefins with H_2O_2. Chem Biochem Eng Quarter 2008;22(1):25–39.

[41] Fraile JM, Garcia JI, Mayoral JA, Vispe E. Effect of the reaction conditions on the epoxidation of alkenes with hydrogen peroxide catalyzed by silica-supported titanium derivatives. J Catal 2001;204:146–56. DOI: 10.1006/jcat.2001.3355

[42] Bhaumik A, Tatsumi T. Intramolecular rearrangement of epoxides generated in situ over titanium silicate molecular sieves. J Catal 1999;182:349–56. DOI: jcat.1998.2355

[43] Chaudhari K, Srinivas D, Ratnasamy P. Reactive oxygen species in titanosilicates TS-1 and TiMCM-41: an in situ EPR spectroscopic study. J Catal 2001;303:25–32. DOI: 10.1006/jcat.2001.3315

[44] W, Wu P, Tatsumi T, Uni. Unique solvent effect of microporous crystalline titanosilicates in the oxidation of 1-hexene and cyclohexene. J Catal 2008;256:62–73. DOI: 10.1016/j.jcat.2008.03.001

[45] Bordiga S, Bonino F, Damin A, Lambert C. Reactivity of Ti(IV) species hosted in TS-1 towards H_2O_2-H_2O solutions investigated by ab initio cluster and periodic approaches combined with experimental XANES and EXAFS data: a review and new highlights. Phys Chem Chem Phys 2007;9:4854–78. DOI: 10.1039/B706637F

[46] Kholdeeva OA, Trukhan NN. Mesoporous titanium silicates as catalysts for the liquid-phase selective oxidation of organic compounds. Russ Chem Rev 2006;75(5):411–32. DOI: 10.1070/RC2006v075n05ABEH001210

[47] Srinivas D, Manikandan P, Laha SC, Kumar R. Reactive oxo-titanium species in titanosilicate molecular sieves: EPR investigations and structure-activity correlations. J Catal 2003;217:160–71. DOI: 1:1016/S0021-9517(03)00060-5

[48] Guidotti M, Ravasio N, Psaro R, Ferraris G, Moretti G. Epoxidation on titanium-containing silicates: do structural feature really affect the catalytic performance?. J Catal 2003;214:242–50. DOI: 10.1016/S0021-9517(02)00152-5

[49] Corma A. State of the art and future challenges of zeolites as catalysts. J Catal 2003;216:298–312. DOI: 10.1016/S0021-9517(02)00132-X

[50] Adam W, Corma A, Reddy TI, Renz M. Diastereoselective catalytic epoxidation of chiral allylic alcohols by the TS-1 and Ti-β zeolites: Evidence for a hydrogen-bonded, peroxy-type loaded complex as oxidizing species. J Org Chem 1997;62:3632–7. DOI: 10.1021/jo970364i

[51] Adam W, Corma A, Martinez A, Mitchell CM, Indrasena T, Renz M, Smerz AK. Diastereoselective epoxidation of allylic alcohols with hydrogen peroxide catalyzed by titanium-containing zeolites or methyltrioxorhenium versus stoichiometric oxidation with dimethyldioxirane: Clues on the active species in the zeolite lattice. J Molecul Catal A: Chem 1997;117:357–66.

[52] Yoon ChW, Hirsekorn KF, Neiding ML, Yang X, Tilley TD. Mechanism of the decomposition of aqueous hydrogen peroxide over heterogeneous TiSBA15 and TS-1 selective oxidation catalysts: insight from spectroscopic and density functional theory studies. ACS Catal 2011;1:1665–78. DOI: 10.1021/cs2003774

[53] Antcliff KL, Murphy DM, Griffiths E, Giamello E. The interaction of H_2O_2 with exchanged titanium oxide sytems (TS-1, TiO_2, [Ti]-APO-5, Ti-ZSM-5). Phys Chem Chem Phys 2003;5:4306–16. DOI: 10.1039/B306398B

[54] Potekhin VV, KLulikova VA, Kochina EG, Potekhin VM. Decomposition of hydrogen peroxide in protic and polar protic solvents on TS-1 heterogeneous catalyst. Russ J Appl Chem 2011;84(7):1195–200.

[55] Guidotti M, Pirovano C, Ravasin N, Lazaro B, Fraile JM, Mayoral JA, Coq B, Galarneau A. The use of H_2O_2 over titanium-grafted mesoporous silica catalysts: step further towards sustainable epoxidation. Green Chem 2009;11:1421–7. DOI: 10.1039/B903302E

[56] Ediati R, Ulfa M, Fansuri H, Ramli Z, Nur H, Prasetyoko D. Influence of TiO_2/TS-1 calcination on hydroxylation of phenol. J Math Fundament Sci 2014;46(1):76–90. DOI: 10.5614/j.math.fund.sci.2014.46.1.7

[57] Fraile JM, Garcia JI, Mayoral JA, Vispe E. Optimization of cyclohexane epoxidation with dilute hydrogen peroxide and silica-supported titanium catalysts. Appl Catal A: Gen 2003;245:363–76. DOI: 10.1016/S0926-860X(02)00643-9

[58] Trukhan NN, Romannikov VN, Shmakov AN, Vanina MP, Paukshtis EA, Bukhtiyarov VI, Kriventsov VV, Yu Danilov I, Kholdeeva OA. H_2O_2 - based selective oxidations over titanium silicates of SBA-15 type. Microporous Mesoporous Mater 2003;59:73–84. DOI: 10.1016/S1387-1811(03)00287-7

[59] Jiang T, Wang W, Han B. Catalytic hydroxylation of benzene to phenol with hydrogen peroxide using catalysts based on molecular sieves. New J Chem 2013;37:1654–64. DOI: 10.1039/c3nj41163j

Chapter 6

Practical Design of Green Catalysts for PET Recycling and Energy Conversion

Arvin Sangalang, Seunghwan Seok and Do Hyun Kim

Additional information is available at the end of the chapter

http://dx.doi.org/10.5772/62041

Abstract

The recycling of chemicals and generation of alternative energy are central topics in the efforts toward sustainable development. Among these, research on plastics recycling and fuel cells has received significant attention, with the aim of designing novel catalysts to improve yield and efficiency. We highlight our work on these areas focusing on the chemical depolymerization of polyethylene terephthalate (PET) to recover its constituent monomer and the development of high-performance anode catalysts for polymer electrolyte membrane fuel cells (PEMFC). We demonstrate various flexible yet practical synthesis strategies (e.g. ultrasound-assisted deposition and biopolymer coating) that were used to obtain catalytic properties optimized for these applications. The effectiveness and simplicity of these methods render the catalysts to be truly green — from synthesis up to process application.

Keywords: PET glycolysis, fuel cell, ultrasound, polydopamine, nanocomposites

1. Introduction

Depleting resources and waste management are the major driving factors for environmental conservation practices, such as recycling and development of alternative energy. Plastics recycling and generation of renewable energy, in particular, are among the direct approaches that aim to curb the consumption of oil-derived commodities. Research efforts on these areas can contribute to the realization of long-term environmental sustainability, aiming for increased productivities and minimal environmental impact. In these efforts, the development of effective catalysts to improve their performance remains a major goal in optimizing processes for resource conservation. Taking both catalytic activity and practical synthesis into account, we present catalyst design strategies that truly conform to principles of green catalysis.

1.1. Chemical recycling of poly(ethylene terephthalate)

Poly(ethylene terephthalate) or PET is currently the world's most widely recycled plastic. Due to its extensive use in various applications, largely in packaging, its worldwide consumption has increased tremendously, adding to the generation of plastic solid wastes. The global market for PET packaging was worth $48.1 billion in 2014, equivalent to about 16 million tons of PET. Furthermore, the projected demand for PET packaging material is estimated to increase further by 4.6% annually, reaching up to 20 million tons by 2019 [1]. The dramatic increase in PET consumption over the past decades has led to significant efforts to recycle the thermoplastic polyester.

The case of PET recycling is currently a logistically well-established, large-scale recycling industry. Bottle recollection rates, for example, have shown a steady increase worldwide, which accounted for at least 2.8 million tons of collected bottles in 2008 (Figure 1). Most of recycled PET flakes produced worldwide are utilized for fiber applications. PET recycling, at present, is largely dominated by physical methods or melt-phase recycling strategies, although various process routes have been studied and remain open to major advancements for better sustainability [2–4].

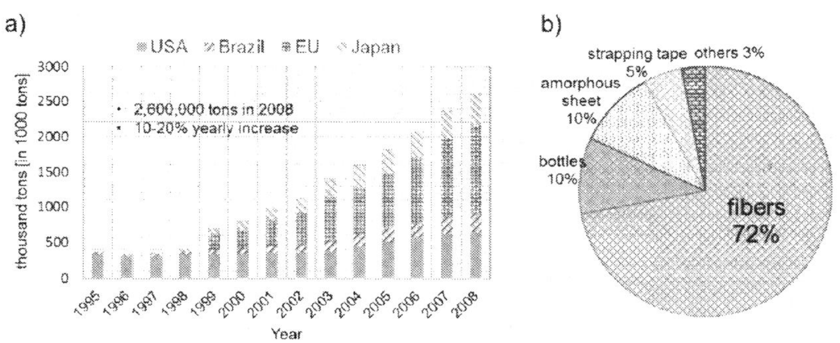

Figure 1. a) Increase in PET bottle recollection rates in various regions worldwide from 1995 to 2008. b) Use distribution of recycled PET flakes.

Significant efforts and renewed interest in chemical recycling routes have resurfaced recently, as stronger calls for long-term sustainability and new process developments in chemical recycling appeared. Extensive reviews on the progress and various aspects of PET chemical recycling are available elsewhere [2–7]. It has been regarded in the recycling literature that chemical recycling offers the most viable approach of sustainable recycling, allowing the recovery of raw materials and other useful chemicals while dealing with long-term waste disposal issues [5]. Figure 2 shows the monomer recovery routes in PET synthesis and chemical recycling. Among them, glycolysis is the simplest and well established, requiring low to moderate process conditions and capital investment (Table 1) [2].

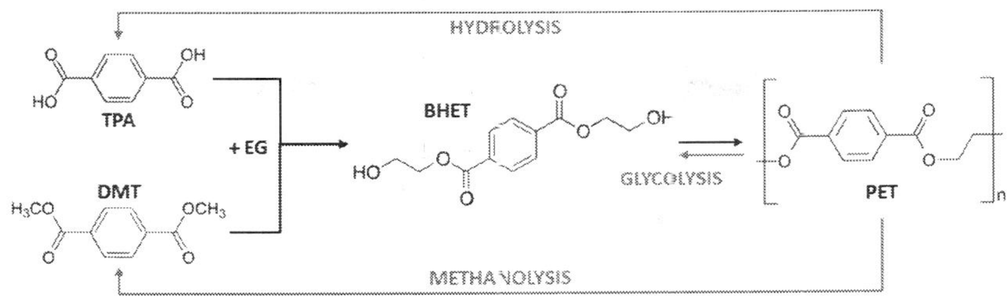

Figure 2. PET monomer recovery routes.

Recycling process	Qualitative plant cost	Economically competitive scale	Safety conditions
Hydrolysis	High	Large plants	High
Methanolysis	High	Large plants	High
Glycolysis	Medium-low	Small plants/batch reactor	Medium

Table 1. Process and economic considerations for PET monomer recovery routes [2]

The products obtained from glycolysis are the polycondensation monomer of PET, bis(2-hydroxyethyl terephthalate) or BHET (Figure 3) and low molecular weight oligomers, which are purified for subsequent use in PET manufacture or conversion into other chemicals. There have been significant issues, however, in process efficiency and product purification that led to recent efforts in the development of novel catalysts.

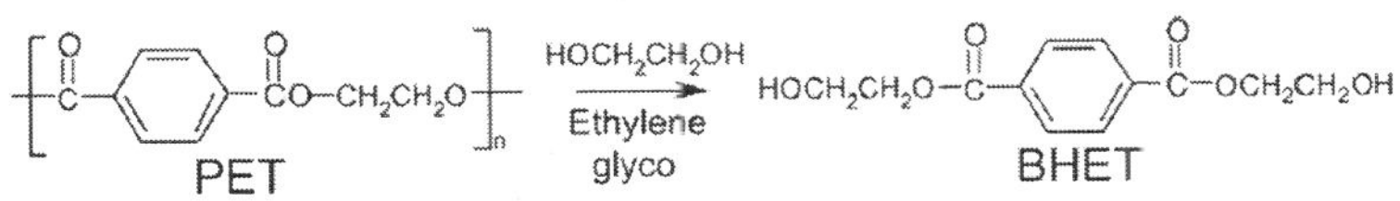

Figure 3. Glycolysis of PET chain releasing the monomer BHET.

Catalyst use is crucial in glycolysis and those used in earlier studies still required long reaction times despite catalyst application. These include metal acetates, alkalis, sulfates, and chlorides, some of which are those of toxic heavy metals [5]. Aside from efficiency, catalyst separation has been a major drawback in these catalysts together with environmental concerns. Despite chemical recycling being touted as the superior viable recycling route in the long-term, significant process developments are needed before it can truly be considered as a practical and competitive alternative to mechanical recycling. Results of life cycle assessments (LCA) show (Figure 4) that not only chemical recycling incurs greater process costs, but it also contributes to an overall negative impact in terms of a larger net consumption of non-renewable energy source and environmental effects [4]. Fortunately, recent developments in novel catalyst design may offer effective solutions to achieve the goal of a truly green and viable PET chemical recycling.

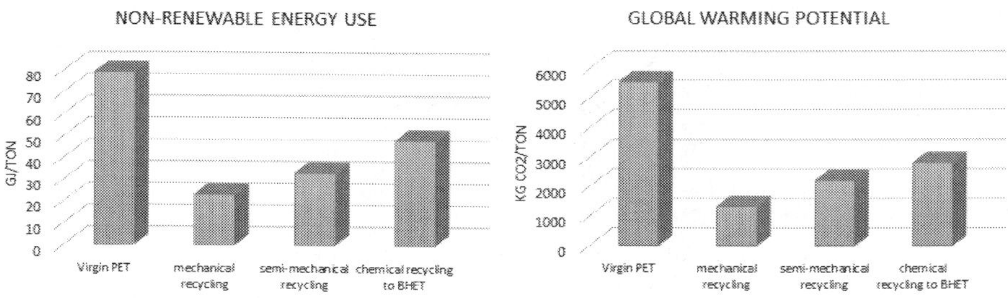

Figure 4. Comparison of life cycle assessment (LCA) results for PET manufacture and recycling in terms of non-renewable energy use and equivalent CO_2 generated per ton of PET fiber [4].

1.2. Hybrid catalysts for polymer electrolyte membrane fuel cells

The development of hybrid catalysts for polymer electrolyte membrane fuel cells (PEMFCs) has received attention recently due to advantageous qualities in its interaction characteristics and physical durability. The combination of conductivity enhancement and water management becomes widely used to improve PEMFC performance without external humidifying system. In this respect, requirements for next promising materials for PEMFCs include such categories as high proton conductivity, membrane electrode assemblies (MEA), and non-toxic and easy fabrication method. The humidifying ability of MEA in fuel cells is crucial for proton conductivity. As such, the approach to enhance conductivity and allow self-humidification at the same time has been considered recently to improve the performance of PEMFCs [8]. Given these requirements, the potential solution is platinum doping for conductivity and using silica as support for its hygroscopic properties. It is a challenge, however, to increase the platinum doping and enhance catalyst conductivity. With our bio-coating strategy for metal doping, we demonstrate a successful implementation of N-doped carbon composite through a silica-based templating technique [9]. In contrast to other reported methods that use chemical vapor deposition (CVD) and plasma treatment of various nitrogen precursors, our method remains on the practical side requiring no expensive equipment and complex treatment processes.

2. Synthesis and performance of metal oxide-based catalysts for PET glycolysis

Clearly, there is a need to consider various factors in designing PET glycolysis catalysts such as catalytic activity, ease of process applicability, and environmental impact. The coordination between a metal cation and the PET carbonyl oxygen is the widely exploited basis of catalytic activity in glycolysis, although other catalytic pathways do exist. Numerous metal compounds have been used as glycolysis catalysts with varying success [6]. Some of these made its way to commercial application through patented technologies [7]. Recently, a significant amount of work in the development of solid catalysts has appeared, potentially addressing the problem of product/catalyst separation.

The first solid catalysts used for glycolysis were zeolites, which showed catalytic performance of only moderate monomer yields (<70%). We have developed various glycolysis catalysts based on metal oxides that were thermally stable and recoverable and provided excellent monomer recovery yields. These are the ideal characteristics of industrially applicable heterogeneous catalysts that could solve issues in existing conventional glycolysis catalysts. We discuss in the following sections several methods used to synthesize them, their properties, and performance as stable and effective glycolysis catalysts. It was possible to obtain free and supported versions of these catalysts using flexible synthesis strategies [10–13] tailored to specific process applications.

2.1. Glycolysis reaction set-up and product analysis

Virgin PET pellets were obtained, mixed with dry ice in a grinder, and reduced to a fine powder with a particle size of <200 µm. The glycolysis reactions were carried out in a 10-mL stainless steel batch-type pressure reactor at 300°C and 1.1 MPa, unless otherwise specified. The reactant mixture of 0.3 g of PET, 1.1 g of EG, and a set amount of catalyst was loaded into the reactor. The reactor was placed in a furnace preheated at the reaction temperature, which, after the given reaction time, was taken out and quenched in cold water to stop the reaction. For quantitative determination of the monomer yield, the reaction products were dissolved in tetrahydrofuran (THF) and analyzed via HPLC. A reverse-phase Zorbax-C8 column and an ultraviolet (UV) detector set at 254 nm were used. The mobile phase used was a 50:50 (v/v) THF/H_2O solution at a flow rate of 1.0 ml/min. The molar yield of BHET was calculated based on the following equation:

$$\text{BHET Molar yield, \%} = \frac{\text{moles of BHET produced}}{\text{moles of PET units}} \times 100\%$$

The monomer BHET was separated from the glycolysis products and purified for qualitative analysis. Boiling water was added to the reaction mixture to dissolve BHET and the hot solution was immediately filtered. Repeated washing was done to extract residual BHET in the filter cake. The filtrate was stored in a refrigerator at 4°C for 24 h, after which white needle-like crystals formed. These were filtered and dried at 70°C for 12 h. The dried crystals were then used for various characterization steps, such as FT-IR, DSC, TGA, and NMR to verify the structure and properties of the recovered monomer, properly distinguishing BHET from its dimer and/or oligomers. Its chemical structure was analyzed by ^{1}H NMR and ^{13}C NMR using DMSO-d_6 as the solvent [10].

2.2. Supported metal oxide composites via ultrasound-assisted synthesis

Metal oxides find important application as catalysts, but agglomeration during their synthesis reduces the active surface area beneficial for catalysis. The combination of solid support and metal has been proposed, since composite materials can isolate the nanoparticle on the supporting material and effectively reduce its size. As a result, such nanocomposites can increase the surface area of metal oxide catalyst [14–16]. Silica, one of the most widely used catalyst supports, is synthesized in this study in the nanoscale and used as a metal oxide support material. Another material considered as support for nanocomposites is graphene oxide (GO). It is compatible with various organic/inorganic nanomaterials, taking advantage of possible chemical modification utilizing the oxygen-containing functional group on its sheet. In addition, it has high chemical stability and specific surface area [17–20]. The synthesis of silica and graphene oxide-based catalysts is presented, where ultrasound was used to reduce reaction time, simplify reaction steps, or perform synthesis in milder conditions.

2.2.1. Ultrasound-assisted catalyst synthesis

Ultrasound has been used in metal deposition, dish cleaning, chemical reactions, and particle dispersion in a solvent [21]. The cavitation phenomenon includes formation, growth, and collapse of bubbles generated in the aqueous solution under ultrasonic irradiation. The phenomenon causes both high temperature (~5000 K) and pressure (~1000 atm) during the sequence of bubble collapse. In a short span of time for bubble formation, it provides very high heating and cooling rate. This unique condition generates intense energy that converts water into H and OH radicals, promoting the formation of metal nanoparticles [22].

Our group applied the ultrasound irradiation method for the fabrication of metal-doped silica nanocomposites [8,11,23,24]. The acoustic cavitation phenomenon facilitates interparticle collision between metal and support material, inducing the binding between metal and support [25,26]. As a result, metal doped on supporting material can be prepared without surfactants or surface modification of support material in a short reaction time and mild reaction conditions.

2.2.2. Ultrasonic deposition of metal oxide catalyst on silica particles

Silica microparticles (SMPs) and silica nanoparticles (SNPs) were used as support for the catalysts. Silica microparticles with size 1–20 μm were purchased from Junsei Chemicals and used without any treatment. Nanosized silica particles were synthesized using the Stöber method, with some modifications. In a sealed round-bottomed flask, 8.0 ml of ammonium hydroxide (28 wt%) and 6.0 ml of deionized water were added to 100 ml of ethanol and stirred for 15 min. Then, 4.7 ml of the silica precursor tetraethyl orthosilicate (TEOS) were added to this solution and stirred at room temperature for 3 h. The resulting precipitate was centrifuged and washed with water and ethanol several times. The product was dried in an oven at 70°C for 8 h followed by calcination at 500°C for 12 h.

To synthesize the silica-supported manganese oxide or zinc oxide catalysts, a predetermined amount of silica support was added to a 1.0 M solution of the precursor [$Mn(NO_3)_2$ xH_2O or

$Zn(NO_3)_2$ $6H_2O$]. The metal oxide loading was set to be 15 wt%. Using a horn-type sonifier, the solution was sonicated for 45 min. A 0.1 M ammonia solution was added at the start of the sonication step in order to keep the pH at around 9.5. The ultrasound-assisted deposition process is illustrated in Figure 5. After separating the particles by centrifugation, they were washed with water and ethanol. The catalyst samples were then dried at 100°C for 8 h and calcined at 350°C for 3 h.

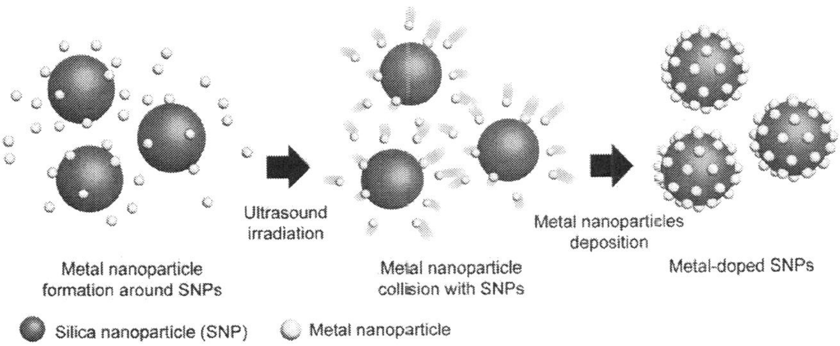

Figure 5. Overall synthesis procedure of metal-doped silica nanoparticle.

The properties of the synthesized silica-supported catalysts are given in Table 2. At 1.0 wt% catalyst-to-PET loading, glycolysis was performed at 300°C for 80 min, after which the product monomer was recovered. Shown in Figure 6 are the FT-IR spectra of the recovered BHET crystals and standard BHET sample. The matching spectra confirm that the product obtained using the silica-supported catalyst was indeed the monomer. The spectrum showed the presence of peaks corresponding to the functional groups in BHET: an -OH band at 3,424 and 1,128 cm^{-1}, an aromatic C-H at 1,456–1,502 cm^{-1}, C-O at 1,712 cm^{-1}, and alkyl C-H at 2,879 and 2,964 cm^{-1} [10].

From the monomer yield versus reaction time data in Figure 7, the order of catalytic activity can be determined as Mn_3O_4/SNPs>ZnO/SNPs>Mn_3O_4 SMPs>ZnO/SMPs. This trend follows the same order as the catalyst surface area and the pore volume given in Table 1. Among the four catalyst samples with different size of support and metal oxide doping, the large amount of active surfaces in the nanoparticle support and the activity of the Mn_3O_4 catalyst could be responsible for the fastest reaction rate and the maximum monomer yield [10]. Although it is evident that using a silica nanoparticle support could improve the catalytic performance, it also has its drawbacks in the practical and industrial perspective. Using a metal oxide catalyst is desired for easier purification of the glycolysis products, but efficient separation of a nanosized catalyst can be challenging. A trade-off between catalytic performance and practical applicability will be inevitable, unless an effective method to separate the catalysts is provided.

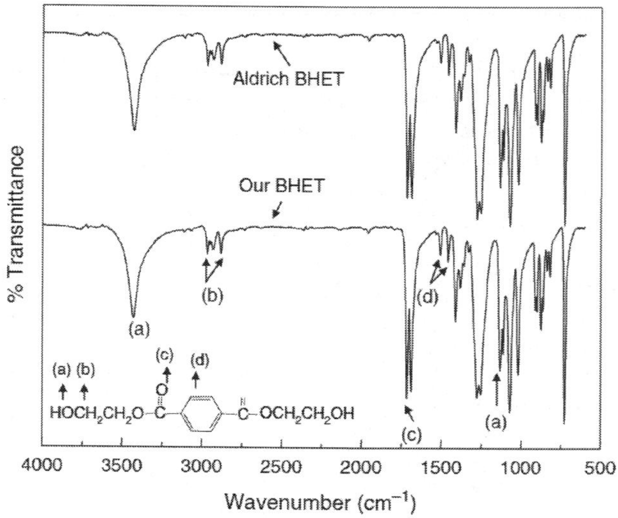

Figure 6. Comparison of FT-IR spectra of standard BHET and BHET recovered from catalyzed glycolysis. Figure from [10].

Catalyst	BET surface area (m^2/g)	Pore volume (cm^3/g)	Average pore diameter (nm)
ZnO/SMPs	2.49	0.014	21.4
Mn_3O_4/SMPs	3.38	0.020	24.4
ZnO/SNPs	22.44	0.154	30.2
Mn_3O_4/SNPs	45.09	0.214	18.9

Table 2. BET surface area, pore volume, and average pore diameter of silica-supported metal oxide catalysts [10]

2.2.3. Sonochemical synthesis of the GO-Mn_3O_4 composites

Mn_3O_4 and its nanocomposites have been utilized as highly effective catalysts for various applications, including PET glycolysis as demonstrated in our previous studies. Some of the known advantages of manganese oxides include excellent catalytic activity, low cost, abundance, and being environmental benign [27–30]. Depending on the oxidation states of manganese, there are several forms of MnO_x (e.g. MnO_2, MnO, Mn_2O_3, and Mn_3O_4), each of which has different applications. Hausmannite (Mn_3O_4), with both Mn^{2+} and Mn^{3+} ions in its crystal structure, has been widely used in catalytic applications. When loaded onto a support to provide a large surface area and prevent aggregation, it could significantly enhance the depolymerization of PET [10].

http://dx.doi.org/10.5772/62041

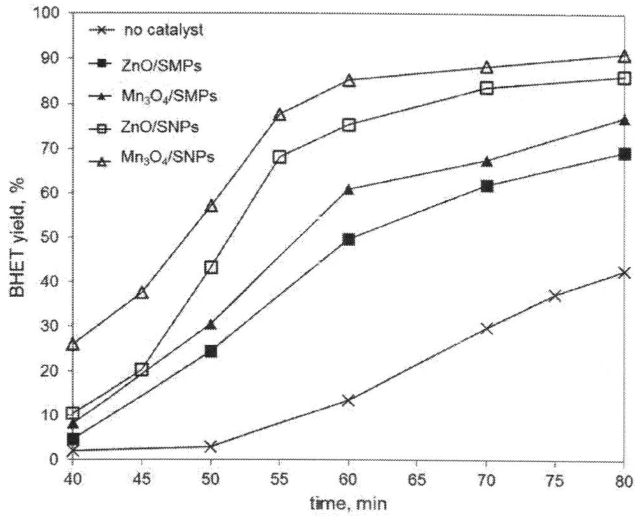

Figure 7. Molar yield of BHET at 300°C and 1.1 MPa using silica-supported catalysts. Figure from [10].

Simultaneous formation and direct deposition of MnO_x nanostructures have been reported by using redox reaction and electrodeposition [21,29–31]. One of the most favorable methods to do this is the reduction of permanganate ions into the insoluble manganese dioxide induced by carbon such as that in a graphene structure. The procedure is simple and the reaction has a self-limiting character [27,28,32]. This method can also be adapted to deposit Mn_3O_4 onto graphene through thermal reduction of MnO_2 over 1000°C [16]. This is an energy-intensive process, over which alternative methods of synthesis using milder conditions would be preferable [21,33]. In our work, sonochemical methods were used to facilitate mild conditions for synthesis and reduce reaction time involved in Mn_3O_4 deposition as illustrated in Figure 8.

Using the modified Hummers method, graphene oxide (GO) was prepared from graphite flakes [34]. One gram of graphite was added to 50 mL of concentrated H_2SO_4 in an ice bath. Then, 3.5 g of $KMnO_4$ were added and stirred for 2 h at 35°C. The suspension was then kept at 98°C to which deionized (DI) water was added dropwise. Then, 25 mL of 3% H_2O_2 aqueous solution were poured into the mixture and filtered with a 0.1 mm pore diameter Anodisc™ membrane. The product was washed with 10% HCl aqueous solution and DI water. By applying ultrasound to the filtered graphite oxide cake suspended in DI water, exfoliated GO was obtained and subsequently dried. Dispersion of GO in DI water at 0.5g/mL concentration was prepared. Mixtures of 10 mg/mL $KMnO_4$ and the GO dispersion at varied volume ratios of 0.01, 0.03, and 0.05 were used for the synthesis of GO-Mn_3O_4 composite samples A, B, and C, respectively. These are then subjected to ultrasonication at a power of 80 W/cm for 30 min using a horn-type sonicator. The resulting suspension was then filtered and washed with DI water and then ethanol [11].

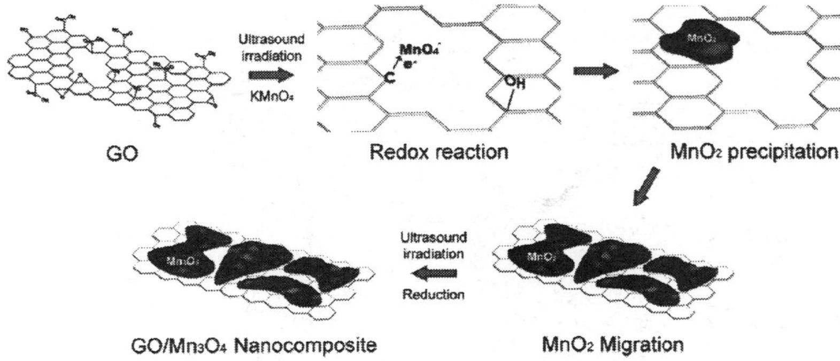

Figure 8. Schematic illustration of ultrasound-assisted synthesis of the GO-Mn_3O_4 composites. MnO_4 is first reduced to MnO_2 and precipitated onto the GO support by oxidizing carbon. The reduction of MnO_2 to Mn_3O_4 then takes place in the following steps.

The formation of GO-Mn_3O_4 nanocomposite was verified by various characterization methods such as XRD, XPS, and Raman spectroscopy. TEM images of pristine GO and the obtained composite are shown in Figures 9a and 9b, indicating the coverage of the GO surface. The high-resolution TEM image in Figure 9c shows the lattice fringes and diffraction pattern of the Mn_3O_4 crystal structure. Compared to the silica-supported composites and conventional metal salt catalysts [10,35], the monomer yield using the GO-Mn_3O_4 nanocomposite was comparable or higher, reaching more than 90% (Figure 10). The yields for the composite were all above 90%, showing improvement from that of bare Mn_3O_4 at 83%. However, the Mn_3O_4 without the support aggregated into micron scale. The GO support could prevent the aggregation of Mn_3O_4 and provide an enlarged and stable active sites [11].

2.3. Magnetically recoverable superparamagnetic γ-Fe_2O_3 nanocatalyst

Efficient recovery of the catalysts in PET glycolysis is an important aspect that several researchers have attempted to address recently. Among the recoverable catalysts studied are various ionic liquids, which were shown to provide molar yields up to 80% [36–39]. Aside from yield, there are several issues that should be considered in using ionic liquids as industrial glycolysis catalysts such as cost, stability, and robustness with respect to process variables such as moisture content, to which ionic liquids are very sensitive [10,11,40]. Urea was also reported to be a reusable glycolysis catalyst at mild temperatures [41]. Vacuum distillation was used to recover the catalyst, however, whose high energy requirements can be counterproductive.

We have studied magnetic nanoparticles as a recoverable glycolysis catalyst, among which γ-Fe_2O_3 was chosen as excellent candidate being known to have good performance in a number of reactions [42–44]. This was the first attempt to utilize magnetic nanomaterials in PET depolymerisation. Nanosized γ-Fe_2O_3 was selected due to its stability, high catalytic activity, and superparamagnetic property [42]. The superparamagnetic behavior allows recovery by application of a magnetic field yet allows good dispersion in the reaction medium, as it has

http://dx.doi.org/10.5772/62041

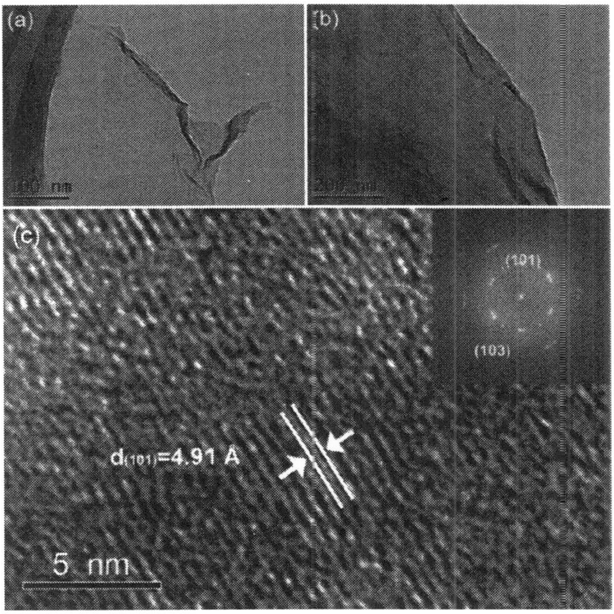

Figure 9. Low-magnification TEM images of (a) GO, (b) GO-Mn_3O_4, and (c) high-resolution TEM image of GO-Mn_3O_4 showing the d-spacing of the (101) crystal plane of Mn_3O_4 and its diffraction rings [11].

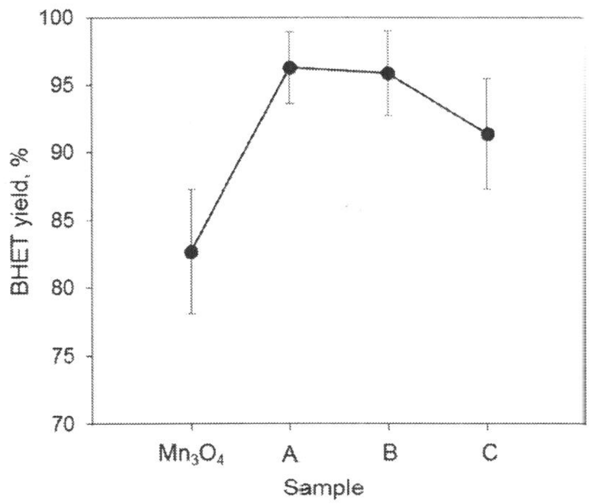

Figure 10. BHET yields for bare Mn_3O_4 and GO-supported Mn_3O_4 catalysts.

zero remanent magnetization. Iron oxides have further advantages being cheap, nontoxic, and abundant [12].

2.3.1. *Synthesis of superparamagnetic γ-Fe_2O_3 catalyst and characterization*

Slight modifications to the conventional co-precipitation method were done to obtain γ-Fe_2O_3 nanoparticles from Fe_3O_4 [45,46]. For the synthesis of Fe_3O_4 nanoparticles, predetermined amounts of $FeCl_2$ and $FeCl_3$ precursors were dissolved in an HCl solution (0.4 M, 25 mL) to prepare aqueous solutions of Fe^{2+}/Fe^{3+} with 1:2 molar ratio. Sodium hydroxide (1.5 M, 250 mL) was then added rapidly and stirred. The addition of NaOH solution instantly produced a black precipitate, characteristic of Fe_3O_4. The surfactant used to control the particle size was citric acid (0.2 M, 50 mL). The synthesis was performed in a nitrogen atmosphere. The formed black precipitate was separated by placing a magnet and decanting the solution. The product was washed with water four times and centrifuged at 4000 rpm for 4 min. The final washing step was done using a 0.01 M HCl solution in order to neutralize the anionic charges on the nanoparticle surface. Calcination of the dried Fe_3O_4 powder at 210°C for 3 h induced the phase transformation into γ-Fe_2O_3 [12].

The particle morphology of the synthesized catalyst was observed by a 200 kV transmission electron microscope (TEM). As shown in Figure 11a, the particles had a size distribution in the range of 8–14 nm, with mean size of 10 nm. The BET surface area measured was 147 m^2/g. The superparamagnetic property of the nanosized γ-Fe_2O_3 was confirmed by the magnetization curve obtained using a vibrating sample magnetometer (VSM). Figure 11b shows the magnetic behavior in the presence of a magnetic field, exhibiting a strong response with saturation magnetization reaching 47 emu/g. The curve does not have a hysteresis loop and retains no magnetization when the magnetic field is removed. The advantageous consequence of this property is redispersability of the catalyst particles when used in subsequent reactions [12].

The obtained X-ray diffraction (XRD) spectra of the catalyst suggest that the material is γ-Fe_2O_3. This is not conclusive, however, because the XRD patterns of Fe_3O_4 and γ-Fe_2O_3 are very similar. X-ray photoelectron spectroscopy (XPS) analysis of Fe2p cores was performed to distinguish the two phases. Higher binding energies before the calcination as shown in Figure 11c are indicative of Fe_3O_4. The shift to lower binding energies after the calcination step confirms the transformation of Fe_3O_4 to γ-Fe_2O_3 [12].

2.3.2. *Catalytic activity, recoverability, and stability with repeated use*

The catalytic performance of the synthesized γ-Fe_2O_3 nanoparticles was compared to previously studied silica nanoparticle-supported metal oxide catalysts [10,40]. Under the same reaction conditions and catalyst/PET weight ratio, γ-Fe_2O_3 delivered comparable performance (Figure 12). The BHET yield reached higher than 90% in 70 min at 1.0% catalyst-to-PET loading. As with the supported nanocatalysts, the excellent catalytic performance of the γ-Fe_2O_3 nanoparticles may be attributed to the high surface area and greater accessibility to active sites [12].

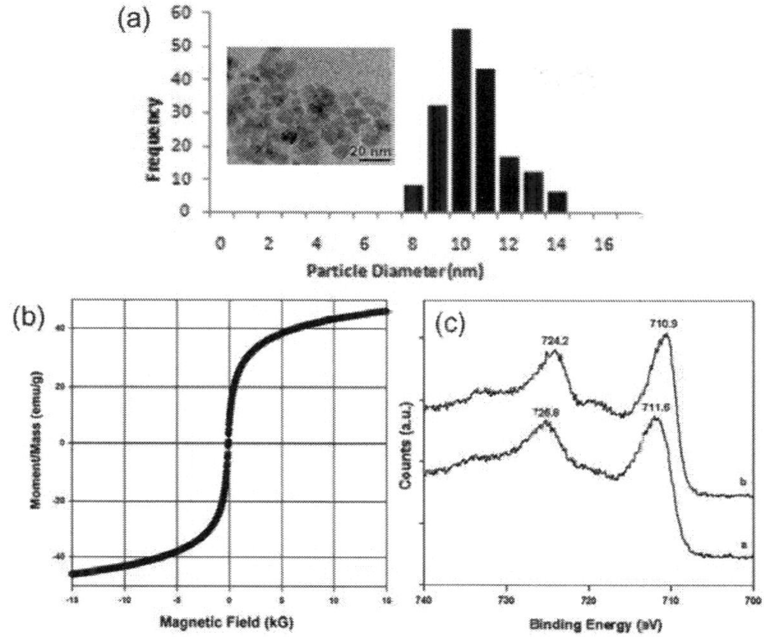

Figure 11. (a) Particle morphology, (b) superparamagnetic property, and (c) phase identification of the synthesized γ-Fe_2O_3 nanoparticles. Figure from [12].

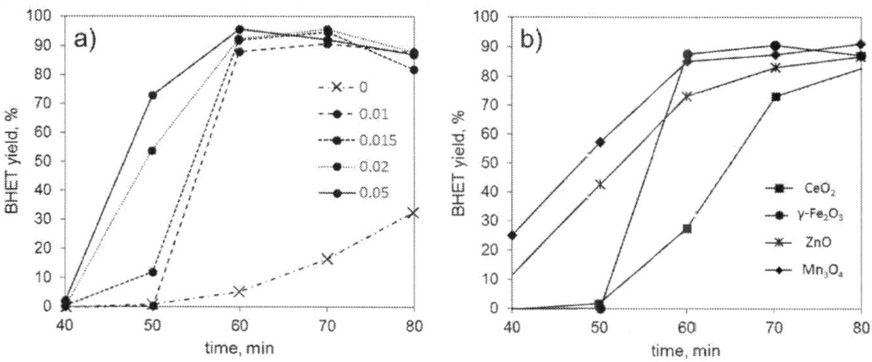

Figure 12. Catalytic performance of γ-Fe_2O_3 nanoparticles at various loadings and comparison to other metal oxide catalysts. Figure from [12].

The main potential of using γ-Fe_2O_3 as candidate for industrial glycolysis catalyst is the easy separation method and stability. In addition to the superparamagnetic property of γ-Fe_2O_3 nanoparticles that was beneficial to catalytic performance and separation, its stability with repeated use was also successfully demonstrated. The catalyst was reused in 10 reaction repeat cycles and did not significantly affect the BHET monomer yield, as shown in Figure 13. The thermal stability was supported by thermogravimetric analysis (TGA), while the XRD spectra of the used catalyst also proved phase stability [12]. These demonstrate the robust characteristics of the catalyst, withstanding repeated use at elevated temperature and pressure without deterioration in performance. Given comparable performance to a homogenous catalyst as shown in Figure 13b, the superparamagnetic catalyst provides a more practical approach to catalyst separation by application of a magnetic field.

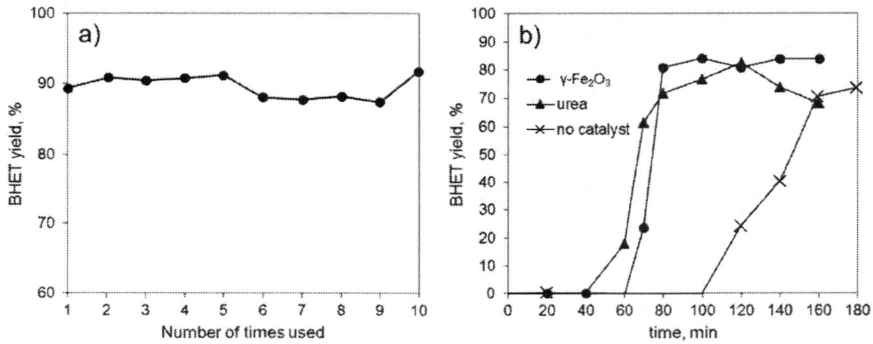

Figure 13. Assessment of catalyst stability with repeated use and comparison to a recoverable homogeneous catalyst at the same reaction conditions. Figure from [12].

2.4. Mesoporous spinel oxide catalysts

Due to several advantages such as high mechanical strength, possibility of regeneration, easier separation, and robust process integration, metal oxides can be considered superior to conventional PET glycolysis catalysts. Moreover, there are numerous possibilities to tailor their physical and chemical properties for the desired catalytic performance and functionality [47]. For the enhancement of catalytic activity, for example, altering the metal composition by introduction of another metal could result in higher catalytic activity [48–50]. The same principle is used to potentially enhance the performance of metal oxide catalysts for PET glycolysis. In this section, pure oxides and mixed-oxide spinel oxides of zinc, manganese, and cobalt were synthesised by simple precipitation or co-precipitation methods.

Simple precipitation method was used to synthesize the pure metal oxides. A 1.0 M solution of the salt precursors ($Mn(NO_3)_2$ xH_2O, $Zn(NO_3)_2$ $6H_2O$, or $Co(NO_3)_2$ $6H_2O$) was mixed with 0.1 M ammonium hydroxide to set the pH at 9.0. Precipitates of the corresponding metal hydroxides formed, which were filtered, washed with water, and dried at 100°C for 8 h. The oxide form was obtained by calcination of the dried powder at 600°C for 4 h. For the mixed metal oxides, a modified co-precipitation method was implemented [51–53]. The molar ratios of the metal precursors were fixed to be 1:2. Similar to the previous synthesis, a 0.1 M ammonium hydroxide solution was stirred into the bimetallic precursor solutions, setting the pH value at 9.0. The same procedures for filtering, washing, drying, and calcining were performed on the mixed metal oxides [13]. The physical and chemical properties of the synthesized catalysts are summarized in Table 3. The analysis of atomic composition of the oxides via EDS was performed with results shown in Table 4.

Catalyst	BET surface area (m^2/g)	Pore volume (cm^3/g)	Average pore diameter (nm)	Acid site concentration (mmol/g)
ZnO	7.67	0.06	31.33	0.041
Co_3O_4	17.50	0.09	18.80	0.032
$ZnCo_2O_4$	21.65	0.08	12.23	0.045
Mn_3O_4	22.57	0.01	14.70	0.060
$CoMn_2O_4$	25.10	0.07	10.16	0.067
$ZnMn_2O_4$	32.40	0.14	14.50	0.088

Table 3. Surface area, pore dimensions, and acid site concentration of the pure and mixed oxide catalysts [53]

Catalyst		Atom percentage			Atomic ratio (Mixed oxides)		
Zn	Mn	Co	O	Mn/Zn	Mn/Co	Co/Zn	
ZnO	56.3	-	-	43.7	-	-	-
Mn_3O_4	-	61.4	-	38.6	-	-	-
Co_3O_4	-	-	75.9	24.1	-	-	-
$ZnMn_2O_4$	21.7	40.9	-	37.4	1.88	-	-
$CoMn_2O_4$	-	36.2	21.0	42.9	-	1.72	-
$ZnCo_2O_4$	24.1	-	38.9	37.0	-	-	1.61

Table 4. Atomic analysis of the synthesized oxide catalysts by EDX [53]

The catalytic activity of metal oxides in PET glycolysis is influenced by the interaction of the metal cation and the carbonyl oxygen in the polyester. The nature of the metal, oxidation state, and crystal structure affect this interaction. In spinel oxides, the metal cations can be located in tetrahedral and octahedral sites. The metal covalency of two or more different metals in the mixed oxides can result in a beneficial interaction that could enhance its redox properties and catalytic activity. In this study, the best catalytic activity for glycolysis was demonstrated by

$ZnMn_2O_4$ catalyst (Figure 14). This catalyst has the ion pair Zn^{2+}/Mn^{3+} in its crystal lattice compared to Co^{2+}/Mn^{3+} and Zn^{2+}/Co^{2+} in the other mixed spinels. The high activity of the catalyst was attributed to the nature of the manganese ions combined with structural effects in the spinel crystal [13].

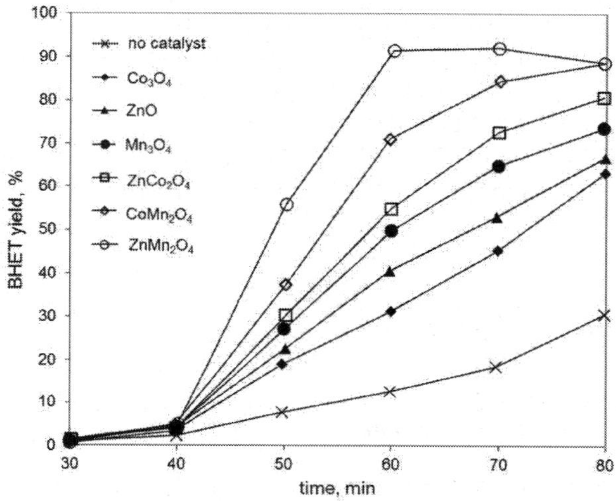

Figure 14. Comparison of BHET yields among the pure and mixed oxide catalysts [13].

2.5. Purity of recovered BHET monomer

For all the glycolysis reactions above using various catalysts, the recycled BHET was recovered via simple recrystallization. Several characterization methods have been employed to verify the structure and purity of the recovered monomer. The FT-IR spectra of the recycled BHET matched that of the standard sample [10], without extra peaks characteristic of contamination. As it is also possible that dimers and oligomers are not effectively separated, thermal and structure analyses have been performed using thermogravimetric analysis (TGA) and nuclear magnetic resonance spectroscopy (NMR). The proton and carbon NMR spectra identified peaks corresponding to distinct groups in the monomer and dimer backbones [13]. Along with TGA thermogram profiles, the structure characterization verified the good separation of the BHET from its oligomers.

3. Dopamine-induced carbon@silica hybrid as fuel cell anode catalyst

Aside from high proton conductivity and self-humidifying properties, principles of practical design for potential materials for polymer electrolyte membrane fuel cell (PEMFC) demand a fabrication method that is non-toxic and requires no complex apparatus for easy and large-scale production. Hybrid catalysts with carbon material and inorganic material have received attention from many researchers for PEMFC [54,55].

In hybrid material synthesis, a variety of methods are used to combine organic and inorganic materials. However, these fabrication methods require expensive equipment and complex processes, thereby limiting mass production [56–58]. To resolve this problem, our group has developed a simple and green fabrication of a hybrid catalyst for PEMFC. The hybrid composite consists of Pt nanoparticle, conductive dopamine coating layer, and silica support. Pt nanoparticle has been used as a typical catalyst for electrode catalyst for energy storage application. We also used Pt nanoparticle as the anode catalyst for PEMFC. To deposit Pt nanoparticles on the surface of silica, we applied dopamine coating as an adhesive layer [9]. The overall synthesis procedure for the metal and dopamine-induced carbon@silica nanocomposite is illustrated in Figure 15.

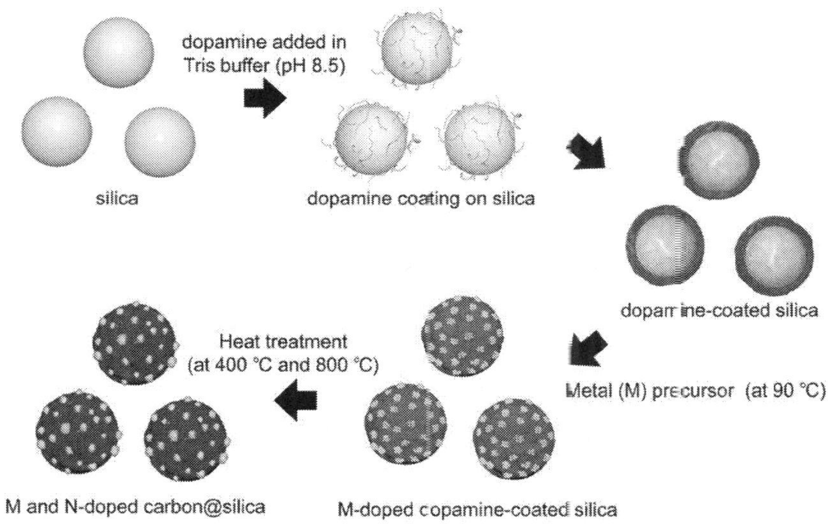

Figure 15. Overall procedure for synthesis of metal (M)- and nitrogen (N)-doped carbon@silica nanocomposite.

For the preparation of a hybrid anode catalyst, silica was fabricated by water-in-oil microemulsion method that produces a narrow size distribution. Silica is generally applied as support material because of its mechanical, chemical, and thermal stability. Its size and surface property can also be easily modified. Additionally, it has hygroscopic property that maintains the humidity needed for the proton conductivity in PEMFC. The humidity dependence of the membrane's conductivity typically required humidity control using an external equipment. However, the application of silica as support for the anode catalyst could provide both humidity and conductivity to the proton conductive membrane [8].

To deposit Pt nanoparticles on the surface of silica, we applied dopamine coating as an adhesive layer and self-reducing agent. Dopamine is a mussel-inspired adhesive based on a protein located at the end of the byssus of mussels. The protein allows mussels to strongly attach on rock surfaces or support in sea. The adhesion property of the protein helps mussels hold onto substrates in wet condition. Mussel-inspired dopamine is similar in chemical structure, with catechol and amine functional group as the protein [59]. Attempting to mimic the adhesive functionality, many researchers coated dopamine on various types of substrates. After dopamine coating, the substrate achieved a more hydrophilic surface and can be well dispersed in water. Dopamine coating also acts as a metal adhesive layer through the self-reducing ability of catechol group in the dopamine chemical structure [60]. Moreover, the coating is a simple and controllable dip-coating process in alkaline water solution (pH 8.5). The coating thickness can be tailored in nanoscale by adjusting the temperature, pH, time, and concentration of dopamine. Because of these advantages, dopamine coating has been applied to a variety of fields, such as surface modification, heavy metal removal, carbon material functionalization, and fabrication of conductive N-doped carbon sources [56–58,61].

We prepared Pt and N-doped carbon@silica with dopamine coating via metal decoration and carbonization steps. The dopamine coating procedure was performed on a Tris buffer solution at pH 8.5. Silica particles were dispersed in the solution. Then, 10 mM dopamine was added and mixed mildly for 6 h. After the coating step, the particle was washed several times with DI water. A thin layer of dopamine was observed to be coated well on silica with a thickness of 11.5 nm as shown in Figures 16c and 16d. The prepared dopamine-coated silica was re-dispersed in DI water. The dopamine-coated silica was then used to serve as catalyst support with metal adhesion properties. The Pt nanoparticles on the coated silica were deposited through the self-reducing ability of catechol groups in dopamine, as shown in Figure 17. This phenomenon releases electrons via the transformation of the R-OH group to a C=O group, which reduce Pt ions into Pt nanoparticles.

The dopamine coating layer naturally prohibits electron transfer (low conductivity). To enhance catalyst conductivity, we conducted carbonization of the Pt and dopamine-coated silica in an inert atmosphere at 800 °C. The carbonized dopamine layer contained nitrogen (N) and had other functional groups removed. N-doped carbon layer from dopamine coating enhanced both mechanical stability and conductivity of the material [57].

We applied Pt and dopamine-induced carbon@silica hybrid composite for polymer membrane electrolyte fuel cell as anode catalyst. The composite was evaluated and compared with commercial Pt/C catalyst under zero relative humidity. The carbonized dopamine coating and silica could be useful in PEMFC under non-humidified conditions. Pt and N-doped carbon@silica-based composites and commercial Pt/C were evaluated by galvanostatic polarization and power density curves as functions of current density for a single PEMFC at zero relative humidity (Figure 18). The Pt and dopamine-induced carbon@silica composite exhibited a maximum power density of 0.55 W cm^{-2}, which exceeds that of commercial Pt/C (0.45

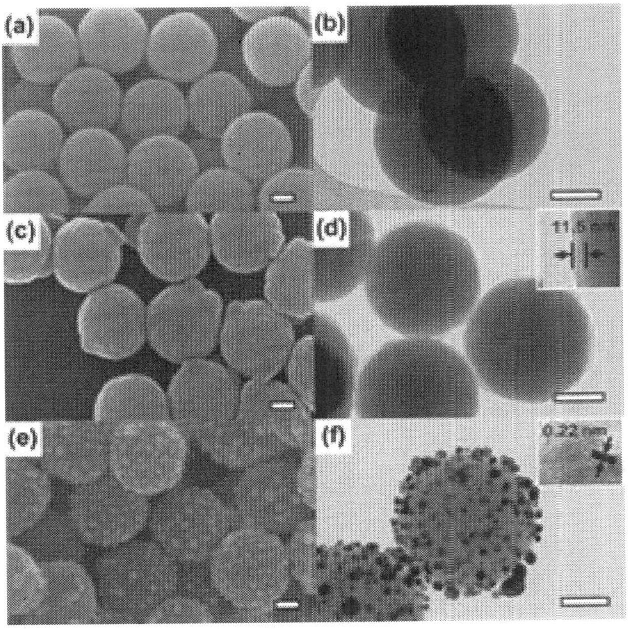

Figure 16. SEM and TEM images of (a and b) silica, (c and d) dopamine-coated silica, and (e and f) Pt and dopamine-induced carbon@silica. Scale bars are 100 nm.

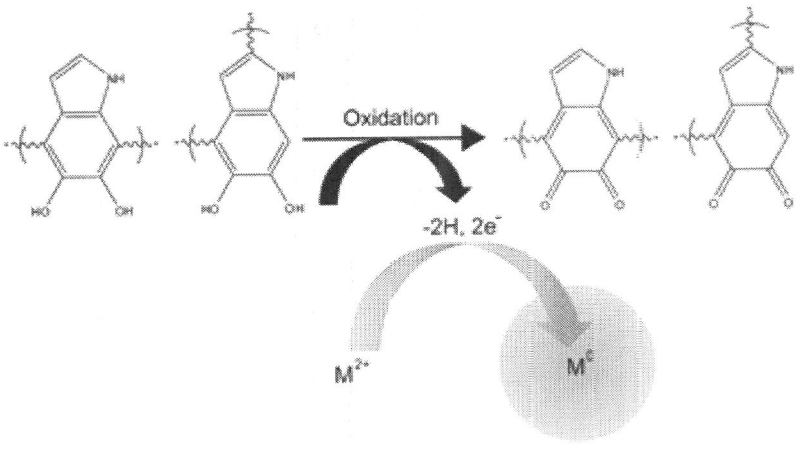

Figure 17. Reduction pathway from Pt ion into Pt nanoparticle using self-reducing ability of dopamine coating layer.

W cm^{-2}). The improved performance of the composite is attributed to the hybrid synergy effect of silica hygroscopic property and N-doped dopamine-induced carbon layer. Additionally, the N-doped carbon obtained after the carbonization step could make the composite more conductive with the enhancement of charge transfer at interfaces. As a result, these advantages of the Pt and dopamine-induced carbon@silica composite could contribute to enhance performance under low humidity conditions compared to the commercial Pt/C catalyst.

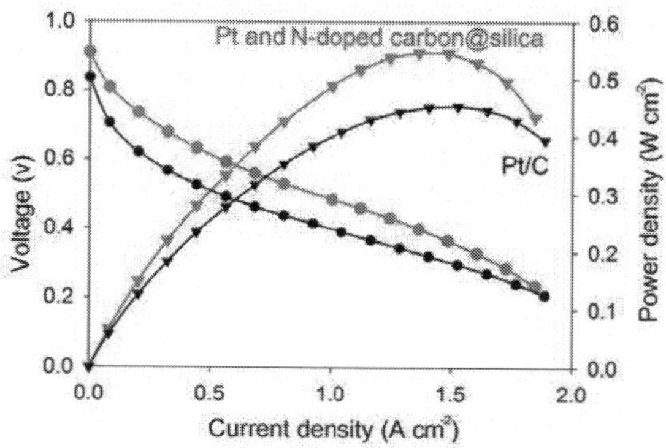

Figure 18. I-V polarization (left axis) and power density (right axis) curve for PEMFC single cell (H_2/O_2) with Pt and N-doped carbon@silica nanocomposites (red triangle) and commercial Pt/carbon (black circle) used as an anode catalyst under relative humidity at 0%.

4. Conclusion

We have presented various techniques to obtain novel catalysts that successfully enhance the efficiency and productivity of PET glycolysis for monomer recycling as well as the performance of polymer electrolyte membrane fuel cells. The catalysts can be considered a practical solution that addresses various issues in these processes, such as catalyst separability, reusability, and performance in limited operating conditions. The use of ultrasound-assisted synthesis provided an efficient alternative synthesis approach to obtain metal-oxide composites of silica and graphene oxide under mild conditions. The use of these catalyst composites offers the potential of industrial-scale use given their high activity in addition to thermal and chemical stability. Another practical and effective approach is the use of a magnetically recoverable catalyst, γ-Fe_2O_3. The superparamagnetic nanocatalyst, offering comparable performance and stability as the other solid catalysts, has further advantages of efficient separation and relatively simplicity of synthesis approach. For fuel cell development, we developed a non-toxic and easy chemical method to fabricate Pt and dopamine-induced carbon@silica composite without complex apparatus. The silica nanoparticles as support material acted as a self-humidifying material, while the dopamine coating played the role of a self-reducing agent and adhesive layer for anchoring the Pt nanoparticles. Additionally, the N-doped carbon from carbonization of dopamine coating enhanced electrical conductivity of the composite. This resulted to overall superior performance of the composite as anode catalyst even under low humidity conditions. The fabrication methods and alternative catalysis strategies presented will be valuable for the commercialization of green catalysts for PET monomer recycling and energy conversion systems.

Acknowledgements

This work was supported by the R&D Center for Valuable Recycling (Global-Top Environmental Technology Development Program) funded by the Ministry of Environment (Project No.: GT-11-C-01-250-0) and by Basic Science Research Program through the National Research Foundation of Korea (NRF) funded by the Ministry of Science, ICT & Future Planning (No. 2014R1A5A1009799).

Author details

Arvin Sangalang, Seunghwan Seok and Do Hyun Kim*

*Address all correspondence to: DoHyun.Kim@kaist.ac.kr

Department of Chemical and Biomolecular Engineering, KAIST, Daejeon, Republic of Korea

References

[1] Smithers P. Demand for PET packaging material in 2019 [Internet]. April 2014. Available from: http://www.smitherspira.com/news/2014/april/demand-for-pet-packaging-material-in-2019 [Accessed: 2015-10-03]

[2] Scheirs J, Long TE, editors. Modern Polyesters: Chemistry and Technology of Polyesters and Copolyesters. England: John Wiley & Sons Ltd; 2003. 750 p.

[3] Welle F. Twenty years of PET bottle to bottle recycling. Resources Conserv. Recycling. 2011;55(11):865–875. DOI: 10.1016/j.resconrec.2011.04.009

[4] Shen L, Worrell E, Patel MK. Open-loop recycling: a LCA case study of PET bottle-to-fibre recycling. Resources Conserv. Recycling. 2010;55(1):34–52. DOI: 10.1016/j.resconrec.2010.06.014

[5] Bartolome L, Imran M, Cho BG, Al-Masry WA, Kim DH. Recent developments in the chemical recycling of PET. In: Achilias D, editor. Material Recycling — Trends and Perspectives. INTECH; 2012. p. 65–84. DOI: 10.5772/33800

[6] George N, Kurian T. Recent developments in the chemical recycling of postconsumer poly(ethylene terephthalate) melts. Ind. Eng. Chem. 2014;53(37):14185–14198. DOI: 10.1021/ie501995m

[7] Lorenzetti C, Manaresi P, Berti C, Barbiroli G. Chemical recovery of useful chemicals from polyester (PET) waste for resource conservation: a survey of state of the art. J. Polym. Environ. 2006;14(1):89–101. DOI: 10.1007/s10924-005-8711-1

[8] Choi I, Lee KG, Ahn SH, Kim DH, Kwon OJ, Kim JJ. Sonochemical synthesis of Pt-deposited SiO_2 nanocomposite and its catalytic application for polymer electrolyte membrane fuel cell under low-humidity conditions. Catal. Commun. 2012;21:86–90. DOI: 10.1016/j.catcom.2012.02.005

[9] Seok S, Choi I, Lee KG, Choi BG, Park KJ, Park JY, Kwon OJ, Lee SJ, Kim DH. Dopamine-induced Pt and N-doped carbon@silica hybrids as a high-performance anode catalyst of polymer electrolyte membrane fuel cell. RSC Adv. 2014;4:42582–42584. DOI: 10.1039/C4RA06819J

[10] Imran M, Lee KG, Imtiaz Q, Kim BG, Han M, Kim DH. Metal oxide-doped silica nanoparticles for the catalytic glycolysis of polyethylene terephthalate. J. Nanosci. Nanotechnol. 2011;11(1):824–828. DOI: 10.1166/jnn.2011.3201

[11] Park G, Bartolome L, Lee KG, Lee SJ, Kim DH, Park TJ. One-step sonochemical synthesis of a graphene oxide-manganese oxide nanocomposite for catalytic glycolysis of poly(ethylene terephthalate). Nanoscale. 2012;4(13):3879–3885. DOI: 10.1039/C2NR30168G

[12] Bartolome L, Imran M, Lee KG, Sangalang A, Ahn JK, Kim DH. Superparamagnetic γ-Fe_2O_3 nanoparticles as easily recoverable catalyst for the chemical recycling of PET. Green Chem. 2014;16(1):279–286. DOI: 10.1039/C3GC41834K

[13] Imran M, Kim DH, Al-Masry WA, Mahmood A, Hassan A, Haider S, Ramay SM. Manganese-, cobalt-, and zinc-based mixed-oxide spinels as novel catalysts for the chemical recycling of poly(ethylene terephthalate) via glycolysis. Polym. Degrad. Stabil. 2013;98(4):904–915. DOI: 10.1016/j.polymdegradstab.2013.01.007

[14] Min BK, Santra AK, Goodman DW. Understanding silica-supported metal catalysts: Pd/silica as a case study. Catal. Today. 2003;85(2):113–124. DOI: 10.1016/S0920-5861(03)00380-8

[15] Juszczyk W, Karpiński Z. Characterization of supported palladium catalysts; II. $PdSiO_2$. J. Catal. 1989;117(2):519–532. DOI: 10.1016/0021-9517(89)90361-8

[16] Xie Z, Liu Z, Wang Y, Yang Q, Xu L, Ding W. An overview of recent development in composite catalysts from porous materials for various reactions and processes. Int. J. Mol. Sci. 2010;11(5):2152–2187. DOI: 10.3390/ijms11052152

[17] Stankovich S, Dikin DA, Dommett GHB, Kohlhaas KM, Zimney EJ, Stach EA, Piner RD, Nguyen ST, Ruoff RS. Graphene-based composite materials. Nature. 2006;442(7100):282–286. DOI: 10.1038/nature04969

[18] Stankovich S, Piner RD, Nguyen ST, Ruoff RS. Synthesis and exfoliation of isocyanate-treated graphene oxide nanoplatelets. Carbon. 2006;44(15):3342–3347. DOI: 10.1016/j.carbon.2006.06.004

[19] Muszynski R, Seger B, Kamat PV. Decorating graphene sheets with gold nanoparticles. J. Phys. Chem. C. 2008;112(14):5263–5266. DOI: 10.1021/jp800977b

[20] Bai H, Li C, Wang X, Shi G. A pH-sensitive graphene oxide composite hydrogel. Chem. Commun. 2010;46(14):2376–2378. DOI: 10.1039/C000051E

[21] Suslick KS, Price GJ. Applications of ultrasound to materials chemistry. Annu. Rev. Mater. Sci. 1999;29(1):295–326. DOI: 10.1146/annurev.matsci.29.1.295

[22] Okitsu K, Bandow H, Maeda Y. Sonochemical preparation of ultrafine palladium particles. Chem. Mater. 1996;8(2):315–317. DOI: 10.1021/cm950285s

[23] Park G, Lee KG, Lee SJ, Park TJ, Wi R, Kim DH. Synthesis of graphene-gold nanocomposites via sonochemical reduction. J. Nanosci. Nanotechnol. 2011;11(7):6095–6101. DOI: 10.1166/jnn.2011.4446

[24] Lee KG, Wi R, Park TJ, Yoon SH, Lee J, Lee SJ, Kim DH. Synthesis and characterization of gold-deposited red, green and blue fluorescent silica nanoparticles for biosensor application. Chem. Commun. 2010;46(34):6374–6376. DOI: 10.1039/C0CC00762E

[25] Doktycz SJ, Suslick KS. Interparticle collisions driven by ultrasound. Science. 1990;247:1067–1069. DOI: 10.1126/science.2309118

[26] Suslick KS. Sonochemistry. Science. 1990;247:1439–1445. DOI: 10.1126/science.247.4949.1439

[27] Wang H, Cui LF, Yang Y, Casalongue HS, Robinson JT, Liang Y, Cui Y, Dai H. Mn_3O_4-graphene hybrid as a high-capacity anode material for lithium ion batteries. J. Am. Chem. Soc. 2010;132(40):13978–13980. DOI: 10.1021/ja105296a

[28] Yan J, Fan Z, Wei T, Qian W, Zhang M, Wei F. Fast and reversible surface redox reaction of graphene-MnO_2 composites as supercapacitor electrodes. Carbon. 2010;48(13):3825–3833. DOI: 10.1016/j.carbon.2010.06.047

[29] Chen S, Zhu J, Wang X. From graphene to metal oxide nanolamellas: a phenomenon of morphology transmission. ACS Nano. 2010;4(10):6212–6218. DOI: 10.1021/nn101857y

[30] Yin J, Gao F, Wu Y, Wang J, Lu Q. Synthesis of Mn_3O_4 octahedrons and other manganese-based nanostructures through a simple and green route. CrystEngComm. 2010;12(11):3401–3403. DOI: 10.1039/C003551N

[31] Liu B, Thomas PS, Williams RP, Donne SW. Thermal characterisation of chemically reduced electrolytic manganese dioxide. J. Thermal Anal. Calorim. 2005;80(3):625–629. DOI: 10.1007/s10973-005-0704-8

[32] Huang H, Wang X. Graphene nanoplate-MnO_2 composites for supercapacitors: a controllable oxidation approach. Nanoscale. 2011;3(8):3185–3191. DOI: 10.1039/C1NR10229J

[33] Qiu LG, Li ZQ, Wu Y, Wang W, Xu T, Jiang X. Facile synthesis of nanocrystals of a microporous metal-organic framework by an ultrasonic method and selective sensing of organoamines. Chem. Commun. 2008;31:3642–3644. DOI: 10.1039/B804126A

[34] Offeman RE, Hummers WS. Preparation of graphitic oxide. J. Am. Chem. Soc. 1958;80:1339.

[35] López-Fonseca R, Duque-Ingunza I, De Rivas B, Arnaiz S, Gutiérrez-Ortiz JI. Chemical recycling of post-consumer PET wastes by glycolysis in the presence of metal salts. Polym. Degrad. Stabil. 2010;95(6):1022–1028. DOI: 10.1016/j.polymdegradstab.2010.03.007

[36] Wang H, Li Z, Liu Y, Zhang X, Zhang S. Degradation of poly (ethylene terephthalate) using ionic liquids. Green Chem. 2009;11(10):1568–1575. DOI: 10.1039/B906831G

[37] Wang H, Liu Y, Li Z, Zhang X, Zhang S, Zhang Y. Glycolysis of poly (ethylene terephthalate) catalyzed by ionic liquids. Eur. Polym. J. 2009;45(5):1535–1544. DOI: 10.1016/j.eurpolymj.2009.01.025

[38] Wang H, Yan R, Li Z, Zhang X, Zhang S. Fe-containing magnetic ionic liquid as an effective catalyst for the glycolysis of poly (ethylene terephthalate). Catal. Commun. 2010;11(8):763–767. DOI: 10.1016/j.catcom.2010.02.011

[39] Yue QF, Wang CX, Zhang LN, Ni Y, Jin YX. Glycolysis of poly (ethylene terephthalate) (PET) using basic ionic liquids as catalysts. Polym. Degrad. Stabil. 2011;96(4):399–403. DOI: 10.1016/j.polymdegradstab.2010.12.020

[40] Wi R, Imran M, Lee KG, Yoon SH, Cho BG, Kim DH. Effect of support size on the catalytic activity of metal-oxide-doped silica particles in the glycolysis of polyethylene terephthalate. J. Nanosci. Nanotechnol. 2011;11(7):6544–6549. DOI: 10.1166/jnn.2011.4393

[41] Wang Q, Yao X, Tang S, Lu X, Zhang X, Zhang S. Urea as an efficient and reusable catalyst for the glycolysis of poly (ethylene terephthalate) wastes and the role of hydrogen bond in this process. Green Chem. 2012;14(9):2559–2566. DOI: 10.1039/C2GC35696A

[42] Koukabi N, Kolvari E, Khazaei A, Zolfigol MA, Shirmardi-Shaghasemi B, and Khavasi HR. Hantzsch reaction on free nano-Fe_2O_3 catalyst: excellent reactivity combined with facile catalyst recovery and recyclability. Chem. Commun. 2011;47(32):9230–9232. DOI: 10.1039/C1CC12693H

[43] Garade AC, Bharadwaj M, Bhagwat SV, Athawale AA, Rode CV. An efficient γ-Fe_2O_3 catalyst for liquid phase air oxidation of p-hydroxybenzyl alcohol under mild conditions. Catal. Commun. 2009;10(5):485–489. DOI: 10.1016/j.catcom.2008.10.044

[44] Picasso G, Quintilla A, Pina MP, Herguido J. Total combustion of methyl-ethyl ketone over Fe_2O_3 based catalytic membrane reactors. Appl. Catal. B: Environ. 2003;46(1):133–143. DOI: 10.1016/S0926-3373(03)00219-4

[45] Kang YS, Risbud S, Rabolt JF, Stroeve P. Synthesis and characterization of nanometer-size Fe_3O_4 and γ-Fe_2O_3 particles. Chem. Mater. 1996;8(9):2209–2211. DOI: 10.1021/cm960157j

[46] Jeong, JR, Lee SJ, Kim JD, Shin SC. Magnetic properties of γ-Fe_2O_3 nanoparticles made by coprecipitation method. Phys. Stat. Solidi B Basic Res. 2004;241(7):1593–1596. DOI: 10.1002/pssb.200304549

[47] Henry CR. Catalysis by nanoparticles. In: Heiz U, Landman U, editors. Nanocatalysis. Heidelberg: Springer Berlin/Heidelberg; 2007. p. 245–268. DOI: 10.1007/978-3-540-32646-5

[48] Ghose J, Murthy K. Activity of Cu^{2+} ions on the tetrahedral and octahedral sites of spinel oxide catalysts for CO oxidation. J. Catal. 1996;162(2):359–360. DOI: 10.1006/jcat.1996.0293

[49] Guilhaume N, Primet M. Catalytic combustion of methane: copper oxide supported on high-specific-area spinels synthesized by a sol-gel process. J. Chem. Soc. Faraday Trans. 1994;90(11):1541–1545. DOI: 10.1039/FT9949001541

[50] Słoczyński J, Ziółkowski J, Grzybowska B, Grabowski R, Jachewicz D, Wcisło K, Gengembre L. Oxidative dehydrogenation of propane on $Ni_xMg_{(1-x)}Al_2O_4$ and $NiCr_2O_4$ spinels. J. Catal. 1999;187(2):410–418. DOI: 10.1006/jcat.1999.2626

[51] Niu X, Du W, Du W. Preparation and gas sensing properties of ZnM2O4 (M=Fe, Co, Cr). Sens. Actuators B: Chem. 2004;99(2):405–409. DOI: 10.1016/j.snb.2003.12.007

[52] Ferraris G, Fierro G, Jacono ML, Inversi M, Dragone R. A study of the catalytic activity of cobalt-zinc manganites for the reduction of NO by hydrocarbons. Appl. Catal. B: Environ. 2002;36(4):251–260. DOI: 10.1016/S0926-3373(01)00289-2

[53] Mokhtar M, Basahel SN, Al-Angary YO. Nanosized spinel oxide catalysts for CO-oxidation prepared via CoMnMgAl quaternary hydrotalcite route. J. Alloys Compounds. 2010;493(1):376–384. DOI: 10.1016/j.jallcom.2009.12.106

[54] Zhang L, Wang L, Holt CMB, Zahiri B, Li Z, Malek K, Navessin T, Eikerling MH, Mitlin D. Highly corrosion resistant platinum-niobium oxide-carbon nanotube electrodes for the oxygen reduction in PEM fuel cells. Energy Environ. Sci. 2012;5(3):6156–6172. DOI: 10.1039/C2EE02689A

[55] Long D, Li W, Qiao W, Miyawaki J, Yoon SH, Mochida I, Ling L. Partially unzipped carbon nanotubes as a superior catalyst support for PEM fuel cells. Chem. Commun. 2011;47(33):9429–9431. DOI: 10.1039/C1CC13488D

[56] Hong D, You I, Lee H, Lee S-G, Choi IS, Kang SM. Polydopamine circle-patterns on a superhydrophobic AAO surface: water-capturing property. Bull. Kor. Chem. Soc. 2013;34(10):3141–3142. DOI: 10.5012/bkcs.2013.34.10.3141

[57] Ryu S, Chou JB, Lee K, Lee D, Hong SH, Zhao R, Lee H, Kim SG. Direct insulation-to-conduction transformation of adhesive catecholamine for simultaneous increases of electrical conductivity and mechanical strength of CNT fibers. Adv. Mater. 2015;27(21):3250–3255. DOI: 10.1002/adma.201500914

[58] Lee M, Rho J, Lee D-E, Hong S, Choi SJ, Messersmith PB, Lee H. Water detoxification by a substrate-bound catecholamine adsorbent. ChemPlusChem. 2012;77(11):987–990. DOI: 10.1002/cplu.201200209

[59] Lee H, Dellatore SM, Miller WM, Messersmith PB. Mussel-inspired surface chemistry for multifunctional coatings. Science. 2007;318(5849):426–430. DOI: 10.1126/science.1147241

[60] Hong S, Lee JS, Ryu J, Lee SH, Lee DY, Kim D-P, Park CB, Lee H. Bio-inspired strategy for on-surface synthesis of silver nanoparticles for metal/organic hybrid nanomaterials and LDI-MS substrates. Nanotechnology. 2011;22(49):494020. DOI: 10.1088/0957-4484/22/49/494020

[61] Kang SM, Park S, Kim D, Ruoff RS, Lee H. Simultaneous reduction and surface functionalization of graphene oxide by mussel-inspired chemistry. Adv. Funct. Mater. 2011;21(1):108–112. DOI: 10.1002/adfm.201001692

Section 3

Recent Progress in Catalytic Materials for Photocatalytic Applications

Chapter 7

Advanced Nanomatericals for Solar Photocatalysis

Le Li and Minqiang Wang

Additional information is available at the end of the chapter

http://dx.doi.org/10.5772/62206

Abstract

Heterogeneous photocatalysis using semiconductors and renewable solar energy has been regarded as one of the most promising processes to alleviate and even solve both the world crises of energy supply and environmental pollution. Recently, numerous semiconducting materials and its composites have been studied for their photocatalytic applications. In this chapter, we briefly summarize recent progress in the binary oxide system (including TiO_2 and α-Fe_2O_3), ternary oxide (Bi system), and the semiconducting materials and their composites which have remarkable applications in photocatalytic degradation of toxic pollutants, hydrogen production and as an adsorbent for wastewater treatment. In addition, we highlight the challenges and opportunities when we implement photocatalytic materials to help on the development of energy research and find ways to approach major problems.

Keywords: Photocatalysis, TiO_2, α-Fe_2O_3, $BiVO_4$, Bi_2WO_4, BiOX, Binary oxide, Ternary oxide

1. Introduction

Ever-increasing environmental issues and consumption of fossil fuels have stimulated extensive research on the utilization of sustainable solar energy [1]. The extensive use of fossil fuels has led to a serious energy crisis and environmental pollution, which are the two major challenges facing the world in the 21st century. Among the many advanced technologies available today, heterogeneous photocatalysis in view of semiconductors taking advantage of regeneration solar energy has been identified as one of the most prospective strategies for resolving both the environmental and energy problems and has thus caused much attention during the recent decades [2–4]. In the past decades, numerous results have been reported for photocatalysis and their applications to produce hydrogen from water (see Fig.1 a) [5–8], convert solar energy into electric energy (see Fig.1 b) [9,10], degrade organic pollutants (see

Fig.1 c) [11–13] and reduce CO_2 into organic fuels (see Fig.1 d) [13–15]. Besides the naturally abundant in nonrenewable energy sources such as solar energy can be renewed into chemical or electrical and thermal energies by using semiconductors having persisting materials in the process of photocatalysis [17-20]. Generally speaking, the mechanism of a typical power-driven photocatalysis process is mainly owing to three critical related synergistic steps: (i) light absorption and charge excitation; (ii) charge separation and transport from the semiconductor particle to its surface active sites; (iii) surface photocatalytic chemical reactions, and this process is similar to the fundamental mechanism of photocatalysis in power systems.

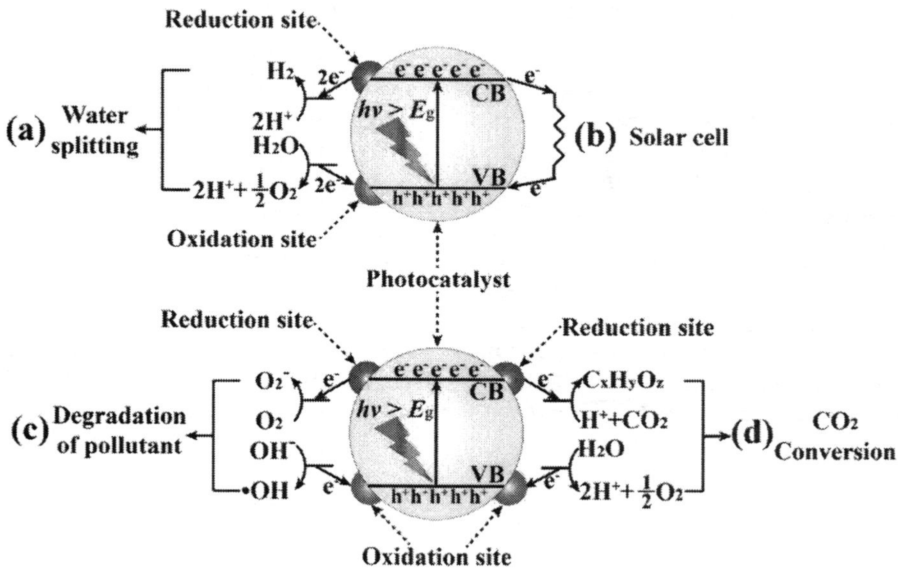

Figure 1. Photocatalytic mechanisms of water splitting, solar cell, degradation of pollutants, CO_2 reduction via one-step photoexcitation. CB and VB represent the conduction and valence bands, respectively [16].

Typically, the electron-hole pair with specific reduction and oxidation potential will be created on its conduction band (CB) and valence band (VB) under the irradiation of incident light with energy greater than the band gap of a given semiconductor. Here, the band gap of the semiconductor determines the utilization rate of the energy of the incident light, and the CB and VB values are the origin of the reduction and oxidation abilities of the photoexcited electrona and holes [21]. However, in practical process, the performance of photocatalysts is mainly related to two conditions: (i) the energy (*hv*) of the incident photon should be larger than the energy gap (*Eg*) of the photocatalyst; (ii) the redox potential of reactants should be located between the CB and VB of the semiconductor photocatalyst. One the one hand, the former condition indicates a narrow band gap, which can facilitate the efficient utilization of incident solar-light. On the other hand, the latter condition demonstrate that a more higher CB potential and a more lower VB potential, which are thermodynamically beneficial for the reduction and oxidation reactions of the reactants, respectively. But a high CB and low VB potential means a broad band gap of a photocatalyst, which leads to the poor solar–light utilization as discussed in condition (i). It is obvious that these conditions above are mutually contradiction, and it is important to find the balance point to design the photocatalyst. However, for a single component photocatalyst, it is difficult to possess both wide light –absorption range and strong redox ability concurrently. Besides, in the single-component structure, the photogenerated electrons in the CB can easily return to the VB or trap in the defect state and recombine with the holes, which seriously reduces the utilization efficiency of solar energy [22-24]. Hence, designing appropriate heterogeneous photocatalytic systems should be an effective way to overcome this problem.

The chapter is divided into four main sections. In the first part, we describe the importance of binary oxide system photocatalytic matericals in the case of two prominent and widely studied matel oxides: Titania (TiO_2), hematite (α-Fe_2O_3). In the second part, we do focus on materials with a specific ternary oxide photocatalytic matericals, such as Bi systems photocatalytic materials. In the third part, we discuss the semiconducting materials and its composites which have promising applications in the area of energy and environment especially in photocatalysis. In the end, we highlight the challenges and opportunities on the way to implement photocatalytic materials to help on the development of energy research and finding ways to approach for the major problems. Hence, we believe that a comprehensive chapter on advanced nanomatericals for solar photocatalytic is desirable for the further development of the novel photocatalytic materials and deeper understanding of photocatalytic mechanisms will be achieved in the near future, through more fundamental interdisciplinary research.

2. Binary oxide system

2.1. TiO_2

TiO_2 has turned out to be one of the most commonly investigated semiconductors due to its low cost, long–term thermodynamic stability in aqueous solution, low toxicity, and high efficiency in the removal of pollutants in water and air as well as hydrogen generation [25-28]. There following are the four commonly known polymorphs of TiO_2 found in nature: anatase (tetragonal), rutile (tetragonal), brookite (orthorhombic), and TiO_2(B) (monoclinic) [29,30]. Rutile TiO_2 has a tetragonal crystal structure and contains six atoms per unit cell as shown in Fig. 2 [31]. Rutile is the most thermodynamically stable polymorph of TiO_2 at all temperatures, exhibiting lower total free energy than metastable phases of anatase and brookite. Anatase TiO_2 has a crystalline structure that corresponds to the tetragonal system but the distortion of the TiO_6 octahedron is slightly larger for the anatase phase. Anatase is the most commonly used in photocatalytic applications due to its inherent superior photocatalytic properties [32-34].

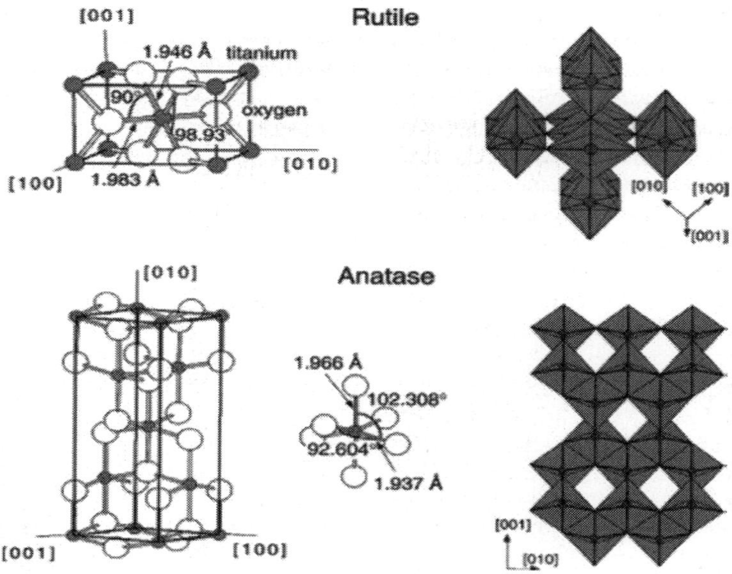

Figure 2. Representations of the TiO_2 anatase and rutile forms [31].

Anatase is the least thermodynamically stable TiO_2 polymorph as a bulk phase, although, from energy calculations, it appears as the most stable phase when the grain size is below 10–20 nm [35,36]. The crystalline structure of the TiO_2 oxides can be described in terms of TiO_6 octahedral chains. These differ by the distortion of each octahedron and the assembly pattern of the resulting octahedral chains. The Ti–Ti distances in the anatase structure are greater than in rutile, while the Ti–O distances are shorter [37]. These structural differences lead to different mass densities as well as different electronic band structures. As a result, the anatase phase is 9% less dense than rutile and it presents more pronounced localization of the Ti 3d states and further a narrower 3d band. Also, the O 2p–Ti 3d hybridization is different in the two structures. Anatase exhibits a valence and conduction band with more pronounced O 2p–Ti 3d character and less nonbonding self-interaction between similar ions (e.g., anion-anion and cation-cation interactions) [38]. The importance of the covalent vs ionic contributions to the metal-oxygen bond has already been discussed in a more general context for Ti oxides [39,40]. Therefore, it could be claimed that differential structural characteristics between anatase and rutile of TiO_2 are possibly attributed to the difference in the mobility of the charge carriers upon light excitation.

In 1972, K. Honda and A. Fujishima discovered the photosensitization effect of a TiO_2 electrode on the electrolysis of H_2O into H_2 and O_2 using a Pt metal electrode as cathode and a TiO_2 photoanode irradiated with UV light. They found that, under UV light irradiation of the TiO_2 electrode, the electrolysis of H_2O proceeded at a much lower bias voltage as compared with normal electrolysis [41]. From then on, the TiO_2-based photocatalyst have been extensively studied in the past few decades due to its proper energy bandgap that matches the UV –visible light irradiation, which favors many light-driven applications [42-47]. Moreover, TiO_2 has got many advantages and the nature of this material is naturally abundant, commercially available, economically viable, chemically stable, non-toxic and environmental eco –friendly [44]. However, TiO_2 has also faced few problems as photocatalysts in applying solar energy processes due to its low sunlight spectrum matching, limited activity and reduced sensitivity [48]. To overcome this shortage, recently many researchers have developed many different modification methods to TiO_2 material to make it as a potential challenging material for highly active photocatalyst [48-55]. Among these works, crystal growth, doping and heterostructuring of semiconductor photocatalysts are commonly used and can substantial tune the light-response range, redox potentials of photoinduced charge carriers, and electron-hole pair separation probability within the photocatalysts. Specifically, crystal growth can be critical in controlling the phase, shape, and size of photocatalysts, as well as their crystallinity and specific surface area. By rationally controlling crystal growth, the intrinsic surface atomic structure and resultant surface states of the derived photocatalysts can be adjusted. For materials design, doping effect can exert a substantial influence on modifying the electronic structure and the construction of heteroatomic surface structures of the aiming material. In particular, nonmetal doping (N [56,57], C [58-60], S [61,62], B [63-65], F [66-68], Br [69], I [70-73], P [74]) in photocatalyst has attracted increasing attention due to its effectiveness in realizing visible–light photocatalytic activity of wide bandgap semiconductor photocatalysts. The chemical states and locations of dopants are considered to be key factors in adjusting the spectral distribution of the induced electronic states of those dopants and reconstructing favorable surface structure for photocatalysis. The hybrids of two or more semiconductor systems, that is, heterostructures, seem to be possess advantageous in more efficiently utilizing solar light by combining different electronic structures when compared with sing-phase semiconductor photocatalysts. Furthermore, an efficient photo-excited electron or hole transfer from one component to another with proper band edge matching can greatly decrease the electron-hole recombination probability and increase the lifetime of charge carriers, which further promoting the photocatalytic efficiency. In addition to the basic requirements of electronic structure for each unit in the integrated photocatalytic systems a favorable interface contact between the two materials is essential in promoting interface charge carrier transfer through different pathways. Fig. 3 demonstrates the connection between crystal growth, doping and hetero-structure of semiconductors for heterogeneous photocatalysis, (CB: conduction band; VB: valence band) [75].

According to the Wulff construction and calculated surface energy, the shape of anatase under equilibrium conditions is a slightly truncated tetragonal bipyramid enclosed with eight isosceles trapezoidal surfaces of {101} and two top squares of {001}, as shown in Fig. 4 [76] It is predicted that the percentage of {101} is as high as 94%. Although the surface energy of {010} (0.53 J m $^{-2}$) was calculated to be between {001} (0.90 J m^{-2}) and {101} (0.44 J m $^{-2}$) [77], it is surprising that no {010} will appears in the equilibrium shape of anatase. Anatase TiO_2 is usually exposed with low-index facets. Theoretical calculations indicate that the (101) surface (0.44 J m^{-2}) is the thermodynamically the most stable surface, the (001) surface (0.90 J m^{-2}) is

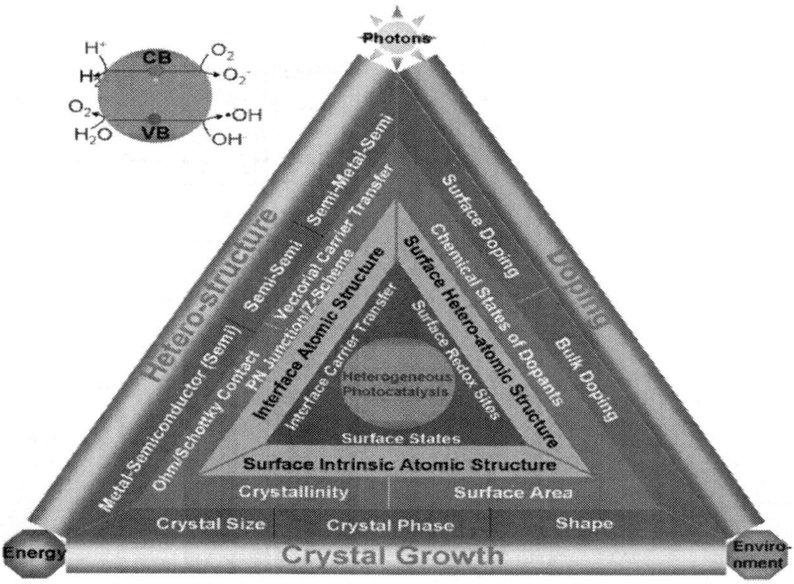

Figure 3. Correlation of key factors in crystal growth, doping and heterostructuring of semiconductors for photocatalysis. (CB: conduction band; VB:valence band) [75].

the most active and the (100) and (110) surfaces are between the (101) and (001) surfaces. As a consequence, facets that have a high surface energy diminish quickly in the minimization of surface energy during the crystal-growth process. Therefore, a large percentage of high active facets has become a popular target in the synthesis of anatase TiO_2 crystals. In the case of rutile, the predicted equilibrium shape of a macroscopic crystal was constructed with (110), (100), (001) and (011) faces (see Fig. 4) [78]. It is found that in the equilibrium shape, the most stable (110) face with the lowest surface energy of 15.6 meV au^{-2} dominates the shape, whereas (001) with the highest surface energy of 28.9 meV au^{-2} does not exist at all. Gong et al. demonstrated the systematic results of the structures and energetics of 10 stoichiometric 1×1 low-index surfaces with different possible terminations of brookite [79]. The determining factors of the relative stabilities of different faces are found to be negatively related to the concentration of exposed coordinatively unsaturated Ti atoms. The equilibrium shape of brookite crystal is shown in Fig. 4, we can observe that the most of it is composed of (111), (210), (010) and reconstructed (001) facets. It is worth noting that brookite (210) is one of the most stable facets, which has a very similar atomic structure to the most stable facet (101) of anatase. However, their electronic states are different, which may result in different chemical reactivities [80].

Usually, different facets of a single–crystalline material possess distinctive adsorption, catalytic reactivity and selectivity, which are caused by its different geometric and electronic structures [81]. Since Lu and his coworkers first reported that the uniform anatase single

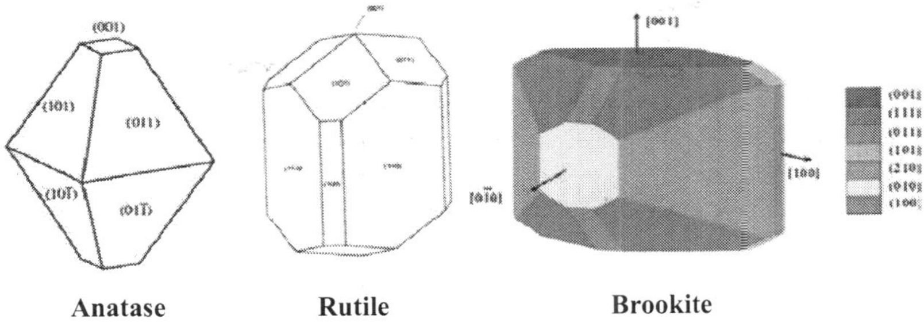

Figure 4. The equilibrium shape of a TiO_2 crystal in the anatase, rutile and brookite, according to the Wulff construction and the calculated surface energies [76, 78,79].

crystals with 47% {001} facets displayed superior photoactivity [82], crystal facet engineering has proven to be an effective strategy to finely tune the efficiency and selectivity of heterogeneous photocatalysts for different applications. Besides, different crystal facets can also facilitate the separation of electrons and holes [83]. To date, many improved synthesis procedures have been successfully developed and lots of exciting advances have been achieved [84-99]. Lu and coworkers demonstrated that, under UV light irradiation, the sheet-like anatase TiO_2 crystal dominated by {001} facets is capable of producing OH that is more than five times higher than that of Degussa P25 TiO_2 [85]. They concluded that the high density unsaturated five-fold Ti and their unique electronic structures of the {001} facets should be responsible for the improved photoactivity. Similar results also have been reported by other groups. For example, Han et al. [84] reported that the photocatalytic ability of TiO_2 nanosheets with {001} facets was higher than that of P25 in the degradation of methyl orange(MO) molecules. Zhang et al. [86] successfully synthesized the a remarkable 80% level of reactive {001} facets microsheet anatase TiO_2 single–crystal photocatalyst, which exhibited much better photocatalytic performance in the oxidative decomposition of organic pollutant. By tuning the percentage of the {001} facets, the photoreactivity was enhanced from 40.0% to 84.5%. The reactive {001} facets played an important role in the photocatalytic reaction owing to their strong ability to dissociatively adsorb water to form hydrogen peroxide and peroxide radicals. Although high-energy {001} facets have been widely studied, anatase TiO_2 crystals with higher-energy {100} facets have been less well developed. Recently, Li and Xu [100] reported a facile hydrothermal route for the synthesis of tetragonal–faceted nanorods (NRs) of anatase TiO_2 with highly exposed higher-energy {100} facets, which exhibited higher reactivity owing to the large percentage of {100} facets compared with crystals that have normal majority {101} facets.

These results discussed above demonstrated that a higher density of surface–unsaturated atoms will lead to a high surface energy of the crystal facets, which generally exhibit better photocatalytic performance. However, recent studies have shown that a high surface energy does not always make the crystal facets highly reactive in photocatalytic reactions. For example, Liu et al. [97] demonstrated a raised conduction band of nanosized single crystals of anatase TiO_2 with 82% {101} facets compared to the crystals with 72% {001} facets, which is determined by UV/Vis adsorption spectroscopy. This different electronic–band difference will further lead to a difference in atomic coordination, and this decrease will result in an enhanced photoactivity in the splitting of water into hydrogen. This example shows that the band-gap of crystal facets or crystal plates will change as the change in the arrangement of surface atoms. As a result, the redox power of the photoexcited electrons and holes will be correspondingly changed.

Generally speaking, the {101} facets are more reductive than {001} facets, which could act as possible tanks of photogenerated electrons, while {001} facets act as oxidation sites, which play a major role in the photooxidative processes [101,102]. For example, Pan et al. reported that low–index facets of anatase TiO_2 follow the photoreactivity order of {001} < {101} < {010} for photocatalytic hydrogen evolution and·OH radical generation [103]. Similarly, a seeded growth technique also demonstrated that the {101} facets of anatase TiO_2 are more active than the {001} facets for photocatalytic water splitting [104]. Surprisingly, it was even found that the photocatalytic activity for H_2 production over the {111} facet exposed anatase TiO_2 is about 5, 9, and 13 times higher than that of the TiO_2 sample exposed with dominant {010}, {101}, and {001} facets, respectively [105]. However, most researchers ignored the synergetic effects of various co-exposed facets in one sample. More attention has to be paid to finding special facets rather than the balanced ratio of different exposed facets for the best photocatalytic efficiency of water splitting.

Recently, Yu's group found that an optimal ratio of the exposed {101} and {001} facets of TiO_2 played a significant role in the enhancement of photocatalytic performance for the reduction of CO_2 [106]. As shown in Fig. 5, the surplus electrons on the {101} facets will overflow onto the {001} facets and then have a fast recombination with the holes on the {001} facets if the percentage of {101} facets is too low to hold all the photoexcited electrons, this process will lead to a decrease in the photocatalytic activity. The results clearly showed that it is of great importance to find the balanced ratio of different exposed facets in achieving the best photo-catalytic efficiency [107]. This finding may shed light on the design and fabrication of advanced nanosheet-based semiconductors for water splitting.

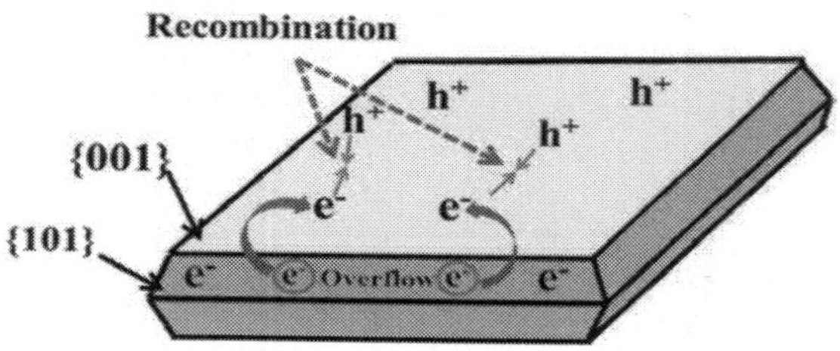

Figure 5. The electron overflow effect on the {101} facets of TiO_2 [107].

In 2011, crystal facet dependence of TiO_2 photocatalysis has been evidenced by using single-molecule imaging and kinetic analysis [108]. Single–particle spectroscopy (microscopy) has been used to explore the structural and kinetic features of "bulk" catalysis because of its high sensitivity and selectivity. This study demonstrated that the reaction sites for the effective reduction of the probe molecules were preferentially located on the {101} facets of the crystal rather than on the high surface energy {001} facets. This preference originated from the unique properties of the {101} facets in terms of their electron-trapping probability induced by the specific facet. This observation emphasizes the important role of the {101} surfaces as the reductive site in TiO_2 photocatalysis and is in agreement with the conclusion that the reactivity of the {001} facets towards oxidation is higher than that of the {101} facets.

This aforementioned investigation shows that the reactivity of a photocatalyst can be controlled by tuning the exposed facets. Another tool-morphology control, which means the formation of different surface facets with different surface atomic structures, also provides an effective method to tune the selectivity of the photocatalysts. For example, Liu et al. [95] reported a fluoride-mediated self-transformation method for fabricating hollow TiO_2 microspheres (HTS) with anatase polyhedra with about 20% exposed {001} facets. The fluorinated HTS exhibited preferential decomposition of methyl orange (MO) compared with methyl blue (MB). In contrast, surface-modified HTS that was either washed with NaOH or calcined at 600°C favored the decomposition of MB over MO. This example demonstrated the importance of the surface structure in modifying the catalytic selectivity of titania. Therefore, it is expected that, by controlling the exposed facets, we can design photocatalysts with both high reactivity and high selectivity. The photocatalysts that are terminated by specific facets allow the same adsorption states of reactant molecules and generate photoexcited electrons of similar energies on the specific facets. It is worth noting that these properties will be beneficial for the solar-induced selective photoconversion of carbon dioxides into specific valuable fuels because this process typically requires an undesired separation process and produces mixed hydrocarbons, including CH_4, CH_3OH, and HCOOH. It is believed that the breakthrough in making specific facets will intensify the development of selective organic transformations that are based on semiconductor photocatalysts.

High-index facets of nanomaterials usually have unique surface atomic structures, such as a high density of atomic steps, dangling bonds, kinks, and ledges, which can all act as active sites. Unfortunately, these unique surface atomic structures always have a high surface energy and high crystal growth rate, which is not naturally preferential growth and is easy to rapidly diminish during the crystal-growth process, so it is quite challenging to synthesize tailor–made crystals.

Yang and coworkers first reported the formation of anatase TiO_2 crystals that are exposed by high–index {105} facets [109]. They produced the product with well–faceted surface by a modified high–temperature gas–phase oxidation route with titanium tetrachloride ($TiCl_4$) as the Ti source. During the $TiCl_4$ oxidization process, the co–adsorption of oxygen, chlorine, or other related species will occur and may specifically lower the Gibbs free energy of the {105} facets thus the typical atomic configuration on the {105} facets can be stabilized and reserved. The unique stepped atomic configuration on the high–index {105} facets makes these materials promising candidates in the areas of renewable clean energy and environmental remediation.

Rutile is the thermodynamically stable phase of TiO_2 polymorphs, which can be obtained by typically three methods: (i) the hydrolysis of Ti precursors and subsequent crystallization; (ii) the post–transformation from anatase/brookite phase via thermal treatment (phase transformation temperature required depends on the particle size of TiO_2) [110] and (iii) mechanical processing [111]. Although rutile is considered to be less active in photocatalytic reactions compared to anatase, nanostructured rutile has also been used photocatalysis applications and in some cases show even higher activity than anatase. Band gap of rutile TiO_2 is 0.2 eV smaller than anatase one and further results in a wider absorption range, which may be the advantage of this phase.

Afterwards, various morphologies of rutile have been developed [112-119], with the nanorod being is a common morphology. The synthesis routes of such rutile nanorods with a high aspect ratio have been well documented in the literature [114,119,120-129]. Generally speaking, the presence of Cl ions as mineralizer in the synthesis system is favourable for rutile TiO_2, regardless of the source of Cl. In the case of the specific synthesis routes of controlling morphology of rutile, there are two representative examples demonstrate the formation of faceted rutile crystals. One is the rapid formation of self-assembled microspheres with rutile nanorods by microwave heating of $TiCl_3$ at 200°C for only 1 min [115]. The nanorods are exposed with {110} and {111} facets, but because of the extremely rapid growth rate, the surface is not smooth. Interestingly, the synthetic rutile nanorods have a smaller bandgap of 2.8 eV compared with the conventional 3.0 eV, which may facilitate the photocatalysis ability under visible light irradiation. The other one is reported by Kakiuchi et al. [116], who observed the dependence of degree of perfection of facets on hydrothermal temperature, where $TiCl_3$ was also used as a precursor together with NaCl additive. For example, at low temperature (80°C), only needle–like nanorods without well–recognized facets were formed. However, when elevating the temperature to 200°C, well–developed lateral {110} and top {111} facets can be observed. Apparently, this result indicates that a higher temperature is favorable for growing crystals with well-developed facets.

Compared with anatase and rutile, brookite phase TiO_2 has attracted little interest due to the generally considered lack of photocatalytic activity. However, increasing literatures have shown that brookite is also photocatalytically active and even has unique photocatalytic properties in some cases [130-133]. However, among the synthetic brookites, crystal facets are usually non-recognizable. Interestingly, Buonsanti et al. [134] developed a nonhydrolytic synthesis route to successfully prepare high–quality anisotropically shaped brookite nanorods with a length of 30–200 nm. These rods are determined to be dominantly enclosed with the longitudinal {210}/{100} and basal {001} facet, which is in agreement with the equilibrium shape of brookite crystals predicted from the Wulff construction.

2.2. α-Fe_2O_3

Hematite (α-Fe_2O_3) is the most thermodynamically stable form of iron oxide under ambient conditions and it is also the most common form of crystalline iron oxide. The iron and oxygen atoms are naturally arranged in the corundum structure, which is trigonal–hexagonal scalenohedral (3 2/m) with space group R–3c, lattice parameters a = 5.0356 Å, c = 13.7489 Å, and six formula units per unit cell [135,136]. It is easy to understand hematite's structure based on the packing of the anions, O^{2-}, which are arranged in a hexagonal closed–packed lattice along the

http://dx.doi.org/10.5772/62206

[001] direction. The cations (Fe^{3+}) occupy the two–thirds of the octahedral interstices (regularly, with two filled followed by one vacant) in the (001) basal planes, and the tetrahedral sites remain unoccupied. The arrangement of cations can also be considered as producing pairs of FeO_6 octahedra that share edges with three neighboring octahedra in the same plane and one face with an octahedron in an adjacent plane in the [001] direction (Fig. 6). The face–sharing is responsible for a trigonal distortion of the octahedra as the proximal iron atoms are repelled to optimize the crystal's Madelung energy. As a result, hematite exhibits a C_{3v} symmetry and there are two different Fe–O bond lengths (Figure 6). However, the electronic structures of the distorted FeO_6 octahedral are thought to be similar to undistorted clusters [133,136]. Hematite is antiferromagnetic at temperatures below 260 K and is a weak (parasitic) ferromagnet at room temperatures. While the magnetic properties of hemitate are not particularly dependent on its photo electrochemical performance, the iron spin configuration does influence the optoelectronic and carrier transport properties of hematite. The absorption of photons by hematite starts from the near–infrared spectral region where weak absorption bands (with absorption coefficients, a, of the order 10^3 cm^{-1}) are due to transition states electrons between two d orbital energy levels of the Fe^{3+} ion, which are split by an intrinsic crystal field [136,138]. Analysis by means of a Tauc plot shows the indirect nature of the band gap for the α-Fe_2O_3 involving d orbital to d orbital transition sand a direct transitions from O (2p) to Fe (3d), which occurs only for band gaps > 3.2 eV [139-141].

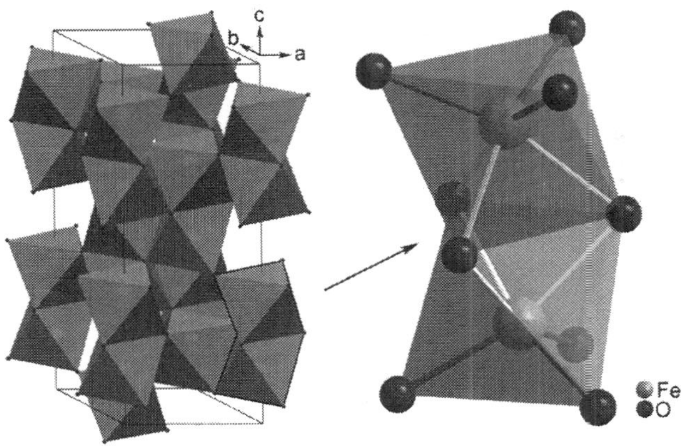

Figure 6. The unit cell (left) of hematite shows the octahedral face–sharing Fe_2O_9 dimers forming chains in the c direction. A detailed view (right) of one Fe_2O_9 dimer shows how the electrostatic repulsion of the Fe^{3+} cations produce long (light grey) and short (dark grey) Fe–O bonds [136,137].

Hematite (α-Fe_2O_3), an environmental friendly n-type semiconductor (Eg = 2.1 eV), has been widely used in many fields such as lithium–ion batteries [142], gas sensors [143–145], photocatalysis [146,147], water treatment [148] and water splitting for generating H_2. Hematite is one promising candidate for photocatalytic applications due to its narrow band gap of about 2.0–2.2 eV. Further, hematite absorbs light up to 600 nm, collects up to 40% of the solar spectrum energy, is stable in most aqueous solutions (pH > 3), and is one of the cheapest semiconductor materials available. Due to the band gap value α-Fe_2O_3 and the fact that it,s valence band edge is substantially lower than the water oxidation potential, it is a promising photoanode material for photoelectrochemical (PEC) water splitting. Photochemical water splitting involves a dispersed material in pure water and accordingly produces hydrogenand oxygen homogeneously throughout the solution [149]. The theoretical photocurrent density of α-Fe_2O_3 is ~12.6 mA/cm^2 under AM 1.5 G solar irradiation, and the solar energy conversion efficiency is ~15.5% in an ideal tandem PEC cell [150,151]. However, the photocatalytic performance of α-Fe_2O_3 is limited by certain factors such as high recombination rate of electrons and holes, low diffusion lengths of holes (2–4 nm), and poor conductivity, which led to both low efficiencies and a larger requisite over potential for photo-assisted water oxidation [152-156]. Many attempts have been made by researchers to overcome these anomalies of α-Fe_2O_3 such as lowering the recombination rate by forming nanostructures, enhancement in conductivity by doping with suitable metals and improving the charge transfer ability [157,158]. Apart from water-splitting applications, the photocatalytic activity of hematite can be used for the elimination of organic compounds in water treatment applications.

As the surface area plays an important role in determining the photocatalytic activity of materials, researchers have attempted to reduce the size of photocatalytic materials and enhance the photocatalytic properties of these materials by producing hematite in a nanoscale powder form. Many methods have been followed to synthesize α-Fe_2O_3 in a nanocrystalline form and in different shapes including hydrolysis [159], co-precipitation [160,161], hydrothermal methods [162-164], solvothermal methods [165,166], ionic liquid-assisted synthesis [167], thermal decomposition [168], combustion methods [169], and a combination reflex condensation and hydrothermal method [170].

Hosseinian et al. [171] synthesized nanostructured iron oxide of different morphologies and different phase compositions (α-Fe_2O_3 and Fe_3O_4) by a solid–state reaction (SSR) route. The photocatalytic activity was checked with respect to degradation of rhodamine B(RhB), and it was observed that the samples containing a mixture of α-Fe_2O_3 and Fe_3O_4 showed better photocatalytic activity than that of the pure α-Fe_2O_3. The higher photocatalytic activity observed for a mixed-phase sample was attributed to the higher transfer of electrons and holes generated during the photoreaction of α-Fe_2O_3 to the valence band of Fe_3O_4, which limits the recombination rates [172]. Yang et al. [173] synthesized α-Fe_2O_3 nanoparticles of uniform size (170nm to 2μm) by a hydrothermal route to study both magnetic as well as photocatalytic properties. The α-Fe_2O_3 powders with the smaller crystallite sizes show the highest photocatalytic degradation efficiency than that of the powders with larger crystallite sizes. Further, all the samples showed higher efficiency for degradation of the dye than that of the commercially available Degussa P25. Apte et al. [169] synthesized nano structured α-Fe_2O_3 powders in size ranging 25–55nm and their photocatalytic activity was analyzed with respect to the decomposition of hydrogen sulfide (H_2S) gas. α-Fe_2O_3 (necked structures) showed good photocatalytic properties and production of H_2. Zhou et al. [174] synthesized nanorods of α-Fe_2O_3 by thermal dehydration and compared the photocatalytic activity with microrods. The authors reported a higher degradation rate for rhodamine B(RhB) for nanodimensional α-Fe_2O_3 than that of the corresponding micron-sized rods. Higher Fe–O bond stretching frequencies were proposed as one of the key factors behind the enhanced photocatalytic activity. Particle size, composition, porosity, and the local structures are also the key factors that affect the photocatalytic properties of materials. Townsend et al. [159] compared the photocatalytic activity of three forms of Fe_2O_3 including bulk (crystallite size 120 nm), ultrasonicated bulk (crystallite size 40 nm), and nanopowders of α-Fe_2O_3 (crystallite size 5.4 nm). They found that the rate of oxygen evolution is higher when the crystallite size becomes smaller, and the highest rate was reported for α-Fe_2O_3 nanopowders (1072 μmol/h g). In the case of α-Fe_2O_3 nanopowders, the hole diffusion length is comparable to the crystallite size, which results in more availability of holes to react with water. Dang et al. [160] reported the effects of calcination temperature, reaction temperature, amount of catalyst, and duration of reaction on the catalytic properties. They reported an increase in photocatalytic activity with increasing calcination temperature, reaction temperature, and catalytic amount up to a certain extent, after which the activity decreases. Similar effects were also reported by Pawar et al. [175] for α-Fe_2O_3 nanoparticles synthesized by a sol-gel technique followed by the heat treatment at different calcination temperatures. The efficiency of the catalyst was analyzed with respect to various experimental variables such as calcination temperature, pH, light intensity, and concentration of dye and catalyst. Samples calcined at 600°C show the highest photocatalytic activity because of the formation of the more dominant α-Fe_2O_3 phase. The photocatalytic properties were analyzed for the 3–10 pH range, and the reactions at higher pH conditions showed better photocatalytic properties. In basic pH conditions, formation of OH· radical is more favored and electrostatic abstractive effects between cationic malachite green dye and negatively charged surface of α-Fe_2O_3 increases, which results in a higher probability of dye degradation Light intensity shows a linear effect on the photocatalytic properties of α-Fe_2O_3 due to the increased availability of photons for the reaction. Similar effects were also reported by Liu et al. [176] for α-Fe_2O_3 nanorods. These authors examined the effect of the amount of catalyst and initial dye concentration on the photocatalytic properties. The optimum catalyst amount was reported to be 50 mg/L to achieve the highest photocatalytic activity. However, the photocatalytic activity degrades with increasing dye concentration. This effect was justified in terms of a decrease in transparency with an increase in dye and catalyst concentration after a particular value.

In a photoreaction, the porosity of the catalyst plays a major role in enhancing the photocatalytic properties. Sundarmurthy et al. [177] synthesized 1D α-Fe_2O_3 nanobraids and nanoporous structures by electrospinning to analyze the photocatalytic properties. The nanostructures show superior photocatalytic activity for the degradation of Congo red dye (CR) in a small fraction of time due to the porous surface and nanosized crystallites of α-Fe_2O_3, which provide more active catalytic centers and allow effective interaction between organic dye and α-Fe_2O_3, thereby enhancing photocatalytic degradation performance. α-Fe_2O_3 porous structures were prepared by Zhang et al. [162] and the photocatalytic activity was analyzed by the degradation of methylene blue(MB). They analyzed the effect of porosity and the amount of catalyst on photocatalytic activity. It has been reported that an optimized amount of catalyst (20 mg) is required for getting the highest rate of degradation of MB, less or more than this amount leads to lower photocatalytic activity. Large amounts of catalyst result in lesser illumination, and when the amount of catalyst is insufficient, the active sites are not sufficient to degrade the organic dye. Geng et al. [178] followed a number of Ni^{2+}/surfactant system routes for synthesizing α-Fe_2O_3 with a porous structure and rough surface which shows better photocatalytic properties than that of the α-Fe_2O_3 nanoparticles in the degradation of MB as a result of higher surface area. Gang et al. [147] prepared α-Fe_2O_3 micro/nano spheres synthesized by hydrothermal synthesis followed by the thermal treatment. The micro/nano spheres show a better dye degradation efficiency than that of the nanopowders. The calculated reaction rate for spherical structures is more than twice than that of the reaction rate of nanopowders and 12 times the reaction rate of the micron-sized powders. The better photocatalytic activityis the result of the higher specific surface area and porous structures.

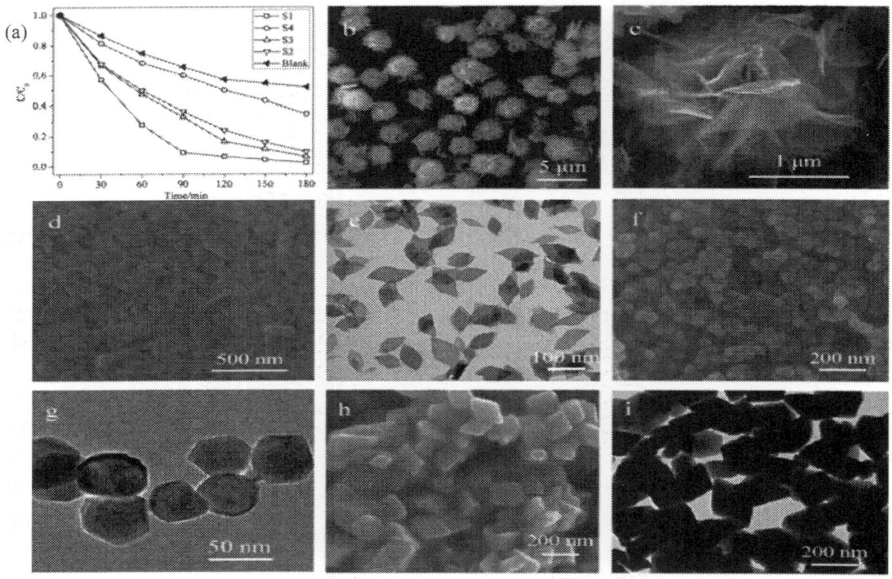

Figure 7. Photocatalytic degradation rate of RhB over the α-Fe_2O_3 nanostructures under visible light illumination in the presence of H_2O_2 additive (a),and SEM/TEM images of the α-Fe_2O_3 nanostructures: (b,c)S1;(d,e)S2;(f,g)S3;and(h,i)S4 [179].

Xu et al. [179], Zhou et al. [180] and Bharathi et al. [170] reported the effect of the surface morphology of α-Fe_2O_3 on its photocatalytic activity. α-Fe_2O_3 nanostructures with different morphologies such as microflowers, nanospindles, nanoparticles and nanorhombohedra were synthesized (Fig. 7) [179]. The photocatalytic activity was analyzed by monitoring the degradation of RhB in the presence of the catalyst. The best photocatalytic activity was observed for the samples with highest surface area and porosity. Similar surface area effects were also reported by Cheng et al. [181] for flower-like α-Fe_2O_3 nanostructures synthesized by a biphasic interfacial reaction route. The photocatalytic properties of α-Fe_2O_3 were evaluated by measuring the degradation of RhB. The results were compared with the commercial α-Fe_2O_3

powders and nanoflowers were found to have a better photocatalytic property than the commercial powders. The enhancement was related to the increase in crystallinity and increase in the surface area, which is also supported by results of other authors for TiO_2 [182] and Fe_2O_3 [183]. Similar surface area effects were also reported by Cao et al. [184], Xu et al. [166], and Li et al. [163] for α-Fe_2O_3 hollow microspheres prepared by solvothermal and hydrothermal methods. The photocatalytic activity was analyzed by the degradation of salicylic acid. The hollow spheres associated with nanosheets show better photocatalytic activity than that of then anoparticles of α-Fe_2O_3. Similar results were also reported by Majiet al. [168], where α-Fe_2O_3 powders prepared at 500°C show better photocatalytic activity for the degradation of rose Bengal dye than that of the powders prepared at 600°C and commercially available TiO_2(Degussa–25) as a result of higher surface area. α-Fe_2O_3 hollow spindles and spheres were prepared by Li et al. [164] and Xu et al. [167], respectively. These authors reported an enhancement in photocatalytic degradation efficiency as a result of the enhancement in specific surface area, which results in more unsaturated surface coordination sites exposed to the solution. The hollow microsphere facilitates more electron-hole transport and lowers the recombination rate. Hollow microspheres allow multiple reflections of visible light within the interior that encourage a more efficient use of the light source and enhance light–harvesting, leading to an increased quantity of $\cdot OH$ available to participate in the photo-catalytic reaction. Along with this, the hollow spheres also provide ideal channels for the dye molecules and increase the probability of interaction.

Apart from crystallite size, the orientation of crystallites also plays a major role in enhancing the photocatalytic properties. This effect has been reported by Wu et al. [185], in which the authors prepared α-Fe_2O_3 nanocubes by a solvothermal method and reported a higher photocatalytic property for the {104} planes than that of the samples with {012} planes. The photocatalytic properties involve Fenton's reaction. The amount of Fe^{3+} on the surfaces of the catalyst play a very important role in the Fenton reaction in which the reduction of Fe^{3+} to Fe^{2+} generates hydroxyl radicals ($\cdot OH$) [186]. It has been reported by Lv et al. [187] that {104} planes of α-Fe_2O_3 contain 10.3 atoms/nm^2 of exposed Fe^{3+} ions, whereas the {012} planes contain 7.33 atoms/nm^2 of exposed Fe^{3+} ions. This explains the higher reactivity of {104} planes than that of the {012} planes. Along with the surface morphology, oxygen pressure and amount of the catalyst also play a major role in enhancing the photocatalytic properties. Isaev et al. [188] reported an enhancement in the photocatalytic activity with an increase in the quantity of Fe_2O_3 up to a certain point, after that, the photocatalytic activity is decreased. Similarly, the authors reported an enhancement in dye degradation with increased oxygen content. The reason behind the enhancement in photocatalytic behavior is due to the formation of more oxygen-containing active species such as $HO\cdot$, $O_2\cdot$, and $HO_2\cdot$ oxidizing species. Zhou et al. [189] investigate visible–light–induced photodegradation of model dye rhodamine B (RhB) in the presence of hydrogen peroxide (H_2O_2) over hematite architectures, namely 1D nanorods, 2D nanoplates, and 3D nanocubes (Fig. 8), and the reactivity trend can be rationalized as exposed facets in the order {110} > {012} >> {001}. This photocatalytic activity order can be well explained by different facets of α-Fe_2O_3 surface atomic and electronic structures.

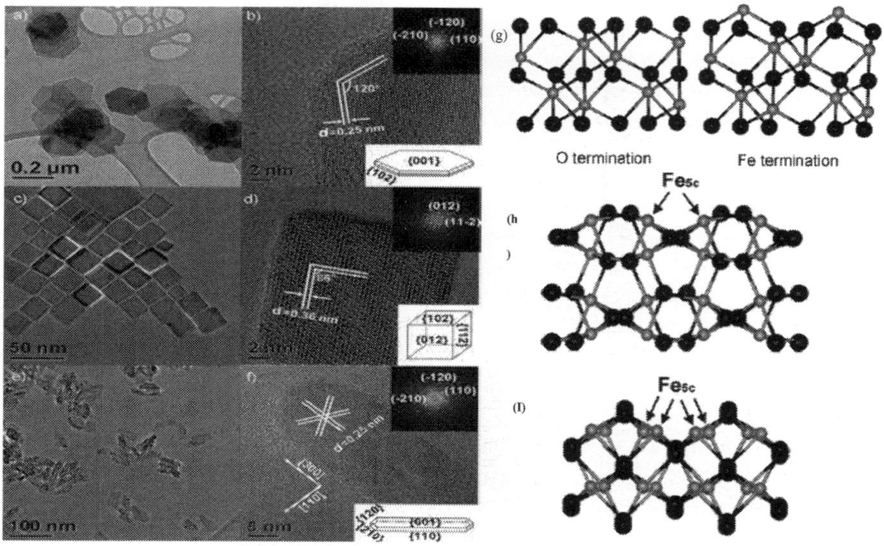

Figure 8. Representative morphologies and structures of α-Fe_2O_3 architectures.(a) TEM image and (b) HRTEM image of 2D α-Fe_2O_3 nanoplates. Insets: FFT pattern and drawing of a plate. (c) TEM image and (d) HRTEM image of 3D α-Fe_2O_3 nanocubes. Insets: FFT pattern and drawing of a cube. (e) TEM image and f) HRTEM image of 1D α-Fe_2O_3 nanorods. Insets: FFT pattern and drawing of a rod. Side views of surface terminations of α-Fe_2O_3. (g) {001}, (h) {012}, and (i) {110}. Large black spheres are oxygen and small gray spheres are iron. The coordinatively unsaturated iron atoms on the {012} and {110} surfaces are shown by arrows [189].

3. Ternary oxide system

3.1. $BiVO_4$

Bismuth vanadate ($BiVO_4$), which is an n−type semiconductor, has been identified as one of the most promising photocatalytic materials. As it is well known, $BiVO_4$ exists in three polymorphs of monoclinic scheelite, tetragonal scheelite, and tetragonal zircon structures, with bandgaps of 2.4, 2.34, and 2.9 eV, respectively. $BiVO_4$ exists naturally as the mineral pucherite with an orthorhombic crystal structure [190]. However, $BiVO_4$ prepared in the laboratory does not adopt the pucherite structure but crystallizes either in a scheelite or a zircon-type structure (Fig. 9) [191,192]. The scheelite structure can have a tetragonal crystal system (space group: $I4_1/a$ with $a = b = 5.1470$ Å, $c = 11.7216$ Å) or a monoclinic crystal system (space group: $I2/b$ with $a = 5.1935$ Å, $b = 5.0898$ Å, $c = 11.6972$ Å, and $b = 90.3871$) [192,193] while the zircon-type structure has a tetragonal crystal system (space group: I_41/a with $a = b = 7.303$ Å and $c = 6.584$ Å) [192,194].

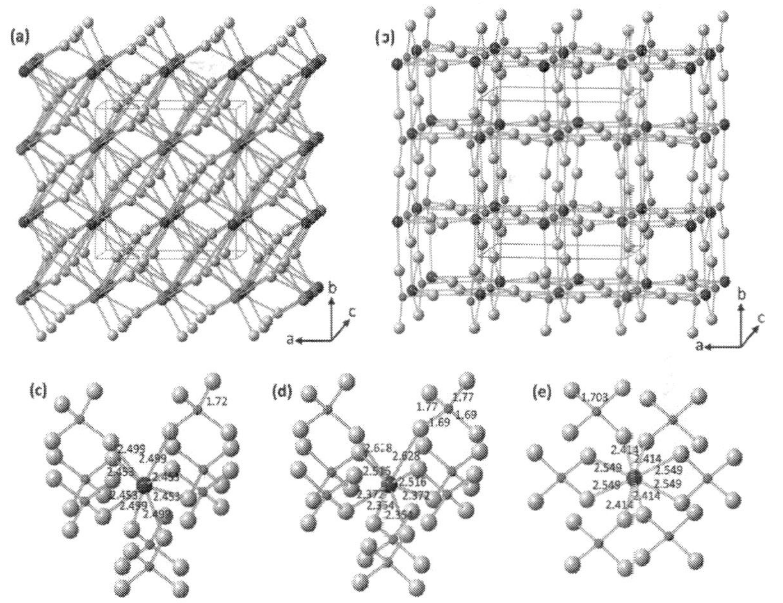

Figure 9. Crystal structures of (a) tetragonal scheelite and (b) zircon-type $BiVO_4$ (red: V, purple: Bi, and gray: O). The crystal structure of monoclinic scheelite is very similar to what is shown in (a) with the exception being the subtle changes in atomic positions of Bi, V, and O. Local coordination of V and Bi ions in (c) tetragonal scheelite, (d) monoclinic scheelite, (e) and zircon–type $BiVO_4$ structure with bond lengths shown in Å [192,194].

In the scheelite structure, four O atoms coordinate each V ion within a tetrahedral site and eight O atoms from eight different VO_4 tetrahedral units coordinate each Bi ion [192,193]. Fig. 9(a) shows the four–coordinated V center and the eight–coordinated Bi center alternating along the [001] direction. Two Bi centers and one V center coordinate each O atom in this structure, and a three–dimensional structure was formed by holding the Bi and V centers. The only difference between the tetragonal and monoclinic scheelite structure is that the local environments of V and Bi ions are more significantly distorted in the monoclinic structure, which removes the fourfold symmetry necessary for a tetragonal system. For example, in the tetragonal scheelite, all four V–O bond lengths were equal (1.72 Å), while in a monoclinic scheelite structure, there are two different V–O bond lengths(1.77 Å and 1.69 Å). In the same manner, in the tetragonal scheelite structure, only two very similar Bi–O distances exist (2.453 Å and 2.499 Å), while in the monoclinic scheelite structure, the Bi–O distances change significantly (2.354 Å, 2.372 Å, 2.516 Å and 2.628 Å) [192,193]. The significant distortion of the Bi–polyhedra indicates that the Bi 6s alone is more sterically expressed in the monoclinic scheelite structure.

It should be noted that the monoclinic scheelite structure of $BiVO_4$ was originally reported with the space group *I*2/b, which is a nonstandard space group [192,193]. Some recent studies of $BiVO_4$ have used a standard space group *C2/c*, which is converted from *I2/b*. Changes in the crystallographic axes via the conversion of a monoclinic *I*–centered (body-centered) cell to a monoclinic C-centered cell are shown in Fig. 10 [195]. With this cell conversion, the new cell parameters for *C2/c* are a' = 7.2472 Å, b' = 11.6972 Å, c' = 5.0898 Å, and β' = 134.225°. The choice of the *I*–centered monoclinic cell has the advantage of easily showing its structural relationship to the tetragonal scheelite structure that was reported in a body–centered space group, $I4_1/a$, using the identical unit cell choice and crystallographic axes. Since both *I2/b* and *C2/c* space groups, which have different unit cell choices and crystallographic axes, are commonly used to describe the monoclinic scheelite structure of $BiVO_4$, it is necessary to clarify the space group used when referring to specific atomic planes or crystal directions as well as the *hkl* indices of X-ray diffraction peaks in order to prevent any possible confusion [192].

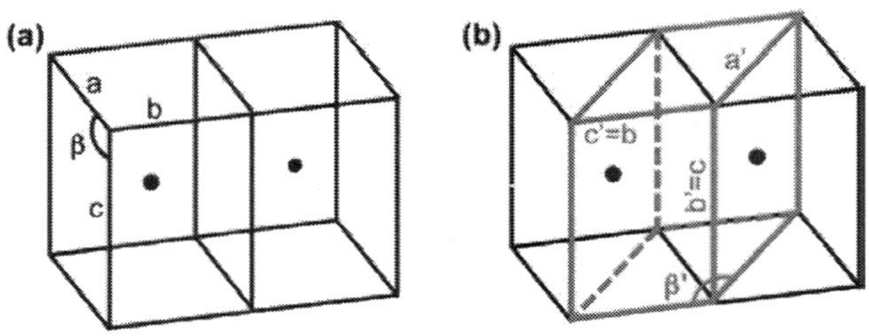

Figure 10. Cell conversion of (a) *I*–centered monoclinic to (b) C–centered monoclinic. *a, b, c,* and *β* represent the unit cell parameters for the *I*–centered cell and a, b0, c0, and b0 for the C–centered cell [195].

In the zircon–type structure, V is still stabilized by four O atoms and Bi is coordinated by eight O atoms. However, since two VO_4 units provide two O atoms to Bi, each Bi is surrounded by only six VO_4 units, as shown in Fig. 9 (e). To form a 3D structure, two Bi centers and one V center are connected by all oxygen atoms, which holds the V and Bi centers together [192].

It was reported that the low temperature synthesis (e.g., precipitation at room temperature) can form a zircon–type structure [192,196,197]. However, in this process, kinetics plays a critical role in the determination of final products, so the structure type obtained at low temperatures may change with different synthesis methods used and detailed conditions. A phase transition from tetragonal zircon to monoclinic scheelite was reported to occur irreversibly at 670–770 K [192,196]. Among scheelite structures, the tetragonal phase is a high temperature phase and the phase transition between monoclinic scheelite $BiVO_4$ and tetragonal scheelite $BiVO_4$ was observed to occur reversibly at 528 K [192,196].

In the zircon–type $BiVO_4$, the charge-transfer transition from O 2p orbitals to empty V 3d is mainly responsible for the bandgap transition. In the scheelite structure, the bandgap is reduced because the 6s state of Bi^{3+} appears above the O 2p and the transition from Bi $6s^2$ (or hybrid Bi $6s^2$–O 2p orbitals) to the V 3d becomes possible. Among scheelite $BiVO_4$ structures, Tokunaga et al. reported that monoclinic scheelite structure shows much higher photocatalysis activity for the photocatalytic water oxidation compared with tetragonal scheelite structure

http://dx.doi.org/10.5772/62206

[192,198]. The bandgap energies of the tetragonal and monoclinic scheelite $BiVO_4$ shows little difference and the more severe distortion of the metal polyhedra present in the monoclinic scheelite $BiVO_4$is the reason why the photocatalytic performance is different [192,198]. As discussed earlier, the local environment of Bi in the monoclinic scheelite structure is much more distorted than that in the tetragonal scheelite structure (Fig. 9 (c) and (d)).

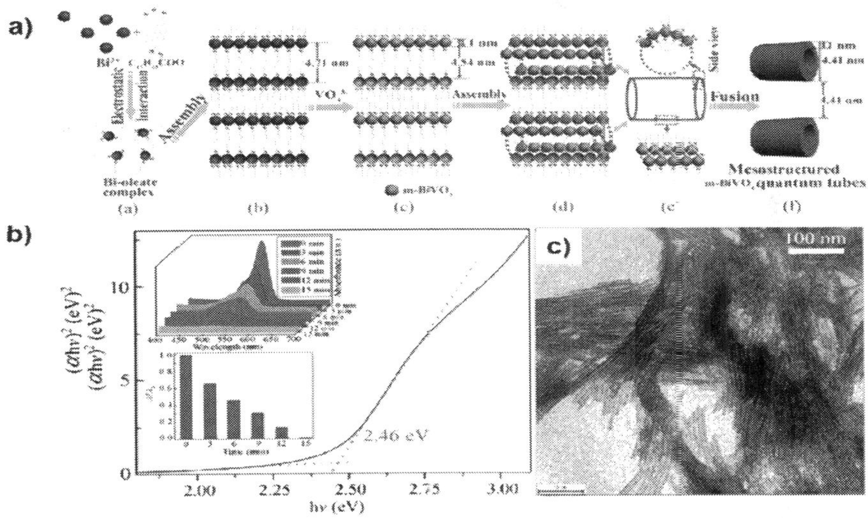

Figure 11. (a) Mechanism for the formation of $BiVO_4$ quantum tubes. (b) Optical absorption edge of $BiVO_4$ quantum tubes (top and bottom insets:photodegradation of RhB vs. irradiation time under visible light). (c) TEM image of $BiVO_4$ quantum tubes after the photodegradation [205]

Morphology control is an efficient method to facilitate carrier transportation and light harvesting, accelerate charge movement within the material structure and assist the collection and separation of electron-hole pairs at the interface of the materials [199-202]. Controllable synthesis of $BiVO_4$ with controlled morphologies has been notable developed (such as nanorods, nanowires, nanotubes (NTs), nanobelts, nanoellipsoids, hollow spheres, and even some hierarchical architectures) and corresponding morphology-dependent photocatalytic properties have also been extensively studied [203-211]. For example, Tada et al. first fabricated $BiVO_4$ nanorods (NRs) using polyethylene glycol (PEG) as a shape–directing agent [203]. Yu et al. developed a template–free solvothermal method to synthesize $BiVO_4$ nanotubes (NTs) [203]. Xie et al. reported a novel assembly–fusion strategy for the synthesis of $BiVO_4$ quantum tubes with an ultra-narrow diameter of 5 nm ultrathin wall thickness down to 1 nm, and exposed {010} facets (Fig. 11 (a), (c)) [206]. As the increase of the reaction time, optical absorption edge and band energy of the $BiVO_4$ quantum tubes are significantly blue–shifted compared with bulk $BiVO_4$, which is due to the well-known quantum size confinement effect (Fig. 11 (b)).

Nanosized building blocks, such as nanowires, nanobelts, nanosheets, and nanotubes possess interesting properties, and the self-assembling of them into hierarchical architectures is much more interesting and has attracted great attention [199]. Liu et al. and Chen et al. synthesized $BiVO_4$ porous hollow microspheres composed of single–crystalline nanosheets using a solvothermal–induced self-assembling method (Fig. 12 (a), (b)) [207,209]. These hollow microspheres exhibited excellent photocatalytic activity due to the increased specific surface area and light harvesting ability. Xie et al. also reported the multi-responsive function of ellipsoidal $BiVO_4$ assembled from many small nanoparticles with major exposed {101} facets [210]. Similarly, Zhao et al. synthesized uniform hyperbranched $BiVO_4$ via a surfactant–free hydrothermal route (Fig. 12 (c)) [211]. The crystal consists of four trunks with branches distributed on opposite sides, this unique structure is beneficial from the different growth rates along a, b, and c axes: preferentially along the [100] direction at the beginning and subsequently along the [010] and [001] directions. The loosely packed building units of the hyperbranched structure exhibits excellent photocatalytic activity, because (i) the small crystal size allows the inside generated electron-hole pairs efficiently transporting from inside out to the surface and (ii) the large surface area provides abundant active sites for the photocatalytic reaction and promotes light harvesting as well as reactant adsorption.

One of the main reasons for the charge recombination in $BiVO_4$ is the long diffusion length of the photo–induced electrons [212-214]. Tailoring porous $BiVO_4$, especially ordered porous structures, can shorten the diffusion length and thus facilitate charge migration, providing a readily accessible channel and increasing the adsorption of reactants and the supply of more surface active sites [214,210]. Yu et al. reported that ordered mesoporous $BiVO_4$ shows a higher photoactivity than conventional $BiVO_4$, and this mesoporous $BiVO_4$ was fabricated by nanocasting using mesoporous silica KIT–6 as the replica parent template (Fig. 13 (a)) [210]. Ordered macroporous $BiVO_4$ with controllable dual porosity was synthesized by Xie et al. for efficient solar water splitting and the relationship between the geometrical characteristics and the charge migration was also demonstrated (Fig. 13 (b)) [215]. There are mainly two factors that determined by the geometrical characteristics of periodically ordered macroporous structures (Fig. 13 (b), (i)): the diameter of the macropores surrounded by the final skeletal walls (denoted as D1) and the diameter of the pores between neighboring macropores (denoted as D2). Previously, Lee et al. observed an efficient photo-induced charge drift mobility within the proper D1 size [216]. Based on this, Xie et al. further synthesized ordered macroporous $BiVO_4$ architectures with controllable dual porosity (aforementioned as D1 and D2) via a modified colloidal crystal templating method (Fig. 13 (b), (i) and (ii)), and verified that charge migration in periodically ordered macroporous architectures has a strong dependence on D1 and D2 (Fig. 13 (b), (iii)) [215]. On the one hand, no matter in the bulk and on the surface, it is believed that a smaller D2 is favorable for charge migration. On the other hand, a smaller D1 blocks bulk charge migration but facilitates surface charge migration.

The specific crystal facet determines the surface active sites and even the electronic structure, as a result crystal facets play a critical role in photocatalysis. [198,217], and consequently, it is of great importance to develop the crystals exposed with highly reactive facets [217-223]. Xi et al. synthesized well-defined $BiVO_4$ nanosheets exposed with {001} facets using a straightfor-

http://dx.doi.org/10.5772/62206

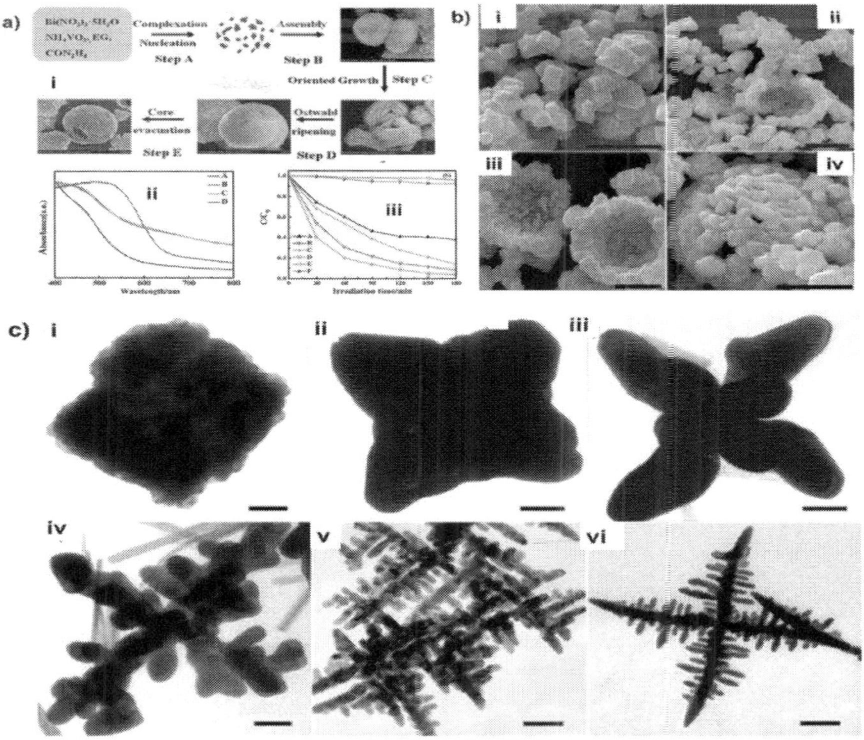

Figure 12. (a) Formation mechanism, UV–Vis absorption, and RhB photodegradation of hollow $BiVO_4$ microspheres [207]. (b) Morphology evolution of $BiVO_4$ hollow spheres via a hydrothermal method using urea as the guiding surfactant (I: 2 h; II: 4 h; III: 8 h; IV: 12 h; V: 24 h; scale bar is 2 μm) [208]. (c) Morphology evolution of hyperbranched $BiVO_4$ at intervals of 10 min (I), 20 min (II), 30 min (III), 45 min (IV), 1 h (V), and 3 h (VI), respectively (the scale bars are 100, 200, 200, 200, 400 and 500 nm, respectively) [210].

ward hydrothermal route without any template or organic surfactant (Fig. 15 (a)) [218]. Typically, $BiVO_4$ crystals show a regular decahedron shape with controllable exposed facets of {010}, {011}, {110} and {111}, as shown in Fig. 14. Li et al. Synthesized $BiVO_4$ with a highly exposed (010) facet using $TiCl_3$ as a directing agent, and correlated this to the high activity in O_2 evolution on $BiVO_4$ (Fig. 15 (b)) [219]. Inspired by this work, facet-dependent photocatalytic activity for water oxidization on $BiVO_4$ was investigated by density functional theory (DFT) calculations, particularly between the (010) and (011) facets (Fig. 14 (c)) [220]. The (010) facet has a higher activity compared with the (011) facet due to its higher charge carriers mobility, easier adsorption of water, and lower overall potential energy of O_2 evolution.

Recently, many studies have reported that photo-induced electrons and holes may be drifted to different crystal facets [217,221-223], which means photo-reduction and oxidation may

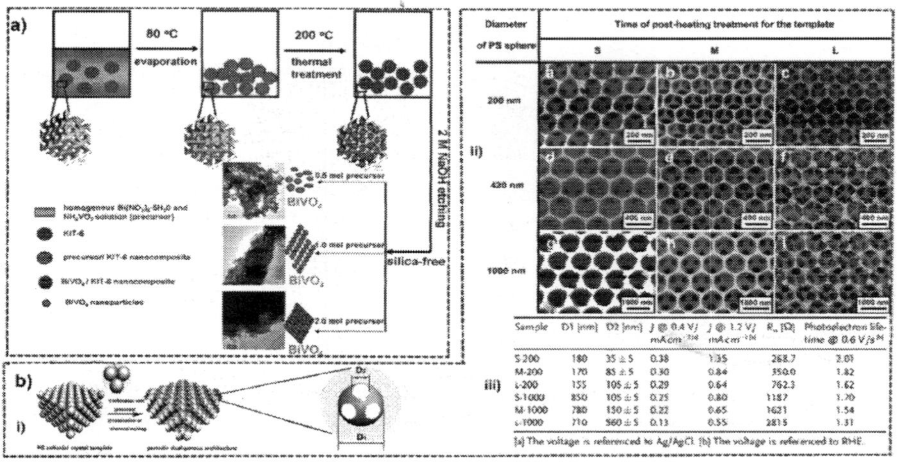

Figure 13. (a) Proposed process for the fabrication of ordered mesoporous $BiVO_4$ [211]. (b) (i)Schematic representation of dual porosity in periodically ordered porous $BiVO_4$; (ii) typical SEM images of corresponding $BiVO_4$; (iii) relationship between PEC performance and dual porosity [216].

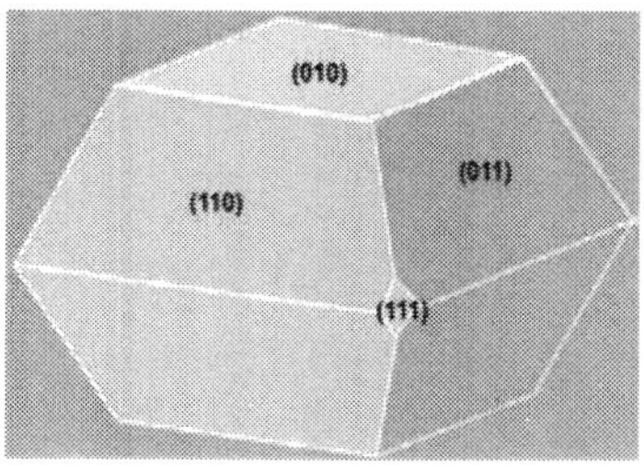

Figure 14. Typical crystal of $BiVO_4$ exposed with the {010}, {011}, {110} and {111} facets.

happen preferentially on different facets. Therefore, the cooperation of different facets is very important to obtain high quantum efficiency. Using photochemical labeling, Li et al. discovered that photo-excited electrons-driven reduction reaction (Pt–photodeposition) and photo-excited holes-driven oxidation reaction (MnO_x-photodeposition) take place on the {010} and {110} facets, respectively (Fig 16a) [222], which implies that the photo-induced electrons and holes move to the {010} and {110} facets, respectively. Notably, it provides a very useful inspiration to selectively deposited co-catalyst on specific facets via photodeposition (Fig. 16 (b)) [223]. Using this concept, the photocatalyst with Pt on the {010} facets and MnO_x on the {110} facets exhibits a much higher activity in both photocatalytic and PEC water oxidation, compared with the counterparts with randomly distributed Pt and PbO_2 co–catalysts (Fig. 16 (c), (d)). The coupling of co–catalysts on selected semiconductor facets may open up a new strategy for developing highly efficient photocatalysts.

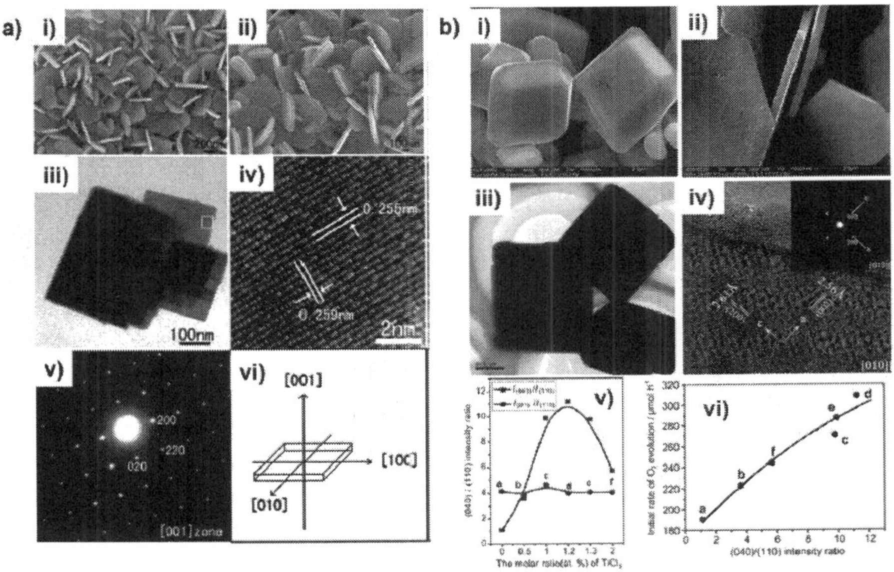

Figure 15. (a) SEM and HRTEM images of $BiVO_4$ nanoplates exposed with the {001} facets [213]. (b) Facet(010/110)-dependent photoactivity of oxygen evolution on $BiVC_4$ [214].

3.2. Bi_2WO_6

The Aurivillius family have a general formula of $Bi_2A_{n-1}B_nO_{3n+3}$ (A = Ca, Sr, Ba, Pb, Bi, Na, K and B = Ti, Nb, Ta, Mo, W, Fe), and Bi_2WO_6 is the simplest member of this family (where n = 1) and usually have the layer structures and unique properties [224]. Fig. 17 shows a schematic structure of the Bi_2WO_6 crystalline with orthorhombic structures constructed by alternating $(Bi_2O_2)_n^{2n+}$ layers and perovskite–like$(WO_4)_n^{2n-}$ layers [225]. More recently, many Aurivillius-based compounds have been reported which exhibit interesting properties suitable for photocatalytic applications. Of these, Bi_2WO_6 is the simplest and probably the most studied example within this family. In this bismuth tungstate, the perovskite–like structure is defined by WO_6 units which form a layer perpendicular to the (100) direction and sandwiched between the $(Bi_2O_2)^{2+}$ units. Layers sandwiched structure favors the efficient separation of photogenerated electron-hole pairs and then improves the photocatalytic activity, which can be ascribed to the formed internal electric fields between the slabs [226,227]. Due to its preferable band composition and unique layered structure, Bi_2WO_6 possesses several advantages as photocatalysts over the competing materials, especially in the view of practical applications, including its desirable visible– light absorption, relatively high photocatalytic activity and good stability.

Bi_2WO_6 consists of accumulated layers of corner-sharing WO_6 octahedral sheets and bismuth oxide sheets [228,229]. The conduction band of Bi_2WO_6 is composed of the W5d orbital; its valence band is formed by the hybridization of the O2p and Bi6s orbitals, which not only makes

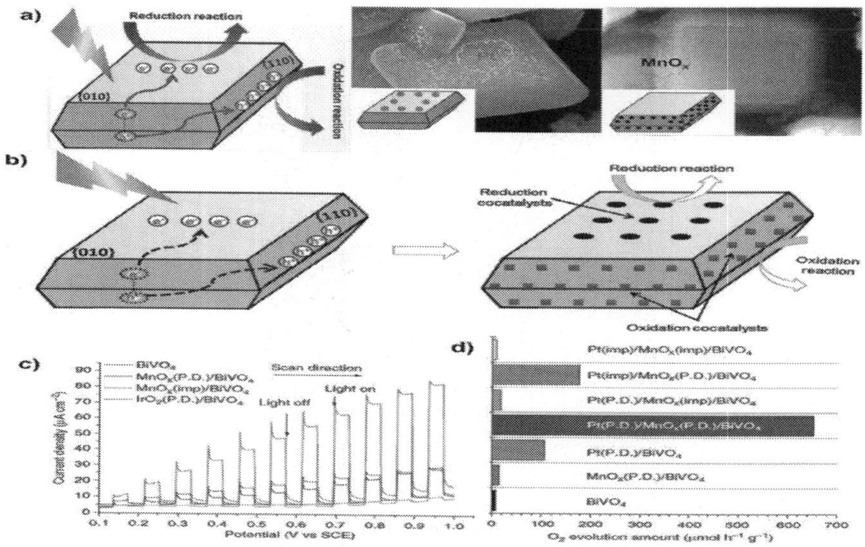

Figure 16. (a) Charge separation between the {010} and {110} facets confirmed by Pt and PbO_2 photodeposition on $BiVO_4$ [218]. (b) Selective deposition of dual redox co-catalysts on specific facets of $BiVO_4$ [219]. (c, d) Photoelectrocatalytic and photocatalytic water oxidation activity of $BiVO_4$ with selectively deposited co-catalysts on specific facets and randomly distributed co-catalysts [220].

the VB largely dispersed and thus results in a narrowed band gap of Bi_2WO_6 (2.8 eV) capable of absorbing visible light ($\lambda > 400$ nm), but also favors the mobility of photogenerated holes for specific oxidation reactions [230]. Such a band structure indicates that charge transfer in Bi_2WO_6 upon photoexcitation occurs from the O2p + Bi6s hybrid orbitals to the empty W5d orbitals, as illustrated in Fig. 18 [231].

As early as in 1999, the solid–state method was first used by Kudo et al. to synthesize Bi_2WO_6 photocatalyst [232], but the particle sizes of the product are in micrometers and the specific surface area is very small, which greatly limit its application in the photocatalysis. In the aim of obtaining micro or nanosized Bi_2WO_6 structures with enhanced photocatalytic activity, several groups have developed many advanced synthetic methods including sol–gel method [233], combustion synthesis method [234], ultrasonic method [235], co-precipitation method [236], sol-gel method/calcining method [237], and hydro/solvothermal method [238-245].

Bi_2WO_6 micro/nano–structures with diverse shapes exhibit different photocatalytic activities, and currently some of them have been used not only for the photodegradation of other organic pollutants but also for the photocatalytic disinfection. In 2005, Zhu's group have developed a Bi_2WO_6 nanoplates [246,247] applied in the photodegradation of rhodamine B (RhB) under visible–light irradiation. Notably, the photocatalytic reaction constant (k) of the best quality Bi_2WO_6 nanoplates is three times higher than that of the sample prepared by solid–state reaction [246]. In addition, they found a significantly pH-dependence of the photo-assisted

http://dx.doi.org/10.5772/62206

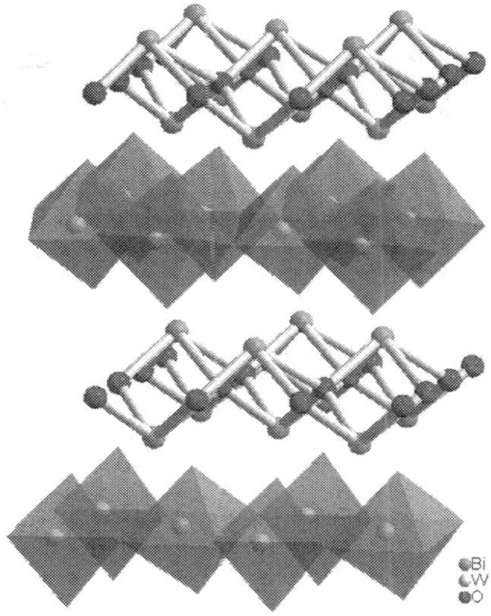

Figure 17. Structure of Bi_2WO_6 showing the WO_4^{2-} and $Bi_2O_2^{2+}$ layers [225].

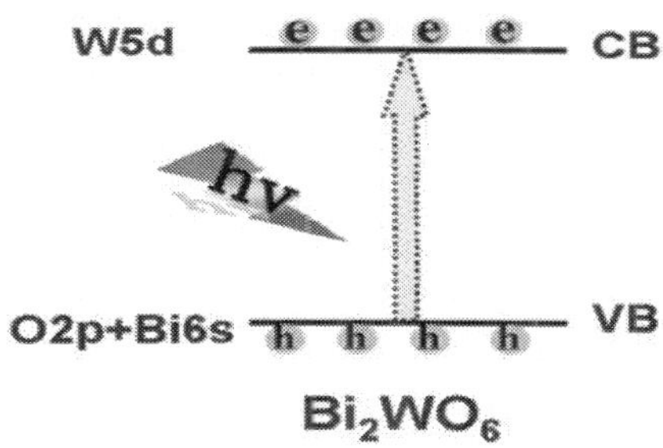

Figure 18. Band structure of the Bi_2WO_6 photocatalyst. [231]

degradation of RhB in aqueous Bi_2WO_6 as the pH varies from 5.03 to 9 89, where the highest degradation rate was achieved at pH 6.53. It is proposed that the pH of the solutions can affect the mode and extent of adsorption of RhB on the Bi_2WO_6 surface and further the transformation rate of RhB indirectly. They further used the total organic carbon measurement to determine the high mineralized degree of RhB [247]. Further studies on the mechanism shows that a photocatalytic process and a photosensitized process is involved in the the Bi_2WO_6–assisted photodegradation of RhB [247]. However, the contribution of RhB photodegradation driven by the light–excited RhB was much slower than by the light–excited Bi_2WO_6. The experimental results show that only 19% of RhB was degraded by photosensitized action, while 81% of RhB was degraded by a photocatalytic process [247].

In order to further improve the photocatalytic activity of Bi_2WO_6, some groups have developed Bi_2WO_6 nanoplates superstructures [248,249-253]. Zhang group also prepared Bi_2WO_6 micro/nanostructures, including nanoplates, tyre/helixlike, disintegrated–flower–like and flower –like superstructures [238,239]. The photodegradation results of RhB show that these Bi_2WO_6 micro/nanostructures exhibit different photocatalytic activities under visible-light ($\lambda > 400$ nm) irradiation, as shown in Fig. 19 (a). Among these photocatalysts, the uncalcined flower-like Bi_2WO_6 superstructure prepared with pH = 1 processes an improved photocatalytic performance, which can degrade 84% of RhB in 60 min [238]. Besides, the photocatalytic performance can be further improved by the calcination process, and the result calcined flower–like Bi_2WO_6 superstructure has a higher photocatalytic activity, which can degrade 97% of RhB in 60 min (Fig. 19 (b)). This performance is also superior to other traditional photocatalysts such as TiO_2 (P25) and bulk SSR-Bi_2WO_6 powder prepared by solid–state reaction [239].

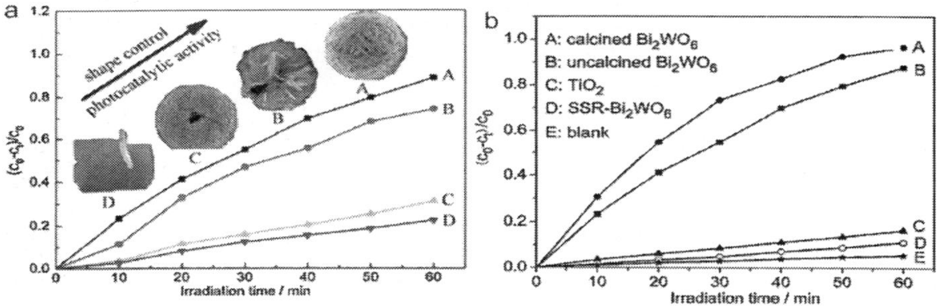

Figure 19. (a) The photodegradation efficiencies of RhB as a function of irradiation time by different Bi_2WO_6 nano/micro–structures: (A) the uncalcined flower–like Bi_2WO_6 superstructure prepared with pH = 1, (B) the uncalcined disintegrated flower-like Bi_2WO_6 superstructure prepared with pH = 2.5, (C) the calcined tyre/helix–like Bi_2WO_6 superstructures prepared with pH = 1 and P123, (D) the uncalcined Bi_2WO_6 nanoplates prepared with pH = 7.5; [239] (b) the photodegradation efficiencies of RhB as a function of irradiation time by photocatalyst samples: (A) the calcined flower-like Bi_2WO_6 superstructure, (B) the uncalcined flower-like Bi_2WO_6 superstructure, (C) TiO_2 (P25), (D)bulk SSR-Bi_2WO_6 powder, and (E) blank [238].

The novel flower-like superstructures of the uncalcined or calcined Bi_2WO_6 is mainly responsible for the highly improved photocatalytic activity. At the same time, as shown in SEM images (Fig. 20), there are plenty of meso– or macro–diameter sized pores in the flower-like superstructures, which can be considered as electron transport paths that also contributes to the photocatalysis process. [254]. It is generally believed that it is an integral part of the architectural design if the reactant molecules can easily move in or out of the nanostructured materials, the efficiency of the photocatalysis can be improved, and here, meso– or macro –diameter sized pores provides the important transport paths [254,255]. They also believe that the introduction of textural transport paths in the uncalcined or calcined Bi_2WO_6 superstructures facilitate the reactant molecules to easily incorporate with the reactive sites on the framework walls of photocatalysts, which leads to excellent photocatalytic performance for the degradation of RhB [239]. On the other hand, fewer defects, which acting as electron-hole recombination centers, can be significantly reduced by improved crystallinity of Bi_2WO_6 through the calcination process of Bi_2WO_6 [238]. This has been proved by Amano et al. [256] who experimentally investigated the influence of crystallization on the lifetime of photoexcited electrons from Bi_2WO_6. The recombination rate of electrons decay with holes can be characterized by the intensity of transient IR absorption after a 355 nm laser pulse [256]. If an appreciable absorbance at 100 μs in Bi_2WO_6 crystalline can be observed, it means a slow recombination rate and a long lifetime of photogenerated carriers, which is beneficial for driving appreciable photocatalytic reactions. However, no transient absorption for amorphous Bi_2WO_6 samples was observed implying a fast recombination of electron-hole pairs, leading to negligible photocatalytic activity. Therefore, the higher photocatalytic activity of the calcined Bi_2WO_6 is explicable in several cases [238,251,256].

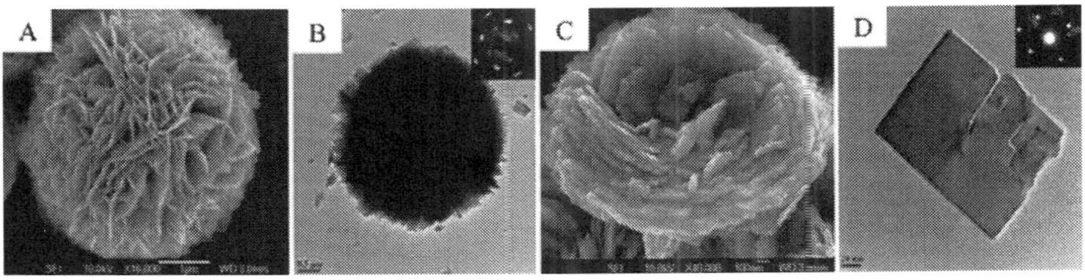

Figure 20. SEM (A) and TEM (B) images of an individual flower–like Bi_2WO_6 superstructure (inset: SEAD pattern recorded at the corner of this individual sphere); (C) SEM image of a broken Bi_2WO_6 sample; (D) TEM image of a peeled fragment (inset: SEAD pattern recorded at this individual nanoplate) (conditions: pH = 1, hydrothermally treated at 160 °C for 20 h, no surfactant, uncalcined) [238].

It is worth noting that Amano et al. [256] have demonstrated a high photocatalytic activity of crystalline Bi_2WO_6 under visible-light (λ > 400 nm) irradiation for oxidative decomposition of gaseous acetaldehyde (AcH) to produce CO_2, however, amorphous Bi_2WO_6 sample exhibits negligible photocatalytic activity under same condition. Because the mineralization of colorless AcH does not involve a dye–sensitized process, this result provides an solid conclusion that crystalline Bi_2WO_6 has excellent visible–light–driven photocatalytic activity. Furthermore, the photocatalytic activity of crystalline Bi_2WO_6 has been finely evidenced by its diffuse reflectance photoabsorption spectrum and action spectrum, that is, 8% apparent quantum efficiency at wavelength of 400 nm [256]. In the future, much more clear evidence needs to be provided to explore the mechanism of visible–light–driven photocatalytic activity of Bi_2WO_6. [256].

Recently, Amano et al. [252] reported the preparation of Bi_2WO_6 superstructures with similar hierarchical architecture, secondary particle size, crystalline shape, exposed crystalline lattice planes, and crystalline content. The only difference is that as the increasing hydrothermal reaction temperature, their specific surface areas of the products were different due to the increase thickness of crystalline rectangular platelets. The specific surface area of the product is very important, because when levels of crystalline content of Bi_2WO_6 flake-ball particles is similar, the higher the specific surface area is, the better the photocatalysis ability it shows.

This proportional relation could be explained by the fact that the initial rate of AcH decomposition was expressed by first-order kinetics with respect to the amount of surface-adsorbed AcH, which is proportional to the specific surface area of Bi_2WO_6 samples.

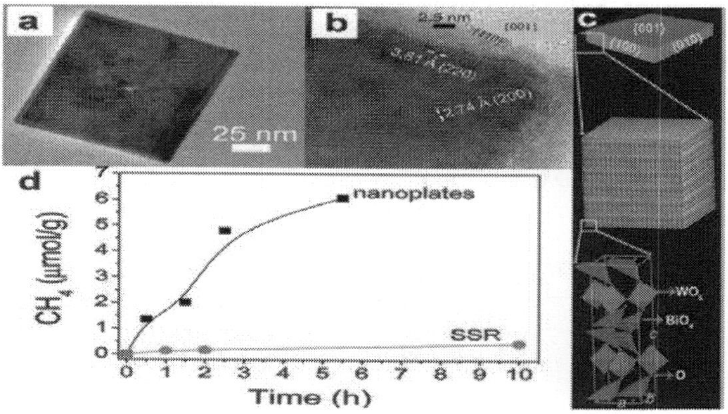

Figure 21. (a) TEM, (b) HRTEM, and (c) structural model of Bi_2WO_6 square nanoplates. (d) CH_4 generation over nanoplates and the SSR sample under visible–light irradiation ($\lambda > 420$ nm) [257].

In 2011, Zou and coworkers reported a remarkable increase in the CO_2 reduction with water to yield CH_4 over Bi_2WO_6 square nanoplates (Fig. 21 (a)–(c)) under visible-light irradiation as compared with yield over Bi_2WO_6 made by solid-state reaction (SSR) [257]. In detail, the CH_4 production rate increases from 0.045 mmol g^{-1} h^{-1} for the SSR sample to 1.1 mmol g^{-1} h^{-1} for the nanoplate catalyst (Fig. 21 (d)) [257]. Considering that the band gap of B_2WO_6 nanoplates and SSR sample is very close, geometrical factors of the photocatalyst is mainly responsible for the photoactivity enhancement. Firstly, reducing lateral dimension of the nanoplate to the nanometer scale offers a higher specific surface area. Secondly, the ultrathin geometry of the nanoplate facilitates the transfer of the charge carriers from the bulk onto the surface, where they participate in the photoreduction reaction. Thirdly, the preferentially exposed (001) crystal plane of the nanoplates is more effective than other crystal planes [257].

3.3. BiOX (X = Cl, Br, and I)

Another new class of interesting layered materials, Bismuth oxyhalides (BiOX; X = Cl, Br, and I), shows promising photocatalytic energy conversion and environment remediation ability, because of their unique layered–structure–mediated fascinating physicochemical properties and suitable band structures, along with their high chemical and optical stability, nontoxicity, low cost, and corrosion resistance [258-260]. The layered BiOX (X = Cl, Br and I) semiconductor materials, as members of the Sillen–Aurivillius family, have a tetragonal PbFCl–type structure (space group *P4/nmm*), which consists of [X–Bi–O–Bi–X] slices stacked together by the nonbonding (van der Waals) interactions through halogen atoms along the *c*-axis [261]. In each

http://dx.doi.org/10.5772/62206

[X–Bi–O–Bi–X] layer, a bismuth center is surrounded by four oxygen and four halogen atoms, creating an asymmetric decahedral geometry. The covalent bonds is the interaction bond within the [Bi_2O_2] layers, whereas the [X] layers are stacked together by van der Waals, forces (nonbonding interactions) between the X atoms along the *c*-axis. The strong intralayer covalent bonding and the weak interlayer van der Waals interaction can induce the formation of layered structures. For BiOX crystals, the valance band maximum mainly comprises of O2p and X np states (n = 3, 4, and 5 for X = Cl, Br and I, respectively) and the Bi 6p states dominate the conduction band minimum [262-266]. As the atomic numbers of X increases, the contribution of X ns states increases remarkably, and the dispersive characteristic of band energy level becomes more and more striking, thereby narrowing the band gap. Taking BiOCl, for example, as illustrated in Fig. 22, BiOX (X = Cl, Br, I) are characterized by the layered structure that are composed of [Bi_2O_2] slabs interleaved with double halogen atom slabs along the [001] direction. A highly anisotropic structural, electrical, optical, and mechanical properties of this material origin from the nature of its strong intralayer covalent bonding and the weak interlayer van der Waals, interaction, this unique structure allows BiOX to apply in many promising potential applications including photocatalytic wastewater and indoor-gas purification, water splitting, organic synthesis, and selective oxidation of alcohol [267-277].

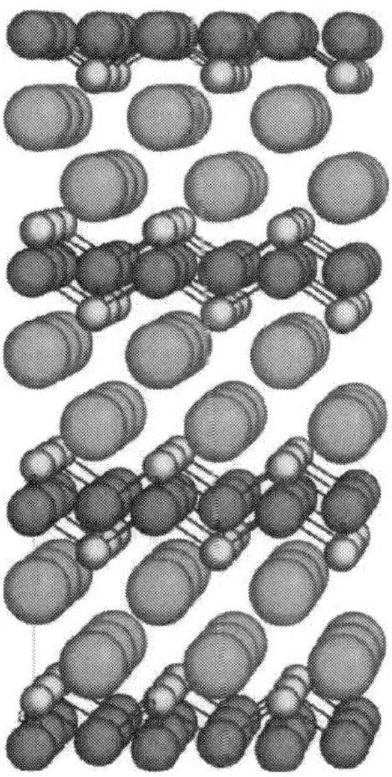

Figure 22. The schematic diagram of crystal structure of BiOCl (green, Cl atoms;yellow, Bi atoms; red, O atoms) [278].

One–dimensional nanostructures (1D), which refers to the materials with nanoscaled thickness and width, while the length can be several micrometers or longer, is considered to be promising in photocatalysis application. The prolonged length scale may allow the 1D nanomaterials to contact the macroscopic world for various measurements [279,280]. Besides, the high aspect ratio of 1D nanostructured semiconductors also facilitates the fast photoexcited electron-hole separation, which is favorable for highly efficient photocatalytic reactions.

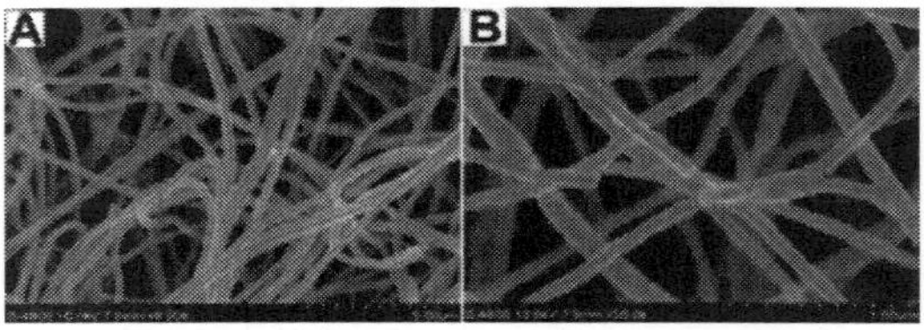

Figure 23. SEM images of the prepared (A) PAN/$BiCl_3$ nanofibers and (B) BiOCl nanofibers [281].

BiOX (X = Cl, Br and I) material is naturally preferentially grow into nanoplates/sheets with 2D features due to its highly anisotropic layered structures. As a result, hard templates for the synthesis of the 1D bismuth oxyhalide nanostructures is commonly used, becaused the template can be easily removed by subsequent thermal or chemical treatments [281-283]. For example, Liu et al. [281] developed the electrospinning method to synthesize BiOCl nanofibers, as shown in Fig. 23. After thermal removal of the polyacrylonitrile (PAN) template at 500°C for 10 h, they can obtain the BiOCl nanofibers with diameters ranging from 80 to 140 nm. Interestingly, the as-prepared BiOCl nanofibers showed high activity towards rhodamine B (RhB) degradation under the UV irradiation, and the photodegradation rate was found to be about three times faster than that of Bi_2O_3 nanofibers obtained in the same way. In addition to PAN, some other templates involving activated carbon fibers (ACFs) [282] and anodic aluminium oxide (AAO) [283] have also been used to prepare BiOCl nanofibers/nanowire arrays, which displayed efficient photocatalytic performance in the degradation of organic dyes.

In the past few years, 2D nanomaterials, such as graphene, transition metal dichalcogenides and layered double hydroxides (LDHs), have gained great attention for their extraordinary physical/chemical features and promising applications in a great deal of applications [284-287]. Intrinsically, van der Waals bonds or electrostatic forces between the layer structure in such 2D nanostructures is the origin of its lamellar structure. Similarly, the layered structure makes BiOX (X = Cl, Br and I) tend to the intrinsic 2D nanostructures, such as nanoplates, nanosheets and nanoflakes. The formed intra-electric field between [Bi_2O_2] layers and halogen atom layers could accelerate the transfer of the photo-induced carriers and enhance the photocatalytic activity of BiOX (X = Cl, Br and I) [288].

To date, numerous synthetic methodologies have been exploited for the preparation of 2D BiOX nanomaterials, such as hydrolysis, [288-292] hydrothermal/solvothermal synthesis, [293-295] and thermal annealing [296]. For instance, recently, Zhang's group [293] has synthesized 2D BiOCl nanosheets with predominantly exposed {001} and {010} facets by selective addition of the mineralizing agent NaOH. Interestingly, BiOCl nanosheets with exposed {001} facets displayed higher UV–induced photocatalytic degradation of MO dye, while the counterpart with exposed {010} facets exhibited higher degradation activity under visible light. On the one hand, the generated internal electric field along the [001] direction is more favorable for direct semiconductor photoexcitation under UV irradiation as shown in Fig. 24, which was also confirmed by the higher photocurrent of {001} facets than that of {010} facets from the transient photocurrent responses. On the other hand, compared with {001} facets, the larger surface area and open channel feature of {010} facets facilitate the adsorption of dye molecules, which further results in its better indirect dye photosensitization performance under visible light irradiation.

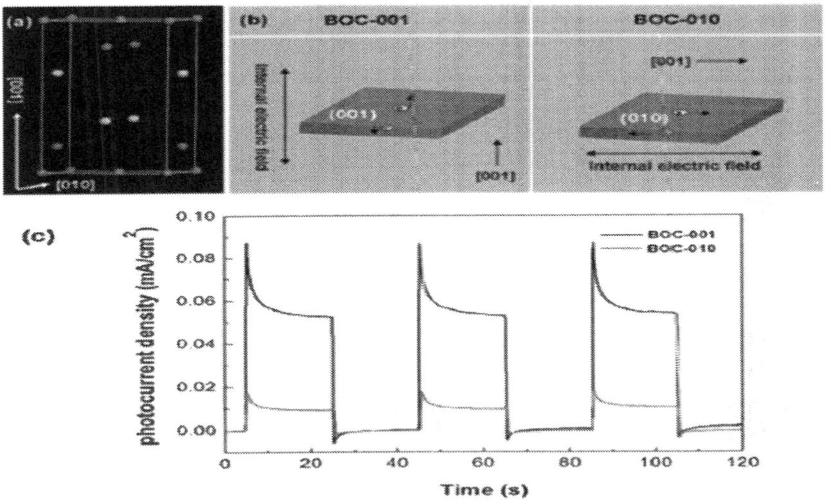

Figure 24. (a) Crystal structure of BiOCl. (b) Model showing the direction of the internal electric field in each of the BiOCl nanosheets. (c) Photocurrent responses of the BiOCl nanosheets in 0.5 M Na_2SO_4 aqueous solutions under UV-vis irradiation [293].

Researches have also using density functional theory (DFT) computations to reveal the nature of such a facet-dependent photocatalytic property in BiOX (X = Cl, Br and I) [297]. The halogen X–terminated {001} facets shows great thermodynamic stability and could efficiently separate photo-generated electron–hole pairs, whereas the formation of deep defect levels in the band gap of BiX–terminated {110} and other facets with surface O vacancies are bad for the carrier separation. This finding reveals the insight into the fundamental facet determined photocatalysis of BiOX (X = Cl, Br and I), which explains the superior photocatalytic performance of BiOX (X = Cl, Br and I) nanosheets with higher percentage of {001} facets than those with lower ones [289,290,296].

Self-assembly is a strong tool in nanotechnology fabrication of making low dimensional (e.g., 1D nanorods, 2D nanosheets, etc.) materials into their higher–dimension (3D) multifunctional superstructures, which plays a major role in material synthesis and device engineering and has been paid much attention recently [298-302]. Comparison with 1D and 2D nanostructures, 3D hierarchical nano/microstructures, which integrate the features of the nanoscale building units and their assembled architectures, are more attractive for solar energy storage and conversion [300-302]. Furthermore, 3D architectures could endow the BiOX (X = Cl, Br, I) semiconductors with improved light harvesting, shortened diffusion pathways, faster interfacial charge separation and more reactive sites, thus enhancing their photocatalytic efficiencies.

Hydro/solvothermal routes are definitely the most robust method among the methodological synthesis of the 3D BiOX (X = Cl, Br and I) hierarchical assemblies, [303-322] which are usually carried out at critical conditions of water or other organic solvents. In 2008, a generalized solvothermal process has been developed by Zhang et al. [303], who use ethylene glycol (EG) to prepare BiOX (X = Cl, Br, and I) hierarchical microspheres from 2D nanoplates. The band gaps of the resulting BiOX (X = Cl, Br and I) samples are calculated to be 3.22, 2.64, and 1.77 eV for BiOCl, BiOBr, and BiOI, respectively. Under visible-light irradiation, the BiOI sample exhibited the best photocatalytic performance with the order of BiOI > BiOBr > BiOCl evaluated by MO dye solution degradation. Almost at the same time, Tang et al. [304] also prepared 3D microspherical BiOBr architectures assembled by nanosheets through EG–assisted solvothermal synthesis. The band gap of the BiOBr architectures is 2.54 eV, so it shows higher photocatalytic activity for MO decomposition under visible–light irradiation than the BiOBr bulk plates.

How to realize the hierarchical architectures in the microstructure modulation of BiOX (X = Cl, Br and I) nano/microstructures with hollow voids draws much attention due to their better penetrability and higher light utilization. Recently, Huang group [313] has developed a method to synthesize uniform BiOBr hollow microspheres in the presence of 2-methoxyethanol solvent a mini-emulsion-mediated solvothermal route. The size of the BiOBr hollow microspheres is in the in the range of 1-2 mm and shell thickness of about 100 nm, which are composed of numerous interlaced 2D nanosheets. As demonstrated in Fig 25, by observing a Tyndall effect of the precursor suspension, the author confirmed that the 1–hexadecyl-3–methylimidazolium bromide ionic liquid ([C16Mim]BrIL) can not only serve as a Br source but also create a colloidal mini-emulsions. The diameter of the BiOBr hollow microsphere is determined by the size of the of the emulsion because the reaction takes place at the phase interface edge of the mini–emulsion rather than in the itself. Under visible-light irradiation, such BiOBr hollow microspheres displayed superior photocatalytic activity in degradation of RhB dye and reduction of Cr^{VI} ions to the samples with micro-flower shape. Xia and co-workers [319] prepared BiOI hollow microspheres by the EG–assisted solvothermal method using 1-butyl-3-methylimidazoliumiodine ([Bmim]I) IL as the reactive templates and I source. Under visible-light irradiation, such 3D BiOI hollow microspheres exhibited higher photocatalytic activity toward MO degradation than that of 2D BiOI nanoplates. Besides the halide ion-containing ILs [304,304,312,314,317,322], surfactants such as poly(vinylpyrrolidone) (PVP) [308] and hexadecyltrimethylammonium bromide (CTAB) [311,312,314,316,318,319] have been used to tailor the self-assembly process of the BiOX (X = Cl, Br, I) hierarchical architectures, and in particular CTAB could act as reactive template to provide Br - ions for BiOBr.

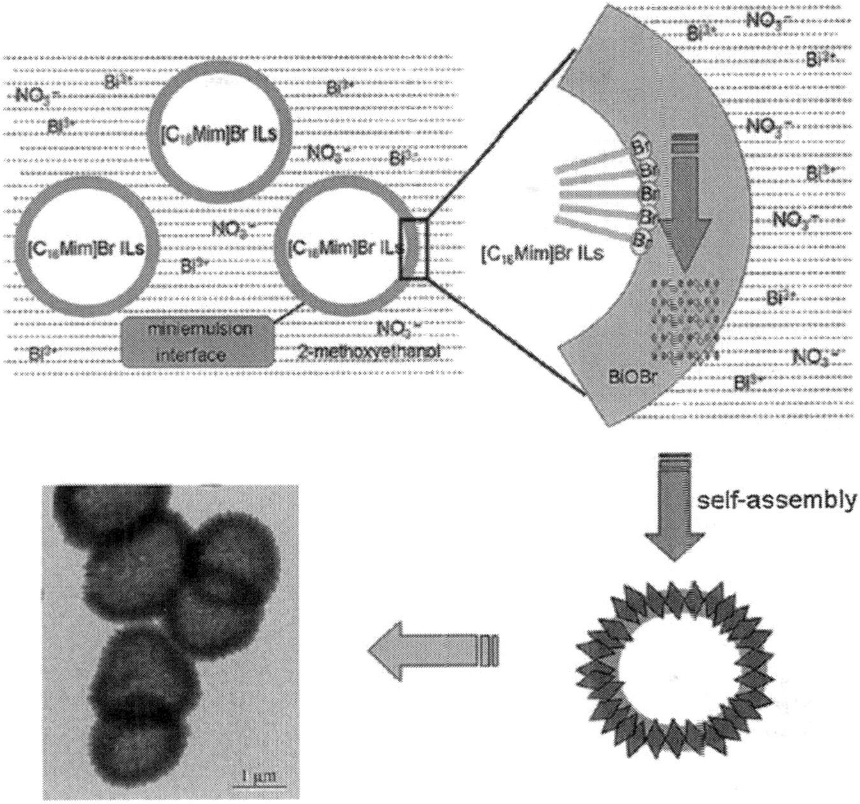

Figure 25. The schematic formation process of the BiOBr hollow microspheres by the mini-emulsionmediated solvothermal route [313].

Apart from the hydro/solvothermal syntheses, other synthetic procedures are also used to synthesize the ordered superstructures of BiOX (X = Cl, Br and I) semiconductors, such as hydrolysis [323,324], direct precipitation [325,326], sonochemical route [327,328], refluxing method [329], chemical bath [330] and solution oxidation process [331]. For example, Xiong and co-workers [331] reported a rapid in situ oxidation process to fabricate 3D flower-like BiOCl hierarchical nanostructures by reacting metallic Bi nanospheres and $FeCl_3$ aqueous solution at room temperature. As illustrated in Fig. 26, in the presence of Cl- ions, the redox potential of Bi species could be reduced from +0.308 V (Bi^{3+}/Bi vs. SHE) to +0.16 V (BiOCl/Bi). Therefore, the high redox potential of Fe^{3+} ($E(Fe^{3+}/Fe^{2+})$ = +0.771 V) could oxidize the surface of Bi nanospheres into the final 3D BiOCl hierarchical nanostructures. Compared with the commercial BiOCl sample, such flower-like BiOCl nanostructures obtained displayed much better RhB photodegradation activity and higher photoelectric conversion performances.

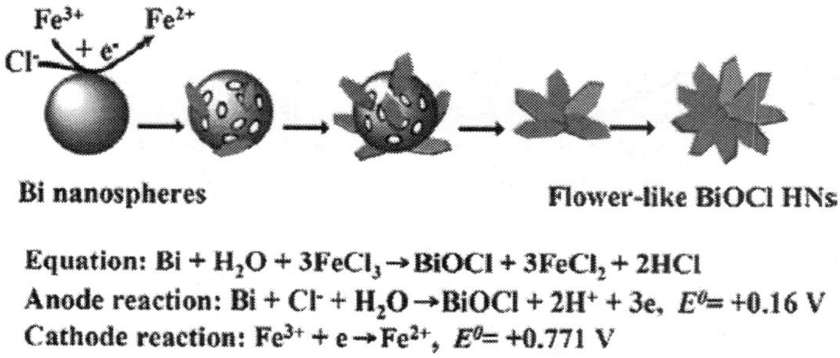

Figure 26. Schematic illustration of the fabrication of flower-like BiOCl hierarchical nanostructures by an in situ oxidation process [331].

Interestingly, Zhang [332] found that the photoactivity of BiOCl nanosheets shows a highly exposed facet-dependent effcet. The BiOCl nanosheets with exposed {001} facets showed higher direct semiconductor photoexcitation activity towards pollutant degradation due to both the surface atomic structure and suitable internal electric fields under UV light irradiation. Under visible light, highly exposed {010} facet BiOCl nanosheets shows superior indirect dye photosensitization activity for methyl orange degradation, which is due to the larger surface area and open channel characteristic of BiOCl nanosheets. It is belived that the enlarged surface area and open channel could enhance the adsorption capacity of methyl orange molecules as well as provide more contact sites between the photocatalyst and dye molecules, thereby facilitating the indirect dye photosensitization process because more efficient electron injection from the photoexcited dye into the conduction band of the catalyst happened (Fig. 27). These findings not only clarified the origin of facet-dependent photoreactivity of BiOCl nanosheets but also provided effective guidance for the design and fabrication of highly efficient bismuth oxyhalide photocatalyst.

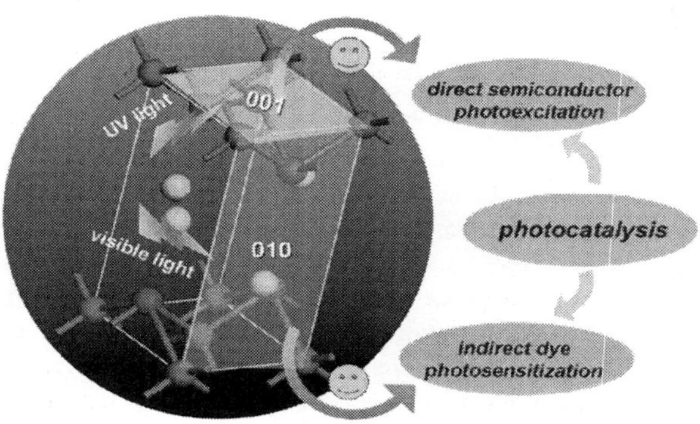

Figure 27. Schematic illustration of facet-dependent photoreactivity of BiOCl single-crystalline nanosheets [332].

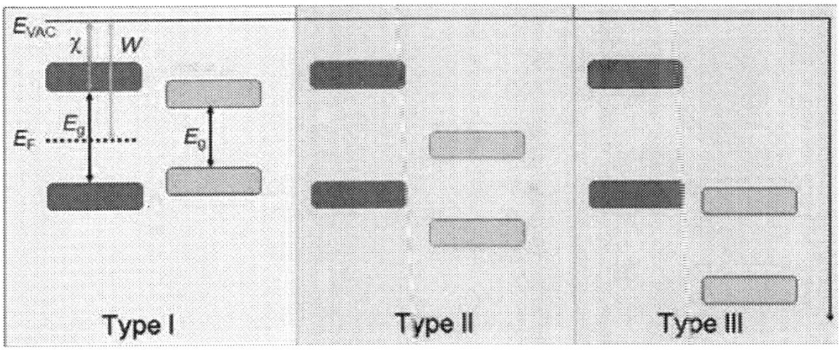

Figure 28. Schematic diagram of three types of semiconductor heterostructures [338].

It is speculated by Ye et al. that oxygen vacancy induced by UV light could yield an intermediate state between the valence and conduction bands to narrow the band gap, which may make oxygen–deficient BiOCl a promising alternative for the visible–light–driven photocatalytic reaction [333,334]. Zhang group recently found that oxygen vacancies of BiOCl can be created by the reductive ethylene glycol because it could easily react with the oxygen –terminated (001) surface at 160°C, which is evidenced by the electron spin resonance (ESR) spectra [335]. The resulting oxygen vacancies not only extended the light-response edge up to 650 nm but also enabled the effective capture of photoinduced electrons and molecular oxygen to generate superoxide anion radicals, both of which are of great important for realizing high photocatalysis efficiency of the photocatalyst. Recently, Xie et al. demonstrated that with the reduced thickness of the {001} facet–dominant BiOCl nanosheets to the atomic scale, the defects mainly change from isolated defects to triple vacancy, which could significantly promote the sunlight–driven photocatalytic activity of BiOCl nanosheets. The enhanced adsorption capability, the separation of electron-hole pairs and the generated reductive photoexcited electrons is mainly responsible for this improvement [261].

4. Composites systems

In contrast to one individual semiconductor photocatalyst, semiconductor composites are more intriguing for their interfacial heterostructures, which are formed at their junctures and have an important effect on their photocatalytic performances. There are usually three types of band positions in semiconductor heterojunctions: straddling gap (type I), staggered gap (type II) and broken gap (type III), as presented in Fig. 28. Among them, semiconductor composites with the staggered gap (type II) have drown much attention in the field of heterogeneous photocatalysis [336,337]. In this system, the photoinduced electrons and holes can be easily separated at the interface of the two semiconductors via effective interfacial charge transfer, thereby enhances the photocatalytic performance of the semiconductor composites.

4.1. Binary composite

In recent years, tremendous efforts have been made in surface modification of TiO_2 nanomaterials with other semiconductors. Most of these systems possess a high dye adsorption capacity, an extended light absorption range, enhanced charge separation, promoted mass-transfer and thus improved photocatalytic efficiency. This semiconductor provides the best compromise between catalytic performance and stability in aqueous media. Therefore, the magnetic iron oxide/TiO_2 composite photocatalysts have become the research focus in recent years. Using the magnetic properties of iron oxide itself for obtaining the magnetic recoverable photocatalyst has become an important issue in the magnetic iron oxide/TiO_2 composite photocatalyst system [339-342]. For instance, Wang and coworkers have reported the fabrication of core–shell $Fe_3O_4@SiO_2@TiO_2$ microspheres through a wet–chemical approach. The microspheres possess both ferromagnetic and photocatalytic properties. The TiO_2 nanoparticles on the surfaces of the microspheres degraded organic dyes under the illumination of UV light. Furthermore, the microspheres were easily separated from the solution after the photocatalytic process due to the ferromagnetic Fe_3O_4 core. The photocatalysts were recycled for further use and the degradation rate of methyl orange still reached 91% after six cycles of reuse [343]. As shown in Fig. 29, Chalasani and Vasudevan have demonstrated water-dispersible photocatalytic $Fe_3O_4@TiO_2$ core–shell magnetic nanoparticles by anchoring carboxy–methyl beta–cyclodextrin (CMCD) cavities to the TiO_2 shell, and photocatalytically destroyed endocrine-disrupting chemicals, bisphenol A (BPA) and dibutyl phthalate, present in water. The particles, which were typically 12 nm in diameter, were magnetic and removed from the dispersion by magnetic separation and then reused. The concentration of BPA solution was determined by liquid chromatography, and then irradiated under UV light for 60 min. After photodegradation of BPA, the CMCD–$Fe_3O_4@TiO_2$ nanoparticles that were separated from the mixtures by a magnet, and can be reused for the photodegradation of newly prepared BPA solutions. The recycle photocatalytic performance of CMCD–$Fe_3O_4@TiO_2$ for the photodegradation of BPA was excellent and stable, retaining 90% efficiency after 10 cycles [345]. For obtaining the magnetically recovered photocatalysts, Fe_3O_4 and γ-Fe_2O_3 were often employed due to their higher saturation magnetization and good magnetic separation ability.

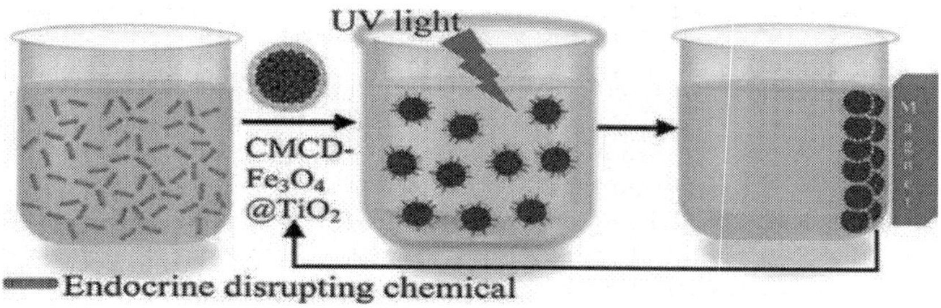

Figure 29. Scheme for the reuse of cyclodextrin-functionalized $Fe_3O_4@TiO_2$ for photocatalytic degradation of endocrine-disrupting chemicals in water supplies [344].

On the other hand, α-Fe_2O_3 has often been introduced into the magnetic iron oxide/TiO_2 composite photocatalyst in order to use its narrow band gap properties and to obtain magnetic iron oxide/TiO_2 composite heterostructures [344-348]. For example, Peng and coworkers have synthesized Fe_2O_3/TiO_2 heterostructural photocatalysts by impregnation of Fe^{3+} on the surface of TiO_2 and annealing at 300°C, the composites possess different mass ratios of Fe_2O_3 vs. TiO_2. The photocatalytic activities of Fe_2O_3/TiO_2 heterocomposites, pure Fe_2O_3 and TiO_2 were studied by the photocatalytic degradation of Orange II dye in aqueous solution under visible-light ($\lambda > 420$ nm) irradiation. The Fe_2O_3/TiO_2 heterogeneous photocatalysts exhibited an enhanced photocatalytic ability for Orange II, higher than either pure Fe_2O_3 or TiO_2. The best photocatalytic performance for Orange II could be obtained when the mass ratio in Fe_2O_3/TiO_2 is 7 : 3. The results illustrate that the generation of heterojunctions between Fe_2O_3 and TiO_2 is key for improving movement and restraining the recombination of photoinduced charge carriers, and finally improving the photocatalytic performance of Fe_2O_3/TiO_2 composites [348]. Recently, Palanisamy and coworkers have prepared Fe_2O_3/TiO_2 (10, 30, 50, 70 and 90 wt% Fe_2O_3) photocatalysts by a sol-gel process. Mesoporous Fe_2O_3/TiO_2 composites exhibited excellent photocatalytic degradation ability for 4–chlorophenol in aqueous solution under sunlight irradiation. The author claimed that the photogenerated electrons in the VB of TiO_2 are transferred to Fe(III) ions resulting in the reduction of Fe(III) ions to Fe(II) ions. Thus, the photoinduced holes in the VB of Fe_2O_3/TiO_2 cause an oxidation reaction and decompose the 4-chlorophenol to CO_2 and H_2O. Meanwhile the transferred electrons in Fe(III) ions could trigger the reduction reaction [349].

Wang's group successfully synthesized Bi_2WO_6-TiO_2 hierarchical heterostructure through a simple and practical electrospinning-assisted route (Fig. 30 (A) and (B)) [350]. As shown in Fig. 30 (A), Bi_2WO_6 nanoplates grew aslant on the primary TiO_2 nanofibers. These three dimensional (3D) Bi_2WO_6–TiO_2 hierarchical heterostructures exhibited enhanced visible–light –driven photocatalytic activity for the decomposition of CH_3CHO, which was almost eight times higher than that of the Bi_2WO_6 sample, and the decomposition rate by the bare TiO_2 could be neglected under visible light irradiation. This high photocatalytic activity was ascribed to the reduced probability of electron–hole recombination and the promoted migration of photogenerated carriers. Similarly, Wang et al. [352] fabricated SnO_2–TiO_2 heterostructured photocatalysts based on TiO_2 nanofibers by combining the electrospinning technique with the hydrothermal method (Fig 30C and D). This SnO_2–TiO_2 composite possessed a high photocatalytic activity for the degradation of rhodamine B (RhB) dye under UV light irradiation, which was almost 2.5 times higher than that of the bare TiO_2. The enhanced photocatalytic efficiency was attributed to the improvement of the separation of photogenerated electrons and holes. Wang's group [352] prepared a graphene–Bi_2WO_6 composite via an in situ hydrothermal reaction (Fig. 31(A) and (B)). This graphene–Bi_2WO_6 photocatalyst showed significantly enhanced photocatalytic activity for the degradation of RhB under visible light ($\lambda > 420$ nm), which was three times greater than that of the pure Bi_2WO_6. The enhanced photocatalytic activity could be attributed to the negative shift in the Fermi level of graphene–Bi_2WO_6 and the high migration efficiency of photoinduced electrons; these electrons may not only be effectively involved in the oxygen reduction reaction but also suppress the charge recombination. Kudo et al. [353] reported the composite of reduced graphene oxide (RGO) with

$BiVO_4$, where an significantly improved PCE activity of a near 10–fold enhancement was observe compared with pure $BiVO_4$ under visible-light irradiation. The longer photoexcited electron lifetime of $BiVO_4$ is mainly responsible for this improvement as the electrons are injected to RGO instantly at the site of generation, leading to a significant reduction in charge recombination.

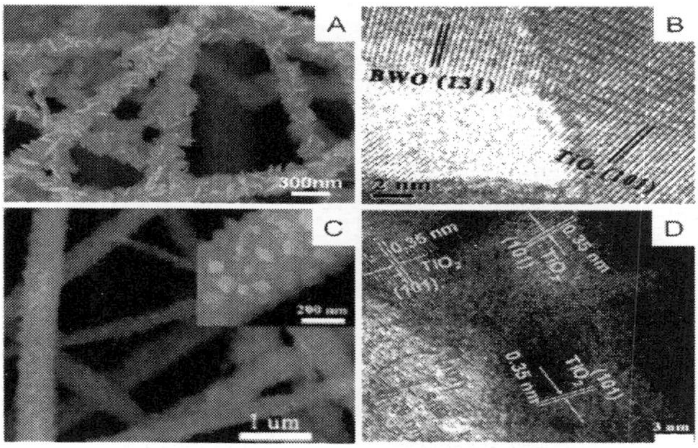

Figure 30. SEM (A) and HRTEM (B) images of Bi_2WO_6/TiO_2 [343]; SEM (C) and HRTEM (D) images of SnO_2/TiO_2 heterostructures [351].

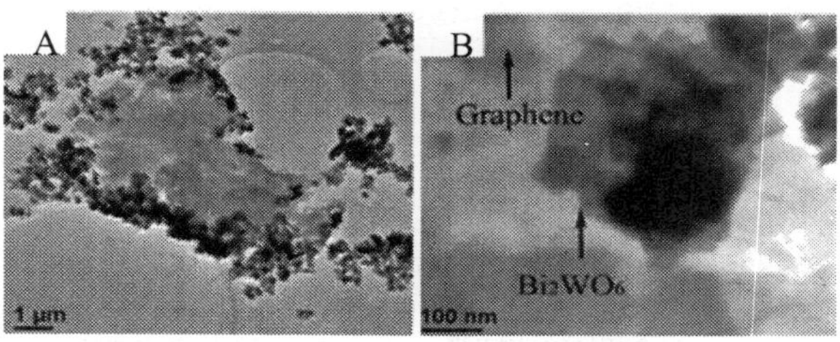

Figure 31. (A and B) TEM images of graphene decorated with Bi_2WO_6 composite [352].

A simple soft-chemical method was used to synthesize the $BiOI/TiO_2$ heterostructures with different Bi to Ti molar ratios at low temperature of 80°C. The degradation of methyl orange under visible-light irradiation (λ > 420 nm) of the material revealed that the $BiOI/TiO_2$ heterostructures exhibited much higher photocatalytic activities than pure BiOI or TiO_2, where 50%$BiOI/TiO_2$ showed the best activity among all these heterostructured photocatalysts [348]. BiOBr–Bi_2WO_6 mesoporous nanosheet composite enhanced photocatalytic activity is attributed to well-matched band edge positions of BiOBr and Bi_2WO_6 and the large specific surface area of the mesoporous nanosheet composites in view of the incorporation of mesopores and the highly exposed BiOBr (001) facet, compared with pure BiOBr and Bi_2WO_6 under exposure to a 3–W LED light [354].

During the past few years, another promising carbon material, graphene, which possess many unique properties, has been used for corporation with BiOX (X = Cl, Br and I) and significantly improved photocatalytic efficiencies was achieved [356-360]. Ai and coworkers [357] have developed a facile solvothermal route tosynthesize BiOBr/graphene hybrids using graphene oxide (GO), bismuth nitrite, and CTAB as the precursors. As shown in Fig. 32, BiOBr nanoplates with hundreds nanometers in size are dispersed randomly on the 2D graphene sheet surface. Evaluated by the removal of gaseous NO under visible-light irradiation, the as-prepared BiOBr/graphene hybrid displays a two times higher removal rate than that of pure BiOBr. It is evidenced that the strong chemical bonding between BiOBr and graphene is mainly responsible for the fast photogenerated electrons transfer from BiOBr to graphene, which further inhibities the unwanted recombination and leading to its enhanced photocatalytic activity.

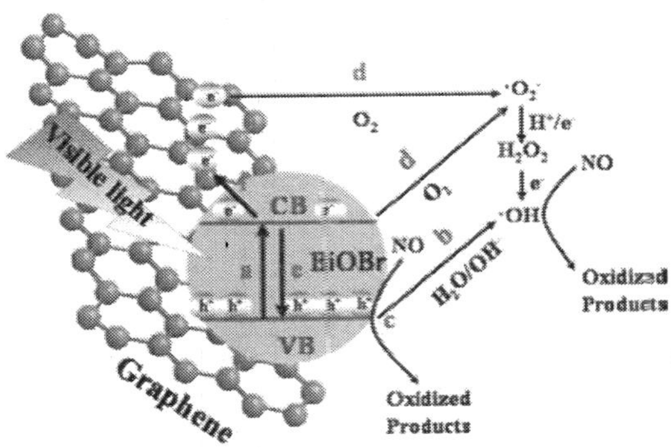

Figure 32. Schematic illustration of the visible–light photocatalytic enhancement of BiOBr/graphene nanocomposites [357].

4.2. Ternary composite

To date, although a variety of approaches have been developed to prepare many kinds of visible–light–driven semiconductor heterojunction photocatalysts, many shortage is still needed to be overcomed, for example, the limited region of visible-light photo–response. To solve these problems, multi–component heterojunction systems have been developed [361,362], in which two or more visible-light active components and an electron-transfer system are spatially integrated as shown in Fig 33 [363].

As demonstrated in Figure 33, since both semiconductor A (S-A) and semiconductor B (S-B) can be excited by UV/visible light and have different photoabsorption ranges, the conjunction of the two materials can overlap and broaden the range of UV/visible–light photoresponse. At the same time, it is well-known that the photocatalytic reaction is initiated by the incident UV/

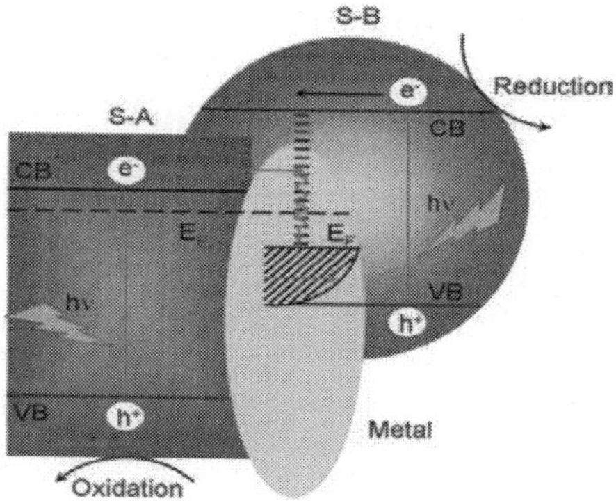

Figure 33. Schematic structure of multicomponent heterojunction systems [363].

visible photons with energy equal or higher than the band-gap in both S–A and S-B, which lead to the creation of photogenerated holes in their VB and electrons in their CB. On the one hand, the electrons in the CB of S–A easily flow into metal (electron transfer I: S–A metal) through the Schottky barrier because the CB (or the Fermi level) of S–A is higher than that of the loaded metal, which is consistent with the previous study on electron transfer from the semiconductor (such as TiO_2) to metal (such as Ag and Au) [361,364]. This process of electron transfer I is faster than the electron-hole recombination between the VB and the CB of S–A. Thus, plenty of electrons in the CB of S–A can be stored in the metal component. As a result, more holes with a strong oxidation power in the VB of S–A escape from the pair recombination and are available to oxidize the pollutants or OH-. On the other hand, since the energy level of metal is above the VB of S–B, holes in the VB of S–B also easily flow into metal (electron transfer II: metal S–B, see Fig. 33), which is faster than the electron-hole recombination between the VB and CB of S–B. More electrons with a strong reduction power in the CB of S–B can escape from the pair recombination and are available to reduce some absorbed compounds (such as O_2 and H^+). Therefore, simultaneous electron transfer I and II (i.e., vectorial electron transfer of S–A metal to S–B in Fig. 33) can occur as a result of UV/visible-light excitation of both S–A and S–B. In these vectorial electron–transfer processes, metal in multicomponent heterojunction systems acts as a storage and/or a recombination center for electrons in the CB of S–A and holes in the VB of S–B, and contributes to enhancing interfacial charge transfer and realizing the complete separation of holes in the VB of S–A and electrons in the CB of S–B. Therefore, the multi-component heterojunction systems can simultaneously and efficiently generate holes with a strong oxidation power in the VB of S–A and electrons with a strong reduction power in the CB of S–B, resulting in greatly enhanced photocatalytic activity, compared with the single semiconductor or semiconductor heterojunctions mentioned above.

In 2006, by using a facile photo–chemical technique, Tada et al. [361] developed a CdS–Au –TiO_2 ternary component nanojunction system (Fig. 34 (A) and (B)). This CdS–Au–TiO_2 triple nanojunction shows significantly improved photocatalytic activity, which was far higher than that of either the single-component or two–components systems. For this photocatalytic CdS

–Au–TiO_2 nanojunction system, 52.2% of methylviologen (MV^{2+}) have been reduced in 100 min, which are 1.6, 1.8 and 2.3 times higher than that of Au/TiO_2, CdS/TiO_2 and TiO_2 [361].

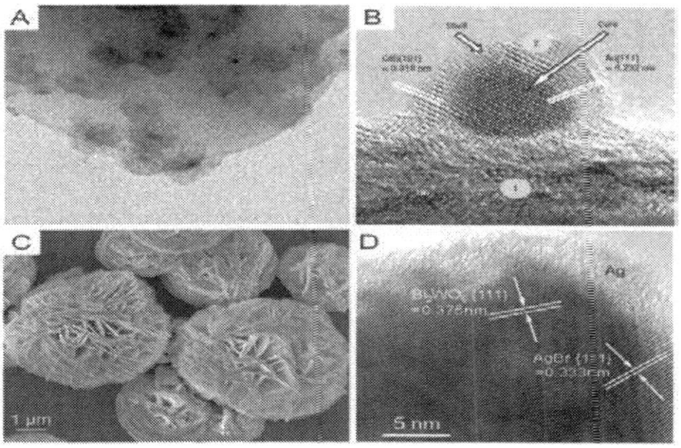

Figure 34. TEM (A) and HRTEM (B) images of Au@CdS–TiO_2 [361]; SEM(C) and HRTEM (D) images of the AgBr–Ag–Bi_2WO_6 nanojunction system [364].

Subsequently, an AgBr–Ag–Bi_2WO_6 nanojunction system was developed by a facile deposition–precipitation method (Fig. 34 (C) and (D)) [364]. This AgBr–Ag–Bi_2WO_5 nanojunction system shows much higher visible–light–driven photocatalytic activity than a photocatalyst with single visible-light response components, such as Bi_2WO_6 nanostructures, Ag–Bi_2WO_6 and AgBr–Ag–TiO_2. For example, with the AgBr–Ag–Bi_2WO_6 nanojunction system as the photocatalyst, the MX–5B could be photocatalytically degraded (42.8 mg L^{-1}) within 60 min under visible-light irradiation, which is higher than that of Bi_2WO_6 nanostructures (2.0 mg L^{-1}), Ag-Bi_2WO_6 (2.9 mg L^{-1}) and AgBr–Ag–TiO_2 (34.1 mg L^{-1}). Furthermore, 65% of pentachlorophenol could be mineralized within 4 h by AgBr–Ag–Bi_2WO_6, which is much higher than that (34.5%) of the AgBr–Ag–TiO_2 composite. This excellent visible–light–driven photocatalytic performance was mainly attributed to the vectorial interparticle electron transfer driven by the two-step excitation of both visible-light-driven components (AgBr and Bi_2WO_6).

A one-step low–temperature chemical bath method was developed to synthesize the flower–like Ag/AgCl/BiOCl composite [365]. The as–prepared Ag/AgCl/BiOCl composite exhibited enhanced visible–light photocatalytic activity on photodegradation of rhodamine B, which was greatly improved in comparison with either pure Ag/AgCl or BiOCl. It is evidenced that the superoxide radical, chlorine radical and the hole play a critical role in the photocatalytic degradation of RhB over the Ag/AgCl/BiOCl. Next, Ag/AgX/BiOX (X = Cl, Br) three-component visible-light-driven photocatalysts were synthesized by a low–temperature chemical bath method (Fig. 35) [366]. The Ag/AgX/BiOX composites showed enhanced visible–light–driven photocatalytic activity for the degradation of rhodamine B, which was much higher than Ag/AgX and BiOX. The photocatalytic mechanisms were analyzed by active species trapping and superoxide radical quantification experiments. The role of metallic Ag in Ag/AgCl/BiOCl and Ag/AgBr/BiOBr were analyzed, and we found that the role of metallic Ag was a surface plasmon resonance and the Z–scheme bridge for Ag/AgCl/BiOCl and Ag/AgBr/BiOBr, respectively. This results suggests that no matter in narrow band gap photocatalysts (Eg < 3.1 eV) or wide band gap photocatalysts (Eg > 3.1 eV), metallic Ag can enhance visible-light-driven photocatalytic activity though the different roles.

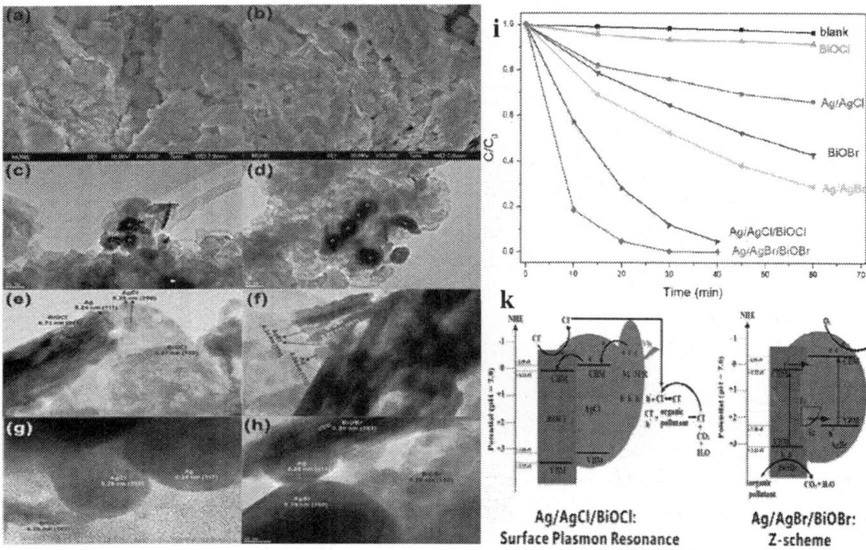

Figure 35. FESEM images of Ag/AgCl/BiOCl (a) and Ag/AgBr/BiOBr (b): the red arrows pointing out the Ag/AgX. TEM images of Ag/AgCl/BiOCl (c) and Ag/AgBr/BiOBr (d): blue rings show the small Ag/AgX, and the red dots point out the large Ag/AgX. HRTEM images of Ag/AgCl/BiOCl (e) and Ag/AgBr/BiOBr (f) with small Ag/AgX and HRTEM images of Ag/AgCl/BiOCl (g) and Ag/AgBr/BiOBr (h) with large Ag/AgX (i)the photocatalytic degradation percentage of RhB under visible-light irradiation ($\lambda \geq 400$ nm) and (k) schematic structure of multicomponent heterojunction systems [366].

Recently, our group has demonstrated a simple and efficient one-pot approach to prepare Ag/r–GO/TiO_2 composites using solvothermal method under atmospheric pressures (Fig. 36) [367], where N,N–dimethylacetamide serves as the reducing agent for Ag and GO reduction. On account of the experimental result, we concluded that the introduction of Ag into classical graphene/TiO_2 system (i) availably expands the absorption range, (ii) improves the photogenerated electron separation and (iii)increases the photocatalysis reaction active sites. The optimized composite sample exhibits outstanding photocatalysis activity compared with pure TiO_2 under simulated sunlight. We further proposed that besides the above three advantages of Ag, different sizes of Ag nanoparticles are also responsible for the improved photocatalysis ability, where small–sized Ag nanoparticles (2~5 nm) could store photoexcited electrons that generated from TiO_2, while large-sized Ag nanoparticles could utilize visible light due to their localized surface plasmon resonance (LSPR) absorption. Our work gives a new insight into the photocatalysis mechanism of noble metal/r-GO/TiO_2 composites and provides a new pathway into the design of TiO_2-based photocatalysts and promote their practical application in various environmental and energy issues.

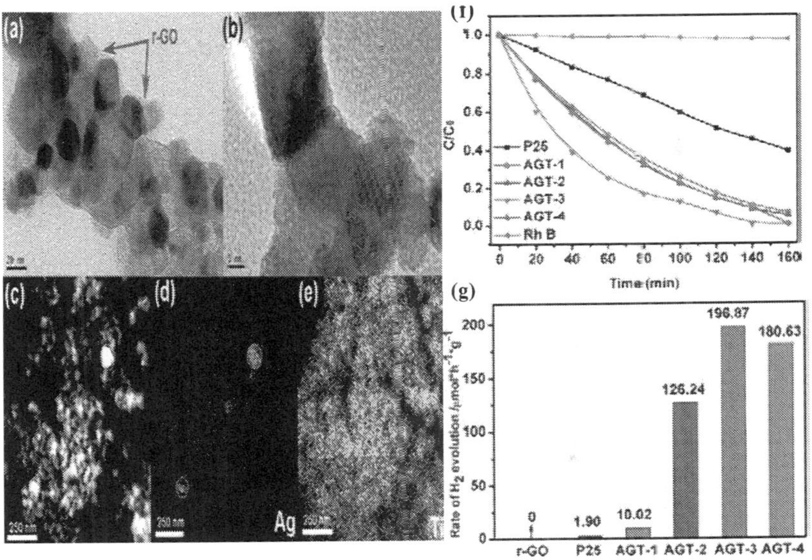

Figure 36. (a) TEM and (b) HRTEM images of sample AGT. (c) STEM model of Ag/r-GO/TiO_2. Elemental mapping of (d) Ag and (e) Ti in the same area in (c). (f)Photocatalytic degradation of Rh B under simulated sunlight irradiation over P25, Ag/r-GO/TiO_2 composites with different $AgNO_3$ contents. (g) Comparison of the photocatalytic activity of r -GO, P25 and Ag/r-GO/TiO_2 composites with different $AgNO_3$ contents for the photocatalytic H_2 production under simulated sunlight irradiation [367].

Next, we have replaced the Ag with MoS_2 quantum dots (QDs) and demonstrated a simple and an efficient one-pot approach to prepare MoS_2 quantum dots-graphene-TiO_2 (MGT) composites using a solvothermal method under obtained atmospheric pressures and at low temperatures (Fig. 37) [368]. The shape of MoS_2 obtained using this method is quantum dot instead of a layered sheet because of the interaction between functional groups on GO sheets and Mo precursors in a suitable solvent environment. In addition, it shows significantly increased photodegradation performance even without a noble-metal cocatalyst, which is due to the increased charge separation, visible-light absorbance, specific surface area and reaction sites upon the introduction of MoS_2 QDs. Besides, the enhancement mainly came from holes left in the TiO_2 crystals rather than electrons transferring to reduced graphene oxide (RGO).

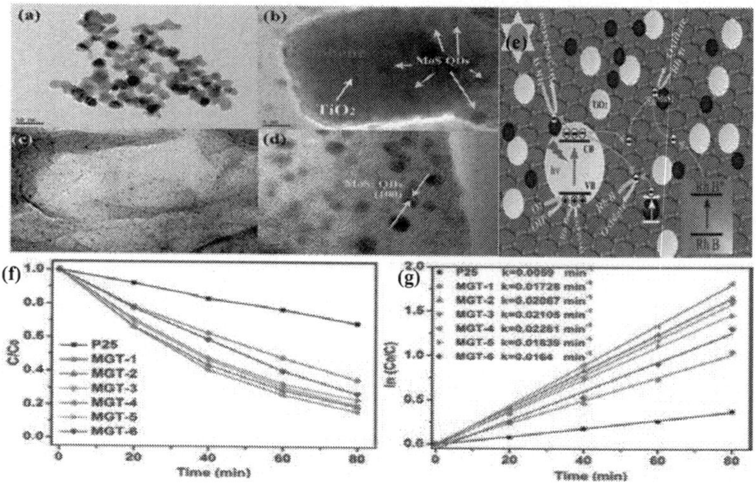

Figure 37. TEM and HRTEM images of the sample (a), (b) MGT–4 and (c), (d)MoS_2–graphene. (e) Proposed mechanism for the photodegradation of RhB by MGT under simulated sunlight irradiation (f) Photocatalytic degradation and (g) photocatalytic degradation reaction of RhB under simulated sunlight irradiation over P25, MGT composites with different MoS_2 contents [368].

5. Summary and outlook

Photocatalysis appears to be a promising avenue to solve environmental and energy issues in the future. Although the photocatalytic processes involve a complicated sequence of multiple synergistic or competing steps, the efficient utilization of solar energy (especial visible-light energy) and improvement in separation and transportation of charge carriers are the main challenges and current trend to design highly effective photocatalysts. Finally, we conclude that this chapter, after discussing with various materials and its composites for photocatalytic process, may be useful for further applications in the area of energy and environment. In summary, we have discussed the general strategies and recent progress in photocatalysis for developing highly efficient and stable photocatalysts, including: (1) Titania (TiO_2), iron oxides (α-Fe_2O_3); (2) ternary oxide photocatalytic matericals, such as Bi systems photocatalytic materials and (3) semiconducting materials and its composites. The achieved progress in photocatalysis indicates a promising route to enhance the photocatalytic efficiencies of photocatalytic semiconductors.

To date, in addition to different kinds of semiconductor materials and its composites significant advances have been reported to improve the photocatalytic efficiencies that range from environmental remediation to clean–energy harvesting by enhancing the utilization of sunlight or improving the separation/transportation of the electron–hole pairs some examples

http://dx.doi.org/10.5772/62206

are highlighted in this chapter. Extending light-response to the visible– or even infrared regions, decreasing the amount of recombination of electrons and holes, and increasing the light–harvesting efficiency have been the major tools that have led to such advances. Although great advancements have been made in investigation of heterostructured photocatalytsts, it is still challenging to design more challenges in high efficiency of photocatalytic systems. First, there is no detailed understanding of the charge generation, separation and transportation across nanoscale interfaces of heterostructured photocatalysts, which are critical for the design and optimization of highly more-efficient and more-reliable photocatalysts. Second, while most available photocatalysts so far can only function in the UV or near–UV regime, the highly effective utilization of visible light is another challenge of heterostructured photocatalysts. Third, photostability of heterostructured photocatalyst is and will still be a major challenge for practical applications. Finally, elucidating and understanding the mechanisms that are involved in various photocatalytic reactions. Therefore, the deepening knowledge of the photocatalytic mechanism and exploration of new materials are indispensible to make substantial breakthroughs for practical application of photocatalysts.

Author details

Le Li and Minqiang Wang*

*Address all correspondence to: mqwang@mail.xjtu.edu.cn

Electronic Materials Research Laboratory (EMRL), Key Laboratory of Education Ministry, International Center for Dielectric Research, Xi'an Jiaotong University, Xi'an, China

References

[1] M. R. Hoffmann, S. T. Martin, W. Y. Choi, D. W. Bahnemann, Chem Rev. 1995, 95, 69–96.

[2] A. Kudo, Y. Miseki, Chem. Soc. Rev. 2009, 38, 253–278.

[3] X. Li, J. Q. Wen, J. X. Low, Y. P. Fang, J G. Yu, Sci. China Mater. 2014, 57,70–100.

[4] Y. Q. Qu, X. F. Duan, Chem. Soc. Rev. 2013, 42, 2568–2580.

[5] H. Tong, S. Ouyang, Y. Bi, N. Umezawa, M. Oshikiri, J. Ye, Adv. Mater. 2012, 24, 229–251.

[6] M. G. Kibria, H. P. T. Nguyen, K. Cui, S. Zhao, D. Liu, H. Guo, M. L.Trudeau, S. Paradis, A. R. Hakima, Z. Mi, ACS Nano. 2013, 7,7886–7893.

[7] Q. J. Xiang, J. G. Yu, M. Jaroniec, J. Am. Chem. Soc. 2012, 134, 6575–6578.

[8] J. A. Christians, R. C. M. Fung, P. V. Kamat, J. Am. Chem. Soc. 2014,136, 758–764.

[9] Q. Zhang, E. Uchaker, S. L. Candelaria, G. Cao, Chem. Soc. Rev. 2013, 42, 3127–3171.

[10] J. G. Yu, X. X. Yu, Environ. Sci. Technol. 2008, 42, 4902–4907.

[11] J. Kim, J. Lee, W. Choi, Chem. Commun. 2008, 756–761.

[12] Y. Bi, S. Ouyang, N. Umezawa, J. Cao, J. Ye, J. Am. Chem. Soc. 2011,133, 6490–6492.

[13] E. V. Kondratenko, G. Mul, J. Baltrusaitis, G. O. Larrazábal, J. Pérez-Ramírez, Energy Environ. Sci. 2013, 6, 3112–3135.

[14] S. C. Roy, O. K. Varghese, M. Paulose, C. A. Grimes, ACS Nano 2010, 4, 1259–1278.

[15] A. Dhakshinamoorthy, S. Navalon, A. Corma, H. Garcia, Energy Environ. Sci. 2012, 5, 9217–9233.

[16] P Zhou, J G Yu, M Jaroniec, Adv. Mater. 2014, 26, 4920–4935.

[17] E. Fujita, J. T. Muckerman, K. Domen, ChemSusChem. 2011, 4, 155–157.

[18] S. Linic, P. Christopher, D. B. Ingram, Nat. Mater. 2011, 10, 911–921.

[19] C. Li, F. Wang, J. C. Yu, Energy Environ. Sci. 2011, 4, 100–113.

[20] H. Zhou, Y. Qu, T. Zeid, X. Duan, Energy Environ. Sci. 2012, 5, 6732–6743.

[21] Q. Li, X. Li, S. Wageh, A. Al-Ghamdi, J. G. Yu, Adv. Energy. Mater. 2015, 17, 10010–10039.

[22] J. M. Elward, A. Chakraborty, J. Chem. Theory Comput. 2013, 9, 4351–4357.

[23] F. L. Formal, S. R. Pendlebury, M. Cornuz, S. D. Tilley, M. Grä tzel, J. R. Durrant, J. Am. Chem. Soc. 2014, 136, 2564–2574.

[24] P. Zhou, J. G. Yu, M. Jaroniec, Adv. Mater. 2014, 26, 4920–4935.

[25] L. L. Tan, S. P. Chai, A. R. Mohamed, Chemsuschem. 2012, 5,1868–1882.

[26] L. Sun, Z. L. Zhao, Y. C. Zhou, L. Liu, Nanoscale. 2012, 4, 613–620.

[27] Q. J. Xiang, J. G. Yu, M. Jaroniec, Chem. Soc. Rev. 2012, 41, 782–796.

[28] M. R. Hoffmann, S. T. Martin, W. Y. Choi, D. W. Bahnemann, Chem. Rev. 1995, 95, 69–96.

[29] O. Carp, C. L. Huisman, A. Reller, Prog. Solid State Chem. 2004, 32, 33–177.

[30] D. Dambournet, I. Belharouak, K. Amine, Chem. Mater. 2010, 22, 1173–1179.

[31] U. Diebold, Surf. Sci. Rep. 2003, 48, 53–229.

[32] O. Carp, C. L. Huisman, A. Reller, Prog. Solid State Chem. 2004, 32, 33–177.

[33] M. A. Fox, M. T. Dulay, Chem. Rev. 1993, 93, 341–357.

[34] A. Fujishima, T. N. Rao, D. A. Tryk, J. Photochem. Photobiol. C. 2000, 1, 1–21.

[35] U. Diebold, N. Ruzycki, G. S. Herman, A. Selloni, Catal.Today. 2003, 85, 93–100.

[36] A. S. Barnard, H. Xu, ACS Nano. 2008, 2, 2237–2242.

[37] J. K. Burdett, T. Hughbands, J. M. Gordon, J. W. Richarson, J. Smith, J. Am. Chem. Soc. 1987, 109, 3639–3646.

[38] R. Asahi, Y. Taga, Y. Mannstadt, A. Freeman, J. Phys. Rev. B. 2000, 61, 7459–7467.

[39] C. Sousa, F. Illas, J. Chem. Phys. 1994, 50, 13974–13983.

[40] A. Kubacka, M. Fernández-García, G Colón, Chem. Rev. 2012, 112, 1555–1614.

[41] A. Fujishima, K. Honda, Nature. 1972, 238, 37–41.

[42] A. L. Linsebigler, G. Lu, J. T. Yates, Chem. Rev. 1995, 95, 735–758.

[43] A. Fujishima, T. N. Rao, D. A. Tryk. J. Photochem. Photobiol. C: Photochem. Rev. 2000, 1, 1–21.

[44] K. Hashimoto, H. Irie, A. Fujishima, Jpn. J. Appl. Phys. 2005, 44, 8269–8285.

[45] U. I. Gaya, A. H. Abdullah, J. Photochem. Photobiol. C: Photochem. Rev. 2008, 9, 1–12.

[46] D. P. Macwan, P. N. Dave, S. Chaturvedi, J. Mater. Sci. 2011, 46, 3669–3686.

[47] X. Guan, J. Du, X. Meng, Y. Sun, B. Sun, Q. Hu, J. Hazard. Mater. 2012, 1, 215–216.

[48] M. D. Hernández-Alonso, F. Fresno, S. Suárez, J. M. Coronado, Energy Environ. Sci. 2009, 2, 1231–1257.

[49] W. Zhou, H. Liu, R. I. Boughton, G. Du, J. Lin, J. Wang, D. Liu, J. Mater. Chem. 2010, 20, 5993–6008.

[50] G. Liu, L. Wang, H. G. Yang, H. M Cheng, G. Q. Lu, J. Mater. Chem. 2010, 20, 831–843.

[51] S. G. Kumar, L. G. Devi, J. Phys. Chem. A. 2011, 115, 13211–13241.

[52] A. A. Ismail, D. W. Bahnemann, J. Mater. Chem. 2011, 21, 11686–11707.

[53] R. Leary, A. Westwood, Carbon. 2011, 49, 741–772.

[54] A. Primo, A. Corma, H. García, Phys. Chem. Chem. Phys. 2011, 13, 886–910.

[55] A. S. Weber, A. M. Grady, R. T. Koodali, Catal. Sci. Technol. 2012, 2, 683–693.

[56] R. Asahi, T. Morikawa, T. Ohwaki, K. Aoki, Y. Taga, Science. 2001, 293, 269–271.

[57] C. Burda, Y. B. Lou, X. B. Chen, A. C. S. Samia, J. Stout, J. L. Gole, Nano Lett. 2003, 3,1049–1051.

[58] S. U. M. Khan, M. Al. Shahry, W. B. Ingler, Science, 2002, 297, 2243–2245.

[59] C. Di Valentin, G. Pacchioni, A. Selloni, Chem. Mater. 2005,17, 6656–6665.

[60] S. Sakthivel, H. Kisch, Angew. Chem. Int. Ed. 2003, 42, 4908–4911.

[61] T. Umebayashi, T. Yamaki, H. Itoh, K. Asai, Appl. Phys. Lett. 2002, 81, 454–456.

[62] W. K. Ho, J. C. Yu, S. C. Lee, J. Solid State Chem. 2006,179, 1171–1176.

[63] W. Zhao, W. H. Ma, C. C. Chen, J. C. Zhao, Z. G. Shuai, J. Am.Chem. Soc. 2004, 126, 4782–4783.

[64] G. Liu, Y. N. Zhao, C. H. Sun, F. Li, G. Q. Lu, H. M. Cheng, Angew. Chem. Int. Ed. 2008, 47, 4516–4520.

[65] S. In, A. Orlov, R. Berg, F. Garcia, S. Pedrosa-Jimenez, M. S. Tikhov, D. S. Wright, R. M. Lambert, J. Am. Chem. Soc. 2007, 129, 13790–13791.

[66] J. G. Yu, W. K. Ho, Z. T. Jiang, L. Z. Zhang, Chem. Mater. 2002, 14, 3808–3816.

[67] D. Li, H. Haneda, N. K. Labhsetwar, S. Hishita, N. Ohashi, Chem. Phys. Lett. 2005, 401, 579–584.

[68] C. Di Valentin, E. Finazzi, G. Pacchioni, A. Selloni, S. Livraghi, A. M. Czoska, M. C. Paganini, E. Giamello, Chem. Mater. 2008, 20, 3706–3714.

[69] H. M. Luo, T. Takata, Y. G. Lee, J. F. Zhao, K. Domen, Y. S. Yan, Chem. Mater. 2004, 16, 846–849.

[70] X. T. Hong, Z. P. Wang, W. M. Cai, F. Lu, J. Zhang, Y. Z. Yang, N. Ma, Y. J. Liu, Chem. Mater. 2005, 17, 1548–1552.

[71] G. Liu, Z. G. Chen, C. L. Dong, Y. N. Zhao, F. Li, G. Q. Lu, H. M. Cheng, J. Phys. Chem. B. 2006, 110, 20823–20828.

[72] W. Y. Su, Y. F. Zhang, Z. H. Li, L. Wu, X. X. Wang, J. Q. Li, X. Z. Fu, Langmuir. 2008, 24, 3422–3428.

[73] S. Tojo, T. Tachikawa, M. Fujitsuka, T. Majima, J. Phys. Chem. C. 2008, 112, 14948–14954.

[74] L. Lin, W. Lin, Y. X. Zhu, B. Y. Zhao, Y. C. Xie, Chem. Lett. 2005, 34, 284–285.

[75] G. Liu, L. Z. Wang, H. G. Yang, H. M. Cheng, G. Q. Lu, J. Mater. Chem. 2010, 20, 831–843.

[76] M. Lazzeri, A. Vittadini, A. Selloni, Phys. Rev. B: Condens. Matter. 2001, 63, 155409–155412.

[77] M. Lazzeri, A. Vittadini, A. Selloni, Phys. Rev. B: Condens. Matter. 2002, 65, 119901–119908.

[78] M. Ramamoorthy, D. Vanderbilt, R. D. Kingsmith, Phys. Rev. B. 1994, 49, 1672–1679.

[79] X. Q. Gong, A. Selloni, Phys. Rev. B: Condens. Matter Mater. Phys. 2007, 76, 235307–235313.

[80] G. Liu, J. C. Yu, G. Q. Lu, H. M. Cheng, Chem. Commun. 2011, 47, 6763–6783.

[81] J. Jiang, K. Zhao, X. Xiao, L. Zhang, J Am. Chem. Soc. 2012, 134, 4473–4476.

[82] H. G. Yang, C. H. Sun, S. Z. Qiao, J. Zou, G. Liu, S. C. Smith, H. M. Cheng, G. Q. Lu, Nature, 2008, 453, 638–641.

[83] M. D'Arienzo, J. Carbajo, A. Bahamonde, M. Crippa, S. Polizzi, R. Scotti, L. Wahba, F. Morazzoni, J. Am. Chem. Soc. 2011, 133, 17652–17661.

[84] X. G. Han, Q. Kuang, M. S. Jin, Z. X. Xie, L. S. Zheng, J. Am. Chem. Soc. 2009, 131, 3152–3153.

[85] H. G. Yang, G. Liu, S. Z. Qiao, C. H. Sun, Y. G. Jin, S. C. Smith, J. Zou, H. M. Cheng, G. Q. Lu, J. Am. Chem. Soc. 2009, 131, 4078–4083.

[86] D. Zhang, G. Li, X. Yang, J. C. Yu, Chem. Commun. 2009, 27, 4381–4383.

[87] J. Zhu, S. H. Wang, Z. F. Bian, S. H. Xie, C. L. Cai, J. G. Wang, H. G. Yang, H. X. Li, CrystEngComm. 2010, 12, 2219–2224.

[88] D. Q. Zhang, G. S. Li, H. B. Wang, K. M. Chan, J. C. Yu, Cryst. Growth Des. 2010, 10, 1130–1137.

[89] F. Amano, O. O. Prieto-Mahaney, Y. Terada, T. Yasumoto, T. Shibayama, B. Ohtani, Chem. Mater. 2009, 21, 2601–2603.

[90] J. S. Chen, Y. L. Tan, C. M. Li, Y. L. Cheah, D. Y. Luan, S. Madhavi, F. Y. C. Boey, L. A. Archer, X. W. Lou, J. Am. Chem. Soc. 2010, 132, 6124–6130.

[91] M. Liu, L. Y. Piao, W. M. Lu, S. T. Ju, L. Zhao, C. L. Zhou, H. L. Li, W. J. Wang, Nanoscale. 2010, 2, 1115–1117.

[92] X. H. Yang, Z. Li, G. Liu, J. Xing, C. H. Sun, H. G. Yang, C. Z. Li, CrystEngComm. 2011, 13, 1378–1383.

[93] X. N. Wang, B. B. Huang, Z. Y. Wang, X. Y. Qin, X. Y. Zhang, Y. Dai, M. H. Whangbo, Chem. Eur. J. 2010, 16, 7106–7109.

[94] X. Y. Ma, Z. G. Chen, S. B. Hartono, H. B. Jiang, J. Zou, S. Z. Qiao, H. G. Yang, Chem. Commun. 2010, 46, 6608–6610.

[95] S. W. Liu, J. G. Yu, M. Jaroniec, J. Am. Chem. Soc. 2010, 132,11914–11916.

[96] M. Liu, L. Piao, L. Zhao, S. Ju, Z. Yan, T. He, C. Zhou, W. Wang, Chem. Commun. 2010, 46, 1664–1666.

[97] G. Liu, C. H. Sun, H. G. Yang, S. C. Smith, L. Z. Wang, G. Q. Lu, H. M. Cheng, Chem. Commun. 2010, 46, 755–757.

[98] Z. K. Zheng, B. B. Huang, X. Y. Qin, X. Y. Zhang, Y. Dai, M. H. Jiang, P. Wang, M. H. Whangbo, Chem. Eur. J. 2009, 15, 12576–12579.

[99] Y. Q. Dai, C. M. Cobley, J. Zeng, Y. M. Sun, Y. N. Xia, Nano Lett. 2009, 9, 2455–2459.

[100] J. M. Li, D. S. Xu, Chem. Commun. 2010, 46, 2301–2303.

[101] M. D'Arienzo, J. Carbajo, A. Bahamonde, M. Crippa, S. Polizzi, R. Scotti, L. Wahba, F. Morazzoni, J. Am. Chem. Soc. 2011, 133, 17652–17661.

[102] T. Tachikawa, S. Yamashita, T. Majima, J. Am. Chem. Soc. 2011, 133, 7197–7204.

[103] J. Pan, G. Liu, G. Q. Lu, H. M. Cheng, Angew. Chem. Int. Ed. 2011, 50, 2133–2137.

[104] T. R. Gordon, M. Cargnello, T. Paik, F. Mangolini, R. T. Weber, P. Fornasiero, C. B. Murray, J. Am. Chem.Soc. 2012, 134, 6751–6761.

[105] H. Xu, P. Reunchan, S. Ouyang, H. Tong, N. Umezawa, T. Kako, J. Ye, Chem. Mater. 2013, 25, 405–411.

[106] J. Yu, J. Low, W. Xiao, P. Zhou, M. Jaroniec, J. Am. Chem. Soc. 2014, 136, 8839–8842.

[107] S. Liu, J. Yu, M. Jaroniec, Chem. Mater. 2011, 23, 4085–4093.

[108] T. Tachikawa, S. Yamashita, T. Majima, J. Am. Chem. Soc. 2011,133, 7197–7204.

[109] H. B. Jiang, Q. A. Cuan, C. Z. Wen, J. Xing, D. Wu, X. Q. Gong, C. Z. Li, H. G. Yang, Angew. Chem. 2011, 123, 3848–3852.

[110] J. Zhang, Q. Xu, M. J. Li, Z. C. Feng, C. Li, J. Phys. Chem. C. 2009, 113, 1698–1704.

[111] G. Liu, F. Li, Z. G. Chen, G. Q. Lu, H. M. Cheng, J. Solid State Chem. 2006, 179, 331–335.

[112] X. Chen, S. S. Mao, Chem. Rev. 2007, 107, 2891–2959.

[113] J. M. Wu, H. C. Shih, W. T. W, Nanotechnology, 2006, 17,105–113.

[114] C. R. Xiong, X. Y. Deng, J. B. Li, Appl. Catal. B. 2010, 94,234–240.

[115] D. Q. Zhang, G. S. Li, F. Wang, J. C. Yu, CrystEngComm, 2010, 12, 1759-1763.

[116] K. Kakiuchi, E. Hosono, H. Imai, T. Kimura, S. Fujihara, J. Cryst. Growth, 2006, 293, 541–545.

[117] X. F. Yang, J. Chen, L. Gong, M. M. Wu, J. C. Yu, J. Am. Chem. Soc. 2009, 131, 12048–12049.

[118] Y. W. Wang, L. Z. Zhang, K. J. Deng, X. Y. Chen, Z. G. Zou, J. Phys. Chem. C, 2007, 111, 2709–2714.

[119] H. M. Cheng, J. M. Ma, Z. G. Zhao, L. M. Qi, Chem. Mater. 1995, 7, 663–671.

[120] Y. Sang, B. Y. Geng, J. Yang, Nanoscale. 2010, 2, 2109–2113.

[121] B. Liu, E. S. Aydil, J. Am. Chem. Soc. 2009, 131, 3985–3390.

[122] T. Y. Ke, C. W. Peng, C. Y. Lee, H. T. Chiu, H. S. Sheu, CrystEngComm, 2009, 11, 1691–1695.

[123] E. Bae, T. Ohno, Appl. Catal. B: Environ. 2009, 91, 634–639.

[124] J. Yu, Y. Chen, A. M. Glushenkov, Cryst. Growth Des. 2009, 9, 1240–1244.

[125] F. Y. Wei, H. L. Zeng, P. Cui, S. C. Peng, T. H. Cheng, Chem. Eng. J. 2008, 144, 119–123.

[126] H. Kaper, F. Endres, I. Djerdj, M. Antonietti, B. M. Smarsly, J. Maier, Y. S. Hu, Small. 2007, 3, 1753–1763.

[127] H. Xu, F. L. Jia, Z. H. Ai, L. Z. Zhang. Cryst. Growth Des. 2007, 7, 1216–1219.

[128] A. Dessombz, D. Chiche, P. Davidson, P. Panine, C. Chaneac, J. P. Jolivet, J. Am. Chem. Soc. 2007, 129, 5904–5909.

[129] M. N. Tahir, P. Theato, P. Oberle, G Melnyk, S. Faiss, U. Kolb, A. Janshoff, M. Stepputat, W. Tremel, Langmuir. 2006, 22, 5209–5912.

[130] B. Zhao, F. Chen, Q. W. Huang, J. L. Zhang, Chem. Commun. 2009, 34, 5115–5117.

[131] M. Addamo, V. Augugliaro, M. Bellardita, A. Di Paola, V. Loddo, G. Palmisano, L. Palmisano, S. Yurdakal, Catal. Lett. 2008, 126, 58–62.

[132] F. Iskandar, A. B. D. Nandiyanto, K. M. Yun, C. J. Hogan, K Okuyama, P. Biswas, Adv. Mater. 2007, 19, 1408–1412.

[133] J. G. Li, C. C. Tang, D. Li, H. Haneda, T. Ishigaki, J. Am. Ceram. Soc. 2004, 87, 1358–1361.

[134] R. Buonsanti, V. Grillo, E. Carlino, C Giannini, T. Kipp, R. Cingolani, P. D. Cozzoli, J. Am. Chem. Soc. 2008, 130, 11223–11233.

[135] R. M. Cornell, U. Schwertmann, The Iron Oxides: Structure, Properties, Reactions, Occurrences, and Uses, 2nd Ed. Wiley-VCH, Weinheim, 2003.

[136] K Sivula, F Le Formal, M Grätzel, ChemSusChem 2011, 4, 432–449.

[137] K. Sivula, R. Zboril, F. Le Formal, R. Robert, A. Weidenkaff, J. Tucek, J. Frydrych, M. Gra°tzel, J. Am. Chem. Soc. 2010, 132, 7436–7444.

[138] A.I. Galuza, A.B. Beznosov, V.V. Eremenko, Low Temp. Phys. 1998, 24, 726–729.

[139] J.H. Kennedy, K.W. Frese, J. Electrochem. Soc. 1978, 125, 709–714.

[140] L.A. Marusak, R. Messier, W.B. White, J. Phys. Chem. Solids. 1980, 41, 981–984.

[141] N.C. Debnath, A.B. Anderson, J. Electrochem. Soc. 1982, 129, 2169–2174.

[142] B. Wang, J.S. Chen, H.B. Wu, Z.Y. Wang, X.W. Liu, J. Am. Chem. Soc. 2011, 133, 17146–17148.

[143] J. Chen, L. Xu, W. Li, X. Gou, Adv. Mater. 2005, 17, 582–586.

[144] Z. Sun, H. Yuan, Z. Liu, B. Han, X. Zhang, Adv. Mater. 2005, 17, 2993–2997.

[145] X. Hu, J.C. Yu, J. Gong, Q. Li, G. Li, Adv. Mater. 2007, 19, 2324–2329.

[146] J. Yu, X. Yu, B. Huang, X. Zhang, Y. Dai, Cryst. Growth Res. 2009, 9, 1474–1480.

[147] G. Liu, Q. Deng, H. Wang, D.H.L. Ng, M. Kong, W. Cai, G. Wang, J. Mater. Chem. 2012, 22, 9704–9713.

[148] C.Y. Cao, J. Qu, W.S. Yan, J.F. Zhu, Z.Y. Wu, W.G. Song, Langmuir 2012, 28, 4573–4579.

[149] K. Sivula, F. Le Formal, M. Gratzel, Chem. Sus. Chem. 2011, 4, 432–449.

[150] Z. Chen, T.F. Jaramillo, T.G. Deutsch, A. Kleiman-Shwarsctein, A.J. Forman, N.Gaillard, R. Garland, K. Takanabe, C. Heske, M. Sunkara, E.W. McFarland, K.Domen, E.L. Miller, J.A. Turner, H.N. Dinh, J. Mater. Res. 2010, 25, 3–16.

[151] J. Brillet, M. Cornuz, F. Le Formal, J.-H. Yum, M. Grätzel, K. Sivula, J. Mater. Res. 2010, 25, 17–24.

[152] C.J. Sartoretti, M. Ulmann, B.D. Alexander, J. Augustynski, A. Weidenkaff, Chem. Phys. Lett. 2003, 376, 194–200.

[153] J.H. Kennedy, J.K.W. Frese, J. Electrochem. Soc. 1978, 125, 709–714.

[154] F.J. Morin, Phys. Rev. 1951, 83, 1005–1010.

[155] F.J. Morin, Phys. Rev. 1954, 93, 1195–1199.

[156] M.P. Dare-Edwards, J.B. Goodenough, A. Hamnett, P.R. Trevellick, J. Chem. Soc.Faraday Trans. 1983, 79, 2027–2041.

[157] K. Sivula, F.L. Formal, M. Grätzel, Chem. Sus. Chem. 2011, 4, 432–449.

[158] M.J. Katz, S.C. Riha, N.C. Jeong, A.B.F. Martinson, O.K. Farha, J.T. Hupp, Coord.Chem. Rev. 2012, 256, 2521–2529.

[159] T.K. Townsend, E.M. Sabio, N.D. Browning, F.E. Osterloh, Energy Environ. Sci. 2011, 4, 4270–4275.

[160] S.N. Dang, S.X. Lu, W.G. Xu, J. Sa, J. Non-Cryst. Solids 2008, 354, 5018–5021.

[161] M.A. Valenzuela, P. Bosch, J. Jiménez-Becerrill, O. Quiroz, A.I. Páez, J. Pho-tochem. Photobiol. A 2002, 148, 177–182.

[162] G. Zhang, Y. Feng, Y. Xu, D. Gao, Y. Sun, Mater. Res. Bull. 2012, 47, 625–630.

[163] L. Li, Y. Chu, Y. Liu, L. Dong, J. Phys. Chem. C. 2007, 111, 2123–2127.

[164] X. Li, X. Yu, J. He, Z. Xu, J. Phys. Chem. C. 2009, 113, 2837–2845.

[165] S. Cao, Y. Zhu, Nanoscale Res. Lett. 2011, 6, 27–36.

[166] J. Xu, Y. Zhu, Cryst. Eng. Comm. 2011, 13, 5162–5169.

[167] L. Xu, J. Xia, K. Wang, L. Wang, H. Li, H. Xu, L. Huang, M. He, Dalton Trans. 2013, 42, 6468–6477.

[168] S.K. Maji, N. Mukherjee, A. Mondal, B. Adhikary, Polyhedron 2012, 33,145–149.

[169] S.K. Apte, S.D. Naik, R.S. Sonawane, B.B. Kale, J. Am. Ceram. Soc. 2007, 90, 412–414.

[170] S. Bharathi, D. Nataraj, K. Senthil, Y. Masuda, J. Nanopart. Res. 2013, 15, 1346–1352.

[171] A. Hosseinian, H. Rezaei, A.R. Mahjoub, World Acad. Sci. Eng. Technol. 2011, 52, 4–21.

[172] C.T. Seip, C. Connor, Nanostruct. Mater. 1999, 12, 183–188.

[173] S. Yang, Y. Xu, Y. Sun, G. Zhang, D. Gao, Cryst. Eng. Comm. 2012, 14, 7915–7921.

[174] X. Zhou, H. Yang, C. Wang, X. Mao, Y. Wang, Y. Yang, G. Liu, J. Phys. Chem. C. 2010, 114, 17051–17061.

[175] M.J. Pawar, A.D. Khajone, M.D. Gaoner, P.S. Chandel, Int. J. Adv. Sci. Res. Tech-nol. 2012, 2, 471–476.

[176] Y. Liu, H. Yu, S. Zhan, Y. Li, Z. Lv, X. Yang, Y. Yu, J. Sol-Gel Sci. Technol. 2011, 58, 716–723.

[177] J. Sundaramurthy, P.S. Kumar, M. Kalaivani, V. Thavasi, S.G. Mhaisalkar, RSC Adv. 2012, 2, 8201–8208.

[178] B. Geng, B. Tao, X. Li, W. Wei, Nanoscale 2012, 4, 1671–1676.

[179] Y. Xu, G. Zhang, G. Du, Y. Sun, D. Gao, Mater. Lett. 2013, 92, 321–324.

[180] H. Zhou, S.S. Wong, ACS Nano 2008, 2, 944–958.

[181] X. Cheng, J. Jiang, M. Hu, G. Mao, F. Bu, C. Lin, Y. Zeng, Q. Zhang, Cryst. Eng.Comm. 2012, 14, 7701–7708.

[182] Y. Mao, S.S. Wong, J. Am. Chem. Soc. 2006, 128, 8217–8226.

[183] J. Bandara, U. Klehm, J. Kiwi, Appl. Catal. B. 2007, 76, 73–81.

[184] S. Cao, Y. Zh, G. Cheng, Y. Huang, J. Phys. Chem. Solids. 2010, 71, 1680–1683.

[185] W. Wu, R. Hao, F. Liu, X. Su, Y. Hou, J. Mater. Chem. A. 2013, 1, 6888–6894.

[186] W. Du, Q. Sun, X. Lv, Y. Xu, Catal. Commun. 2009, 10, 1854–1858.

[187] B. Lv, Z. Liu, H. Tian, Y. Xu, D. Wu, Y. Sun, Adv. Funct. Mater. 2010, 20, 3987–3996.

[188] A.B. Isaev, Z.M. Aliev, N.K. Adamadzieva, N.A. Alieva, G.A. Magomedova, Nanotechnol. Russia. 2009, 4, 475–479.

[189] Zhou X, Lan J, Liu G, Deng K, Yang Y, Nie G, Yu J, Zhi L, Angew Chem Int Ed. 2012, 51, 178–182.

[190] J. D. Bierlein, A. W. Sleight, Solid State Commun. 1975,16, 69–76.

[191] R. S. Roth, J. L. Waring, Am. Mineral. 1963, 48, 1348–1356.

[192] Y Park, KJ McDonald, KS Choi, Chem. Soc. Rev. 2013, 42, 2321–2337.

[193] A. W. Sleight, H.-y. Chen, A. Ferretti, D. E. Cox, Mater. Res. Bull. 1979, 14, 1571–1577.

[194] G. Dreyer, E. Tillmanns, Neues Jahrb. Mineral. Monatsh.e, 1981, 81, 151–156.

[195] Y Park, K J. McDonald, Kyoung-Shin Choi, Chem. Soc. Rev. 2013, 42, 2321–2337.

[196] A. K. Bhattacharya, K. K. Mallick, A. Hartridge, Mater. Lett. 1997, 30, 7-10.

[197] A. Kudo, K. Omori, H. Kato, J. Am. Chem. Soc. 1999, 121, 11459–11461.

[198] S. Tokunaga, H. Kato, A. Kudo, Chem. Mater. 2001, 13, 4624–4628.

[199] H. Tong, S. Ouyang, Y. Bi, N. Umezawa, M. Oshikiri, J. Ye, Adv. Mater. 2012, 24, 229–251.

[200] H. L. Zhou, Y. Q. Qu, T. Zeid, X. F. Duan, Energy Environ. Sci. 2012, 5, 6732–6743.

[201] Y. Ma, X. Wang, Y. Jia, X. Chen, H. Han, C. Li, Chem. Rev. 2014, 114, 9987–10043.

[202] A. Kubacka, M. Fernandez-Garcia, G. Colon, Chem. Rev. 2011, 112, 1555–1614.

[203] S.-i. Eda, M. Fujishima, H. Tada, Appl. Catal. B. 2012, 125, 288–293.

[204] H. He, S. P. Berglund, A. J. E. Rettie, W. D. Chemelewski, P. Xiao, Y. Zhang, C. B. Mullins, J. Mater. Chem. A, 2014, 2, 9371–9379.

[205] W. Wang, Y. Yu, T. An, G. Li, H. Y. Yip, J. C. Yu, P. K. Wong, Environ. Sci. Technol. 2012, 46, 4599–4606.

[206] Y. Sun, Y. Xie, C. Wu, S. Zhang, S. Jiang, Nano Res. 2010, 3, 620–631.

[207] X. Chen, J. Liu, H. Wang, Y. Ding, Y. Sun, H. Yan, J. Mater. Chem. A. 2013, 1, 877–887.

[208] J. Sun, G. Chen, J. Wu, H. Dong, G. Xiong, Appl. Catal. B. 2013, 132, 304–314.

[209] Y. Sun, C. Wu, R. Long, Y. Cui, S. Zhang, Y. Xie, Chem. Commun. 2009, 30, 4542–4544.

[210] Y. Zhao, Y. Xie, X. Zhu, S. Yan, S. Wang, Chem. Eur. J. 2008, 14, 1601–1606.

[211] G. Li, D. Zhang, J. C. Yu, Chem. Mater. 2008, 20, 3983–3992.

[212] Y. Park, K. J. McDonald, K.-S. Choi, Chem. Soc. Rev. 2013, 42, 2321–2337.

[213] F. F. Abdi, N. Firet, R. van de Krol, ChemCatChem, 2013, 5, 490–496.

[214] K. J. McDonald, K.-S. Choi, Energy Environ. Sci. 2012, 5, 8553–8557.

[215] M. Zhou, H. B. Wu, J. Bao, L. Liang, X. W. Lou, Y. Xie, Angew. Chem. Int. Ed. 2013, 52, 8579–8583.

[216] E. S. Kwak, W. Lee, N.-G. Park, J. Kim, H. Lee, Adv. Funct. Mater. 2009, 19, 1093–1100.

[217] G. Liu, J. C. Yu, G. Q. Lu, H. M. Cheng, Chem. Commun. 2011, 47, 6763–6783.

[218] G. Xi, J. Ye, Chem. Commun. 2010, 46, 1893–1895.

[219] D. Wang, H. Jiang, X. Zong, Q. Xu, Y. Ma, G. Li, C. Li, Chem. Eur. J. 2011, 17, 1275–1282.

[220] J. Yang, D. Wang, X. Zhou, C. Li, Chem. Eur. J. 2013, 19, 1320–1326.

[221] J. Pan, G. Liu, G. Q. Lu, H. M. Cheng, Angew. Chem. Int. Ed. 2011, 50, 2133–2137.

[222] R. Li, F. Zhang, D. Wang, J. Yang, M. Li, J. Zhu, X. Zhou, H. Han, C. Li, Nat. Commun. 2013, 4, 1432–1437.

[223] R. Li, H. Han, F. Zhang, D. Wang, C. Li, Energy Environ. Sci. 2014, 7, 1369–1376.

[224] Y.Y. Li, J.P. Liu, X.T. Huang, Nanoscale Res. Lett. 2007, 3, 365–371.

[225] N. Kim, R.N. Vannier, C.P. Grey, Chem. Mater. 2005, 17, 1952–1958.

[226] L. Zhang, H. Wang, Z. Chen, P. K. Wong, J. Liu, Appl. Catal. B. 2011, 106, 1–13.

[227] S. H. Chen, Z. Yin, S. L. Luo, X. J. Li, L. X. Yang, F. Deng, Appl. Surf. Sci. 2012, 259, 7–12.

[228] A. Kudo, S. Hijii, Chem. Lett. 1999, 10, 1103–1104.

[229] J. Tang, Z. Zou, J. Ye, Catal. Lett. 2004, 92, 53–56.

[230] C. Zhang, Y. Zhu, Chem. Mater. 2005, 17, 3537–3545.

[231] N. Zhang, R. Ciriminna, M. Pagliaro, Y. J. Xu, Chem Soc Rev. 2014, 43, 5276–5287.

[232] A. Kudo, S. Hijii, Chem. Lett. 1999, 10, 1103–1104.

[233] G. K. Zhang, F. Lü, M. Li, J. L. Yang, X. Y. Zhang, B. B. Huang, J. Phys. Chem. Solids. 2010, 71, 579–582.

[234] Z. J. Zhang, W. Z. Wang, M. Shang, W. Z. Yin, J. Hazard. Mater. 2010, 177, 1013–1018.

[235] L. Zhou, W. Z. Wang, L. S. Zhang, J. Mol. Catal. 2007, 268, 195–200.

[236] S. O. Alfaro, A. Martínez-de la Cruz, Appl. Catal. A. 2010, 383, 128–133.

[237] S. C. Zhang, C. Zhang, Y. Man, Y. F. Zhu, J. Solid State Chem. 2005, 178, 3823–3830.

[238] L. S. Zhang, W. Z. Wang, Z. G. Chen, L. Zhou, H. L. Xu, W. Zhu, J. Mater. Chem. 2007, 17, 2526–2532.

[239] L. S. Zhang, W. Z. Wang, L. Zhou, H. L. Xu, Small. 2007, 3, 1618–1625.

[240] M. Shang, W. Z. Wang, S. M. Sun, L. Zhou, L. Zhang, J. Phys. Chem. C. 2008, 112, 10407–10411.

[241] J. X. Xia, H. M. Li, Z. J. Luo, H. Xu, K. Wang, S. Yin, Y. S. Yan, Mater. Chem. Phys. 2010, 121, 6–9.

[242] C. X. Xu, X. Wei, Y. M. Guo, H. Q. Wu, Z. H. Ren, G. Xu, G. Shen, G. R. Han, Mater. Res. Bull. 2009, 44, 1635–1641.

[243] C. Y. Wang, H. Zhang, F. Li, L. Y. Zhu, Environ. Sci. Technol. 2010, 44, 6843–6848.

[244] H. B. Fu, W. Q. Yao, L. W. Zhang, Y. F. Zhu, Mater. Res. Bull. 2008, 43, 2617–2625.

[245] H. B. Fu, L. W. Zhang, W. Q. Yao, Y. F. Zhu, Appl. Catal. B. 2006, 66, 100–110.

[246] C. Zhang, Y. F. Zhu, Chem. Mater. 2005, 17, 3537–3545.

[247] H. B. Fu, C. S. Pan, W. Q. Yao, Y. F. Zhu, J. Phys. Chem. B. 2005, 109, 22432–22439.

[248] Y. Tian, G. M. Hua, W. Xu, N. Li, M. Fang, L. D. Zhang, J. Alloys Compd. 2011, 509, 724–730.

[249] F. Amano, K. Nogami, R. Abe, B. Ohtani, Chem. Lett. 2007, 36, 1314–1315.

[250] F. Amano, K. Nogami, R. Abe, B. Ohtani, J. Phys. Chem. C. 2008, 112, 9320–9326.

[251] F. Amano, K. Nogami, R. Abe, B. Ohtani, J. Phys. Chem. C. 2009, 113, 1536–1542.

[252] F. Amano, K. Nogami, R. Abe, M. Tanaka, B. Ohtani, Langmuir 2010, 26, 7174–7180.

[253] S. W. Liu, J. G. Yu, J. Solid State Chem. 2008, 181,1048–1055.

[254] L. Z. Zhang, J. C. Yu, Chem. Commun. 2003, 16, 2078–2079.

[255] D. R. Rolison, Science. 2003, 299, 1698–1701.

[256] F. Amano, A. Yamakata, K. Nogami, M. Osawa, B. Ohtani, J. Am. Chem. Soc. 2008, 130, 17650–17651.

[257] Y. Zhou, Z. Tian, Z. Zhao, Q. Liu, J. Kou, X. Chen, J. Gao, S. Yan, Z. Zou, ACS Appl. Mater. Interfaces. 2011, 3, 3594–3601.

[258] J. Rotmensch, J. L. Whitlock, M. L. Dietz, J. J. Hines, R. C. Reba, E. P. Horwitz, P. V. Harper, Abstr. Pap. Am. Chem. Soc. 1998, 216, U926.

[259] G. G. Briand, N. Burford, Chem. Rev. 1999, 99, 2601–2657.

[260] S. K. Poznyak, A. I. Kulak, Electrochim. Acta. 1990, 35, 1941-1948.

[261] K. L. Zhang, C. M. Liu, F. Q. Huang, C. Zheng, W. D. Wang, Appl. Catal. B. 2006, 68, 125–129.

[262] L. J. Zhao, X. C. Zhang, C. M. Fan, Z. H. Liang, P. D. Han, Phys. B. 2012, 407, 3364–3370.

[263] X. C. Zhang, L. J. Zhao, C. M. Fan, Z. H. Liang, P. D. Han, Comput. Mater. Sci. 2012, 61, 180–184.

[264] H. J. Zhang, L. Liu, Z. Zhou, RSC Adv. 2012, 2, 9224–9229.

[265] W. C. Wang, W. J. Yang, R. Chen, X. B. Duan, Y. L. Tian, D. W. Zeng, B. Shan, Phys. Chem. Chem. Phys. 2012, 14, 2450–2454.

[266] H. J. Zhang, L. Liu, Z. Zhou, Phys. Chem. Chem. Phys. 2012, 14, 1286–1292.

[267] M. L. Guan, C. Xiao, J. Zhang, S. J. Fan, R. An, Q. M. Cheng, J. F. Xie, M. Zhou, B. J. Ye, Y. Xie, J. Am. Chem. Soc. 2013, 135, 10411–10417.

[268] Z. H. Ai, W. K. Ho, S. C. Lee, L. Z. Zhang, Environ. Sci. Technol. 2009, 43, 4143–4150.

[269] K. Li, Y. L. Xu, Y. He, C. Yang, Y. L. Wang, J. P. Jia, Environ. Sci. Technol. 2013, 47, 3490–3497.

[270] Y. F. Fang, Y. P. Huang, J. Yang, P. Wang, G. W. Cheng, Environ. Sci. Technol. 2011, 45, 1593–1600.

[271] H. T. Tian, J. W. Li, M. Ge, Y. P Zhao, L. Liu, Catal. Sci. Technol. 2012, 2, 2351–2355.

[272] F. Chen, H. Q. Liu, S. Bagwasi, X. X. Shen, J. L. Zhang, J. Photochem. Photobiol. A. 2010, 215, 76–80.

[273] J. Xu, W. Meng, Y. Zhang, L. Li, C. S. Guo, Appl. Catal. B. 2011, 107, 355–362.

[274] Z. Y. Yu, B. Detlef, D. Ralf, L. Song, L. Q. Lu, J. Mol. Catal. A: Chem. 2012, 365, 1–7.

[275] L. Zhang, W. Z. Wang, S. M. Sun, Y. Y. Sun, E. P. Gao, J. Xu, Appl. Catal. B. 2013, 132, 315–320.

[276] R. S. Yuan, S. L. Fan, H. X. Zhou, Z. X. Ding, S. Lin, Z. H. Li, Z. Z. Zhang, C. Xu, L. Wu, X. X. Wang, X. Z. Fu, Angew. Chem. Int. Ed. 2013, 52, 1035–1039.

[277] N. Kijima, K. Matano, M. Saito, T. Oikawa, T. Konishi, H. Yasuda, T. Sato, Y. Yoshimura, Appl. Catal. A. 2001, 206, 237–244.

[278] Kubacka A, Fernández-García M, Colón G. Chem. Rev. 2012, 112. 1555–1614.

[279] Y. Xia, P. Yang, Y. Sun, Y. Wu, B. Mayers, B. Gates, Y. Yin, F. Kim, H. Yan, Adv. Mater. 2003, 15, 353–389.

[280] R. S. Devan, R. A. Patil, J. H. Lin, Y. R. Ma, Adv. Funct. Mater. 2012, 22, 3326–3370.

[281] C. H. Wang, C. L. Shao, Y. C. Liu, L. N. Zhang, Scr. Mater. 2008, 59, 332–335.

[282] R. S. Yuan, C. Lin, B. C. Wu, X. Z. Fu, Eur. J. Inorg. Chem. 2009, 24, 3537–3540.

[283] S. J. Wu, C. Wang, Y. F. Cui, T. M. Wang, B. B. Huang, X. Y. Zhang, X. Y. Qin, P. Brault, Mater. Lett. 2010, 64, 115–118.

[284] Q. H. Wang, K. Kalantar-Zadeh, A. Kis, J. N. Coleman, M. S. Strano, Nat. Nanotechnol. 2012, 7, 699–712.

[285] M. Chhowalla, H. S. Shin, G. Eda, L. Li, K. P. Loh, H. Zhang, Nat. Chem. 2013, 5, 263–275.

[286] M. Q. Zhao, Q. Zhang, J. Q. Huang, F. Wei, Adv. Funct. Mater. 2012, 22, 675–694.

[287] J. H. Han, S. Lee, J. Cheon, Chem. Soc. Rev. 2013, 42, 2581–2591.

[288] K. L. Zhang, C. M. Liu, F. Q. Huang, C. Zheng, W. D. Wang, Appl. Catal. B. 2006, 68, 125–129.

[289] L. Q. Ye, L. Zan, L. Tian, T. Peng, J. Zhang, Chem. Commun. 2011, 47, 6951-6953.

[290] D. Zhang, J. Li, Q. G. Wang, Q. S. Wu, J. Mater. Chem. A. 2013, 1, 8622–8629.

[291] X. F. Chang, J. Huang, C. Cheng, Q. Sui, W. Sha, G. B. Ji, S. B. Deng, G. Yu, Catal. Commun. 2010, 11, 460–464.

[292] X. Chang, M. A. Gondal, A. A. Al-Saadi, M. A. Ali, H. Shen, Q. Zhou, J. Zhang, M. Du, Y. Liu, G. Ji, J. Colloid Interface Sci. 2012, 377, 291–298.

[293] J. Jiang, K. Zhao, X. Y. Xiao, L. Z. Zhang, J. Am. Chem. Soc. 2012, 134, 4473–4476.

[294] J. Y. Xiong, G. Cheng, G. F. Li, F. Qin, R. Chen, RSC Adv. 2011, 1, 1542–1553.

[295] M. Shang, W. Z. Wang, L. Zhang, J. Hazard. Mater. 2009, 167, 803–809.

[296] L. Q. Ye, L. H. Tian, T. Y. Peng, L. Zan, J. Mater. Chem. 2011, 21, 12479–12484.

[297] H. J. Zhang, L. Liu, Z. Zhou, RSC Adv. 2012, 2, 9224–9229.

[298] G. M. Whitesides, B. Grzybowski, Science, 2002, 295, 2418–2421.

[299] M. A. Snyder, M. Tsapatsis, Angew. Chem. Int. Ed. 2007, 46, 7560–7573.

[300] D. R. Rolison, J. W. Long, J. C. Lythe, A. F. Fischer, C. P. Rhodes, T. M. McEvoy, M. E. Bourg, A. M. Lubers, Chem. Soc. Rev. 2009, 38, 226–252.

[301] Y. Li, Z. Y. Fu, B. L. Su, Adv. Funct. Mater. 2012, 22, 4634–4667.

[302] Q. F. Zhang, E. Uchaker, S. L. Candelaria, G. Z. Cao, Chem. Soc. Rev. 2013, 42, 3127–3171.

[303] X. Zhang, Z. H. Ai, F. L. Jia, L. Z. Zhang, J. Phys. Chem. C. 2008, 112, 747–753.

[304] X. Y. Qin, H. F. Cheng, W. J. Wang, B. B. Huang, X. Y. Zhang, Y. Dai, Mater. Lett. 2013, 100, 285–288.

[305] J. X. Xia, J. Zhang, S. Yin, H. M. Li, H. Xu, L. Xu, Q. Zhang, J. Phys. Chem. Solids. 2013, 74, 298–304.

[306] J. M. Song, C. J. Mao, H. L. Niu, Y. H. Shen, S. Y. Zhang, CrystEngComm, 2010, 12, 3875–3881.

[307] L. P. Zhu, G. H. Liao, N. C. Bing, L. L. Wang, Y. Yang, H. Y. Xie, CrystEngComm, 2010, 12, 3791–3796.

[308] S. J. Peng, L. L. Li, P. N. Zhu, Y. Z. Wu, M. Srinivasan, S. G. Mhaisalkar, S. Ramakrishna, Q. Y. Yan, Chem. Asian J. 2013, 8, 258–268.

[309] D. H. Wang, G. Q. Gao, Y. W. Zhang, L. S. Zhou, A. W. Xu, W. Chen, Nanoscale. 2012, 4, 7780–7785.

[310] J. Zhang, F. Shi, J. Lin, D. Chen, J. Gao, Z. Huang, X. Ding, C. Tang, Chem. Mater. 2008, 20, 2937–2941.

[311] Z. H. Ai, W. Ho, S. Lee, L. Z. Zhang, Environ. Sci. Technol. 2009, 43, 4143–4150.

[312] J. Xu, W. Meng, Y. Zhang, L. Li, C. S. Guo, Appl. Catal. B. 2011, 107, 355–362.

[313] H. F. Cheng, B. B. Huang, Z. Y. Wang, X. Y. Qin, X. Y. Zhang, Y. Dai, Chem. Eur. J. 2011, 17, 8039–8043.

[314] Y. J. Chen, M. Wen, Q. S. Wu, CrystEngComm, 2011, 13, 3035–3039.

[315] J. X. Xia, S. Yin, H. M. Li, H. Xu, L. Xu, Y. G. Xu, Dalton Trans. 2011, 40, 5249–5258.

[316] L. Zhang, X. F. Cao, X. T. Chen, Z. L. Xue, J. Colloid Interface Sci. 2011, 354, 630–636.

[317] D. Q. Zhang, M. C. Wen, B. Jiang, G. S. Li, J. C. Yu, J. Hazard. Mater. 2012, 211, 104–111.

[318] Y. C. Feng, L. Li, J. W. Li, J. F. Wang, L. Liu, J. Hazard. Mater. 2011, 192, 538–544.

[319] Y. N. Huo, J. Zhang, M. Miao, Y. Jin, Appl. Catal. B. 2012, 111. 334–341.

[320] H. T. Tian, J. W. Li, M. Ge, Y. P. Zhao, L. Liu, Catal. Sci. Technol. 2012, 2, 2351–2355.

[321] Y. Y. Li, J. S. Wang, H. C. Yao, L. Y. Dang, Z. J. Li, J. Mol. Catal. A: Chem. 2011, 334, 116–122.

[322] J. X. Xia, S. Yin, H. M. Li, H. Xu, Y. S. Yan, Q. Zhang, Langmuir. 2011, 27, 1200–1206.

[323] L. Chen, S. F. Yin, R. Huang, Y. Zhou, S. L. Luo, C. T. Au, Catal. Commun. 2012, 23, 54–57.

[324] K. Zhang, J. Liang, S. Wang, J. Liu, K. X. Ren, X. Zheng, H. Luo, Y. J. Peng, X. Zou, X. Bo, J. H. Li, X. B. Yu, Cryst. Growth Des. 2012, 12, 793–803.

[325] Y. Q. Lei, G. H. Wang, S. Y. Song, W. Q. Fan, M. Pang, J. K. Tang, H. J. Zhang, Dalton Trans. 2010, 39, 3273–3278.

[326] R. Hao, X. Xiao, X. X. Zuo, J. M. Nan, W. D. Zhang, J. Hazard. Mater. 2012, 209, 137–145.

[327] Y. Q. Lei, G. H. Wang, S. Y. Song, W. Q. Fan, H. J. Zhang, CrystEngComm, 2009, 11, 1857–1862.

[328] C. H. Deng, H. M. Guan, Mater. Lett. 2013, 107, 119–122.

[329] G. Cheng, J. Y. Xiong, F. J. Stadler, New J. Chem. 2013, 37, 3207–3213.

[330] X. Xiao, W. D. Zhang, J. Mater. Chem. 2010, 20, 5866–5870.

[331] J. Y. Xiong, Z. B. Jiao, G. X. Lu, W. Ren, J. H. Ye, Y. P. Bi, Chem. Eur. J. 2013, 19, 9472–9475.

[332] J. Jiang, K. Zhao, X. Y. Xiao, L. Z. Zhang, J. Am. Chem. Soc. 2012, 134, 4473–4476.

[333] L. Q. Ye, L. Zan, L. H. Tian, T. Y. Peng, J. J. Zhang, Chem. Commun. 2011, 47, 6951–6953.

[334] L. Q. Ye, K. J. Deng, F. Xu, L. H. Tian, T. Y. Peng, L. Zan, Phys. Chem. Chem. Phys. 2012, 14, 82–85.

[335] J. Jiang, L. Z. Zhang, H. Li, W. W. He, J. J. Yin, Nanoscale. 2013, 5, 10573–10581.

[336] A. Kudo, Y. Miseki, Chem. Soc. Rev. 2009, 38, 253–278.

[337] X. B. Chen, S. S. Mao, Chem. Rev. 2007, 107, 2891–2959.

[338] H. F. Cheng, B. B. Huang, Y Dai, Nanoscale, 2014, 6, 2009–2026

[339] S. Xuan, W. Jiang, X. Gong, Y. Hu, Z. Chen, J. Phys. Chem. C, 2008, 113, 553–558.

[340] Q. Yuan, N. Li, W. C. Geng, Y. Chi, X. T. Li, Mater. Res. Bull. 2012, 47, 2396–2402.

[341] Y. Z. Wang, X. B. Fan, S. L. Wang, G. L. Zhang, F. B. Zhang, Mater. Res. Bull. 2013, 48, 785–789.

[342] B. Cui, H. X. Peng, H. Q. Xia, X. H. Guo, H. L. Guo, Sep. Purif. Technol. 2013, 103, 251–257.

[343] Z. Wang, L. Shen, S. Zhu, Int. J. Photoenergy. 2012, 202, 519–524.

[344] R. Chalasani, S. Vasudevan, ACS Nano, 2013, 7, 4093–4104.

[345] H. Liu, L. Gao, J. Am. Ceram. Soc. 2006, 89, 370–373.

[346] W. Zhou, H. Fu, K. Pan, C. Tian, Y. Qu, P. Lu, C. C. Sun, J. Phys. Chem. C. 2008, 112, 19584-19589.

[347] F. X. Chen, W. Q. Fan, T. Y. Zhou, W. H. Huang, Acta Phys. Chim. Sin. 2013, 29, 167–175.

[348] W. S. Tung, W. A. Daoud, ACS Appl. Mater. Interfaces. 2009, 1, 2453–2461.

[349] B. Palanisamy, C. M. Babu, B. Sundaravel, S. Anandan, V. Murugesan, J. Hazard. Mater. 2013, 23, 233–242.

[350] M. Shang, W. Wang, L. Zhang, S. Sun, L. Wang, L. Zhou, J. Phys. Chem. C, 2009, 113, 14727–14731.

[351] C. Wang, C. Shao, X. Zhang, Y. Liu, Inorg. Chem. 2009, 48, 7261–7268.

[352] E. Gao, W. Wang, M. Shang, J. Xu, Phys. Chem. Chem. Phys. 2011, 13, 2887–2893.

[353] Y. H. Ng, A. Iwase, A. Kudo, R. Amal, J Phys Chem Lett, 2010, 1, 2607–2612.

[354] X. Zhang, L. Z. Zhang, T. F. Xie, D. J. Wang, J. Phys. Chem. C. 2009, 113, 7371–7378.

[355] Y. Li, Y. Liu, E. Uchaker, Q. Zhang, S. Sun, J. Wang, Y. Huang, J. Li, G. Cao, J. Mater. Chem. A. 2013, 1, 7949–7956.

[356] F. D. Gao, D. W. Zeng, Q. W. Huang, S. Q. Tian, C. S. Xie, Phys. Chem. Chem. Phys. 2012, 14, 10572–10578.

[357] Z. H. Ai, W. Ho, S. Lee, J. Phys. Chem. C. 2011, 115, 25330–25337.

[358] X. M. Tu, S. L. Luo, G. X. Chen, J. H. Li, Chem. Eur. J. 2012, 18 14359–14366.

[359] S. Y. Song, W. Gao, X. Wang, X. Y. Li, D. P. Liu, Y. Xing, H. J. Zhang, Dalton Trans. 2012, 41, 10472–10476.

[360] H. Liu, W. R. Cao, Y. Su, Z. Chen, Y. Wang, J. Colloid Interface Sci. 2013, 398, 161–167.

[361] H. Tada, T. Mitsui, T. Kiyonaga, T. Akita, K. Tanaka, Nat. Mater. 2006, 5, 782–786.

[362] M. R. Elahifard, S. Rahimnejad, S. Haghighi, M. R. Gholami, J. Am. Chem. Soc. 2007, 129, 9552–9553.

[363] H. L. Wang, L. S. Zhang, Z. G. Chen, J. Q. Hu, S. J. Li, Z. H. Wang, J. S. Liu, X. C. Wang, Chem. Soc. Rev. 2014, 43, 5234–5244.

[364] L. S. Zhang, K. H. Wong, Z. G. Chen, J. C. Yu, J. C. Zhao, C. Hu, C. Y. Chan, P. K. Wong, Appl. Catal. A. 2009, 363,221–229.

[365] W. Xiong, Q. D Zhao, X. Y. Li, D. K. Zhang, Catalysis Communications. 2011, 16, 229–233.

[366] L. Q. Ye, J. Y. Liu, C. Q. Gong, L. H. Tian, T. Y. Peng, L. Zan. ACS Catal. 2012, 2, 1677–1683.

[367] W. Y. Gao, M. Q. Wang, C. X. Ran, X Yao, H Yang, J Liu, D He, J. B. Bai. Nanoscale. 2014, 21, 5498–508.

[368] W. Y. Gao, M. Q. Wang, C. X. Ran, L Li, Chem. Commun. 2015, 51, 1709–1712.

Chapter 8

ZnO/Sepiolite Catalysts – Characterization and Photoactivity Measurements

A. Neren Ökte

Additional information is available at the end of the chapter

http://dx.doi.org/10.5772/61973

Abstract

Sepiolite-supported ZnO catalysts (ZnO-SEP) were prepared and characterized using X-ray diffraction (XRD), surface area (BET) measurements, scanning electron microscopy (SEM) with energy-dispersive X-ray analysis (EDX), X-ray photoelectron spectroscopy (XPS) and UV-vis diffuse reflectance spectra (UV-vis DRS) techniques. XRD analysis supplied information about the generation of ZnO nanoparticles. SEM images and elemental mapping scans revealed variations in the surface morphology of the SEP after ZnO loading. Supported catalysts possessed higher surface areas and pore volumes in comparison to unsupported catalyst (0.25 M ZnO). They also revealed suitable band-gap energies in the UV-A region. XPS analysis confirmed the build-up of ZnO nanoparticles on the SEP matrix with the form of Zn^{2+} oxidation state. Photocatalytic performances were evaluated in terms of methyl orange (MO) decolorization process following pseudo-first-order kinetics. The repeatability of photocatalytic activity was also tested.

Keywords: Sepiolite, ZnO, Supported catalysts, Methyl orange

1. Introduction

The complete degradation of dye molecules is not possible by the application of precipitation, adsorption, oxidation, reduction and biological and electrochemical types of conventional methods. These methods may either end up with less efficiency or create a secondary sludge. Semiconductor photocatalysis is an advanced oxidation process (AOP) for the treatment of air and water streams and emerged as an important technology for the degradation of dye molecules. In summary, the mechanism of photocatalytic processes is initiated by the band-gap illumination and this leads to production of electron-hole pairs. After separation, electrons and holes migrate to the catalyst surface, induce redox reactions with adsorbed pollutants and eventually result in the degradation of the dye pollutants.

ZnO is a wide band-gap (~3.3 eV) semiconductor that has been extensively used because of its catalytic and photochemical properties along with its low cost [1]. There are many reports of ZnO having higher photocatalytic activities than other semiconductors in both air and aqueous media [2,3]. As a photocatalyst, the surface area plays an important role since reactions mainly occur between catalyst surfaces and pollutants. The nanoscale ZnO crystals have shown larger surface areas and higher photocatalyic performances than that of bulk materials [4]. Therefore, recent studies have focused on the synthesis of nanostructured ZnO with tunable size and shape [5,6]. However, ZnO nanoparticles are not stable in acidic and alkaline conditions and also show rapid deactivation in bulk due to increased tendency of aggregation [7]. Hence, the development of industrially viable, cost-effective, eco-friendly adsorbents with attractive multiple functions such as adsorption, and decomposition becomes important. Silicate adsorbents engineered with photocatalytic ingredients, for example, tailoring the aluminosilicate layers and their surfaces with nanostructured semiconducting photocatalysts, can make them multifunctional composites and heterogeneous catalysts [8]. ZnO/clay system reveals the potential of this composite for various applications [9–12]. Among clay minerals, the usage of sepiolite as a support material is rarely reported [13–18]. Natural sepiolite is a very cheap, fibrous and hydrated magnesium silicate with a relatively high surface area. The presence of alkalescent [MgO_6] and acidic [SiO_4] centers in the sepiolite structure enhances the adsorption of reactant molecules and their degradation possibility. Moreover, silicate layers appear as an attractive support for the assembly of small-sized metals and metal-oxide aggregates (clusters and nanoparticles) that have been mainly employed for catalytic purposes. The immobilization of nanoparticles on the inner and outer surfaces of inorganic supports results in the formation of nanocomposite materials. The synergy established among nanoparticles and support systems makes them attractive options for the degradation of pollutants. In the study of Xu et al., quantum-sized ZnO particles supported on sepiolite were prepared using a sol-gel method with the sepiolite of acid activation as carrier and zinc acetate dihydrate ($Zn(CH_3COO)_2\,2H_2O$) and lithium hydroxide monohydrate ($LiOH\,H_2O$) as raw materials [13]. They found that the nano-ZnO supported on sepiolite can not only solve the dispersing problem but also has a positive synergistic effect on the ZnO photocatalysis. Bautista et al. prepared TiO_2-Sep supports of vanadium oxide in order to obtain a new TiO_2 support with a high and thermostable surface area [14]. According to this study, vanadium oxides supported on TiO_2-coated sepiolite and sepiolite characterization studies indicated that well-dispersed vanadium in both types of supports was achieved in the systems with vanadia loading below the theoretical monolayer. Above this vanadia loading, the formation of V_2O_5 nanoparticles with a mainly crystalline character took place as well as the formation of V-Mg mixed metal-oxide phases, especially in systems supported on sepiolite. The hydrophilic character and more open structure of the sepiolite was underlined in the study of Arques et al. [15]. Accordingly, sepiolite appeared to be a convenient support for pyrylium salts to be employed as a heterogeneous solar photocatalyst. Also, promising results have been obtained testing the performance of the new material, with ferulic acid as target pollutant, and important percentages of photo-oxidation were achieved. In another study, monolithic catalysts based on Rh/TiO_2-sepiolite were developed and tested in the decomposition of N_2O traces [16]. The system was found to be extremely sensitive to the amount of rhodium and is an attractive alternative for the elimination of N_2O traces from stationary sources due to the combination of high catalytic activity with a low pressure drop and optimum textural/mechanical properties. The structural and photocatalytic properties of TiO_2-supported sepiolite and sodium ion-treated sepiolite catalysts were also investigated for the degradation of β-naphthol molecule [17]. Sodium chloride treatment enhanced the attraction of sepiolite support through TiO_2 nanoparticles. This study explored that sepiolite can be employed as a catalyst support for the photocatalytic degradation reactions in solution. Such supported catalysts can be readily separated from the suspension without filtration since they decant in minutes, while TiO_2 (Degussa P-25) sample could not sediment over hours. This provided an important, practical advantage in the usage of TiO_2-supported sepiolite catalysts.

In this chapter, a mixed structure of ZnO and sepiolite (ZnO-SEP) is examined for the degradation of an azo dye, methyl orange (MO). Such supported catalyst systems are prepared using a coprecipitation method with highly dispersed active ZnO nanoparticles and investigated in terms of the dark adsorption capacities and photoactivities. X-ray diffraction (XRD), surface area (BET) measurements, scanning electron microscopy (SEM) with energy-dispersive X-ray analysis (EDX), X-ray photoelectron spectroscopy (XPS) and UV-vis diffuse reflectance spectra (UV-vis DRS) techniques are used for the characterization of these catalysts. The decolorization of MO is followed under UV irradiation.

2. Experimental section

The raw SEP used as a support in this study was obtained from the Eskisehir region of Anatolia (Turkey) and characterized by X-ray diffraction and SEM-EDX analyses. The chemical composition was found as SiO_2 (69.25%), MgO (28.92%), Al_2O_3 (1.12%), Na_2O (0.55%) and TiO_2 (0.16%). The clay was ground to approximately 200-mesh size powder. Zinc nitrate hexahydrate ($Zn(NO_3)_2$ $6H_2O$) (99.0%, Merck), sodium carbonate (Na_2CO_3) (analytical grade, Merck) and (4-[[(4-dimethylamino)phenyl]-azo]benzenesulfonic acid sodium salt] (methyl orange, Merck) were used as provided by the suppliers without further purification. Deionized water, purified with an Elga-Pure Water Purification (UHQ II) system, was used for preparing solutions in the experiments.

The following procedures and techniques were applied for the synthesis and characterization of the photocatalysts:

ZnO catalysts in the absence of SEP were prepared by a coprecipitation method using $Zn(NO_3)_2$ $6H_2O$ and Na_2CO_3 precursors [19]. Briefly, the precursors with 0.25 M concentrations were separately dissolved in deionized water. Then, 100 mL of $Zn(NO_3)_2$ $6H_2O$ were added gradually to 100 mL of Na_2CO_3 solution under continuous stirring. The resulting white suspension was stirred for 2 h at room temperature followed by several washings and centrifugations at 4000 rpm. Then, the resulting precipitate was dried at 100°C for 12 h and calcined at 500°C with a heating rate of 10°C min^{-1} for 5 h. Finally, the catalysts were ground into fine powder and named as 0.25 M ZnO.

Depending on the loading of ZnO on the support, precursor concentrations as 0.125 M ZnO, 0.25 M ZnO or 0.5 M ZnO (prepared following the above procedure) were added to the SEP solution. After stirring the mixed suspension for about 12 h at room temperature, several washings and centrifugations (at 4000 rpm) were applied. The resulting catalysts were dried at 100°C for 12 h and then calcined at 500°C with a rate of 10°C min^{-1} for 5 h. Finally, the catalysts were ground into fine powder and named as 0.125 M ZnO-SEP, 0.25 M ZnO-SEP and 0.5 M ZnO-SEP.

X-ray diffraction was used to monitor the formation of crystal planes and measure the crystalline size of ZnO nanoparticles. The analyses were recorded on a Rigaku-D/MAX-Ultima diffractometer using Cu-Kα radiation (λ=1.54 Å) operating at 40 kV and 40 mA and scanning rate of 2 min^{-1}. Nitrogen sorption analysis was used to measure the surface areas, pore volumes and pore diameters of the catalysts.

The nitrogen adsorption/desorption isotherms were obtained at 77 K using Quantachrome Nova 2200e automated gas adsorption system. The specific surface areas were determined using multi-point BET analysis and the pore sizes were measured by the BJH method of adsorption.

The surface morphologies were determined using scanning electron microscopy (SEM) in combination with energy-dispersive X-ray analysis on an ESEM-FEG/EDAX Philips XL-30 instrument operating at 20 kV. The catalysts were fixed with carbon tape prior to the metallization process with gold (Sputter Coater-Balzer SCD050).

In X-ray photoelectron spectroscopy (XPS) tests, Thermo Scientific K-Alpha X-ray photoelectron spectrometer equipped with electron analyzer and Al-Kα micro-focused monochromator was used. The areas of peaks were estimated by calculating the integral of each peak after subtracting a Shirley background and fitting the experimental curve to a combination of Lorentzian/Gaussian lines. Binding energy shifts were observed in the samples and the instrument was calibrated using the carbon peak (C-1s) at 285 eV as in the other studies [20,21].

The UV-vis diffuse reflectance spectra (UV-vis DRS) of the supports, 0.25 M ZnO and supported catalysts were obtained using UV-vis spectrophotometer (UV-2450, Shimadzu) equipped with an integrating sphere reflectance accessory. The baseline correction was done by $BaSO_4$. The spectra were recorded in the range 200–600 nm for the catalysts and the FA using $BaSO_4$ as a reference.

For the photocatalytic experiments, the details of reaction systems were given in our previous studies [22,23]. Briefly, a Pyrex flask reactor was located in an "irradiation box" equipped with eight black light lamps (Philips TL 15W/5BLB) with an emission maximum at λ=365 nm. The lamps were positioned to surround the flask with an incident photon flux of 4.7×10^{15} photons s^{-1}. Then, 0.2 g of catalysts with 200 mL of 3.27 mg L^{-1} MO (unless the concentration effect of MO was controlled) was used as reaction solutions. The flask had an inlet for the air circulation and an outlet for the collection of aliquots. Prior to irradiation, suspensions were magnetically stirred in the dark for 30 min. UV-vis spectrophotometer (UV-2450, Shimadzu) was used to monitor the absorbance spectra of MO at 464 nm as a function of irradiation time. All experiments were performed at room temperature and without concerning the degradation intermediates. Also, measurements were conducted at least twice and the average value was recorded. MO decolorization percentages were calculated by the following equation:

$$\text{Decolorization }\% = \frac{C_0 - C}{C_0} \times 100$$

where C_0 is the initial concentration of MO and C is the concentration of MO after "t" minutes irradiation.

3. Results and discussion

3.1. Characterization of ZnO-SEP catalysts

3.1.1. XRD analysis

The phase identification of the raw SEP and the supported catalysts were performed by XRD analysis from 0° to 70° (2θ) (Figure 1). The characteristic d 110 reflection of the SEP was noticed with a basal spacing of 12.19 Å at 7.24° (2θ). ZnO loadings decrease the intensities of all SEP reflections. For the supported catalysts, alteration in ZnO concentrations did not modify the position of the SEP peaks due to the non-expandable nature of the SEP. Supported catalysts also exhibited wurtzite ZnO structure with d 100, d 002, d 101, d 102, d 110, d 103 and d 200 crystal planes at 31.9°, 34.6°, 36.4°, 47.7°, 56.7°, 63.1° and 66.6° (2θ), respectively. The broad and less intense ZnO peaks got sharper and more intense with the increments in ZnO loadings.

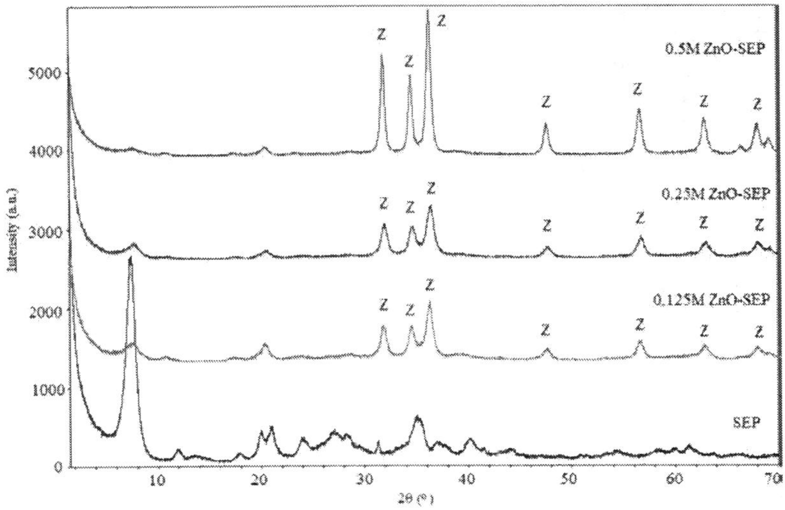

Figure 1. XRD patterns of SEP and supported catalysts (Z: ZnO).

The crystalline sizes (D_{ZnO}) calculated using Scherrer's equation for the broadening of d 101 reflection of ZnO were found as 16.1 nm (for 0.25 M ZnO), 13.4 nm (for 0.125 M ZnO-SEP), 12.3 nm (for 0.25 M ZnO-SEP) and 15.1 nm (for 0.5 M ZnO-SEP) (Table 1). The reduction in the ZnO crystalline sizes and the decrements in the intensities of SEP reflections may suggest highly dispersed ZnO nanoparticles over the surface and bulk.

Catalysts	D_{ZnO} (nm)[a]	BET ($m^2 g^{-1}$)[b]	V_{pore} ($cm^3 g^{-1}$)[c]	R_{pore} (Å)[d]
0.25M ZnO	16.1	7.58	0.012	17.7
SEP	-	104.5	0.140	14.9
0.125M ZnO-SEP	13.4	91	0.151	19.1
0.25M ZnO-SEP	12.3	47	0.077	20.2
0.5M ZnO-SEP	15.1	51	0.091	14.8

a Calculated from the (101) diffraction peak of ZnO using the Scherrer equation.
b Determined from nitrogen adsorption-desorption isotherms using BET equation.
c Determined from cumulative adsorption pore volume using BJH method.
d Determined from adsorption pore size using BJH method.

Table 1. Crystalline sizes (D_{ZnO}), surface areas (BET), total pore volumes (Vpore) and pore radius (Rpore) of 0.25 M ZnO, SEP and supported catalysts.

3.1.2. Nitrogen adsorption desorption isotherms

The textural characterization points out that 0.25 M ZnO had type II isotherms, which refer to non-porous materials (Figure 2A and 2B). However, the raw SEP and the supported catalysts had type IV nitrogen adsorption-desorption isotherms typical of mesoporous structures with hysteresis loops [24]. Almost similar pore radius was detected for all catalysts and the SEP (Table 1). The capillary condensation taking place within the mesopores in the high P/P_0 range became more difficult with the formation of narrower loops, owing to the higher ZnO loading concentrations. Accordingly, 0.125 M ZnO-SEP shows almost similar surface area and pore volume with respect to the SEP, whereas 0.25 M ZnO-SEP and 0.5 M ZnO-SEP reveal lower values. The partial blockage of the pores by the ZnO nanoparticles on the supports' surface sites, edges and corners was expected to induce an easier attraction among MO molecules and the catalyst particles.

3.1.3. SEM (EDX) analysis

Figure 3 illustrates the SEM and elemental mapping images of 0.25 M ZnO-SEP. The characteristic fibrous structure of SEP and pseudo-spherical shape of the as-prepared ZnO catalyst were reported previously [17,22]. The heterogeneously dispersed ZnO nanoparticles within the SEP matrix resulted in a different morphology in comparison to the raw SEP and unsupported catalyst. In the EDX analysis of 0.25 M ZnO-SEP (not shown), the reduction in the peak intensities and percentages of SiO_2 and MgO (from 69.25% to 16.08% for SiO_2 and from 28.92% to 10.93% for MgO) and also simultaneous detection of ZnO peak signified the in situ build-

http://dx.doi.org/10.5772/61973

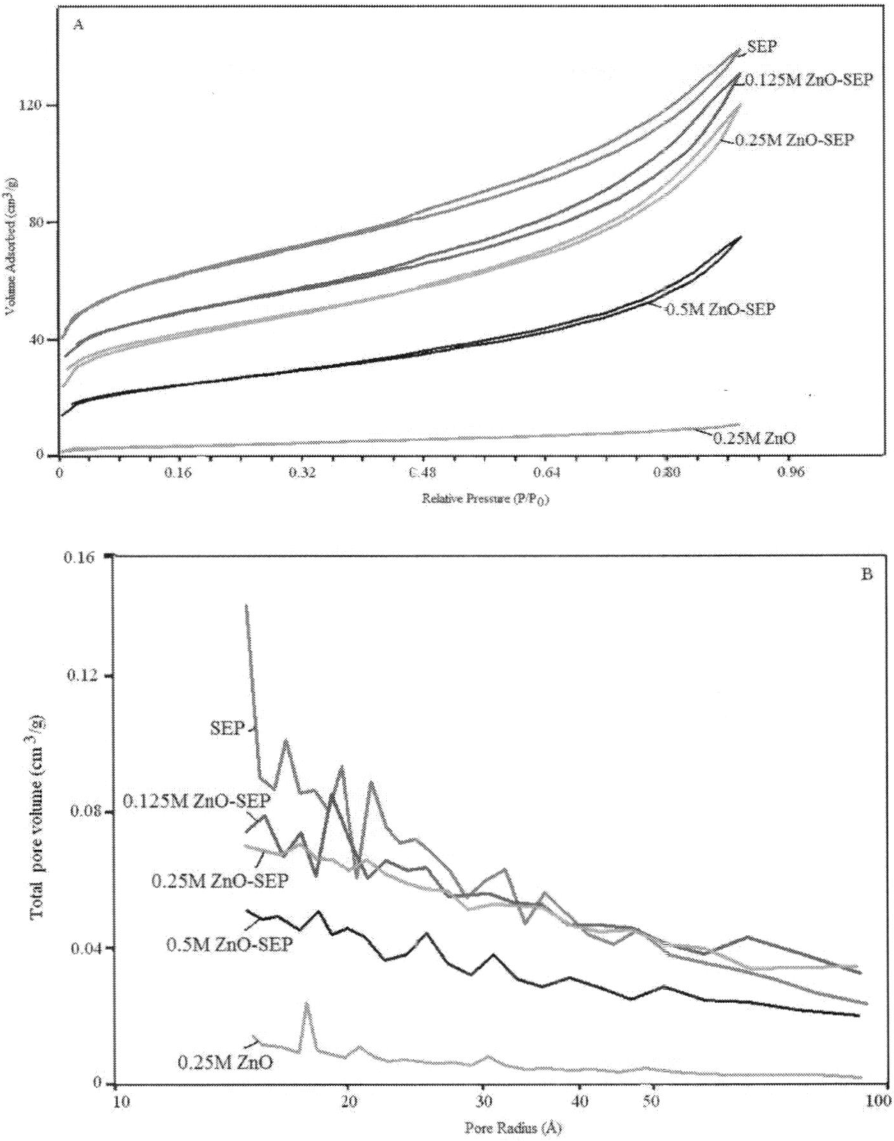

Figure 2. (A) Nitrogen adsorption/desorption isotherms and (B) pore size distribution plots of 0.25 M ZnO, SEP and supported catalysts.

up of ZnO nanoparticles in the SEP structure. In mapping images, ZnO existence was proven by the dominating Zn signals within less dense regions of Si and Mg constituents.

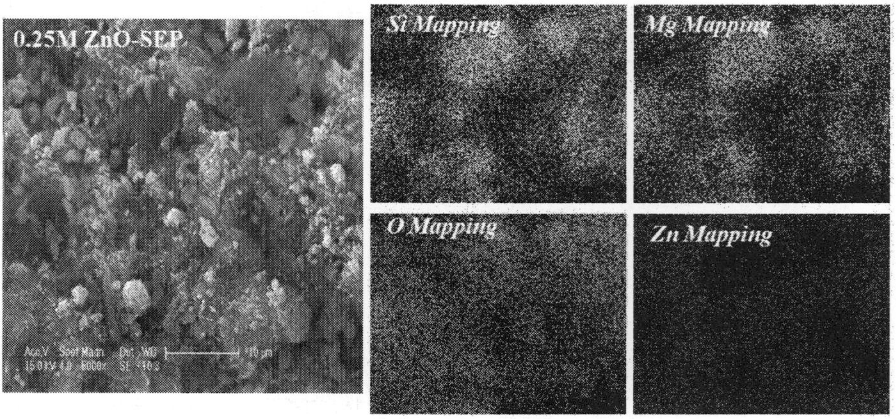

Figure 3. SEM and mapping images of 0.25 M ZnO-SEP.

3.1.4. XPS analysis

In order to control the presence of ZnO on the SEP host, the surface chemical composition was analyzed by the XPS method, focusing in particular on the binding energies of the typical lines of Zn and O (Figure 4). The survey scan of 0.5 M ZnO-SEP reveals Zn peaks in addition to the Mg (49.5 eV), Al (75.6 eV), Si (103 eV), O (531.47 eV) and some Auger peaks of the SEP (Figure 4A). Figure 4B depicts the presence of Zn ion with a doublet matching to Zn 2p3/2 and 2p1/2 core levels (Figure 4B). This shows the presence of Zn^{2+} ions in an oxide environment [25,26]. The O1s photoelectron peak is deconvoluted by three subspectral parts (Figure 4C). The peak at 527 eV is attributed to the lattice oxygen in a Zn-O-Zn network with 8.7% spectral area. The shift in the peak position with respect to ZnO binding energy of 530.5 eV can be due the complex configuration in the ZnO-SEP matrix. For the other two components, 64.7% spectral area refers to MgO and Al_2O_3 at 531 eV and 26.5% spectral area corresponds to SiO_2 at 532.81 eV.

3.1.5. UV-vis DRS analysis

UV-vis absorption spectroscopies of the SEP and supported catalysts are presented in Figure 5A. The raw SEP shows an extended profile in between 200 and 600 nm. In contrast, 0.25 M ZnO reveals its distinctive edge below 400 nm. The supported catalysts resemble this characteristic edge, which is more pronounced for the catalyst having the highest ZnO concentration. The UV activities of the supported catalysts are evidenced by the evaluated band-gap energies based on Kubelka-Munk transformed reflectance spectra as 3.08 eV for 0.125 M ZnO-SEP, 3.09 eV for 0.25 M ZnO-SEP and 3.16 eV for 0.5 M ZnO-SEP (Figure 5B).

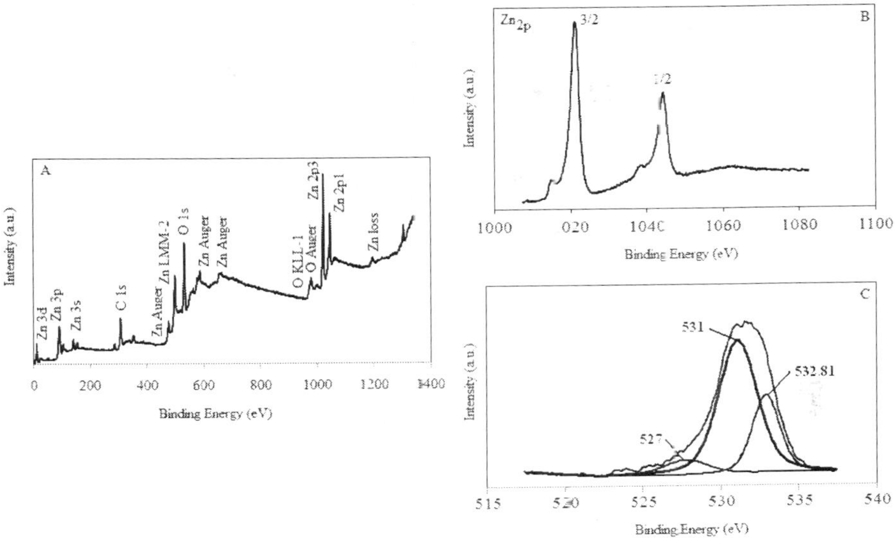

Figure 4. XPS analysis of 0.5 M ZnO-SEP: (A) survey spectrum, (B) Zn 2p spectra and (C) O 1s spectra.

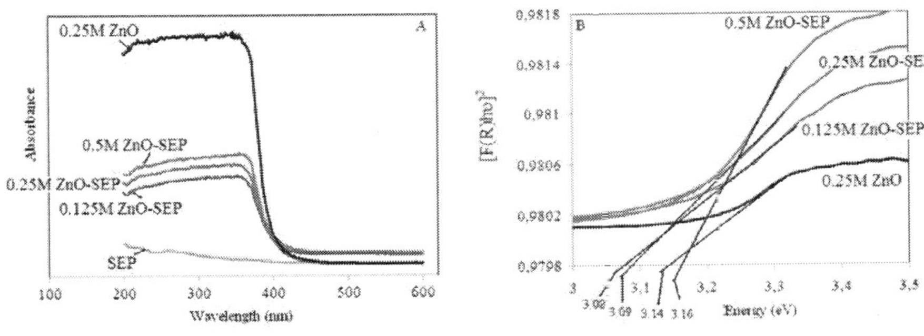

Figure 5. A) Diffuse reflectance spectra of SEP, 0.25 M ZnO and supported catalysts and (B) Kubelka-Munk transformed reflectance spectra of 0.25 M ZnO and supported catalysts.

3.2. Decolorization of MO by ZnO-SEP catalysts

Control experiments were performed under dark conditions for 0.25 M ZnO and 0.25 M ZnO-SEP (Figure 6). The remaining MO percentage was found as 83% in the presence of 0.25 M ZnO after 60 min, and then no significant variation was detected. The supported catalyst, however, shows a lower percentage (73%) in 80 min. The interaction between the SEP matrix and ZnO nanoparticles facilitated the adsorption of MO molecules. The low isoelectric point of the SEP (IEP_{SEP}~2) increased the contact possibility of positively charged Zn^{2+} ions or ZnO (with high isoelectric point ~9) [18]. The attraction between positively charged Zn^{2+} ions and negatively charged MO moiety decreased the intensity of absorption band at 464 nm, which is responsible from the destruction of the chromophoric (-N=N-) group. The negligible photolysis of MO proves the active role of ZnO nanoparticles within the supported catalyst system. It is also noted that the raw SEP does not participate in the decolorization of MO (not shown).

The photoactivities of the supported catalysts were examined under irradiation (Figure 6, inset). The highest MO remaining percentage (43%) was obtained in the presence of 0.125 M ZnO-SEP within 120 min. The percentage of MO decreased upon the loading concentration of ZnO. The synergy established on the mixed structures of ZnO and SEP improved the catalyst performance. Accordingly, 0.5 M ZnO-SEP shows the best activity with the lowest MO remaining percentage (19.8%) in solution.

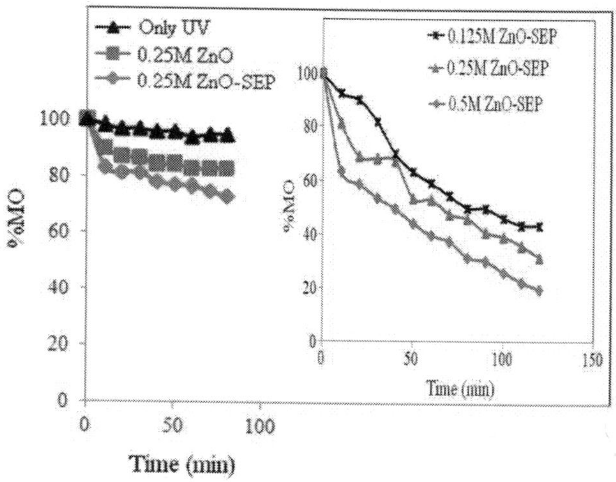

Figure 6. Photolysis and dark adsorption experiments. Inset: Photocatalytic activities of the supported catalysts.

The whole mechanism starts with the illumination of ZnO nanoparticles and production of electron-hole pairs (Eq. (1)).

$$ZnO + h\gamma(UV) \rightarrow e_{CB}^{-} + h_{VB}^{+} \quad (1)$$

The main active species in such processes are known as h_{vb}^{+}, $^{\cdot}OH$ and $O_2^{\cdot-}$ [27,28]. The radicals may form either by the reactions of photogenerated holes or electrons. The photogenerated electron (e_{cb}^{-}) can be easily transferred to the O_2 molecules promoting the $O_2^{\cdot-}$ formation and then converted to the active $^{\cdot}$OH via hydrogen peroxide and hydroperoxyl radical generations (Eqs. (2)–(5)).

$$O_2 + e^- \rightarrow O_2^{\cdot-} \tag{2}$$

$$O_2^{\cdot-} + H^+ \rightarrow HO_2^{\bullet} \tag{3}$$

$$2HO_2^{\bullet} \rightarrow H_2O_2 + O_2 \tag{4}$$

$$H_2O_2 + O_2^{\cdot-} \rightarrow {}^{\bullet}OH + O_2 + OH^- \tag{5}$$

At the same time, the photogenerated h_{vb}^+ can be captured on the catalyst surface undergoing charge transfer with adsorbed water molecules or with surface-bound hydroxide species to generate active $^{\cdot}$OH (Eqs. (6) and (7)) and results in the degradation of MO molecule (Eq. (8)).

$$h_{vb}^+ + H_2O \rightarrow H^+ + {}^{\bullet}OH \tag{6}$$

$$h_{vb}^+ + OH^- \rightarrow {}^{\bullet}OH \tag{7}$$

$${}^{\bullet}OH + MO \rightarrow MO\ (\text{decolorization and degradation}) \tag{8}$$

Thus, the separation of the charge carriers also enhanced the yield of hydroxyl radicals (highly reactive electrophilic oxidants) and improved the photocatalytic activity of the supported catalysts. The addition of the hydroxyl radical to the double bond of the azo group is referred to as the main reaction pathway, with the disappearance of the color. The second route followed the addition of hydroxyl radical to the aromatic rings [29,30]. Alternatively, hydroxyl radicals may attack the carbon atom bearing the azo bond [30]. Further attacks caused the formation of sulfonated intermediates, aromatic amine and phenolic compounds and ring open fragments [31].

Kinetic analysis was performed by varying the initial MO concentration from 16.2 to 3.27 mg L^{-1} in the presence of 0.5 M ZnO-SEP (not shown). The linearity obtained between $\ln(C_0/C)$ versus t plot indicates pseudo-first-order kinetics, where C_0 is the initial concentration of MO (mg L^{-1}) and C is the concentration of MO (mg L^{-1}) at irradiation time t (min) (Figure 7). The rate constants k (min^{-1}) hence calculated (slopes of the lines) were found to decrease with increasing concentration of MO (from 0.01 (3.27 mg L^{-1}) to 0.005 min^{-1} (16.2 mg L^{-1})). This can be a result of blocking of the photocatalytically active sites on the supported catalyst and reducing the interaction of photons with these sites.

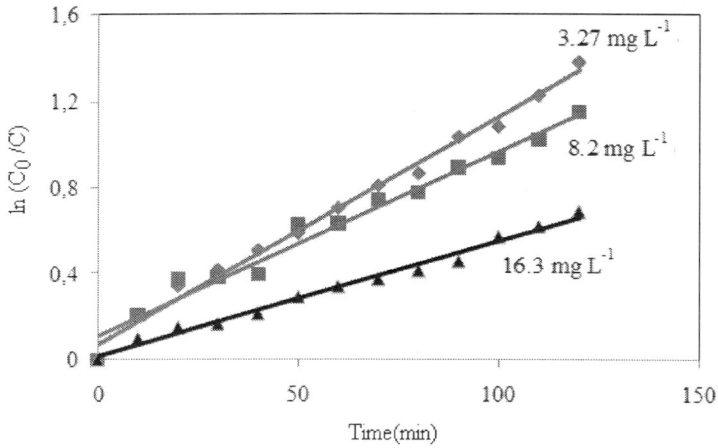

Figure 7. Effect of initial MO concentration on the photoactivity of 0.5 M ZnO-SEP.

The stability of 0.5 M ZnO-SEP catalyst was examined by recycling experiments (Figure 8). For each run, 0.5 M ZnO-SEP was filtrated, washed and calcined at 500°C for 2 h. After four cycles, the percentage of MO remaining in solution was found to increase by only approximately 3% (from 19.8% to 22%). The slight increment in the percentage can be attributed to the catalyst loss during each collection and rinsing steps.

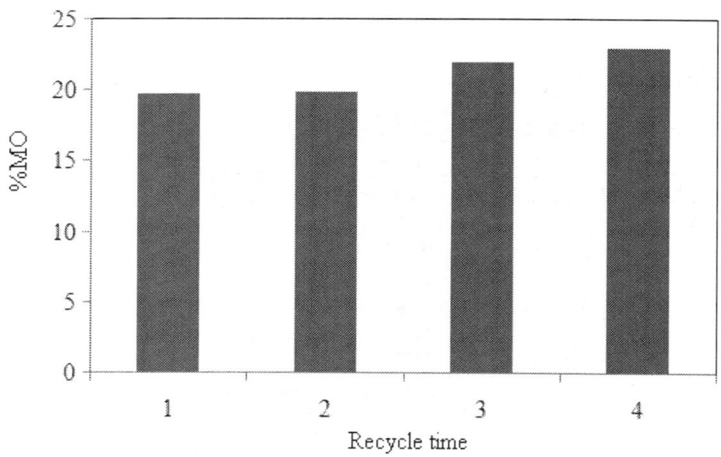

Figure 8. Reuse properties of 0.5 M ZnO-SEP.

4. Conclusions

This chapter has summarized the recent study of the author in an attempt to learn more about supported catalyst framework where clay mineral sepiolite is used as a support material for ZnO nanocatalysts. These photocatalysts were prepared and characterized by XRD, BET, SEM (EDX), XPS and DRUV techniques. EDX, mapping and XPS analysis evidence the existence of ZnO nanoparticles. D_{ZnO} sizes do not exhibit significant differences depending on the ZnO loading concentrations. Moreover, supported catalysts exhibit an absorption edge similar to 0.25 M ZnO. The pronounced performance of the supported catalysts is attributed to the highly dispersed ZnO nanoparticles within the ZnO-SEP matrix. Also, such systems can be readily separated from the suspension without filtration, providing an important, practical advantage in the usage of sepiolite supported catalysts.

Future studies will focus on the synthesis of new supported catalyst systems and their applications in energy and environmental issues.

Acknowledgements

This study was supported by the Boğaziçi University Research Foundation (Project No. 14B05P11/8820).

Author details

A. Neren Ökte*

Address all correspondence to: okteayse@boun.edu.tr

Department of Chemistry, Boğaziçi University, Bebek, Istanbul, Turkey

References

[1] Lam S. M., Sin J. C., Abdullah A. Z., Mohamed A. R., Degradation of wastewaters containing organic dyes photocatalysed by zinc oxide: A review. Desalin. Water Treat. 2012; 41: 131–169. DOI: 10.1080/19443994.2012.664698

[2] Kansal S. K., Ali A. H., Kapoor S., Photocatalytic decolorization of beibrich scarlet dye in aqueous phase using different nanophotocatalysts. Desalination. 2010; 259: 147–155. DOI: 10.1016/j.desal.2010.04.017

[3] Wahab R., Tripathy S. K., Shin H. S., Mohapatra M., Musarrat J., Al-Khedhairy A. A., Kaushik N. K., Photocatalytic oxidation of acetaldehyde with ZnO-quantum dots. Chem. Eng. J. 2013; 226: 154–160. DOI: 10.1016/j.cej.2013.02.128

[4] Kajbafvala A., Ghorbani H., Paravar A., Samberg J. P., Kajbafvala E., Sadrnezhaad S. K., Effects of morphology on photocatalytic performance of zinc oxide nanostructures synthesized by rapid microwave irradiation methods. Superlattice Microst. 2012: 51: 512–522. DOI: 10.1016/j.spmi.2012.01.015

[5] Sin J. C., Lam S. M., Lee K. T., Mohamed A. R., Preparation and photocatalytic properties of visible light-driven samarium-doped ZnO nanorods. Ceram. Int. 2013; 39: 5833–5843. DOI: 10.1016/j.ceramint.2013.01.004

[6] Lam S. M., Sin J. C., Abdullah A. Z., Mohamed A. R., Green hydrothermal synthesis of ZnO nanotubes for photocatalytic degradation of methylparaben. Mater. Lett. 2013; 93: 423–426. DOI: 10.1016/j.matlet.2012.12.008

[7] Mihai G. D., Meynen V., Mertens M., Bilba N., Cool P., Vansant E. F., ZnO nanoparticles supported on mesoporous MCM-41 and SBA15: a comparative physicochemical and photocatalytic study. J. Mater. Sci. 2010; 45: 5786–5794. DOI: 10.1007/s10853-010-4652-8

[8] Mahesh K. P. O., Kuo D. H., Huang B. R., Facile synthesis of heterostructured Ag-deposited SiO_2@TiO_2 composite spheres with enhanced catalytic activity towards the photodegradation of AB 1 dye. J. Mol. Catal. A: Chem. 2015; 396: 290–296. DOI: 10.1016/j.molcata.2014.10.017

[9] Németh J., Rodríguez-Gattorno G., Díaz D., Vázquez-Olmos A. R., Dékány I., Synthesis of ZnO nanoparticles on a clay mineral surface in dimethyl sulfoxide medium. Langmuir. 2004; 20: 2855–2860. DOI: 10.1021/la035097s

[10] Hur S. G., Kim T. W., Hwang S. J., Hwang S. H., Yang J. H., Choy J. H., Heterostructured nanohybrid of zinc oxide-montmorillonite clay. J. Phys. Chem. B. 2006; 110: 1599–1604. DOI: 10.1021/jp0543633

[11] Fatimah I., Wang, S., Wulandari D., ZnO/montmorillonite for photocatalytic and photochemical degradation of methylene blue. Appl. Clay Sci. 2011; 53: 553–560. DOI: 10.1016/j.clay.2011.05.001

[12] Meshram S., Limaye R., Ghodke S., Nigam S., Sonawane S., Chikate R., Continuous flow photocatalytic reactor using ZnO-bentonite nanocomposite for degradation of phenol. Chem. Eng. J. 2011; 172: 1008–1015. DOI: 10.1016/j.cej.2011.07.015

[13] Xu W. G., Liu S. F., Lu S. X., Kang S. Y., Zhou Y., Zhang H. F., Photocatalytic degradation in aqueous solution using quantum-sized ZnO particles supported on sepiolite. J. Colloid Interf. Sci. 2010; 351: 210–216. DOI: 10.1016/j.jcis.2010.07.052

[14] Bautista F. M., Campelo J. M., Luna D., Luque J., Marinas J. M., Vanadium oxides supported on TiO_2-sepiolite and sepiolite: Preparation, structural and acid characteri-

zation and catalytic behavior in selective oxidation of toluene. Appl. Catal. A: Gen. 2007; 325: 336–344. DOI: 10.1016/j.apcata.2007.02.033

[15] Arques A., Amat A. M., Santos-Juanes L., Vercher R. F., Marín M. L., Miranda M. A., Sepiolites as supporting material for organic sensitisers employed in heterogeneous solar photocatalysis. J. Mol. Catal. A: Chem. 2007; 27: 221–226. DOI: 10.1016/j.molcata.2007.02.038

[16] Suárez S., Yatez M., Petre A. L., Martin J. A., Avila P., Blanco J., Development of a new Rh/TiO_2-sepilote monolithic catalyst for N_2O decomposition. Appl. Catal. B: Environ. 2006; 64: 302–311. DOI: 10.1016/j.apcatb.2005.12.006

[17] Ökte A. N., Sayinsöz E., Characterization and photocatalytic activity of TiO_2 supported sepiolite catalysts. Separ. Purif. Technol. 2008; 62: 535–543. DOI: 10.1016/j.seppur.2008.03.011

[18] Portela R., Rubio-Marcos F., Leret P., Fernandez J. F., Banares M. A., Avilla P., Nanostructured ZnO/sepiolite monolithic sorbents for H_2S removal. J. Mater. Chem. A. 2015; 3: 1306–1316. DOI: 10.1039/C4TA04440A

[19] Anandan S., Vinu A., Sheeja Lovely K. L. P., Gokulakrishnan N., Srinivasu P., Mori T., Murugesan V., Sivamurugan V., Ariga K., Photocatalytic activity of La-doped ZnO for the degradation of monocrotophos in aqueous suspension. J. Mol. Cat. A. Chem. 2007; 266: 149–157. DOI: 10.1016/j.molcata.2006.11.008

[20] Lupan O., Chow L., Chai G., Roldan Cuenya B., Naitabdi A., Schulte A. A., Heinrich H., Nanofabrication and characterization of ZnO nanorod arrays and branched microrods by aqueous solution route and rapid thermal processing. Mater. Sci. Eng. B. 2007; 145: 57–66. DOI: 10.1016/j.mseb.2007.10.004

[21] Lupan O., Emelchenko G. A., Ursaki V. V., Chai G., Redkin A. N., Gruzintsev A. N., Tiginyanu I. M., Chow L., Ono L. K., Roldan Cuenya B., Heinrich H., Yakimov E. E., Synthesis and characterization of ZnO nanowires for nanosensor applications. Mater. Res. Bull. 2010; 45: 1026–1032. DOI: 10.1016/j.materresbull.2010.03.027

[22] Ökte A. N., Karamanis D., A novel photoresponsive ZnO-flyash nanocomposite for environmental and energy applications. Appl. Catal. B: Environ. 2013; 142–143: 538–552. DOI: 10.1016/j.apcatb.2003.12.008

[23] Ökte A. N., Karamanis D., Tuncel D., Dual functionality of TiO_2-flyash nanocomposites: Water vapor adsorption and photocatalysis. Catal. Today. 2014; 230: 205–213. DOI: 10.1016/j.cattod.2014.01.031

[24] Brunauer S., Deming L. S., Deming W. E, Teller E., On a theory of the van der Waals adsorption of gases. J. Am. Chem. Soc. 1940; 62: 1723–1732. DOI: 10.1021/ja01864a025

[25] Jing L., Xu Z., Sun X., Shang J., Cai W., The surface properties and photocatalytic activities of ZnO ultrafine particles. Appl. Surf. Sci. 2001; 180: 308–314. DOI: 10.1016/S0169-4332(01)00365-8

[26] Liqiang J., Dejun W., Baiqi W., Shudan L., Baifu X., Honggang F., Jiazhong S., Effects of noble metal modification on surface oxygen composition, charge separation and photocatalytic activity of ZnO nanoparticles. J. Mol. Catal. A. 2006; 244: 193–200. DOI: 10.1016/j.molcata.2005.09.020

[27] Sun H. Q., Feng X. H., Wang S. B., Ang H. M., Tadé M. O., Combination of adsorption, photochemical and photocatalytic degradation of phenol solution over supported zinc oxide: Effects of support and sulphate oxidant. Chem. Eng. J. 2011; 170: 270–277. DOI: 10.1016/j.cej.2011.03.059

[28] Lam S. M., Sin J. C., Abdullah A. Z., Mohamed A. R., Degradationof wastewaters containing organic dyes photocatalysedby zinc oxide: A review. Desalin. Water Treat. 2012; 41: 131–169. DOI: 10.1080/19443994.2012.664698

[29] Joseph J. M., Destaillats H., Hung H., Hoffmann M. R., The sonochemical degradation of azobenzene and related azo dyes:rate enhancements via Fenton's reactions. J. Phys. Chem. A. 2000; 104: 301–307. DOI: 10.1021/jp992354m

[30] Spadaro J. T., Isabelle L., Renganathan V., Hydroxyl radical mediated degradation of azo dyes: evidence for benzene generation. Environ. Sci. Technol. 1994; 28: 1389–1393. DOI: 10.1021/es00056a031

[31] Konstantinou K. I., Albanis T. A., TiO_2-assisted photocatalytic degradation of azo dyes in aqueous solution: kinetic and mechanistic investigations: A review. Appl. Catal. B: Environ. 2004; 49: 1–14. DOI: 10.1016/j.apcatb.2003.12.008

Chapter 9

Preparation of Functionalized Graphene and Gold Nanocomposites – Self-assembly and Catalytic Properties

Jingxin Zhou, Tifeng Jiao, Qingrui Zhang and Jie Hu

Additional information is available at the end of the chapter

http://dx.doi.org/10.5772/62166

Abstract

Nanocomposites and nanomaterials have been attracting more attention in various fields. Nanocomposites can be prepared with a variety of special physical, thermal and other unique properties. They have better properties than conventional microscale composites and can be synthesized using simple and inexpensive techniques. A composite material consists of an assemblage of two materials of different natures completing and allowing us to obtain a material of which the set of performance characteristics is greater than that of the components taken separately. In our recent research, some functionalized nanocomposites and nanomaterials have been prepared and investigated. In addition, some of the analytical methods, theoretical treatments, and synthetic tools, which are being applied in the area of self-assembly and supramolecular chemistry, will be highlighted. In this chapter, we summarize our main research contributions in recent years in two sections: (1) preparation and catalytic properties of some functionalized graphene nanocomposites; (2) preparation and catalytic properties of some functionalized gold nanocomposites. These works not only provided important inspirations for developing graphene-hybridized materials but also opened new possibilities to improve the photocatalytic activity of photocatalyst.

Keywords: nanocomposites, nanomaterials, catalytic property, self-assembly

1. Introduction

Nanocomposites and nanomaterials have been attracting increasing attention in various fields [1–6]. Nanocomposites can be prepared with a variety of special physical, thermal, and other unique properties. They have better properties than conventional microscale composites and can be synthesized using simple and inexpensive techniques. In addition, as the assemblage of various separate components, the designed composite materials would demonstrate greater characteristics and performances. At the same time, nanomaterials and nanocomposites often have unique properties that could enable composite materials with multiple unique properties simultaneously; however, it is often challenging to achieve these properties in large-scale nanocomposites and functionalized nanomaterials. In order to obtain the stated objective composites, it is important to design different models of the interactions in nanocomposites and next controllable properties. Moreover, it is well known that various self-assembly techniques are important to fabricate well-defined hierarchical and organized nanostructures with special properties and controlled capacity [7–10]. For different research systems, such as inorganic–organic hybrids, various nanoparticles, colloidal microspheres, and supramolecular nanostructures, the self-assembly process can demonstrate obvious advantages, especially in preparing special nanostructures. The incorporation of nanocomposites with self-assembly shows new performances for designed composite nanomaterial with different organized style and processing routes.

In addition, as a special sp^2-bonded single layer carbon material, graphene has been attracting more attentions in various application fields, especially as photocatalytic material due to its natural characteristics, such as high chemical stability, large surface area, and excellent adsorption capacity [11]. And some new nanocomposites from graphene materials as photocatalysts have been developed in recent years. For example, the research group of Ullah prepared ternary Pt-graphene/TiO_2 nanocomposites by microwave-assisted method and investigated the photocatalytic degradation for dye rhodamine B [12]. Bai and coworkers reported the preparation of new $ZnWO_4$/graphene hybrid materials by in situ reduction of graphene oxide and characterized the degradation capacity of methylene blue as photocatalysts [13]. Sun's group reported the synthesis of $ZnFe_2O_4$/ZnO nanocomposites anchored on graphene substrates via an ultrasound-aided method [14]. Xu and coworkers prepared reduced graphene oxide/Bi_2WO_6 composite materials as photocatalysts, and compared the photocatalytic activities of composites with pure Bi_2WO_6 [15]. In addition, more research works have been reported to enhance the photocatalytic activity of graphene-based composites as photocatalysts [16–19]. Moreover, in the last several decades, the preparations of gold nanoparticles have been attracting much attention [20–22]. Up to now, various processes [23–25] and designed capping agents [26–29] have demonstrated the synthesis of functional gold nanoparticles with designed nanostructures and tailored behaviors. It is well known that many factors, such as the particle size, shape, colloid stability, and surface functionalized groups, seem to be critical for the applications fields [30–32]. In addition, because the self-assembled nanostructures are important and closely related to molecular skeletons in designed template compounds, the design and synthetic routes are the key step in the whole process.

In recent research work reported by our group, various research systems about graphene and gold nanocomposites/nanomaterials have been designed and characterized. And the involved analytical methods and self-assembly techniques have been also highlighted. So in the present chapter, we would like to demonstrate our recent research contributions in two parts: (1) preparation and catalytic properties of some functionalized graphene nanocomposites; (2) preparation and catalytic properties of some functionalized gold nanocomposites. These works not only provided important inspirations for developing graphene-hybridized materials but also opened new possibilities to improve the photocatalytic activity of photocatalyst.

2. Preparation and catalytic properties of some functionalized graphene nanocomposites

Recently, our group has reported the design and fabrication of hybrid organogels by self-assembly of composites containing cationic compounds and GO [33]. It is interesting to note that the obtained gelation performances can be regulated via different functional substituted headgroups in used compounds. The experimental data indicated that ammonium headgroup in molecular skeletons seemed more favorable for the composite gelation than pyridinium segment. The obtained results suggested that the self-assembly modes in present GO-based composites could be manipulated by controlling efficient headgroup effect. In addition, various weak forces between present building blocks seemed also responsible for the formation of different nanostructures. Based on the obtained results data in present composite gel system, a reasonable mechanism about self-assembly modes in gels is shown in Figure 1. For the prepared CTAB-GO gel, various organized building blocks are obtained in different solvents because of the van der Waals force of substituent chains and the strong electrostatic interaction of ammonium headgroups with oxygen-containing functional groups at GO surface. In addition, for the cases of C16Py-GO and BPy-GO composite gels, the strong π–π stacking between carbon net in GO plane and additional pyridine headgroups showed as more competitive with other forces, such as electrostatic interaction and van der Waals force. Thus, the present research work demonstrates new exploration for the design of GO-based composite gels and self-assembled soft matters.

In addition, we have also demonstrated the formation of organogels by self-assembly of cationic gemini amphiphile–GO composites [34]. Their gelation behaviors in various organic solvents can be controlled by regulating molecular symmetry. The obtained data indicated that the designed functional groups and molecular symmetry could change the self-assembly modes and produce different self-assembled nanostructures. It seemed that longer alkyl chains in molecular skeletons could be helpful to enhance the intermolecular hydrophobic force in the process of self-assembly. So the changes of building blocks and stacking modes between present GO-based composites and different solvents are responsible for the formation of various nanostructures, as seen in Figure 2. It clearly indicated that the formed nanostructures in present composite materials were obviously different, such as nano-wrinkle, nano-lamella, and nano-belt. The obtained different morphologies in self-assembled gels can be mainly due to various formation mechanisms upon special self-assembly modes via weak interactive forces between building blocks. Finally, a reasonable self-assembled mechanism for symmetry effects in formation of present organized nanostructures is demonstrated. Thus, the present GO-based composite gel materials will give a helpful clue for the design and preparation of functional GO composite nanomaterials.

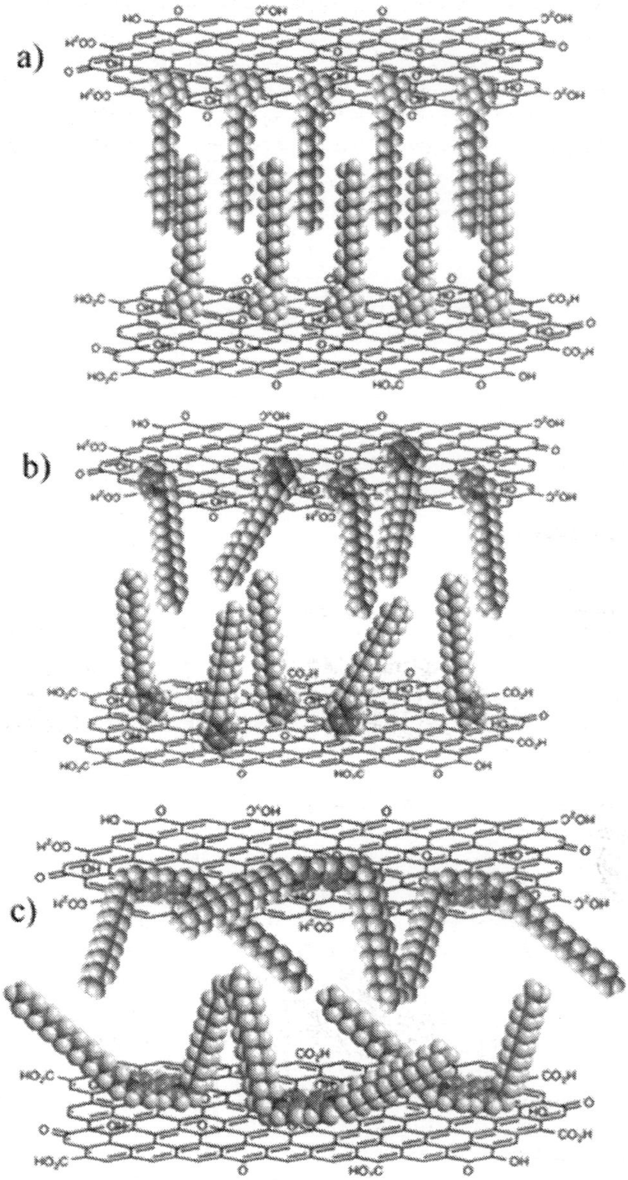

Figure 1. Scheme of different assembly modes in cationic amphiphiles–graphene oxide gels. CTAB-GO (a), C16Py-GO (b), and BPy-GO (c).

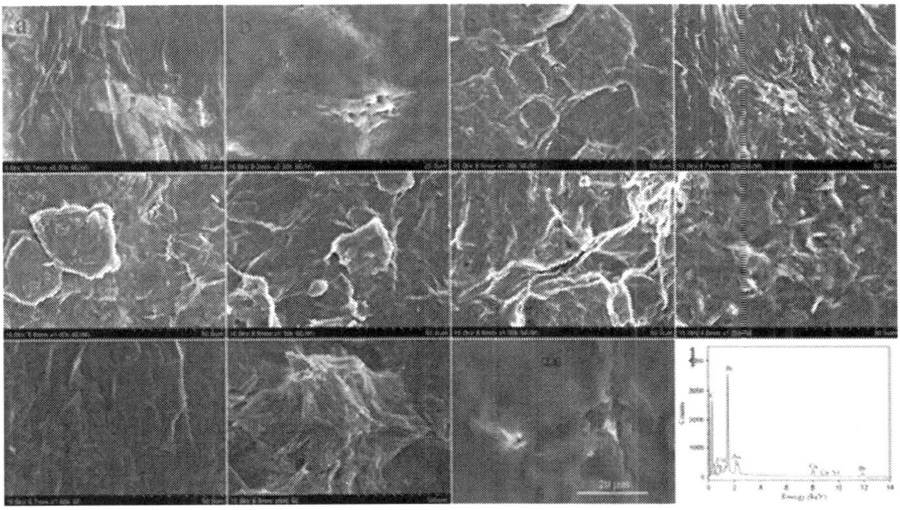

Figure 2. SEM images of xerogels for GO sheet, C18-6-6/GO gels, C18-6-12/GO gels, and C18-6-18/GO gels in different solvents.

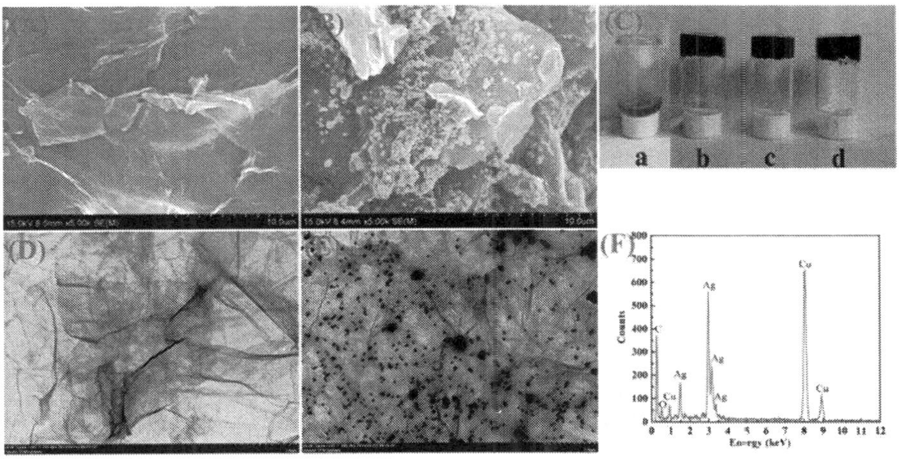

Figure 3. SEM and TEM images for the RGO/PEI and RGO/PEI/Ag hydrogels.

In another composite system, we have reported the design and preparation of silver nanoparticle-containing RGO-based composite hydrogel materials via an in situ reduction process [35]. The obtained experimental data indicated that the prepared composite gels were composed of 3D net-like nanostructures, as seen in Figure 3. In addition, the used preparation method included the in situ reduction of GO and silver acetate in hydrogel structures to fabricate present RGO-based composite hydrogel. So the formed silver nanoparticles were uniformly anchored on RGO surface in composite gel. Moreover, the photocatalytic behaviors for removing dye pollutants are also characterized for the silver nanoparticle-containing RGO-based composite hydrogel, as seen in Figure 4. It is interesting to note that the obtained photocatalytic composite materials can be reused from an aqueous degradation system, indicating the important and potential applications for dye removal and wastewater treatment.

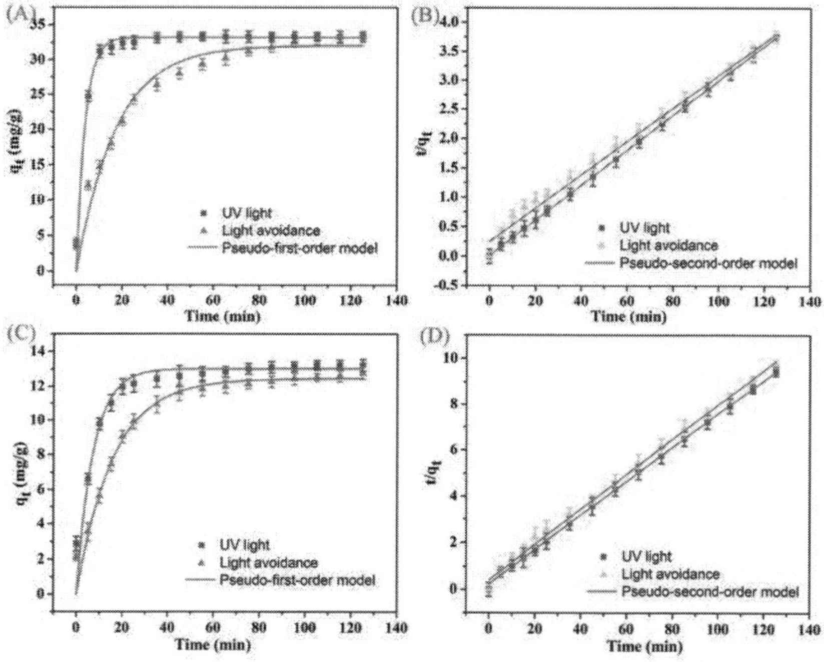

Figure 4. Degradation kinetics curves of as-prepared RGO/PEI/Ag nanocomposites on MB (a, b) and RhB (c, d) at 298 K.

In order to investigate the mechanism of hybrid graphene composites, we have also synthesized some $LaMnO_3$–graphene composites as photocatalysts by a sol–gel method [36]. It is found that $LaMnO_3$ perovskite phase was successfully fixed on graphene surface with well-dispersion capacity, as seen in Figure 5. The data indicated that the photocatalytic capacity of as-formed $LaMnO_3$–graphene composite materials were better than pristine $LaMnO_3$ material, with the detailed results in Figure 6. The enhancement of photocatalytic properties can be mainly due to the high separation efficiency of photo-induced electron–hole pairs originated from the excellent conductivity of graphene in composite and the large interfacial contact between components, which is helpful to increase the dyes adsorption and improve the transfer efficiency in photocatalytic process. This research demonstrated new inspiration for designing photocatalytic graphene-based hybrid materials.

In another work, organized $La_{1-x}Sr_xMnO_3$/graphene thin films were prepared by both sol–gel and spin-coating methods [37]. In experimental process, the formed sol nanoparticles were adsorbed on graphene surface via electrostatic force in aging time. Then, the formed $LaMnO_3$ nanoparticles increase sizes and form crystal domains on graphene surface in the next calcination step. Present obtained nanostructures and morphology were investigated during various characterization techniques. Figure 7 shows the XRD patterns of graphene, $LaMnO_3$, $LaMnO_3$/graphene, and $La_{0.9}Sr_{0.1}MnO_3$/graphene thin film. The diffraction peak of graphene appeared in the vicinity of 23°, which was similar to the diffraction peak position of graphite. However, this peak broadened and weakened because the size of the graphite sheet decreased, the integrity of the crystal structure declined, and the degree of disorder increased. The pattern of $LaMnO_3$ is in agreement with PDF33-0713, indicating its perovskite structure with complete crystal shape. In the process of acid red 3GN photodegradation, $LaMnO_3$/graphene thin film had sound stability and better photocatalytic ability than $LaMnO_3$ thin film. As shown in Figure 8, a mild photodecomposition effect was observed in the degradation of dye. The absorption peak of acid red 3GN dye solution at 509 nm was from the initial 0.5446 to 0.5138 after irradiating for 48 h without photocatalysts. The experimental results indicated that graphene enhanced the dye adsorption, inhibiting the reunion of light-induced e^-–h^+ and improving photocatalytic capacity. In addition, it should be noted that a red shift of absorption edge was found by doping Sr, which seemed helpful to increase the photocatalytic performance of the obtained composite film.

In addition, another new $LaMn_{1-x}Co_xO_3$/graphene composite material as photocatalyst had been designed and prepared by sol–gel method [38]. The experimental data indicated that $LaMnO_3$ perovskite phase was anchored on graphene surface with the special perovskite structure. In addition, the photocatalytic capacity was characterized by the degradation of diamine green B. In the photodegradation process, the graphene component in composite can accelerate the dye adsorption, while doping Co component improves the photocatalytic performance. Thus, the reasonable charge transfer mechanism that occurred in the obtained $LaMnO_3$/graphene composite during photocatalytic process is demonstrated in Figure 9. Firstly, diamine green B molecules could shift to the active surface of prepared composites from solution and organized in self-assembled face-to-face mode via π–π stacking with aromatic graphene net. Due to the effect of these holes and electron transfers, charge recombination is pushed in obtained $LaMnO_3$/graphene composite and improves the efficiency of photocatalytic capacity.

Moreover, we have also reported the preparation of some graphene-based $LaNiO_3$ composite films by both sol–gel method and spin-coating technique [39]. The obtained experimental results indicated that the size of formed $LaNiO_3$ nanoparticles was about 20 nm, well dispersed on graphene surface. The photocatalytic capacity of present hybrid films had been demonstrated by degradation of acid red A. In comparison with pure $LaNiO_3$ films, the designed $LaNiO_3$/graphene composite films showed better photocatalytic behavior. It is interesting to

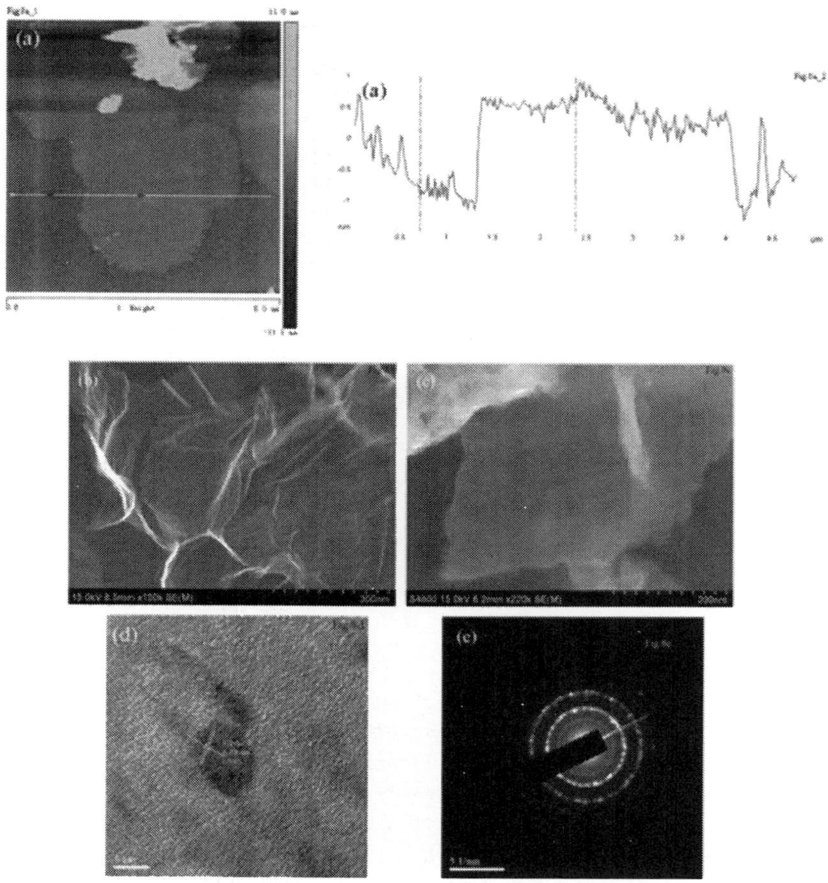

Figure 5. (a) AFM image of the as-synthesized graphene; (b) SEM image of graphene; (c, d, e) SEM, HRTEM, and SA-ED images of $LaMnO_3$–graphene composites.

note that when the content of graphene shifted to the value of about 4%, the photocatalytic efficiency of the obtained composite films was double that of pure $LaNiO_3$ films.

In another system, $La_{1-x}Ca_xMnO_3$ perovskite–graphene composites are synthesized as catalysts for Zn–air cell cathodes [40]. The results indicated that perovskite phase adhered on the surface of graphene sheets, and adding graphene significantly improved the electrochemical performance of $LaMnO_3$. The XPS spectrum of $La_{0.6}Ca_{0.4}MnO_3$–graphene composite is shown in Figure 10. The peak contained all the elements of $La_{0.6}Ca_{0.4}MnO_3$. In addition, the obtained graphene showed gauze-like fold nanostructures, mainly originated from the oxygenic functional groups and the surface defects during preparation process. So the formed porous

http://dx.doi.org/10.5772/62166

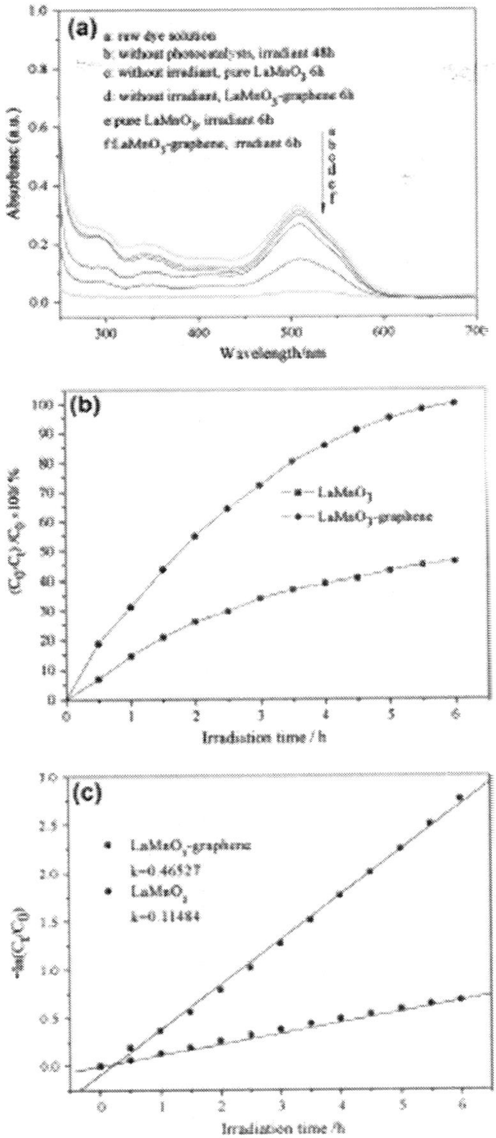

Figure 6. (a) UV–Vis spectral changes of the degradation of acid red A by $LaMnO_3$–graphene and $LaMnO_3$; (b) photocatalytic activities of $LaMnO_3$–graphene composite and $LaMnO_3$; (c) kinetics of photocatalytic degradation by the $LaMnO_3$ and $LaMnO_3$–graphene.

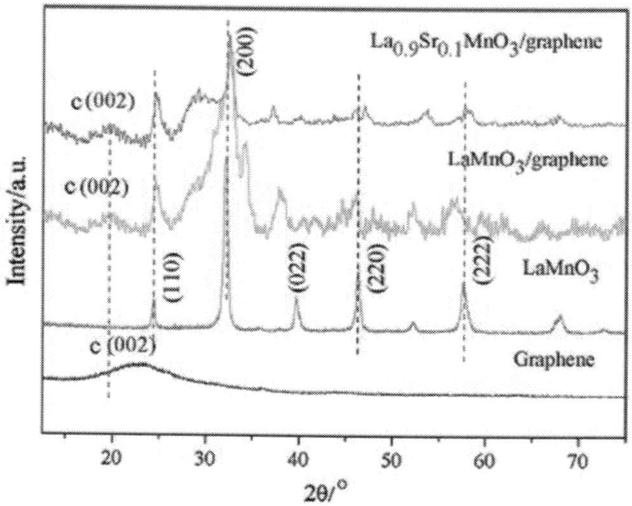

Figure 7. XRD pattern of graphene, $LaMnO_3$, $LaMnO_3$/graphene, and $La_{0.9}Sr_{0.1}MnO_3$/graphene.

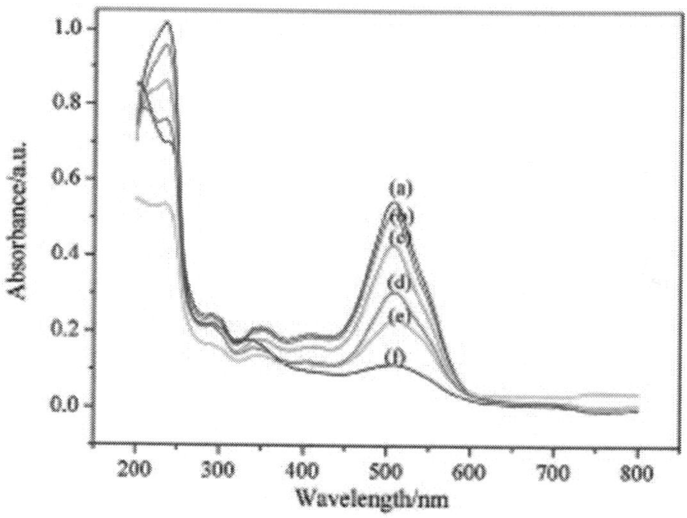

Figure 8. UV–Vis absorb spectrums of acid red-3GN under different conditions: (a) raw dye solution, (b) irradiation 48 h without catalyst, (c) $LaMnO_3$ thin film adsorb dye 4 h without irradiation, (d) $LaMnO_3$/graphene or $La_{0.9}Sr_{0.1}MnO_3$/graphene thin film adsorb dye 4 h without irradiation, (e) irradiation 4 h with $LaMnO_3$ thin film as catalyst, (f) irradiation 4 h with $LaMnO_3$/graphene thin film as catalyst.

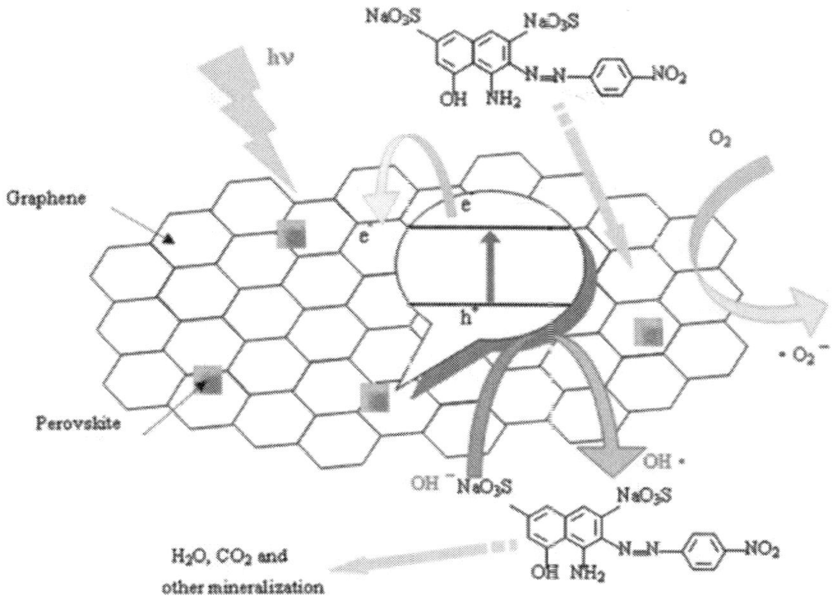

Figure 9. Proposed mechanism for photocatalytic degradation of diamine green B over graphene-based perovskite photocatalysts under light irradiation.

3D nanostructure can obviously enhance the three-phase domains and improve the mass transfer process as catalyst materials of air electrode. The experimental data indicated that the voltage plateau was superior with 10 wt% ratio value of graphene. Moreover, Ca doping maintained the perovskite structure and obviously enhanced the electrocatalytic activity for ORR, and $La_{0.6}Ca_{0.4}MnO_3$–graphene composite demonstrated the best catalytic capacity. Thus, the obtained research work indicates that the prepared graphene-based $La_{1-}xCaxMnO_3$ composites are important material for design of air electrodes catalysts.

Recently, we have also reported the preparation of $LaMnO_3$/graphene thin films with the perovskite-type as new photocatalyst via sol–gel process and spin-coating method [41]. The obtained results indicated that the addition of graphene did not change the perovskite structure, with formed $LaMnO_3$ particles at about 22 nm well dispersed on graphene surface. Figure 11 displays the nitrogen adsorption–desorption isotherms and pore size distribution curves calculated by BJH method for $LaMnO_3$ and $LaMnO_3$/graphene powders. The larger surface area can effectively absorb the dye, thus increasing the contact probability of pollutant molecular and catalyst. Determination of contact angle indicated that the contact angle of glass substrate decreased and the hydrophilicity improved after treating with H_2SO_4 and APTES. The UV–Vis photocatalytic activity of the photocatalysts was evaluated by the degradation of diamine green B. $LaMnO_3$/graphene thin films had better photocatalytic ability than $LaMnO_3$ and TiO_2 films.

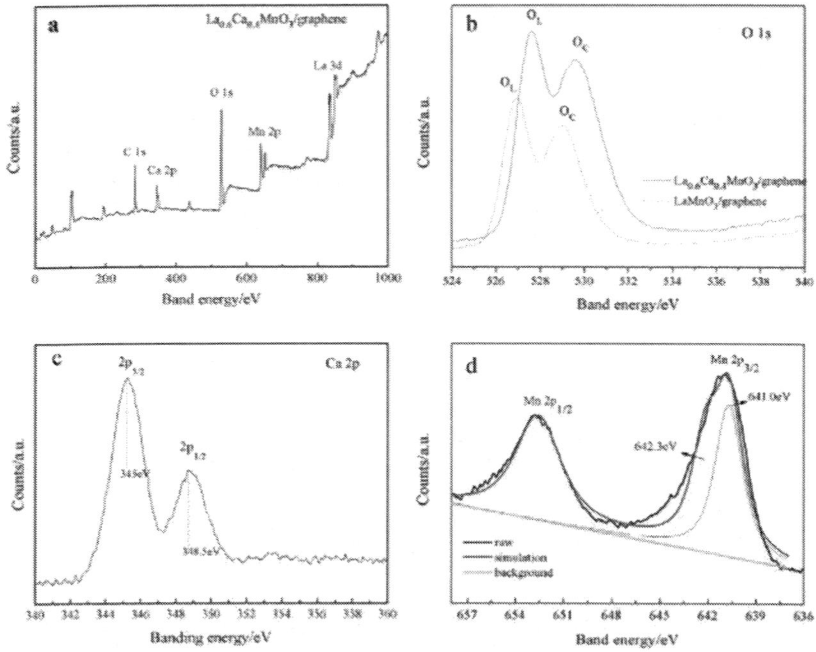

Figure 10. XPS spectrum of $La_{0.6}Ca_{0.4}MnO_3$–graphene composite. (a) Overall spectrum, high-resolution curves of (b) O 1s region, (c) Ca 2p region, and (d) Mn 2p region.

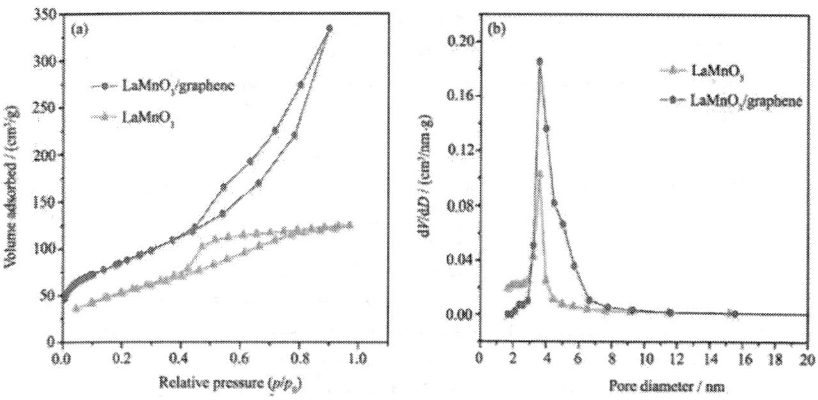

Figure 11. N_2 adsorption–desorption isotherms (a) and pore size distributions of samples (b).

3. Preparation and catalytic properties of some functionalized gold nanocomposites

In this section, many examples have been investigated, including the synthesis of gold nanoparticles by some compounds with hydrophilic spacers and aromatic headgroups at different interface. The obtained experimental data indicated that various gold nanostructures could be prepared by controlling different substituted headgroups in template compounds. In addition, the photocatalytic capacities of as-prepared gold nanoparticles on the degradation of organic dyes were also demonstrated.

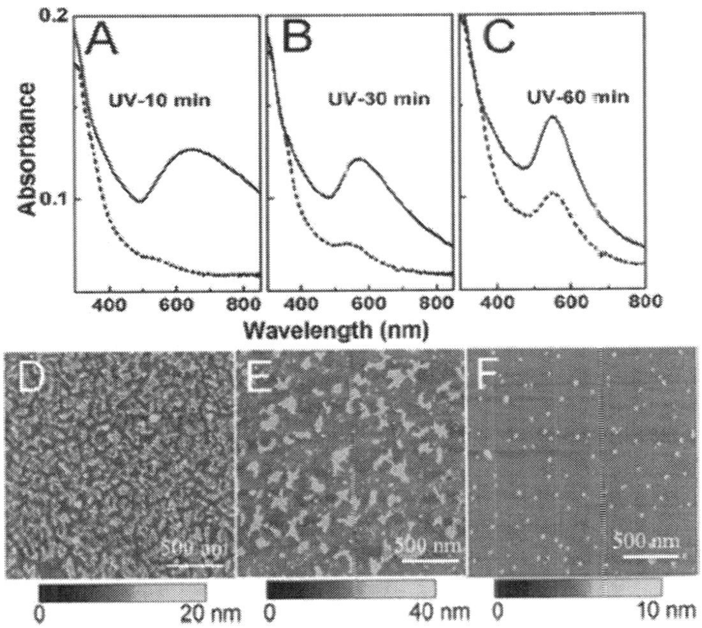

Figure 12. UV–Vis spectra and AFM images of organized LB films containing gold nanoparticles.

Firstly, the effect of gemini compounds on in situ self-assembly and fabrication of gold nanoparticles in organized molecular films have been demonstrated [42]. In comparison with traditional compounds, gemini molecules demonstrated well the capacity to fix chloroaurate ions onto various solid substrates, suggesting an important route to prepare metallic nanoparticles in organized films by chemical reduction method. In addition, gemini compounds could produce a 2D confined environment in LB films to accommodate the formed gold nanoparticles. Thus, the size, shape, and interparticle distances can be regulated by changing various reducing methods. Various nanostructures could be fabricated, such as nanoparticles, nanowires, and the tree-branched domains. Based on the obtained experimental results, different gold nanostructures could be synthesized by regulating various reductants or reducing processes. Moreover, we tried to control gold nanostructures by optical method. The obtained composite film was irradiated with UV light and then reduced by $AuCl_4^-$ ions with hydroquinone, as seen in Figure 12. The present research work provided the new clue for the synthesis of metallic particles in films from special compounds, showing important exploration in designing various gold nanostructures.

In addition, we presented a facile synthetic method to the gold nanostructures using a series of gemini amphiphiles through liquid–liquid biphase method [43]. The gemini amphiphiles themselves could serve as both capping and reducing agents. The spacer and concentration of the gemini amphiphiles played an important role in the formation of gold nanoprisms. It is predicted that gold and other novel metal nanostructures may be produced by gemini amphiphiles whose properties can be well controlled by designing different headgroups, spacers, or alkyl chains. In order to make clear the gold nanostructures, the chloroform solution was cast onto copper grid for TEM measurement, as shown in Figure 13. For gold nanostructures generated by GN1, a few polygon gold nanoparticles were found. In contrast, uniform gold nanoparticles with a size of 11.4 ± 1.2 nm were observed in GN2 chloroform solution. Interestingly, in GN3 chloroform solution, triangular nanoplates and nanoparticles were both observed. The obvious difference suggests that the spacer has an influence on the shapes of gold nanostructures. On the other hand, no gold nanoparticle can be observed in aqueous phase no matter how long the reaction proceeded.

Considering the above research background, we have prepared new kinds of gold nanoparticles via some bolaform Schiff base compounds with hydrophilic spacers and aromatic headgroups in molecular skeletons [44]. By stirring the mixed solution of aqueous $AuCl_4^-$ ions with chloroform solution of used Schiff base molecules, the metal ions shifted to the chloroform phase and reduced to the formation of different gold nanoparticles. The data indicated that different gold nanostructures could be obtained by regulating the molecular skeletons of used bolaform compounds, including spacers and headgroups, as well as the relative ratios of compounds to metal ions. In addition, the characterization of morphologies and spectra indicated that the present designed bolaform amphiphilic compounds could act as both capping agents and reducing agents. So the UV–Vis spectra in different conditions are demonstrated, as seen in Figure 14. From the obtained UV–Vis data, it clearly indicated that the spacer and azomethine segments in bolaform molecules could be positively charged at interface in transferred process. In addition, the photocatalytic capacity of the obtained gold nanoparticles on dye degradation was demonstrated in Figure 15, showing the influence of molecular skeletons in the used compounds on the regulation of prepared gold nanoparticles and next catalytic behaviors. The present obtained data suggested that various gold nanostructures could be designed and synthesized by changing substituted skeletons in used template compounds

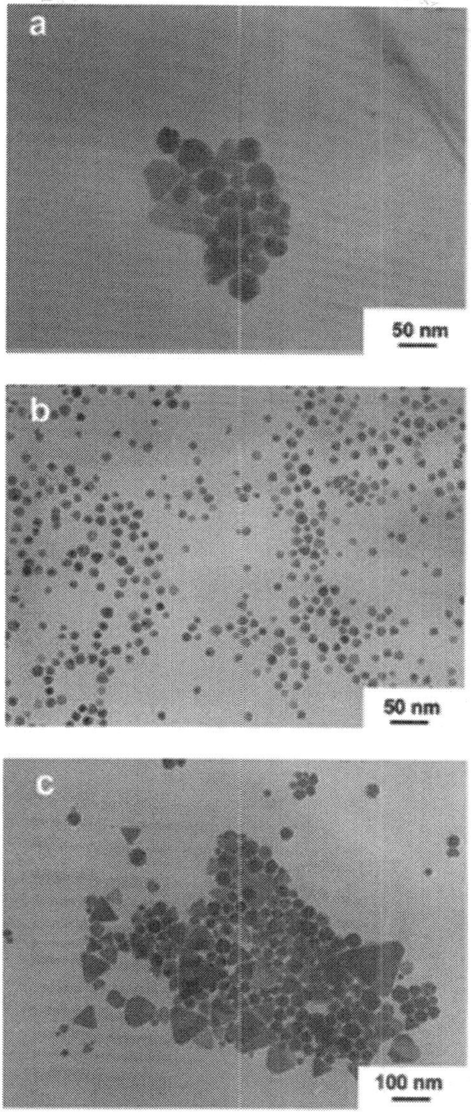

Figure 13. TEM images of reduced gold in GN1 (a), GN2 (b), and GN3 (c) chloroform solution after 36 h of reaction respectively.

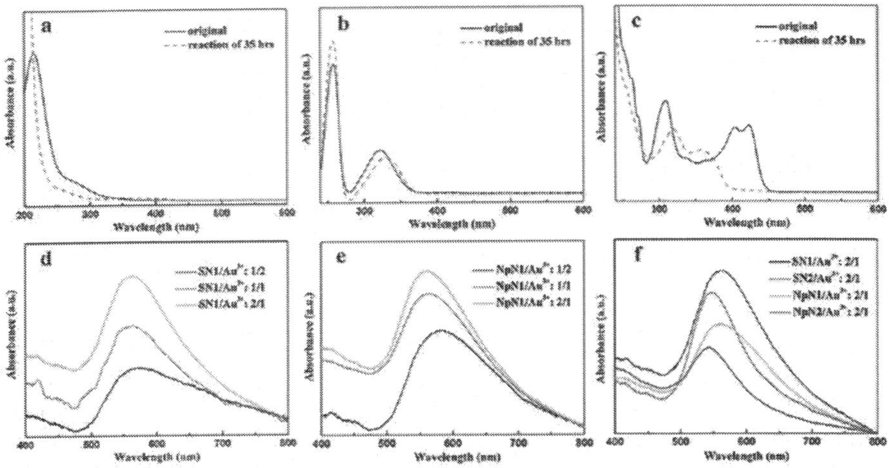

Figure 14. UV–Vis spectra of 0.1 mM $AuCl_4^-$ ions and prepared gold nanostructures in aqueous solution.

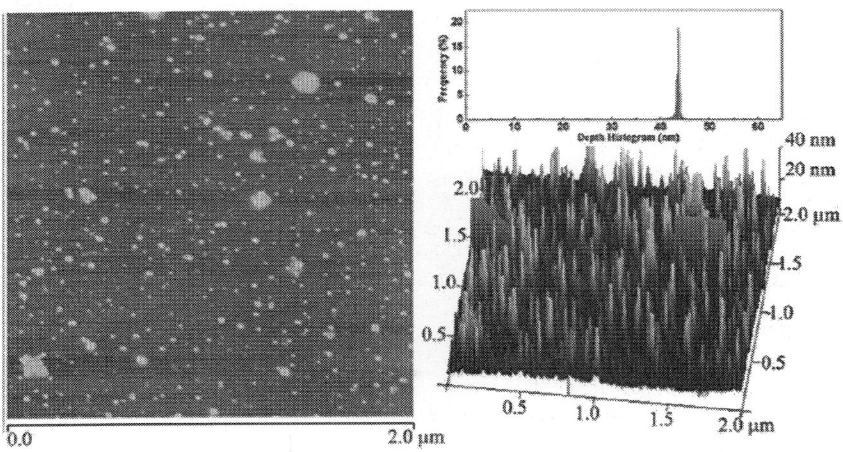

Figure 15. AFM images of height and 3D surface plot view with depth histogram of gold nanoparticles on mica prepared from instant ultrasonic ethanol solution with SN2 to chloroaurate ion ratio of 2:1 after 35 h of reaction.

In addition, some other gold nanoparticles were synthesized by two bolaform cholesteryl imide derivatives with different lengths of ethyleneamine spacers at a liquid–liquid interface [45]. Spectral and morphological measurements indicated that both bolaform amphiphiles could act as both capping agents and reducing agents. To further characterize the prepared gold nanostructures, TEM measurements have been demonstrated, as seen Figure 16. The images indicated that the size distribution of obtained gold nanostructures could be regulated by changing various spacers in used molecular skeletons. In addition, the effect of molar ratio of the template compound to $AuCl_4^-$ ions was also investigated in details. The experimental data indicated that various nanostructures, such as hexagonal, polygon nanoparticles, and nanoplates, could be synthesized. Moreover, the photocatalytic capacity of prepared gold nanostructures on dye degradation was also characterized, suggesting the importance of compounds' skeletons in regulating the formation of gold nanoparticles and changing relative catalytic properties, as demonstrated in Figure 17. The obtained research data would give new clue for the preparation of gold nanostructures by designing special template compounds.

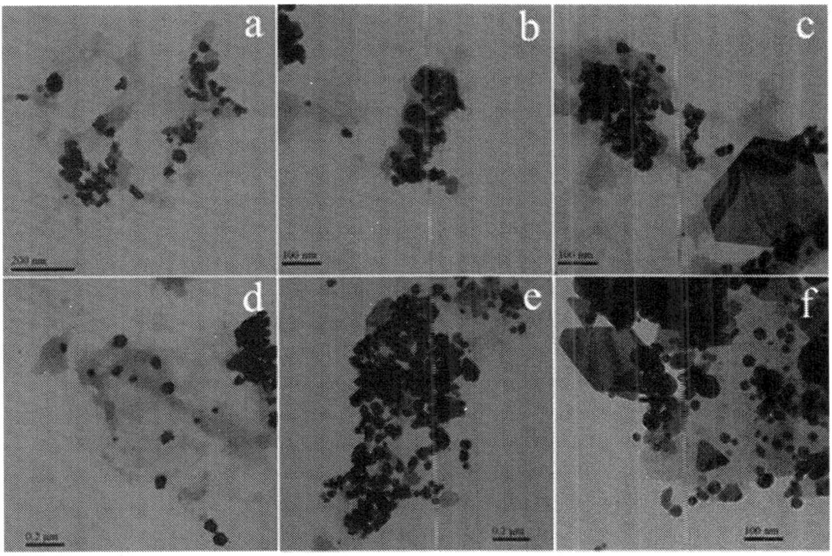

Figure 16. TEM images of gold nanoparticles by using different compounds to chloroaurate ion ratio.

4. Conclusion and perspectives

We are working on the preparation, self-assembly, and application of functionalized nano-composites and nanomaterials. In this chapter, various kinds of nanocomposites including graphene nanocomposites and gold nanoparticles have been designed and characterized. For preparation of graphene composites, hybridization with functional little organic molecules, polymers, and various metal oxides/salts have been achieved, and the self-assembly process and photocatalytic capacities have also been demonstrated. As for the synthesis of gold

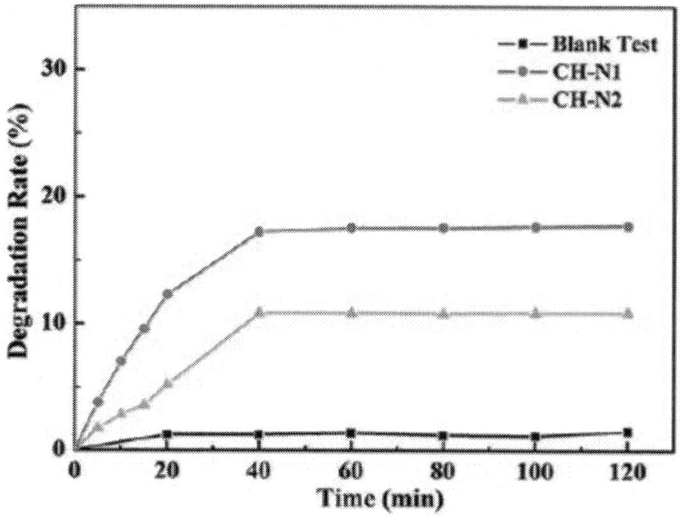

Figure 17. Photocatalytic activities of gold nanoparticles by using two amphiphiles with amphiphile to chloroaurate ion ratio of 2:1 on the degradation of methyl orange.

nanoparticles, the different prepared conditions, such as LB films/bulk, molecular skeletons/ substituted headgroups, and kinds of template compounds, have been investigated to control and regulate the designed gold nanostructures. Thus, the photocatalytic behaviors of as-obtained gold nanoparticles on dyes degradation were also characterized. The above-mentioned research work may provide new and potential perspective for the preparation and analysis of nanocomposite and nanomaterials. In closing, in recent several decades, preparation of novel nanocomposites are promising and emerging as attentive research platforms based on special properties and application fields. In addition, the self-assembly techniques belonging to supramolecular chemistry seem to be key subject in physical chemistry, related to various fundamental and scientific fields such as electron and energy transfer, organic electronics, catalysis engineering, and so on. Thus, the present preliminary research works only give a cursory browse of recent progress in this field. Future research orientation on synthesis of nanocomposites and nanomaterials would mainly focus on the special nanostructures with less-expensive fabrication process to gain high-performance and low-cost nanocomposite and nano-devices.

Acknowledgements

The authors would like to extend their thanks to the National Natural Science Foundation of China (Nos. 21473153, 21207112, and 51402253); the Natural Science Foundation of Hebei

Province (No. B2013203108); the Science Foundation for Excellent Youth Scholars from Universities and Colleges of Hebei Province (No. YQ2013026); the Support Program for the Top Young Talents of Hebei Province; the Scientific and Technological Research and Development Program of Qinhuangdao City (No. 201502A006); the China Postdoctoral Science Foundation (No. 2015M580214); and Open Foundation of National Key Laboratory of Biochemical Engineering (Institute of Process Engineering, Chinese Academy of Sciences) for providing funds for this research.

Author details

Jingxin Zhou[1], Tifeng Jiao[1,2*], Qingrui Zhang[1] and Jie Hu[1]

*Address all correspondence to: tfjiao@ysu.edu.cn

1 Hebei Key Laboratory of Applied Chemistry, School of Environmental and Chemical Engineering, Yanshan University, Qinhuangdao, P. R. China

2 National Key Laboratory of Biochemical Engineering, Institute of Process Engineering, Chinese Academy of Sciences, Beijing, P. R. China

References

[1] Mutiso RM, Winey KI. Electrical properties of polymer nanocomposites containing rod-like nanofillers. *Prog Polymer Sci* 2015;40:63–84.

[2] Campbell TA, Ivanova OS. 3D printing of multifunctional nanocomposites. *Nano Today* 2013;8(2):119–20.

[3] Vashist SK, Luong JHT. Recent advances in electrochemical biosensing schemes using graphene and graphene-based nanocomposites. *Carbon* 2015;84:519–50.

[4] Hu K, Kulkarni DD, Choi I, Tsukruk VV. Graphene-polymer nanocomposites for structural and functional applications. *Prog Polymer Sci* 2014;39(11):1934–72.

[5] Jiao T, Hu J, Zhang Q, Xiao Y. Preparation and self-assembly of functionalized nanocomposites and nanomaterials-relationship between structures and properties. In: Larramendy M. (ed.) *Nanomaterials–Toxicity and Risk Assessment*. Rijeka: InTech; 2015, pp. 177–210.

[6] Jiao T, Wang S, Zhou J. Molecular design and supramolecular assemblies of novel amphiphiles with special molecular structures in organized molecular films. In: Rahman M. (ed.) *Nanomaterials*. Rijeka: InTech: 2011, pp. 315–46.

[7] Lee YS. *Self-Assembly and Nanotechnology Systems: Design, Characterization, and Applications*. New Jersey: John Wiley & Sons, Inc.; 2011, pp. 111–40.

[8] Jiao T, Wang Y, Zhang Q, Zhou J, Gao F. Regulation of substituent groups on morphologies and self-assembly of organogels based on some azobenzene imide derivatives. *Nanoscale Res Lett* 2013;8:160.

[9] Jiao T, Huang Q, Zhang Q, Xiao D, Zhou J, Gao F. Self-assembly of organogels via new luminol imide derivatives: diverse nanostructures and substituent chain effect. *Nanoscale Res Lett* 2013;8:278.

[10] Jiao T, Gao F, Zhang Q, Zhou J, Gao F. Spacer effect on nanostructures and self-assembly in organogels via some bolaform cholesteryl imide derivatives with different spacers. *Nanoscale Res Lett* 2013;8:406.

[11] Liang D, Cui C, Hu H, Wang Y, Xu S, Ying B, Li P, Lu B, Shen H. One-step hydrothermal synthesis of anatase TiO_2/reduced graphene oxide nanocomposites with enhanced photocatalytic activity. *J Alloys Compd* 2014;582:236–40.

[12] Ullah K, Zhu L, Meng ZD, Ye S, Sun Q, Oh WC. A facile and fast synthesis of novel composite Pt-graphene/TiO_2 with enhanced photocatalytic activity under UV/Visible light. *Chem Eng J* 2013;231:76–83.

[13] Bai XJ, Wang L, Zhu YF. Visible photocatalytic activity enhancement of $ZnWO_4$ by graphene hybridization. *ACS Catal* 2012;2:2769–78.

[14] Sun L, Shao R, Tang LQ, Chen ZD. Synthesis of $ZnFe_2O_4$/ZnO nanocomposites immobilized on graphene with enhanced photocatalytic activity under solar light irradiation. *J Alloys Compd* 2013;564:55–62.

[15] Xu JJ, Ao YH, Chen MD. A simple method for the preparation of Bi_2WO_6-reduced graphene oxide with enhanced photocatalytic activity under visible light irradiation. *Mater Lett* 2013;92:126–8.

[16] Li T, Shen JF, Li N, Ye MX. Hydrothermal preparation, characterization and enhanced properties of reduced graphene-$BiFeO_3$ nanocomposite. *Mater Lett* 2013;91:42–4.

[17] Lv T, Pan LK, Liu XJ, Lu T, Zhu G, Sun Z. Enhanced photocatalytic degradation of methylene blue by ZnO-reduced graphene oxide composite synthesized via microwave-assisted reaction. *J Alloys Compd* 2011;509:10086–91.

[18] Jiang LX, Li KX, Yan LS, Dai YH, Huang ZM. Preparation of Ag(Au)/graphene-TiO_2 composite photocatalysts and their catalytic performance under simulated sunlight irradiation. *Chin J Catal* 2012;33(12):1974–81.

[19] Zhang DF, Pu XP, Ding GQ, Shao X, Gao YY, Liu JX, Gao MC, Li Y. Two-phase hydrothermal synthesis of TiO_2-graphene hybrids with improved photocatalytic activity. *J Alloys Compd* 2013;572:199–204.

[20] Ghosh SK, Pal T. Interparticle coupling effect on the surface plasmon resonance of gold nanoparticles: from theory to applications. *Chem Rev* 2007;107:4797–862.

[21] Katz E, Willner I. Integrated nanoparticle-biomolecule hybrid systems: synthesis, properties, and applications. *Angewandte Chemie International Edition* 2004;43(45): 6042–108.

[22] Daniel MC, Astruc D. Gold nanoparticles: assembly, supramolecular chemistry, quantum-size-related properties, and applications toward biology, catalysis, and nanotechnology. *Chem Rev* 2004;104:293–346.

[23] Liao J, Bernard L, Langer M, Schoenberger C, Calame M. Reversible formation of molecular junctions in 2D nanoparticle arrays. *Adv Mater* 2006;18(18):2444–7.

[24] Klajn R, Bishop KJM, Grzybowski BA. Light-controlled self-assembly of reversible and irreversible nanoparticle suprastructures. *Proc Nat Acad Sci USA* 2007;104:10305–9.

[25] Koplin E, Niemeyer CM, Simon U. Formation of electrically conducting DNA-assembled gold nanoparticle monolayers. *J Mater Chem* 2006;16:1338–44.

[26] Meister A, Drescher S, Mey I, Wahab M, Graf G, Garamus VM, Hause G, Mo1gel HJ, Janshoff A, Dobner B, Blume A. Helical nanofibers of self-assembled bipolar phospholipids as template for gold nanoparticles. *J Phys Chem B* 2008;112:4506–11.

[27] Jadzinsky PD, Calero G, Ackerson CJ, Bushnell DA, Kornberg RD. Structure of a thiol monolayer-protected gold nanoparticle at 1.1 Å resolution. *Science* 2007;318:430–3.

[28] Wen Y, Jiang X, Yin G, Yin J. Multi-responsive amphiphilic gold nanoparticles (AuNPs) protected by poly(ether amine) (PEA). *Chem Commun* 2009;43:6595–7.

[29] Song WJ, Du JZ, Sun TM, Zhang PZ, Wang J. Gold nanoparticles capped with polyethyleneimine for enhanced siRNA delivery. *Small* 2010;6:239–46.

[30] Ko S, Park TJ, Kim HS, Kim JH, Cho YJ. Directed self-assembly of gold binding polypeptide-protein: a fusion proteins for development of gold nanoparticle-based SPR immunosensors. *Biosens Bioelectron* 2009;24:2592–7.

[31] George J, Thomas KG. Surface plasmon coupled circular dichroism of Au nanoparticles on peptide nanotubes. *J Am Chem Soc* 2010;132:2502–3.

[32] Cho EC, Au L, Zhang Q, Xia Y. The effects of size, shape, and surface functional group of gold nanostructures on their adsorption and internalization by cells. *Small* 2010;6:517–22.

[33] Jiao T, Wang Y, Zhang Q, Yan X, Zhao X, Zhou J, Gao F. Self-assembly and headgroup effect in nanostructured organogels via cationic amphiphile-graphene oxide composites. *PLoS ONE* 2014;9(7):e101620.

[34] Jiao T, Wang Y, Zhang Q, Yan X, Zhao X, Huo Q, Zhou J, Gao F. Organogels via gemini amphiphile-graphene oxide composites: self-assembly and symmetry effect. *Sci Adv Mater* 2015;7(9):1677–85.

[35] Jiao T, Guo H, Zhang Q, Peng Q, Tang Y, Yan X, Li B. Reduced graphene oxide-based silver nanoparticle-containing composite hydrogel as highly efficient dye catalysts for wastewater treatment. *Sci Rep-UK* 2015;5:11873.

[36] Hu J, Ma J, Wang L, Huang H. Synthesis and photocatalytic properties of $LaMnO_3$-graphene nanocomposites. *J Alloys Compd* 2014;583:539–45.

[37] Hu J, Ma J, Wang L, Huang H. Preparation of $La_{1-x}Sr_xMnO_3$ /graphene thin films and their photocatalytic activity. *Mater Sci Eng B* 2014;180:46–53.

[38] Hu J, Ma J, Wang L, Huang H, Ma L. Preparation, characterization and photocatalytic activity of Co-doped $LaMnO_3$/graphene composites, *Powder Technol* 2014;254:556–62.

[39] Hu J, Wang L, Ma J, Huang H. Study on the preparation and photocatalytic performance of the $LaNiO_3$/graphene composite film. *Rare Metal Mater Eng* 2014;43(7):1736–41.

[40] Hu J, Wang L, Shi L, Huang H. Preparation of $La_{1-x}Ca_xMnO_3$ perovskite-graphene composites as oxygen reduction reaction electrocatalyst in alkaline medium. *J Power Sources* 2014;269:144–51.

[41] Hu J, Men J, Ma J, Huang H. Preparation of $LaMnO_3$/graphene thin films and their photocatalytic activity. *J Rare Earths* 2014;32(12):1127–35.

[42] Zhong L, Jiao T, Liu M. Synthesis and assembly of gold nanoparticles in organized molecular films of gemini amphiphiles. *Langmuir* 2008;24:11677–83.

[43] Zhong L, Jiao T, Liu M. A facile method to the synthesis of gold nanoprisms using a series of gemini amphiphiles. *J Nanosci Nanotechnol* 2009;9:2726–30.

[44] Jiao T, Wang Y, Guo W, Zhang Q, Yan X, Chen J, Wang L, Xie D, Gao F. Synthesis and photocatalytic property of gold nanoparticles by using a series of bolaform Schiff base amphiphiles. *Mater Res Bull* 2012;47(12):4203–9.

[45] Jiao T, Wang Y, Zhang Q, Yan X, Chen J, Zhou J, Gao F. Preparation and photocatalytic property of gold nanoparticles by using two bolaform cholesteryl imide derivatives. *J Dispers Sci Technol* 2013;34(12):1675–82.

Chapter 10

Photochemical Decomposition of Hydrogen Sulfide

Shan Yu and Ying Zhou

Additional information is available at the end of the chapter

http://dx.doi.org/10.5772/61823

Abstract

Hydrogen sulfide is an extremely toxic gas which is generated from both nature factors and human factors. A proper method for the efficient decomposition of hydrogen is of great importance. Using traditional Claus process, hydrogen sulfide could be decomposed into hydrogen oxide and sulfur. One drawback of this process is that the energy stored in hydrogen sulfide is partially wasted by the formation of hydrogen oxide. In fact, the energy could be utilized for the generation of hydrogen, a potential energy source in future, or other chemical products. Various methods that could possibly make better use of hydrogen sulfide have been studied in recent years, like thermal decomposition, plasma method, electrochemical method, and photochemical method. In particular, there have been high hopes in photochemical method due to the possible direct solar energy conversion into chemical energy. Unlike traditional photocatalytic water splitting, hydrogen sulfide decomposition is more accessible from the thermodynamic point of view. Photocatalytic hydrogen sulfide decomposition could occur in both gas phase and solution phase and various systems have been reported. Besides, the photoelectrochemical decomposition of hydrogen sulfide is also highlighted. In this chapter, we will simply introduce the current situation for photochemical decomposition of hydrogen sulfide.

Keywords: photocatalysis, hydrogen sulfide

1. Introduction

Hydrogen sulfide (H_2S) is an extremely toxic and corrosive gas with an odor of rotten eggs. Usually, H_2S could be released due to nature factors like microbial metabolism in the absence of oxygen and volcanic eruptions. However, in modern society, the main source of H_2S should be more attributed to human activities like the refinery of crude oil (desulfurization) and the sweetening of natural gas. For example, due to the exhaustion of high-quality natural gas reservoirs and our continued growth in energy demand, people has to turn to some sour nature gas reservoirs with a large amount of H_2S. As a matter of fact, the H_2S content of sour natural gas at some locations could be as high as 70–80% (like Harmatten, Alberta in Canada) that they are considered unusable [1]. A concentration of H_2S above 320 ppm in air could lead to pulmonary edema with the possibility of death [2], and H_2S must be carefully removed in related human activities.

Classically, the Claus process is the industrial standard to remove hydrogen sulfide. With this process, gaseous hydrogen sulfide could be decomposed into hydrogen oxide and sulfur (see Eq. (1)) with first thermal step at temperature above 850°C (Eq. (2)) and subsequently catalytic step (Eq. (3)) with activated aluminum(III), titanium (IV) oxide and so on [3].

$$2\,H_2S + O_2 \rightarrow 2\,S + 2\,H_2O \tag{1}$$

$$10\,H_2S + 5\,O_2 \rightarrow 2\,H_2S + SO_2 + 7\,S + 8\,H_2O \tag{2}$$

$$2\,H_2S + SO_2 \rightarrow 3\,S + 2\,H_2O \tag{3}$$

Although this process is very mature and yields elemental sulfur as a by-product, one big drawback of it is that the energy stored in hydrogen sulfide is partially wasted by the formation of hydrogen oxide. In fact, the energy stored in H_2S could be utilized for the generation of hydrogen, a potential energy source in future, or other chemical products like H_2O_2. Other disadvantages of Claus treatment include additional tail gas treatment and inflexibility to adjust to changes [4].

Various methods that could possibly make better use of hydrogen sulfide have been studied in recent years, like thermal decomposition, electrochemical method, plasmachemical method, and photochemical method [5]. For thermal decomposition, high temperature above 1000 K for significant conversion of H_2S is often required. Besides, high pressure and proper catalyst like molybdenum sulfide and other metal sulfide are commonly suggested, too. Interestingly, solar furnace was also suggested as the thermal source from the energy source point of view. Electrochemical method like direct electrolysis is often carried out in basic solutions where H_2S is absorbed. Anode poisoning by sulfur is a big challenge. In addition, chemical redox couples such as I^{3-}/I^- and Fe^{3+}/Fe^{2+} are also introduced for indirect electrolysis of H_2S. The main problem of electrochemical method is the high electricity costs today. Plasma generated from microwave, ozonizer, and glow discharge was also reported to be an active species to induce the decomposition of H_2S into H_2 and S. In comparison, the plasma method is relatively clean and effective. However, similar to electrochemical method, the big obstacle of the plasma-chemical method is the use of electricity.

In contrast to others like thermal and electrochemical methods, the photodecomposition of H_2S is much less mature. Nevertheless, it is a very attracting method, as it offers us one possible approach to directly harness solar energy and convert them into chemical energy, in a period that we are under the pressure of both exhaustion of fossil fuel and increase in energy demand worldwide.

2. The principles for photochemical method

As early as 1912, photochemist Giacomo Ciamician has drawn us a picture of the future [6]:

> "On the arid lands there will spring up industrial colonies without smoke and without smokestacks; forests of glass tubes will extend over the plains and glass buildings will rise everywhere; inside of these will take place the photochemical processes that hitherto have been the guarded secret of the plants, but that will have been mastered by human industry which will know how to make them bear even more abundant fruit than nature, for nature is not in a hurry and mankind is. And if in a distant future the supply of coal becomes completely exhausted, civilization will not be checked by that, for life and civilization will continue as long as the sun shines! If our black and nervous civilization, based on coal, shall be followed by a quieter civilization based on the utilization of solar energy, that will not be harmful to progress and to human happiness."

However, even after 100 years later of this vision, human civilization is still "made use almost exclusively of fossil solar energy. Would it not be advantageous to make better use of radiant energy?"

In this chapter, we will mainly focus on photocatalysis (photochemical reaction carried out in the presence of catalyst), which has risen during the last half century. Ever since the discovery that TiO_2 could split water into hydrogen and oxygen with the assistance of light and electricity in 1972, photocatalysis has aroused great interest of people [7]. Usually, photocatalysis is a chemical process triggered by photogenerated electrons and holes from light-responsive materials. Like photosynthesis happening in nature, the light energy could be converted into chemical energy with photocatalysis. Therefore, some photocatalytic reaction like water splitting for hydrogen and oxygen evolution is called *artificial photosynthesis* and has given high hopes.

Both molecule and inorganic semiconductor systems could be constructed for photocatalysis. Typically, three processes are necessary to complete the photocatalyis (Figure 1) : (1) absorption of photons and subsequent generation of free electrons and holes (see Eq. (4) and the enlarged section on the top right of Figure 1) ; (2)charge transfer and separation of photogenerated carriers (pathway C and D), accompanied with the competitive charge recombination processes (pathways A and B); (3) reduction of reaction substrates by electrons; and (4) oxidation of adsorbents by holes (Eq. (5)).

$$\text{semiconductor} + hv \rightarrow e_{CB}^{-} + h_{VB}^{+} \tag{4}$$

$$\mathrm{A} \rightarrow \mathrm{A}^{-},\ \mathrm{D} \rightarrow \mathrm{D}^{+} \tag{5}$$

For molecular systems, these three steps are often occurred on different materials. Taken photocatalytic hydrogen evolution as an example, step 1 is often carried out by one kind molecule (like ruthenium complexes), and step 3 is finished with the help of another molecule (such as recently popular cobalt and nickel complexes), while step 2 occurs both intra and intermolecularly. For semiconductor systems, all three steps could happen on one material (TiO_2 for instance), although sometimes cocatalyst (like Pt nanoparticles) is introduced for a higher light-to-chemical energy conversion. Molecular systems could be easily modified and could help us better observe the underlying catalytic mechanism from molecular level; nevertheless, such systems usually lacks long-term stability and we will mainly focus on semiconductor-based photocatalytic systems in this chapter.

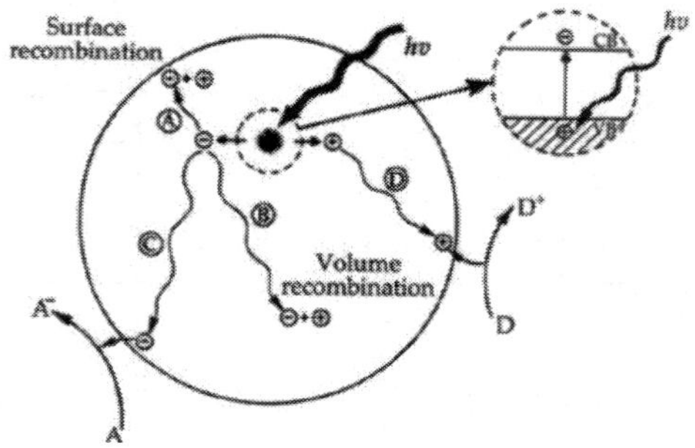

Figure 1. Schematic photoexcitation in a solid followed by deexcitation events. Adapted with permission from reference [8]. Copyright 1995 American Chemical Society.

Various kinds of semiconductors have been developed for photocatalysis. Due to its nontoxicity, low cost, and high stability, TiO_2 is the most studied semiconductor ever since its big sensation in 1972, and it is still very popular today. However, the crystal symmetry of TiO_2 allows only indirect interband transitions, and TiO_2 suffers from serious recombination of charge carriers [9]. Most importantly, the wide band gap of TiO_2 (3.2 eV for anatase and 3.0 eV for rutile) only makes it response to UV light (with wavelength below 398 nm for anatase and 413 nm for rutile), which only accounts for about 4% of the full solar spectrum [10]. Sensitization and doping are two common methods for modification of TiO_2 to increase its responsibility to visible light. Recently, it has been reported that with disorder engineering by hydrogenation, the onset of optical absorption of TiO_2 could be shifted to about 1200 nm (corresponding to 1.0 eV), and no obvious loss of photocatalytic activity of TiO_2 is observed [11].

In addition to TiO_2, many other binary and ternary oxides are also studied, such as d^0 metal oxides ($SrTiO_3$, ZrO_2, Nb_2O_5, Ta_2O_5, $Bi_2W_2O_9$, etc.), d^{10} metal oxides (ZnO, In_2O_3, etc.), and f^0 metal oxides (like CeO_2). Metal sulfides are another important category of photocatalysts. Among them, CdS has attracted large attentions. The main advantage of CdS is its responsibility to visible light (with a direct band gap of 2.4 eV), while one big disadvantage is its instability (mainly the oxidation of S^{2-} in the absence of hole scavenger) under light illumination. Other sulfides like ZnS, $CuInS_2$, $AgIn_2S_2$, and their solid solution have also been well studied for photocatalysis [12]. In particular, carbon materials, like graphene carbon nitride and carbon quantum dots, have lately aroused people's great interests due to their metal-free property and easy preparation [13,14]. Figure 2 shows the band gap and conduction and valence band levels of several typical semiconductors at pH 0. A more comprehensive presentation of the band structure of oxides and sulfide semiconductors was reported by Schoonen and Xu [15]. From the thermodynamic point of view, the conduction and valence band edge of semiconductors is an indication of their reducing and oxidizing ability, respectively. For instance, oxides often have deep valence band edge and hence strong oxidizing ability.

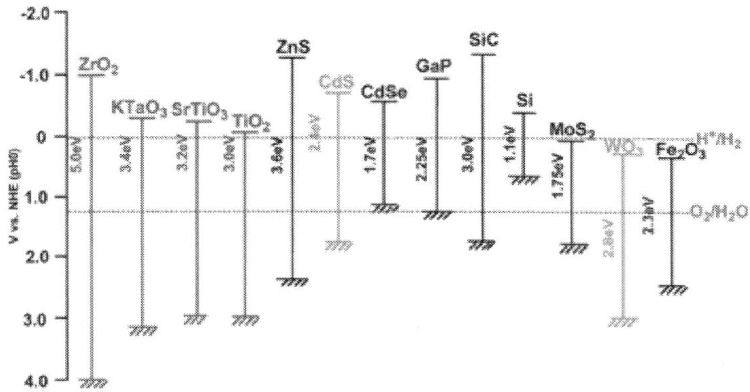

Figure 2. Relationship between band structure of semiconductor and redox potentials of water splitting. Reproduced by permission from the Royal Society of Chemistry from reference [12].

In all photocatalytic reactions were studied, water splitting is considered to be the Holy Grail of solar energy conversion. Over the last 40 years, scientists have been committed to find ideal photocatalytic systems that could turn water into hydrogen and oxygen by solar light. For a semiconductor qualified for water splitting, the conduction band edge should be more negative than the redox potential of H^+/H_2 (0 V vs NHE at pH 0), and the valance band edge should be more positive than the redox potential of O_2/H_2O (1.23 V vs NHE at pH 0). Nevertheless, overpotential and large kinetic barriers are also needed to be considered in practice. Several semiconductor systems have been reported for the stoichiometry water splitting for hydrogen and oxygen evolution (with mole ration of 2:1), such as $In_{1-x}Ni_xTaO_4$ (x = 0–0.2) [16], NiO (0.2 wt%)/$NaTaO_3$:La (2%) [17], and the lately reported visible light-responsive carbon dot/C_3N_4 nanocomposite [14].

As a matter of fact, the photocatalytic decomposition of H_2S is similar to that of water splitting. To some extent, the direct decomposition of H_2S into H_2 and elemental S is much easier than that of H_2O from the thermodynamic point of view: the energy needed for H_2O decomposition is about 237.2 kJ/mol [18], while that for H_2S is only 39.3 kJ/mol [4,10]. The reductive reaction that occurs in the decomposition process of H_2S is still hydrogen evolution from protons in most cases (with exception mentioned below), but the oxidative reaction changes from O_2

evolution to oxidation of S^{2-}. Therefore, for a semiconductor qualified for H_2S decomposition, the conduction band edge should still be more negative than the redox potential of H^+/H_2, but the valance band edge only needs to be more positive than the redox potential of H_2S/S^{2-} (0.14 V vs NHE at pH 0). This means that for semiconductors that are capable of water splitting are all qualified for H_2S decomposition. Besides, for some semiconductor, even if they may be not proper for water splitting due to the less positive valance band edge, they still have the potential for H_2S decomposition. One example is silicon. As seen from Figure 2, the valence band edge of silicon is far more negative than the redox potential of H_2O/O_2, which determines its inability for oxygen evolution. Nevertheless, it could be used in the system of H_2S decomposition (see below).

Like water splitting could occur in both gas phase (water vapor) and liquid phase, H_2S, as an acid gas, could be decomposed in gas phase directly and disposed in liquid phase indirectly after being absorbed by solution. Moreover, here we will have a review of these two cases, respectively.

3. Photocatalytic hydrogen sulfide decomposition by gas phase reaction

Jardim et al. studied the gas phase destruction of H_2S with a low concentration range of hundreds of ppm using TiO_2 as the catalyst and black light lamp as the light source [19]. In the existence of oxygen and water vapor, H_2S could be effectively decomposed (about 99% efficiency) and the main product is determined as SO_4^{2-}. The deactivation of TiO_2 would happen with a H_2S concentration larger than 600 ppm, and it was mainly caused by the adsorption of by-product on its surface. No elemental S was detected by the color change of TiO_2 from white to yellow, and hydrogen evolution was not considered in this study. Notably, if oxygen is absent in the system, H_2S could be barely removed.

In a similar experiment, with the assistance of in situ FT-IR, Anderson et al. confirmed that no other gaseous products like SO_2 or SO, and SO^{2-} adsorbed on TiO_2 may be one intermediate during the "eight electron transfer" process [20]. Furthermore, Sano et al. have found that the photodeposition of Ag on TiO_2 would promote the adsorption of H_2S on the sample, possibly due to the partially oxidized silver surface, and the deposited Ag could act as a cocatalyst for removal of H_2S. Both factors made Ag-deposited TiO_2 more efficient for H_2S degradation [21].

In addition, Sánchez et al. have tried glass "Raschig" ring, poly(ethylene terephthalate) (PET), and cellulose acetate (CA) as the supports to load TiO_2 for photocatalytic treatment of H_2S gas [22]. Glass rings supported TiO_2 (which has underwent fire treatment) outperforms PET and CA supported TiO_2. For PET and CA supports with low temperature treatment, PET supports displayed the higher photocatalytic activity, and TiO_2 caused the degradation of CA supports under illumination. Different from reports before, although SO_4^{2-} is one main product of H_2S removal, SO_2 was detected from these systems.

The interaction of H_2S with the semiconductor surfaces has also been investigated. Two adsorption modes of H_2S with high defect density rutile TiO_2 (110) surfaces were suggested:

dissociative adsorption with both H and S atom attached to the Ti atom at low H_2S concentration and molecular adsorption at high H_2S concentration [23]. Moreover, the preadsorption of H_2S would significantly block O_2 adsorption on TiO_2 surfaces even in the presence of large Ti^{3+} cations. Using Langmuir isotherm, Sopyan further discovered that H_2S adsorbed more strongly on rutile (0.7 molecules / nm^2) rather than anatase (0.4 molecules/nm^2). This is in sharp contrast with other molecules like acetaldehyde and ammonia [24]. Consequently, photocatalytic activity of anatase film is only 1.5 times higher than that of rutile for degradation of H_2S.

In all the above systems, H_2S in gas phase is studied within low concentration (tens to hundreds of ppm) and people mainly concerns with the oxidation product of H_2S. Little attention is paid to the reductive reaction of H_2S. Nevertheless, the reduction of H_2S (which is often the conversion of H_2S into H_2) is more attractive from an energy point of view.

Early in 1990s, Naman has combined thermal and photocalytic decomposition of H_2S together and studied the influence of light influx on the thermal decomposition of H_2S by V_xS_y on different substrates (TiO_2, Al_2O_3, and ZnO) [25]. Under light irradiation, the conversion of H_2S to H_2 was increased by 27.6%, 44.6%, and 16.5% at 500°C, respectively. The Arrhenius activation energy for H_2S decomposition has also calculated to be 50% of that in darkness. The author tentatively attributes this photoactivation effect on thermal decomposition to the photoexcitation of semiconductors (including V_xS_y) and the subsequent generated charge carriers.

In 2008, Li et al. have compared the activity of five typical semiconductors TiO_2, CdS, ZnS, ZnO, and $ZnIn_2S_4$ for the direct decomposition of H_2S in gaseous phase [26]. With illumination of Xe lamp and Pt loading (0.2 wt%), the efficiency of the decomposition of 5% H_2S in argon decreases as a sequence of ZnS > TiO_2 > $ZnIn_2S_4$ > ZnS > CdS under the gas flow rate of 6 ± 0.5 mL/min. Various noble metal loadings on ZnS have been compared, and it turns out that Ir is superior than others (Pd, Pt, Ru, Rh, and Au), which improves the hydrogen evolution efficiency from 1.2 to 4.5 μmol/h. Doping ZnS was also carried out, and transition metal Cu^{2+} doping (0.5% mol) could greatly promote the decomposition process and improve efficiency of the hydrogen evolution by about 20 times in contrast to blank ZnS. In addition, the absorption edge of ZnS shift from 400 to 450 nm after Cu doping, and this contributes to a photocatalytic H_2 production rate of 17 μmol/h under visible light irradiation (λ > 420 nm). Similarly, one limitation of this research is that only the reduction product, H_2, is detected in the system and the oxidative products are ignored.

Although systematic experimental studies of the photocatalytic decomposition of H_2S in gaseous phase are scarce, thermodynamic analysis of solar-based photocatalytic H_2S decomposition has recently been reported, which may be instructive for further studies on experiments [27]. Analysis indicates that energy efficiency of this process is not significantly affected by the intensity of solar irradiation. Exergy efficiency (the second law efficiency) will decrease with the increase of solar intensity, while the hydrogen yield will increase. Although the exergy efficiency value of current catalyst is calculated to be less than 1%, the author envisioned that an exergy efficiency of 10% could be achieved in the near future, and a maximum exergy efficiency of 27% may be obtained for a chemical conversion ratio of 0.6 if close to optimum cases of the quantum efficiency and the catalyst band gap can be obtained.

4. Photocatalytic hydrogen sulfide decomposition in solution

4.1. H_2S decomposition in aqueous solution

In comparison with solid gas phase photocatalysis, more often H_2S is first absorbed in solution. Under these circumstances, H_2S mainly participate in the photocatalytic reaction in the form of S^{2-} or HS^-, depending on the pH of the system. (The dissociative constants for the first and the second dissociation of H_2S at 298 K are 1.02×10^{-7} and 1.3×10^{-12}, respectively.) Hydrogen (in most cases) is generated in these systems as a result of the proton reduction. H_2S has a high solubility in pure water (ca. 0.1 M at 298 K); however, due to the limited availability of S^{2-} and HS^-, the hydrogen evolution efficiency is low [28]. More often, H_2S is absorbed by alkaline solution like NaOH and KOH solution, and sulfide solution is often used to replace the gaseous H_2S for photocatalytic H_2 evolution.

In 1976, Wrighton et al. reported that when using Na_2S as the sacrificial reagents in the presence of NaOH, the photocorrosion of CdS or CdSe photoelectrodes could be effectively inhibited [29]. The added S^{2-} in solution is oxidized, judging from the color change of the solution from transparent to yellow. H_2 was evolved at the Pt counter electrode. Later, Nozik proved that when the Schottky-type n-CdS/Pt photochemical diodes was suspended in the solution containing 1 M Na_2S and 1 M NaOH, hydrogen evolution could be observed with the illumination of simulated sunlight [30].

Early exhaustive studies of such work were conducted by Grätzel et al. [31]. When loaded with RuO_2 (0.1 wt%), CdS shows a high H_2 evolution rate of 0.128 mL g^{-1} h^{-1} in the presence of 0.1 M Na_2S (pH 3). With the S^{2-} ions present in the photocatalytic solution, a H_2S to H_2 conversion efficiency of 90% was calculated. Also, the concomitantly formed oxidation product S would not interfere with the water reduction (hydrogen evolution). The reaction mechanisms are shown in Eqs. (6–8):

$$\mathrm{CdS} + h\nu \rightarrow e^- + h^+ \tag{6}$$

$$2\,\mathrm{H}^+ + 2\,e^- \rightarrow \mathrm{H}_2 \tag{7}$$

$$\mathrm{H_2S} + 2h^+ \rightarrow 2\mathrm{H}^+ + \mathrm{S} \tag{8}$$

Moreover, the overall reaction corresponds to H_2S splitting into H_2 and S with the assistance of two photons (Eq. (9)):

$$\mathrm{H_2S} + 2h\nu \rightarrow \mathrm{H_2} + \mathrm{S} \tag{9}$$

During the photocatalytic process, although oxygen reduction can compete with hydrogen reduction, the existence of O_2 has little effect on the system's efficiency. Furthermore, the study shows that a basic solution and a higher RuO_2 loading (but no more than 0.5 wt%) could obviously improve the efficiency, while increase of the concentration of S^{2-} in solution and Pt loading on CdS seems has no significant influence on hydrogen evolution. Under optimal conditions, a quantum yield of 0.35 is obtained. In a similar CdS involved photocatalytic system with S^{2-} as the electron donor, Reber et al. pointed out that in contrast to an acidic environment, only disulfide instead of elemental sulfur is formed in an alkaline medium [32].

Photocatalytic hydrogen evolution with in situ H_2S absorption in alkaline solution has been carried out with various kinds of semiconductor photocatalysts, too (Table 1). The mechanism of such systems is similar to that contains sulfide solution. One recent example with relatively high efficiency for hydrogen evolution was reported by Kale et al. with nanostructure Bi_2S_3, which has a direct band gap of 1.3–1.7 eV [33]. Both nanorod and hierarchical nanoflower Bi_2S_3 were synthesized by hydrothermal method. With continuous H_2S bubbling into KOH solution, a hydrogen evolution efficiency of 8.38 and 7.08 mmol g^{-1} h^{-1} was observed for nanoflower and nanorod, respectively, under solar irradiation (from 11:30 a.m. to 2:30 p.m.).

photocatalyst	light source	aqueous reaction solution	Cocatal./H_2 activity (μmol·h^{-1} g^{-1})	Quantum yiled (%)	Ref.
$CdIn_2S_4$	450-W Xe	H_2S + KOH	6960	17.1 (500 nm)	[40]
$ZnIn_2S_4$	300-W Xe	H_2S + KOH	10574	-	[41]
N-doped TiO_2	300-W Xe	H_2S + KOH	8800	-	[42]
N-doped ZnO	300-W Xe	H_2S + KOH	19785	-	[43]
Bi_2S_3	Sunlight	H_2S + KOH	8880	-	[33]
6,13-Pentacenequinone	450-W Xe, > 420 nm	H_2S + KOH	48480	-	[44]
$CdSe_{0.5}S_{0.5}$ in GeO_2 glass	450-W Xe, > 420 nm	H_2S + KOH	8165	26 (> 420 nm)	[45]
CdSe in GeO_2 glass	450-W Xe, > 420 nm	H_2S + KOH	7257	21 (> 420 nm)	[45]
CdS in GeO_2 glass	450-W Xe, > 420 nm	H_2S + KOH	7560	-	[46]
Bi QD in GeO_2 glass	450-W Xe, > 420 nm	H_2S + KOH	11541	-	[47]
$Cd_{0.1}Zn_{0.9}S$	450-W Xe, > 420 nm	H_2S + KOH	8320	-	[48]
$FeGaO_3$	450-W Xe,	H_2S + KOH	NiO_x/5890	9.3 (550 nm)	[49]

photocatalyst	light source	aqueous reaction solution	Cocatal./H_2 activity ($\mu mol \cdot h^{-1} g^{-1}$)	Quantum yiled (%)	Ref.
	> 420 nm				
$FeGaO_3$	450-W Xe, > 420 nm	H_2S + KOH	NiO_x/4730	7.5 (550 nm)	[49]
$CuGa_2O_4$	450-W Xe, > 420 nm	H_2S + KOH	3212	5.3 (550 nm)	[50]
$CuGa_{1.4}Fe_{0.6}O_4$	450-W Xe, > 420 nm	H_2S + KOH	RuO_2/9548	15.0 (550 nm)	[50]
$CuGaO_2$	450-W Xe, > 420 nm	H_2S + KOH	7316	11.4 (550 nm)	[51]
$CuGa_{0.065}In_{0.935}O_2$	450-W Xe, > 420 nm	H_2S + KOH	RuO_2/8656	13.6 (550 nm)	[51]
$Nb_2Zr_6O_{17-x}N_x$	450-W Xe, > 420 nm	H_2S + KOH	8566	13.5 (550 nm)	[52]
CdS-TiO_2	500-W Hg, > 420 nm	H_2S + NaOH	Pt/9800	41 (> 420 nm)	[53]
CdS in HY zeolite	250-W Hg, > 400 nm	H_2S + NaOH / Na_2SO_3	24000	-	[54]

Table 1. Photocatalytic systems directly using H_2S gas dissolved in alkaline solution for hydrogen evolution.

4.2. H_2S decomposition through S^{2-}/SO_3^{2-} solution

One challenge often encounters with alkaline sulfide solution for photocatalytic hydrogen evolution is the interference of by-product. Disulfide and polysulfide ions usually form in alkaline sulfide solution by reaction between S^{2-} and elemental S immediately after the photooxidation (see Eqs. (10–12)). These ions are yellow and can act as an optical filter, which reduces the absorption of photocatalyst. In addition, polysulfide would compete with protons for reduction. Therefore, with the accumulation of disulfide, the hydrogen evolution efficiency of related systems is slowed down. A common solution for this is the addition of SO_3^{2-} into the system. The additional sulfite could react with sulfur and avoid the generation of polysulfide; meanwhile, colorless thiosulfate is formed, which is thermodynamically less easily reduced than protons:

$$S^{2-} + 2h^{+} \rightarrow S \tag{10}$$

$$S^{2-} + S \rightarrow S_2^{2-} \tag{11}$$

$$S_2^{2-} + 2\,S \rightarrow S_4^{2-} \tag{12}$$

$$SO_3^{2-} + S \rightarrow S_2O_3^{2-} \tag{13}$$

Then the net oxidative reaction that occurs in such a photocatalytic system is

$$SO_3^{2-} + S^{2-} + 2h^{+} \rightarrow S_2O_3^{2-} \tag{14}$$

and the whole photocatalytic hydrogen evolution reaction corresponds to

$$S^{2-} + SO_3^{2-} + 2\,H_2O + 2\,h\upsilon \rightarrow S_2O_3^{2-} + 2\,OH^{-} + H_2 \tag{15}$$

Photocatalytic hydrogen evolution systems based on S^{2-}/SO_3^{2-} solution is widely reported, and some typical reports are given in Table 2 [10]. As a matter of fact, the S^{2-}/SO_3^{2-} solution is one of the most famous sacrificial donors for photocatalytic hydrogen evolution under basic environment; this is especially true for metal sulfide photocatalysts. CdS, ZnS, $CuInS_2$ $ZnInS_2$, and their solid solution are all well studied for photocatalytic hydrogen evolution with such system. Metal sulfide often suffers from instability in photocatalytic processes as a result of the self-oxidation of sulfide with other sacrificial donors, but this could be effectively inhibited in the presence of sulfide in solution. This may be one important reason for the wide use of S^{2-}/SO_3^{2-} solution in photocatalysis. In contrast, metal oxide is less popular in such system, probably due to their small response in the visible light region.

Using S^{2-}/SO_3^{2-} solution, Kudo et al. have developed a series of visible light-responsive ZnS–$CuInS_2$–$AgInS_2$ solid solution photocatalysts for hydrogen evolution under irradiation from the solar simulator [34]. With increasing the proportion of $CuInS_2$ and $AgInS_2$ in the solid solution, the absorption spectrum of the photocatalyst could be extended to near-infrared region; however, hydrogen evolution was only observed with light absorption of wavelength less than 650 nm. When loaded with 0.75 wt% Ru, the initial hydrogen rate of 8.2 L m^{-2} h^{-1} and a quantum yield of 7.4% (at both 480 and 520 nm) could be observed with irradiation of solar simulator. Furthermore, they have demonstrated that with this photocatalytic system, a solar hydrogen evolution rate of about 2 L m^{-2} h^{-1} could be obtained for a reactor of 1 m^2 in November in Tokyo [12]. In addition, CdS loaded with RuO_2 (0.25 wt%) is also evaluated for its potential for commercial application. When the CdS-RuO_2 concentration is 2.0 mg mL^{-1} in 500 mL solution of 0.1 M Na_2S and 0.1 M Na_2SO_3 (with a surface area of 112 cm^2), a hydrogen generation rate of 28 mL h^{-1} could be achieved under solar light irradiation [35].

photocatalyst	light source	Cocatalyst / H_2 activity ($\mu mol \cdot h^{-1} g^{-1}$)	Quantum yiled (%)
CdS	300-W Xe, > 420 nm	Pt-PdS/29 233	93 (420 nm)
CdS/ZnS	350-W Xe, > 430 nm	900	10.2 (420 nm)
CdS/TiO_2	350-W Hg, > 420 nm	Pt/6400	-
CdS/ZnO	300-W Xe	Pt/3870	3.2 (300-600 nm)
$CdS/LaMnO_3$	300-W Xe, > 420 nm	375	-
c-CdS/Pt/hex-CdS	500-W Hg-Xe, > 420 nm	13 360	-
$CdS/Na_2Ti_2O_4(OH)_2$	300-W Xe, > 420 nm	Pt/2680	43.4 (420 nm)
$CdS/Zr_{0.25}Ti_{0.75}PO_4$	300-W Xe, > 430 nm	Pt/2300	27.2 (420 nm)
$CdS/AgGaS_2$	450-W Hg, > 420 nm	Pt/4730	19.7 (> 420 nm)
CdS:Ag	900-W Xe	Pt/33480	~25 (450 nm)
CdS-ZnS:Ag	900-W Xe	Pt/40957.5	37 (450 nm)
CdS:In/Cu	300-W W-H, > 420 nm	Pt/2456	26.5 (420 nm)
CdS:Mn	500-W Xe, > 420 nm	RuO_x/1935	7 (> 420 nm)
$Cd_{0.1}Zn_{0.9}S$:Ni	350-W Xe, > 420 nm	Pt/585.5	15.9 (420 nm)
$(Zn_{0.95}Cu_{0.05})_{0.67}Cd_{0.33}S$	300-W Xe, > 420 nm	Pt/3633.3	31.8 (420 nm)
ZnS:C	500-W Hg, > 420 nm	Pt/~90	-
ZnS:Ni	300-W Xe, > 420 nm	160	1.3 (420 nm)
ZnS:Pb/Cl	300-W Xe, > 420 nm	93	-
CdSe	700-W Hg, > 400 nm	436	13.4 (> 400 nm)
In_2S_3	300-W Xe, > 400 nm	Pd/960.2	2.1 (430 nm)
$CuInS_2$	500-W Xe, > 420 nm	Pt/84	-
$ZnIn_2S_4$	300-W Xe, > 430 nm	Pt/562	18.4 (420 nm)
$ZnIn_2S_4$:Cu	300-W Xe, > 430 nm	Pt/757.5	14.2 (420 nm)
$AgGaS_2$	500-W Hg, > 420 nm	Pt/2960	12.4 (> 420 nm)
$CuGa_3S_5$	300-W Xe, > 420 nm	NiS/~2800	1.3 (420-520 nm)
$AgIn_5S_8$	300-W Xe, > 420 nm	Pt/200	5.3 (411.2 nm)
Ag_2ZnSnS_4	300-W Xe, > 420 nm	Ru/1607	3 (500 nm)
Cu_2ZnGeS_4	300-W Xe, > 420 nm	Ru/1233	-
$CuGa_2In_3S_8$	300-W Xe, > 420 nm	Rh/10667	15 (560 nm)
$AgGa_2In_3S_8$	300-W Xe, > 420 nm	Rh/3433	15 (490 nm)
$AgInZn_7S_9$	300-W Xe, > 420 nm	Pt/3164.7	20 (420 nm)
$ZnS-In_2S_3-CuS$	300-W Xe, > 400 nm	360 000	22.6 (420 nm)

photocatalyst	light source	Cocatalyst / H_2 activity (μmol·h^{-1} g^{-1})	Quantum yiled (%)
$ZnS\text{-}In_2S_3\text{-}Ag_2S$	300-W Xe, > 400 nm	220 000	19.8 (420 nm)
$Cu_{0.25}Ag_{0.25}In_{0.5}ZnS_2$	300-W Xe, > 420 nm	Ru/7666.7	7.4 (520 nm)
$In(OH)_3$:S/Zn	300-W Xe, > 420 nm	Pt/223.3	0.59 (420 nm)
$ZnS_{1-x-0.5y}O_x(OH)_y$	400-W H, > 420 nm	~460	3.0 (400-700 nm)
$AgGaS_2/TiO_2$	450-W Hg, > 420 nm	Pt/4200	17.5 (> 420 nm)
$TiO_{2-x}N_x/WO_3$	300-W Hg, > 400 nm	Pd/1005	0.45 (> 400 nm)
$Cr_2O_3/Na_2Ti_2O_4(OH)_2$	350-W Xe, > 400 nm	36.4	-

Table 2. Photocatalysts for hydrogen evolution using S^{2-}/SO_3^{2-} related solution as the sacrificial donor under visible-light irradiation.

4.3. Thiosulfate cycle for H_2S decomposition

To make more efficient use of solution with mixed sulfide and sulfite for photocatalytic hydrogen evolution, Grätzel et al. further propose the concept "thiosulfate cycle" [36]. Under light illumination, $S_2O_3^{2-}$ could be disproportionated into S^{2-} and SO_3^{2-} with the assistance of TiO_2 (see specific reaction in Eqs. (16–19) and overall reaction in Eq. (20)). Oxidation products like SO_4^{2-} and $S_2O_6^{2-}$ are excluded from the system, and the 1:2 stoichiometric ratio of S^{2-} and SO_3^{2-} is maintained during the whole irradiation time:

$$TiO_2 + 2\,hv \rightarrow 2\,h^+ + 2\,e^- \tag{16}$$

$$2\,S_2O_3^{2-} + 2\,h^+ \rightarrow S_4O_6^{2-} \tag{17}$$

$$S_4O_6^{2-} + 3\,OH^- \rightarrow SO_3^{2-} + \frac{3}{2}\,S_2O_3^{2-} + \frac{3}{2}\,H_2O \tag{18}$$

$$S_2O_3^{2-} + 2\,e^- \rightarrow S^{2-} + SO_3^{2-} \tag{19}$$

$$\frac{3}{2}\,S_2O_3^{2-} + 3\,OH^- + 2\,hv \rightarrow 2\,SO_3^{2-} + S^{2-} + \frac{3}{2}\,H_2O \tag{20}$$

Therefore, if a system contains both photocatalytic hydrogen generation (Eq. (15)) and sulfite generation (Eq. (20)) compartments and one coordinates to the other well, three molecules of

H_2 would be produced with the oxidation of one mol of S^{2-} into SO_3^{2-} through the thiosulfate cycle (Eq. (21)):

$$S^{2-} + 3\,H_2O + 10\,hv \rightarrow SO_3^{2-} + 3\,H_2 \tag{21}$$

With such a cycle, no sulfur or thiosulfate would accumulate in such system. Figure 3 shows such a possible two-compartment system composed of CdS and TiO_2. Ideally, for the generation of 1 mol of H_2, 2 mol and 4/3 mol photons are needed to be absorbed by CdS and TiO_2, respectively. However, whether the efficiency of the two half cycles could match each other effectively is one important question unveiled to us, and there is no further clear report of this system till now.

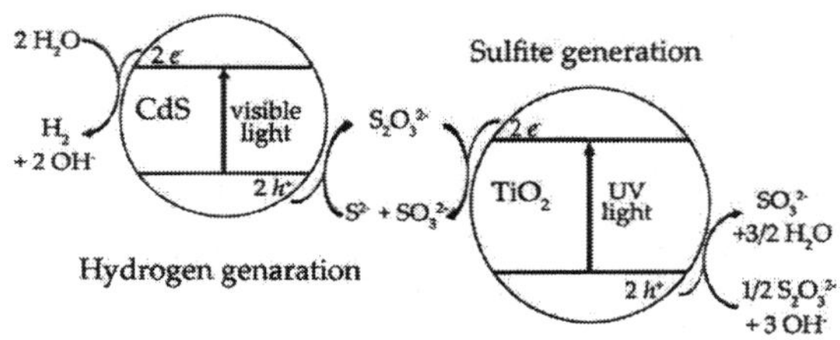

Figure 3. Schematic illustration of H_2S decomposition by two photosystems, linked through the $S_2O_3^{2-}/S^{2-}/SO_3^{2-}$ redox system.

4.4. H_2S decomposition in ethanolamine solution

In addition to hydroxide alkaline solution, some other additives are introduced to promote the absorption of H_2S in solution. For example, ethanolamine solution is frequently used in gas sweetening industry. Naman and Grätzel have dissolved H_2S in aqueous solution of alkanolamines (including monoethanolamine (MEA), diethanolamine (DEA), and triethanolamine (TEA)) and studied the photocatalytic efficiency of such system with vanadium sulfide as the photocatalyst [37]. Taking monoethanolamine for instance, one monoethanolamine was able to dissolve one molecule of H_2S (see Eqs. (22 and 23)). However, one big disadvantage of this method is that ethanolamines themselves could be decomposed under light illumination and the amount of ammonia detected from the photocatalytic system could even be higher than that of H_2:

$$2HOCH_2CH_2NH_2 + H_2S \rightleftharpoons (HOCH_2CH_2NH_3)_2S \tag{22}$$

$$(HOCH_2CH_2NH_3)_2S + H_2S \rightleftharpoons (HOCH_2CH_2NH_3)_2(SH)_2 \tag{23}$$

http://dx.doi.org/10.5772/61823

Furthermore, Li et al. used anhydrous ethanolamine solution to absorb H_2S [38]. Different from early report that system containing aqueous MEA outperforms that contains DEA and TEA, nonaqueous DEA solution is best for H_2S decomposition with CdS-based photocatalyst. Such a system is also better than system with NaOH-Na_2S solution from both the point of lifetime and rate for photocatalytic hydrogen evolution. In addition, the reduction of polysulfide in H_2S-DEM system is effectively depressed and could hardly compete with the proton reduction, which commonly occurs in NaOH-Na_2S system.

4.5. Extraction of elemental sulfur

Although numerous kinds of catalysts have been reported for the decomposition of H_2S through the above-mentioned method and hydrogen indeed evolves from solution, one problem is that S^{2-} often transforms into polysulfide, thiosulfate, or sulfite. How to deal with these by-products is another big challenge for us. Elemental S is more favored as the by-product; nevertheless, it could not be recovered from such photocatalytic system. To obtain pure sulfur, people have developed several ideas.

One simple method is to take advantage of the limited acid stability of complex sulfur species. Both polysulfide and thiosulfate would produce S when the pH value of the system decreases to a certain extent. That is, if the outlet reaction solution after photocatalysis (containing polysulfide or thiosulfate) encounters the inlet acidic gas H_2S, elemental S could possibly be precipitated from the system with a proper drop of pH:

$$S_2O_3^{2-} + H_2S \rightarrow S + HS^- + HSO_3^- \tag{24}$$

$$S_2^{2-} + H_2S \rightarrow S + 2HS^- \tag{25}$$

In this regard, Linkous et al. have designed a circulating photoreactor for H_2S decomposition (Figure 4) [39]. The feasibility of this system was conducted. In the photoreactor, hydroxide would be generated along with H_2 evolution, and the pH of the solution would increase. Nevertheless, when this solution flows into the scrubber tower, pH would decrease due to the input H_2S gas. For the fresh reaction solution constituted of both S^{2-} and SO_3^{2-}, $S_2O_3^{2-}$ would be generated after photocatalysis, and pH must be lowered to 4.2 (by neutralization with H_2S) for sulfur release from $S_2O_3^{2-}$. Then S could be collected as precipitates and the remaining solution (enriched with HS^- and HSO_3^-) would be sent back to the photoreactor for another round of photocatalysis, with a low pH (≤ 4.2). For fresh reaction solution only constituted of S^{2-}, polysulfide would be generated and the pH of the solution need to be lower than 10 for precipitation of sulfur. Normally, the photocatalytic systems using S^{2-} or SO_3^{2-} as the electron donor are more efficient for hydrogen evolution under basic conditions (pH ≥10); in some cases, the system could not even work under a relatively acidic environment (like pH 4). This urges us to reconsider the effect of SO_3^{2-} under such circumstances: as described above, SO_3^{2-} are widely used in S^{2-} involved hydrogen evolution system to avoid the generation of polysufide (which competes not only with catalyst from light absorption but also with protons for reduction by electrons), but the acidity necessary for the release of S from the obtained thiosulfate would greatly reduce the photocatalytic activity of the catalysts. In their study, Linkous pointed out that if the depth of reaction solution in photoreactor is less than 1 cm, in order to reduce the light absorption of polysufide, S^{2-} alone as the electron donor for photocatalytic hydrogen evolution is probably more suitable for the cyclic sulfur release in a CdS/Pt involved system. Additionally, another problem of this design is that if the commonly studied suspension system is used for photoreaction, photocatalyst could not be easily separated with the solution. Therefore, catalyst may need to be immobilized for circulating.

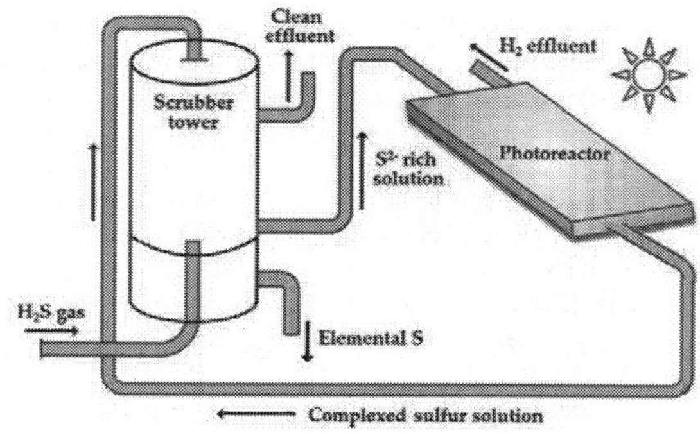

Figure 4. Generalized scheme for light-driven H_2S decomposition using an immobilized photocatalyst. Reprinted from reference [39], Copyright (1995), with permission from Elsevier.

5. Photoelectrochemical decomposition of hydrogen sulfide

In addition to the photocatalytic decomposition of H_2S alone, sometimes photochemical method is combined with electrochemical method for the decomposition of H_2S, that is, the photoelectron-chemical (PEC cell) decomposition of H_2S. Actually, the colloidal semiconductor photocatalyst system mentioned above could also be seen as some kind of short-circuit PEC cell, in which both anodic and cathodic reaction occurs on the surface of semiconductors at the same time (similar to the "photochemical diodes" developed by Nozik [30]). In this section, traditional PEC cells (with separated anode and cathode connected by wires) would be mainly focused. These cells could not only decompose hydrogen sulfide but also generate electricity. In addition, voltage bias could be applied to the cells if the drive force of light is not enough for hydrogen sulfide decomposition.

In 1987, Kainthla and Bockris reported a PEC cell for the decomposition of H_2S based on CdSe anode and Pt cathode [55]. CdSe film was directly grown on Ti substrate. Using polysulfide (prepared by H_2S dissolution in NaOH and subsequent addition of sulfur) as the electrolyte, an open circuit voltage of 0.62 V and short circuit current of 8.82 mA cm^{-2} could be achieved. H_2 bubbles could be observed to leave the Pt cathode when photocurrent flows through the cell and a Faraday efficiency of 0.97 is calculated. With the gradual accumulation of polysufide during the reaction, elemental sulfur would precipitates from the solution when polysulfide reaches its solubility limit. Stability of the cell is also tested and short circuit drops less than 10% with continuous illumination of 2 weeks. The total cell conversion efficiency (ε) given as the ratio of the recoverable energy to the input energy is calculated based on Eq. (26):

$$\varepsilon(\%) = (0.171I + IV_{cell}) \times 100 / W_{light} \tag{26}$$

where $0.171I$ is the chemical energy storage for H_2S decomposition into H_2 and S, IV_{cell} is the electrical energy generated from the cell, and W_{light} is the intensity of light. Maximum light to chemical energy storage, light to electrical energy, and total cell conversion efficiency occurs at cell voltage of 0, 0.3, and 0.275 V, with the corresponding efficiency to be 1.5%, 1.8%, and 2.85%, respectively. In this regard, the PEC cell can be operated in a manner that electrical energy or chemical energy can be selectively collected.

It is noteworthy that for eliminating the competition of polysulfide with proton for reduction in this cell, which is also a big problem in suspension systems, anode and cathode are placed in two compartments, and Nafion membrane is used to prevent the contact of polysulfide with cathode. If there is no Nafion membrane, only polysulfide is reduced into sulfide, and no H_2 could be detected from the system. Under this circumstance, no net chemical reaction happens in the cell, and light energy could only be converted into electrical energy.

Another advantage for PEC cells is that some strategies for the electrochemical decomposition of H_2S could be extended to PEC cell. One strategy is the indirect decomposition of H_2S with the assistance of redox couple like I^-/I_3^-(or I^-/IO_3^-) and Fe^{3+}/Fe^{2+}, in which the electrical energy or solar energy is first stored in the redox intermediate species, and then the intermediate could drive the following chemical reactions. Although indirect strategy may consume more additional energy for H_2S decomposition from a thermodynamical point of view, it is kinetically more favored and is beneficial for the extraction of elemental sulfur from the system.

Lately, Li and Wang et al. have adopted this strategy in PEC cells for H_2S decomposition and achieved good results. PEC cell with p-type Si deposited with protective TiO_2/Ti n+ doping layer and H_2 evolution cocatalyst Pt (Pt/TiO_2/Ti/n+p-Si) as the photocathode and Pt plates as anode was reported for the decomposition of H_2S [56]. In a two-compartment cell separated by Nafion membrane, freshly prepared 0.2 M of $FeSO_4$ (or KI) in 0.5 M of H_2SO_4 solution and 0.5 M of H_2SO_4 was used as the anodic and cathodic electrolyte, respectively. After H_2S bubbling into the anode compartment, S and H_2 could be separately produced from the anode and the cathode under light illumination at an applied potential of 0.2 V vs RHE. In this system, the chemical redox couple is significant for the conversion of H_2S into H_2 and S (Eqs. (27–33)):

$$\text{Photoelectrode} + 2\,h\nu \rightarrow 2\,h^+ + 2\,e^- \tag{27}$$

$$\text{Anode: } Fe^{2+} + 2\,h^{+} \rightarrow Fe^{3+} \tag{28}$$

$$H_2S + 2\,Fe^{3+} \rightarrow 2\,Fe^{2+} + 2\,H^{+} + S \text{ (chemical reaction)} \tag{29}$$

$$\text{Or anode: } 3\,I^{-} + 2\,h^{+} \rightarrow I_3^{-} \tag{30}$$

$$H_2S + I_3^{-} \rightarrow 3\,I^{-} + 2\,H^{+} + S \text{ (chemical reaction)} \tag{31}$$

$$\text{Cathode: } 2\,H^{+} + 2\,e^{-} \rightarrow H_2 \tag{32}$$

$$\text{Overall reaction: } H_2S + 2\,hv \rightarrow H_2 + S \tag{33}$$

Control experiment shows that if there is no existence of Fe^{2+} or I^- in the electrolyte, such experiment is unsuccessful due to the low solubility of H_2S in acidic solution. Besides, n-type Si coated with 3,4-ethylenedioxythiophene (PEDOT) as the anode was also tested in this system, and it turns out that Fe^{2+} and I^- could be easily oxidized on it. Nevertheless, due to the low stability of the n-type Si anode, further study in this report is unclear.

Notably, they further developed this indirect strategy in PEC cell and have made H_2O_2 and S from H_2S in the presence of oxygen [57]. This is quite novel because most study related to H_2S decomposition is limited to H_2 as the only reduced product now. In addition to the redox couple I^-/I_3^- in the anode compartment of the cell for S production, another redox couple anthraquinone/anthrahydroquinone (AQ/H_2AQ) was introduced to the cathode cell for H_2O_2 production. In fact, AQ is also an important reaction substrate in Hysulf process, one indirect strategy related to the thermal decomposition of H_2S. The anode reaction is still the same as Eqs. (30 and 31), but the cathode reaction and the overall reaction change as follows (Eqs. (34–36)):

$$\text{Cathode: } 2\,H^{+} + AQ + 2\,e^{-} \rightarrow H_2AQ \tag{34}$$

$$H_2AQ + O_2 \rightarrow AQ + H_2O_2 \text{ (chemical reaction)} \tag{35}$$

$$\text{Overall reaction: } H_2S + O_2 + 2\,hv \rightarrow H_2O_2 + S \tag{36}$$

At zero bias, Pt/p+ n Si photoanode and Pt cathode can simultaneously oxidize I^- to I_3^- and reduce AQ to H_2AQ, respectively. Solar to chemical conversion efficiency was estimated to be 1.1%. If Pt cathode is replaced with carbon plate, a higher photocurrent could be observed.

6. Conclusion

In general, H_2S is a highly polluted gas that must be carefully handled and removed. The traditional Claus process suffers from high-energy consumption and waste of potential energy, H_2. The photochemical decomposition of H_2S, which emerges with the rise of photocatalysis in the last century, could be one improved method for H_2S disposal. Lots of progress in the field of the photochemical decomposition of H_2S has been made in both gaseous phase and liquid phase. The mechanism of such reaction has been studied, and the efficiency of these systems has been calculated. Most often, the photochemical decomposition of H_2S is indirectly carried out in the form of photocatalytic H_2 production from aqueous sulfide solution. Details of the photochemical decomposition of H_2S, such as extraction of elemental sulfur from reaction system and the cyclic operation, were also of preliminary consideration. In addition, photochemistry was combined with electrochemistry for H_2S conversion: photoelectrochemical cells were built to extract H_2 (or H_2O_2) and S from H_2S with the assistance of redox couples.

In 2009, Li et al. reported CdS loaded with PdS and Pt dual cocatalyst can effectively generate H_2, with a quantum yield of 93% at 420 nm in the presence of S^{2-}/SO_3^{2-} solution and no deactivation was observed within illumination of 100 h for H_2 generation [58]. This is probably the most efficiency system reported relevant to the photocatalytic decomposition of H_2S till now. However, a lot of scientific problems are still unsolved, and there is a long, long way to go for the real application of the photocatalytic decomposition of H_2S in large scale chemical processing. In present, problems below may be considered in priority:

In gaseous phase systems, the concentration studied for H_2S decomposition is often low (with a volume concentration on ppm level); they are not practical in real industrial process. Also, people tend to focus on half of the reaction (oxidation of S^{2-} to SO_4^{2-} or H_2 generation). This is especially true in solution phase system with S^{2-}/SO_3^{2-} or S^{2-} as the electron donor: most reports only consider how to improve the efficiency of hydrogen evolution. Without the thorough consideration of both oxidizing and reducing reactions, the photochemical decomposition of H_2S is not persuasive. Moreover, in solution phase system for H_2S decomposition, along with H_2 evolution, the simultaneously generated polysulfide or thiosulfate is also a pollutant to environment; subsequent processing of such reaction solution should be cared for meaningful utilization of H_2S. Although systems have been designed for sulfur generation from polysulfide or thiosulfate solution, successful trials are limited and the subsequent separation of sulfur from solution is also a challenge.

Current catalysts with high efficiency of photochemical H_2S decomposition are mainly metal sulfide loading with noble metal cocatalyst like Pt, RuO_2, and so on. Although CdS is considered one of the most efficient photocatalyst for H_2 generation under visible light, the high toxicity of CdS should be taken seriously. New materials are needed to be exploited, and carbon materials may be alternative photocatalysts in consideration of cost, stability, and toxicity. Besides, noble metal poisoning by sulfide is another problem could happen sometimes and new earth abundant (low cost) cocatalyst resistive to sulfide poisoning is necessary. Transition metals like Fe, Co, and Ni and their compounds could be promising from the current available data. Similar in PEC cells for H_2S decomposition, stability and cost could be big problems, too.

To conclude, the photochemical decomposition of H_2S is still in a relatively early stage. New photocatalytic H_2S decomposition systems with low cost, high quantum efficiency, and long stability should be further developed, especially those responsive to the visible light region, which account for 43% in the full solar spectra. (Taking similar photocatalytic water spitting as a reference, a quantum yield of 30% at 600 nm is the starting point for practical application, which corresponds to about 5% solar energy conversion.) This may be fulfilled with optimized structure design, including chemical composition, electron and band structure, crystal structure and crystallinity, surface state, morphology, and so on, which is currently highlighted in nanoscience and technology. Moreover, people should keep in mind that oxidation and reduction of H_2S is equally important for H_2S decomposition if we want to handle H_2S in a really green way.

Author details

Shan Yu and Ying Zhou*

*Address all correspondence to: yzhou@swpu.edu.cn

State Key Laboratory of Oil and Gas Reservoir Geology and Exploitation, School of Materials Science and Engineering, Southwest Petroleum University, Chengdu, China

References

[1] Kappauf T, Murray JP, Palumbo R, Diver RB, Fletcher EA. Hydrogen and sulfur from hydrogen sulfide: IV. quenching the effluent from a solar furnace. Energy. 1985;10:1119–1137. DOI: 10.1016/0360-5442(85)90027-1

[2] Hydrogen sulfide [Internet]. 2015. Available from: https://en.wikipedia.org/wiki/Hydrogen_sulfide [Accessed: 2015-07-01], and reference therein.

[3] Claus process [Internet]. 2015. Available from: https://en.wikipedia.org/wiki/Claus_process [Accessed: 2015-07-05], and reference therein.

[4] Tambwekar SV, Subrahmanyam M. Photocatalytic generation of hydrogen from hydrogen sulfide: an energy bargain. International Journal of Hydrogen Energy. 1997;22:959–965. DOI: 10.1016/s0360-3199(97)00002-5

[5] Zaman J, Chakma A. Production of hydrogen and sulfur from hydrogen sulfide. Fuel Processing Technology. 1995;41:159–198. DOI: 10.1016/0378-3820(94)00085-8

[6] Ciamician G. The photochemistry of the future. Science. 1912;36:385–394. DOI: 10.1126/science.36.926.385

[7] Fujishima A, Honda K. Electrochemical photolysis of water at a semiconductor electrode. Nature. 1972;238:37–38. DOI: 10.1038/238037a0

[8] Linsebigler AL, Lu G, Yates JT. Photocatalysis on TiO_2 surfaces: principles, mechanisms, and selected results. Chemical Reviews. 1995;95:735–758. DOI: 10.1021/cr00035a013

[9] Habisreutinger SN, Schmidt-Mende L, Stolarczyk JK. Photocatalytic reduction of CO_2 on TiO_2 and other semiconductors. Angewandte Chemie. 2013;52:7372–7408. DOI: 10.1002/anie.201207199

[10] Chen X, Shen S, Guo L, Mao SS. Semiconductor-based photocatalytic hydrogen generation. Chemical Reviews. 2010;110:6503–6570. DOI: 10.1021/cr1001645, and references therein.

[11] Chen X, Liu L, Yu PY, Mao SS. Increasing solar absorption for photocatalysis with black hydrogenated titanium dioxide nanocrystals. Science. 2011;331:746–750. DOI: 10.1126/science.1200448

[12] Kudo A, Miseki Y. Heterogeneous photocatalyst materials for water splitting. Chemical Society Reviews. 2009;38:253–278. DOI: 10.1039/b800489g

[13] Wang X, Maeda K, Thomas A, Takanabe K, Xin G, Carlsson JM, Domen K, Antonietti M. A metal-free polymeric photocatalyst for hydrogen production from water under visible light. Nature Materials. 2009;8:76–80. DOI: 10.1038/nmat2317

[14] Liu J, Liu Y, Liu N, Han Y, Zhang X, Huang H, Lifshitz Y, Lee S-T, Zhong J, Kang Z. Metal-free efficient photocatalyst for stable visible water splitting via a two-electron pathway. Science. 2015;347:970–974. DOI: 10.1126/science.aaa3145

[15] Xu Y, Schoonen MAA. The absolute energy positions of conduction and valence bands of selected semiconducting minerals. American Mineralogist. 2000;85:543–556. DOI: 10.2138/am-2000-5110

[16] Zou Z, Ye J, Sayama K, Arakawa H. Direct splitting of water under visible light irradiation with an oxide semiconductor photocatalyst. Nature. 2001;414:625–627. DOI: 10.1038/414625a

[17] Kato H, Asakura K, Kudo A. Highly efficient water splitting into H_2 and O_2 over lanthanum-doped $NaTaO_3$ photocatalysts with high crystallinity and surface nanostructure. Journal of the American Chemical Society. 2003;125:3082–3089. DOI: 10.1021/ja027751g

[18] Walter MG, Warren EL, McKone JR, Boettcher SW, Mi Q, Santori EA, Lewis NS. Solar water splitting cells. Chemical Reviews. 2010;110:6446–6473. DOI: 10.1021/cr1002326

[19] Canela MC, Alberici RM, Jardim WF. Gas-phase destruction of H_2S using TiO_2/UV-VIS. Journal of Photochemistry and Photobiology A: Chemistry. 1998;112:73–80. DOI: 10.1016/x1010-6030(97)00261-x

[20] Kataoka S, Lee E, Tejedor-Tejedor MI, Anderson MA. Photocatalytic degradation of hydrogen sulfide and in situ FT-IR analysis of reaction products on surface of TiO_2. Applied Catalysis B: Environmental. 2005;61:159–163. DOI: 10.1016/j.apcatb.2005.04.018

[21] Kato S, Hirano Y, Iwata M, Sano T, Takeuchi K, Matsuzawa S. Photocatalytic degradation of gaseous sulfur compounds by silver-deposited titanium dioxide. Applied Catalysis B: Environmental. 2005;57:109–115. DOI: 10.1016/j.apcatb.2004.10.015

[22] Portela R, Sánchez B, Coronado JM, Candal R, Suárez, S. Selection of TiO_2-support: UV-transparent alternatives and long-term use limitations for H_2S removal. Catalysis Today. 2007;129:223–230. DOI: 10.1016/j.cattod.2007.08.005

[23] Smith KE, Henrich VE. Interaction of H_2S with high defect density TiO_2(110) surfaces. Surface Science. 1989;217:445–458. DOI: 10.1016/0039-6028(89)90440-8

[24] Sopyan I. Kinetic analysis on photocatalytic degradation of gaseous acetaldehyde, ammonia and hydrogen sulfide on nanosized porous TiO_2 films. Science and Technology of Advanced Materials. 2007;8:33–39. DOI: 10.1016/j.stam.2006.10.004

[25] Naman SA. Comparison between thermal decomposition and photosplitting of H_2S over V_xS_y supported on oxides at 450–550°C in a static system. International Journal of Hydrogen Energy. 1992;17:499–504. DOI: 10.1016/0360-3199(92)90148-p

[26] Ma G, Yan H, Zong X, Ma B, Jiang H, Wen F, Li C. Photocatalytic splitting of H_2S to produce hydrogen by gas-solid phase reaction. Chinese Journal of Catalysis. 2008;29:313–315. DOI: 10.1016/s1872-2067(08)60029-7

[27] Shamim RO, Dincer I, Naterer G. Thermodynamic analysis of solar-based photocatalytic hydrogen sulphide dissociation for hydrogen production. International Journal of Hydrogen Energy. 2014;39:15342–15351. DOI: 10.1016/j.ijhydene.2014.07.094

[28] Jang JS, Gyu Kim H, Borse, PH, Lee JS. Simultaneous hydrogen production and decomposition of dissolved in alkaline water over composite photocatalysts under visible light irradiation. International Journal of Hydrogen Energy. 2007;32:4786–4791. DOI: 10.1016/j.ijhydene.2007.06.026

[29] Ellis, AB, Kaiser, SW Wrighton, MS. Visible light to electrical energy conversion. Stable cadmium sulfide and cadmium selenide photoelectrodes in aqueous electrolytes. Journal of the American Chemical Society. 1976;98:1635–1637. DOI: 10.1021/ja00422a087

[30] Nozik AJ. Photochemical diodes. Applied Physics Letters. 1977;30:567–569. DOI: 10.1063/1.89262

[31] Borgarello E, Kalyanasundaram K, Grätzel M, Pelizzetti E. Visible light induced generation of hydrogen from H_2S in CdS-dispersions, hole transfer catalysis by RuO_2. Helvetica Chimica Acta. 1982;65:243–248. DOI: 10.1002/hlca.19820650123

[32] Buehler N, Meier K, Reber JF. Photochemical hydrogen production with cadmium sulfide suspensions. Journal of Physical Chemistry. 1984;88:3261–3268. DOI: 10.1021/j150659a025

[33] Kawade UV, Panmand RP, Sethi YA, Kulkarni MV, Apte SK, Naik SD, Kale BB. Environmentally benign enhanced hydrogen production via lethal H_2S under natural sunlight using hierarchical nanostructured bismuth sulfide. RSC Advances. 2014;4:49295–49302. DOI: 10.1039/c4ra07143c

[34] Tsuji I, Kato H, Kudo, A. Visible-light-induced H_2 evolution from an aqueous solution containing sulfide and sulfite over a ZnS–$CuInS_2$–$AgInS_2$ solid-solution photocatalyst. Angewandte Chemie International Edition. 2005;44:3565–3568. DOI: 10.1002/anie.200500314

[35] Grzyll LR, Thomas JJ, Barile RG. Photoelectrochemical conversion of hydrogen sulfide to hydrogen using artificial light and solar radiation. International Journal of Hydrogen Energy. 1989;14:647–651. DOI: 10.1016/0360-3199(89)90040-2

[36] Borgarello E, Desilvestro J, Grätzel M, Pelizzetti E. Photoreduction of thiosulfate in semiconductor dispersions. Helvetica Chimica Acta. 1983;66:1827–1834. DOI: 10.1002/hlca.19830660622

[37] Naman SA, Grätzel M. Visible-light generation of hydrogen from hydrogen sulphide in aqueous solutions of ethanolamines containing vanadium sulphide dispersions. Journal of Photochemistry and Photobiology A: Chemistry. 1994;77:249–253. DOI: 10.1016/1010-6030(94)80050-2

[38] Ma G, Yan H, Shi J, Zong X, Lei Z, Li C. Direct splitting of H_2S into H_2 and S on CdS-based photocatalyst under visible light irradiation. Journal of Catalysis. 2008;260:134–140. DOI: 10.1016/j.jcat.2008.09.017

[39] Linkous CA, Muradov NZ, Ramser SN. Consideration of reactor design for solar hydrogen production from hydrogen sulfide using semiconductor particulates. International Journal of Hydrogen Energy. 1995:20:701–709. DOI: 10.1016/0360-3199(94)00127-l

[40] Kale BB, Baeg JO, Lee SM, Chang H, Moon SJ, Lee, CW. $CdIn_2S_4$ nanotubes and "marigold" nanostructures: a visible-light photocatalyst. Advanced Functional Materials. 2006;16:1349–1354. DOI: 10.1002/adfm.200500525

[41] Chaudhari NS, Bhirud AP, Sonawane RS, Nikam LK, Warule SS, Rane VH, Kale, BB. Ecofriendly hydrogen production from abundant hydrogen sulfide using solar light-driven hierarchical nanostructured $ZnIn_2S_4$ photocatalyst. Green Chemistry. 2011;13:2500–2506. DOI: 10.1039/c1gc15515f

[42] Chaudhari NS, Warule SS, Dhanmane SA, Kulkarni MV, Valant M, Kale BB. Nanostructured n-doped TiO_2 marigold flowers for an efficient solar hydrogen production from H_2S. Nanoscale. 2013;5:9383–9390. DOI: 10.1039/c3nr02975a

[43] Bhirud AP, Sathaye SD, Waichal RP, Nikam LK, Kale BB. An eco-friendly, highly stable and efficient nanostructured p-type n-doped ZnO photocatalyst for environmentally benign solar hydrogen production. Green Chemistry. 2012;14:2790–2798. DOI: 10.1039/c2gc35519a

[44] Pandit VU, Arbuj SS, Mulik UP, Kale BB. Novel functionality of organic 6,13-pentacenequinone as a photocatalyst for hydrogen production under solar light. Environmental Science and Technology. 2014;48:4178–4183. DOI: 10.1021/es405150p

[45] Apte SK, Garaje SN, Naik SD, Waichal RP, Baeg J-O, Kale, BB. Quantum confinement controlled solar hydrogen production from hydrogen sulfide using a highly stable $CdS_{0.5}Se_{0.5}/CdSe$ quantum dot-glass nanosystem. Nanoscale. 2014;6:908–915. DOI: 10.1039/c3nr04898e

[46] Apte SK, Garaje SN, Valant M, Kale BB. Eco-friendly solar light driven hydrogen production from copious waste H_2S and organic dye degradation by stable and efficient orthorhombic CdS quantum dots-GeO_2 glass photocatalyst. Green Chemistry. 2012;14:1455–1462. DOI: 10.1039/c2gc16416g

[47] Apte SK, Garaje SN, Naik SD, Waichal RP, Kale BB. Environmentally benign enhanced H_2 production from abundant copious waste H_2S using size tuneable cubic bismuth (Bi^0) quantum dots-GeO_2 glass photocatalyst under solar light. Green Chemistry. 2013;15:3459–3467. DOI: 10.1039/c3gc41940a

[48] Garaje SN, Apte SK, Naik SD, Ambekar JD, Sonawane RS, Kulkarni MV, Vinu A, Kale BB. Template-free synthesis of nanostructured CdxZn1-xS with tunable band structure for h-2 production and organic dye degradation using solar light. Environmental Science and Technology. 2013;47:6664–6672. DOI: 10.1021/es3045955

[49] Subramanian E, Baeg J-O, Lee, SM, Moon, S-J, Kong K-j. Dissociation of H_2S under visible light irradiation ($\lambda \geq 420$ nm) with $FeGaO_3$ photocatalysts for the production of hydrogen. International Journal of Hydrogen Energy. 2008;33:6586–6594. DOI: 10.1016/j.ijhydene.2008.07.016

[50] Gurunathan K, Baeg J-O, Lee SM, Subramanian E, Moon S-J, Kong K-j. Visible light active pristine and doped spinel photocatalysts for solar hydrogen production. International Journal of Hydrogen Energy. 2008;33:2646–2652. DOI: 10.1016/j.ijhydene.2008.03.018

[51] Gurunathan K, Baeg JO, Lee SM, Subrarnanian E, Moon SJ, Kong KJ. Visible light assisted highly efficient hydrogen production from H_2S decomposition by $CuGaO_2$ and $CuGa_{1-x}In_xO_2$ delafossite oxides bearing nanostructured co-catalysts. Catalysis Communications. 2008;9:395–402. DOI: 10.1016/j.catcom.2007.07.021

[52] Kanade KG, Baeg J-O, Kale BB, Mi Lee S, Moon S-J, Kong, K-j. Rose-red color oxynitride : a visible light photocatalyst to hydrogen production. International Journal of Hydrogen Energy. 2007;32:4678–4684. DOI: 10.1016/j.ijhydene.2007.07.040

[53] Bao N, Shen L, Takata T, Domen K, Gupta A, Yanagisawa K, Grimes CA. Facile Cd –thiourea complex thermolysis synthesis of phase-controlled Cds nanocrystals for photocatalytic hydrogen production under visible light. Journal of Physical Chemistry C. 2007;111:17527–17534. DOI: 10.1021/jp076566s

[54] Bai XF, Cao Y, Wu W. Photocatalytic decomposition of H2S to produce H_2 over CdS nanoparticles formed in HY-zeolite pore. Renewable Energy. 2011;36:2589–2592. DOI: 10.1016/j.renene.2010.04.037

[55] Kainthla RC, Bockris JOM. Photoelectrolysis of H_2S using an n-CdSe photoanode. International Journal of Hydrogen Energy. 1987;12:23–26. DOI: 10.1016/0360-3199(87)90122-4

[56] Zong X, Han J, Seger B, Chen H, Lu G, Li C, Wang L. An integrated photoelectrochemical–chemical loop for solar-driven overall splitting of hydrogen sulfide. Angewandte Chemie International Edition. 2014;53:4399–4403. DOI: 10.1002/anie.201400571

[57] Zong X, Chen H, Seger B, Pedersen T, Dargusch MS, McFarland EW, Li C, Wang L. Selective production of hydrogen peroxide and oxidation of hydrogen sulfide in an unbiased solar photoelectrochemical cell. Energy and Environmental Science. 2014;7:3347–3351. DOI: 10.1039/c4ee01503g

[58] Yan H, Yang J, Ma G, Wu G, Zong X, Lei Z, Shi J, Li C. Visible-light-driven hydrogen production with extremely high quantum efficiency on Pt–PdS/CdS photocatalyst. Journal of Catalysis. 2009;266:165–168. DOI: 10.1016/j.jcat.2009.06.024

Chapter 11

Metal/Semiconductor Hybrid Nanocrystals and Synergistic Photocatalysis Applications

Jiatao Zhang, Muwei Ji, Jiajia Liu and Meng Xu

Additional information is available at the end of the chapter

http://dx.doi.org/10.5772/61888

Abstract

This review focuses on recent research efforts to synthesize metal/semiconductor hybrid nanocrystals, understand and control the photocatalytic applications. First, we summarize the synthesis methods and recent presented metal/seminconductor morphologies, including heterodimer, core/shell, and yolk/shell etc. The metal clusters and nanocrystals deposition on semiconductor micro/nano substrates with well-defined crystal face exposure will be clarified into heterodimer part. The outline of this synthesis part will be the large lattice mismatch directed interface, contact and morphologies evolution. For detailed instructions on each synthesis, the readers are referred to the corresponding literature. Secondly, the recent upcoming photocatalysis applications and research progress of these hybrid nanocrystals will be reviewed, including the photocatalytic hydrogen evolution (water splitting), photo-reduction of CO_2 and other newly emerging potential photosynthesis applications of metal/semiconductor hybrid nanocrystals. Finally, we summarize and outlook the future of this topic. From this review, we try to facilitate the understanding and further improvement of current and practical metal/semiconductor hybrid nanocrystals and photocatalysis applications.

Keywords: Metal/semiconductor, hybrid nanocrystals, photocatalysis, sythesis, interface, lattice mismatch, Plasmon, exciton, photooxidation, photoreduction

1. Introduction

The shape and size effect, the controllable doping, heterocomposite, and interface are the prerequisite of colloidal nanocrystals for exploring their optoelectronic properties, such as fluorescence, plasmon–exciton coupling, efficient electron/hole separation, and enhanced photoelectric conversion [1–4]. The use of photoexcited electrons and holes in semiconductor nanocrystals as reduction and oxidation reagents is an intriguing way of harvesting photon energy to drive chemical reactions since increasing energy demand and environmental pollution create a pressing need for clean and sustainable energy solution. The high efficient separation and collection of photoexcited electrons (e-) and holes (h+) are the key points to get high efficient photocatalysis applications. Hybrid nanocrystals composed of semiconductor and metal components are receiving extensive attention in recent years due to their high efficient separation of photoexcited electrons and holes and potentional photocatalysis applications [1,4–5]. Furthermore, hybrid nanocrystals composed of semiconductor and plasmonic metal components are receiving extensive attention. The simultaneous existence and coupling of localized surface plasmon resonance induced plasmon and excitons in semiconductors, as well as the synergistic interactions between the two components [1, 6–7].

This review focuses on recent research efforts to synthesize metal/semiconductor hybrid nanocrystals to understand and control the photocatalytic applications. First, we summarize the synthesis methods and recent presented metal/semiconductor morphologies, including heterodimer, core/shell, and yolk/shell. The metal clusters and nanocrystals deposition on semiconductor micro/nanosubstrates with well-defined crystal face exposure will be clarified into heterodimer part. The outline of this synthesis part will be the large lattice mismatch-directed interface, contact, and morphology evolution. For detailed instructions on each synthesis, the readers are referred to the corresponding literature.

Second, the recent upcoming photocatalysis applications and research progress of these hybrid nanocrystals will be reviewed, including the photocatalytic hydrogen evolution (water splitting), photoreduction of CO_2, and other newly emerging potential photosynthesis applications of metal/semiconductor hybrid nanocrystals. Finally, we provide a summary and outlook on the future of this topic. From this review, we try to facilitate the understanding and further improvement of current and practical metal/semiconductor hybrid nanocrystals and photocatalysis applications.

2. Metal/semiconductor hybrid nanocrystal synthesis

According to the classical hetero epitaxial growth theory [8], when a secondary material (referred to as "2") has to be deposited over a preexisting seed substrate of a different material (denoted as "1"), total Gibbs free surface energy change function ΔG_s can be described as follows: $\Delta G_s = \gamma_1 - \gamma_2 + \gamma_{1,2}$, where γ_1 and γ_2 are the surface energies of seed substrate and the deposited materials, respectively, the solid/solution interfacial energies of colloidal nanostructure in the liquid medium, and $\gamma_{1,2}$ is strain-related solid/solid interfacial energy that depends on the bonding strength and degree of crystallographic compatibility between materials 1 and 2. As shown in Figure 1, if material 2 exposes lower energy surfaces ($\gamma_2 < \gamma_1$) and/or attains good lattice matching with the material 1 ($\gamma_{1,2}$ is small), then it is easy to get continuous and uniform core/shell nanostructure ($\Delta G_s > 0$, Frank—van der Merwe (FM) growth mode). On the contrary, if material 2 is featured by higher energy surfaces ($\gamma_2 > \gamma_1$) and/or $\gamma_{1,2}$ is high owing to the large lattice mismatch, then it will tend to deposit as a discontinuous island-like domain to be heterodimer structure (ΔG_s <0, Volmer—Weber (VW)growth mode) or even separate from each other. Therefore, the metal/semiconductor hybrid nanocrystal synthesis should follow the mechanism of Figure 1. However, in colloidal phase, γ_1 and γ_2 would be strongly influenced by adhesion of solvent, capping ligands and precursors, thus changing $\gamma_{1,2}$ too [9,10]. Considering of the significant impact of binding of organic stabilizers or other solution species to the surface energy terms and therefore altering the ultimate ΔG_s, the colloidal phase synthesis will be potential for the control of large lattice mismatch-directed shape evolution and morphology control when choosing different solvent, capping ligands, and reactant precursors.

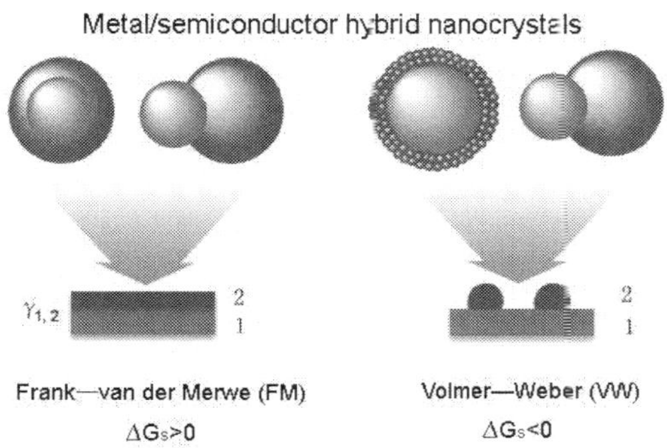

Figure 1. General metal/semiconductor hybrid nanocrystal synthesis sketches by traditional heterogeneous epitaxy growth mechanism: (A) Franck–van der Merwe (FM) mode; (B) Volmer–Weber (VM) mode.

2.1. Metal/semiconductor heterodimers by direct heterogeneous deposition

A broad family of metal/semiconductor hybrid nanocrystals has been obtained by accomplishing direct heterogeneous nucleation and growth of one or more secondary material layers onto preformed metal NC seeds serving as starting "cores" [9,10]. Synthetic strategies aim, on one side, at inhibiting homogeneous self-nucleation of isolated NCs made of the shell material and, on the other side, at achieving size- and shape-mediated deposition of the shell beyond limitations imposed by misfit strain at the metal/semiconductor interface. Practical techniques to realize these objectives rely on the colloidal synthesis, the "active" surface of "core" and the slow-nucleation of another material to deposit on "core" homogeneously, depending on the inherent chemical accessibility of the "core" or "seeds" as well on the reactivity of the shell molecular precursors (generally nonaqueous) solution media. The regulation of the thermodynamics and kinetics of reactions, namely, the temperature and selection of suitable surface-adhering organic ligands or surfactants and the optimal reactant injection rate all critically influence on the temporal evolution of the supersaturation degree.

To date, disparate combinations of metals, semiconductors, and oxides have been addressed by manipulating direct heterogeneous deposition pathways [11–25]. Here, we typically review most recent achievement in the field, the metal/semiconductor heterodimers (as scheme in Figure 1). Representative transmission electron microscopy (TEM) and high-resolution TEM examples of core–shell NCs can be found in Figure 2, respectively.

The one-pot synthesis of metal/semiconductor heterodimers mostly focus on the bifunctional heterodimers of nanoparticles, such as the conjugate of quantum dots (QDs) and magnetic nanoparticles reported by Gu et al. [20], the noble metal/metal oxide heterodimers by Wang et al. [21] and Wu et al. [22], and the bimagnetic FePt–iron oxide heterodimer nanocrystals by Figuerola et al. [23] and Cozzoli et al. [24] (see also Figure 2).

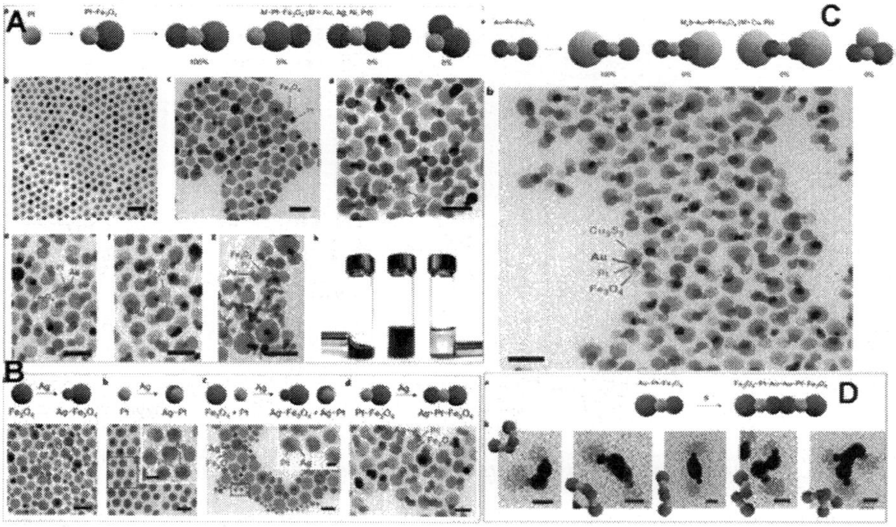

Figure 2. (A) FePt/CdS heterodimers by one-pot synthesis: (a) the scheme of synthesis process, (b) a TEM image, and (c) a high-resolution TEM image of FePt/CdS heterodimers. Adapted from Gu et al. [20] with permission; copyright American Chemical Society. (B) Seed-mediated high temperature growth of M/metal oxide heterodimers: (a) the scheme of synthesis, (b) 5–12 nm Au–Fe_3O_4, (c) 5–12 nm Ag-Fe_3O_4, and (d) 5–12 nm Pt–Fe_3O_4. Adapted from Wang et al. [21] with permission; copyright American Chemical Society. FePt/In_2O_3 heterodimers were adapted from Wu et al. [22] with permission; copyright American Chemical Society. (C) One-pot, two-step colloidal strategy to prepare FePt/ iron oxide heterodimer nanocrystals: (a) the scheme of synthesis method; (b) the TEM image of FePt/ Iron Oxide heterodimers; and (c) the HRTEM image of FePt/ Iron Oxide heterodimer. Adapted from Figuerola et al. [23] and Cozzoli et al. [24] with permission; copyright American Chemical Society.

The Au/ CdSe or Au/CdS heterodimer nanocrystals, because of the noble metal and II-VI QDs, are potential for photocatalysis applications. Banin et al. used surface nucleation and growth of a second phase to get Au/CdSe nanorod (NR) heterodimers, the first time to realize site-selective growth of Au NPs on CdSe NRs tips (see also Figure 3A) [13, 25, 26]. Du et al. used similar way to directly deposit the Au NPs on CdS NRs to get heterodimers (see also Figure 3B) [27]. The as-obtained Au/CdS and Au/CdSe NR heterodimers promote the electron/hole separation and exhibit promising photocatalytic activity for the water-splitting reaction in photoelectrochemical cells and photodegradation applications.

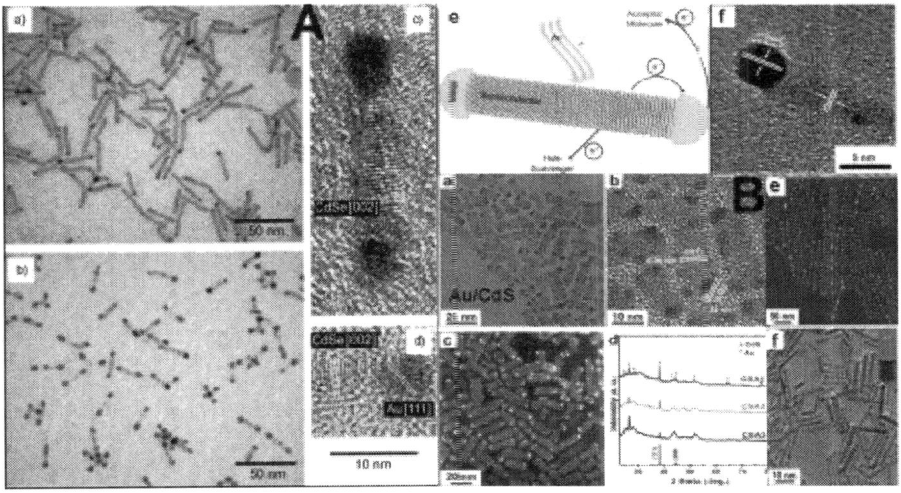

Figure 3. (A) TEM images showing controlled growth of gold onto the tips of CdSe nanorods. (a, b) The size of the gold tips can be controlled by varying the amount of gold precursor added during growth. (c, d) HRTEM images of a single nanodumbbell (c) and a nanodumbbell tip (d); the CdSe lattice for the rod in the center and gold tips at the rod edges can be identified. (From [25, 26], reprinted with permission from the AAAS.) (B) (a) TEM image, (b) HRTEM image, and (c) HAADF-STEM image of Au/CdS NRs hybrid nanocrystals. (d) XRD patterns of hybrid nanocrystals of Au/CdS NRs. (e) HAADF-STEM images in which the size of Au NPs is ~8 nm. (f) TEM images of Au/CdS hybrid nanocrystals in which the size of Au NPs is ~3 nm. Adapted from Du et al. [27] with permission; copyright Wiley-VCH Verlag GmbH & Co. KGaA.

Colloidal metal/semiconductor hybrid nanoparticles contain multiple nanoscale domains fused together by solid-state interfaces. They represent an emerging class of multifunctional lab-on-a-particle architectures that underpin future advances in solar energy conversion, photocatalysis. Buck et al. reported that the known direct heterogeneous deposition could be applied in a predictable and stepwise manner to build complex hybrid nanoparticle architectures that include M–Pt–Fe_3O_4 (M=Au, Ag, Ni, Pd) heterotrimers, M_xS–Au–Pt–Fe_3O_4 (M=Pb, Cu) heterotetramers, and higher-order oligomers based on the heterotrimeric Au–Pt–Fe_3O_4 building block [18]. This synthetic framework conceptually mimics the total synthesis approach used by chemists to construct complex organic molecules (see Figure 4).

In other words, the advantage of direct heterogeneous deposition colloidal method is it could get heterodimers with high monodispersity. The disadvantages are as follows: (1) The metal part or semiconductor part could not be mediated in more larger size range, such as tens of nanometers to hundreds of nanometers due to the LaMer growth mechanism of high

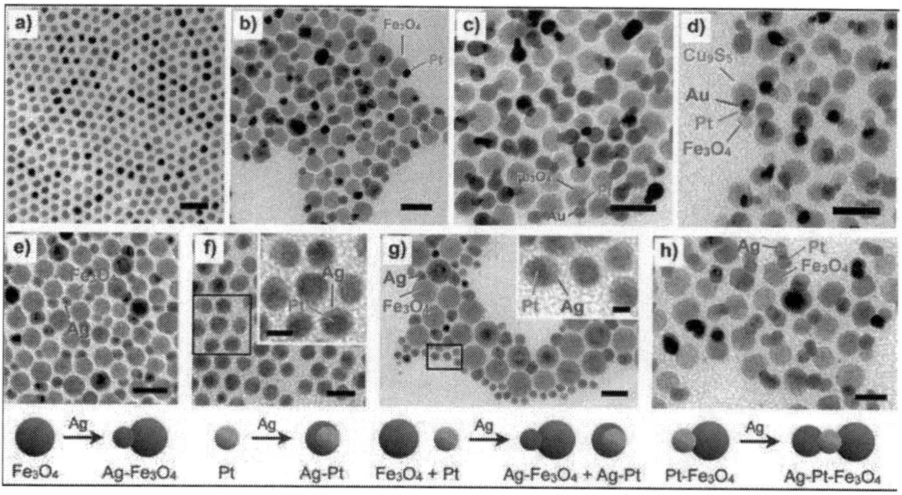

Figure 4. (A–D) TEM images showing the stepwise direct heterogeneous deposition of linear nanoparticle heterotetramers: (A) Pt nanoparticles, (B) Pt–Fe_3O_4 heterodimers, (C) Au–Pt–Fe_3O_4 heterotrimers, and (D) Cu_9S_5–Au–Pt–Fe_3O_4 heterotetramers. (E–H) TEM images and schematic representations showing a series of metal/semiconductor heterodimers (same synthetic conditions for all steps) aimed at understanding the site-selective deposition of Ag onto Pt–Fe_3O_4 heterodimers: (E) Ag grown off Fe_3O_4 nanoparticles, (F) Ag grown off Pt nanoparticles, (G) Ag grown indiscriminately off both Fe_3O_4 and Pt nanoparticles when both are present as a physical mixture, and (H) Ag grown exclusively off the Pt domain when Pt and Fe_3O_4 are directly attached as heterodimers. Reprinted from Buck et al. [18] with permission (Copyright of Nature group 2011, Macmillan Publishers Limited).

temperature organic phase synthesis. (2) The interface between metal and semiconductor is still not controlled well to be clear enough because of the size of two parts mostly still limited at <10 nm range. The lattice mismatch-induced stain energy here is usually too large to get more defects on the interface, which is not helpful for the highly efficient photoinduced electron/hole separation and the following photocatalysis applications.

2.2. Large lattice mismatch-directed shape evolution and morphology control

The growth of monocrystalline semiconductor-based metal/semiconductor hybrid nanostructures with modulated composition, morphology, and interface strain are the prerequisite for exploring their plasmon–exciton coupling, efficient electron/hole separation, and enhanced photocatalysis properties. As the schematic process in Figure 1, different from generally used one-pot epitaxial growth on performed metal nanoparticle seeds, Ouyang et al. and Zhang et al. unprecedentedly took two steps of facile chemical thermodynamics processes to maximally adjust the surface energies γ_2, $\gamma_{1,2}$, and then ΔG_s to change their overgrowth modes from FM mode to VW mode gradually. Furthermore, Ouyang et al. and Zhang et al. unprecedentedly used the cation exchange initiated the topotactic in situ conversion from amorphous to be single-crystalline structure [28–33]. In this case, the more flexible size range (from smaller than

10 nm to tens of nanometers, to hundreds of nanometers, and to micrometer scale), morphology (isotropic and anisotropic), interface, and single-crystallinity of metal and semiconductor part could be tailored synergistically.

By controlling soft acid–base coordination reactions between molecular complexes and colloidal nanostructures, Zhang et al. and Ouyang et al. showed that chemical thermodynamics could drive nanoscale monocrystalline growth of the semiconductor shell on metal nanosubstrates and then enhanced light–matter-spin interactions in these judiciously engineered nanostructures could be achieved [28, 3].

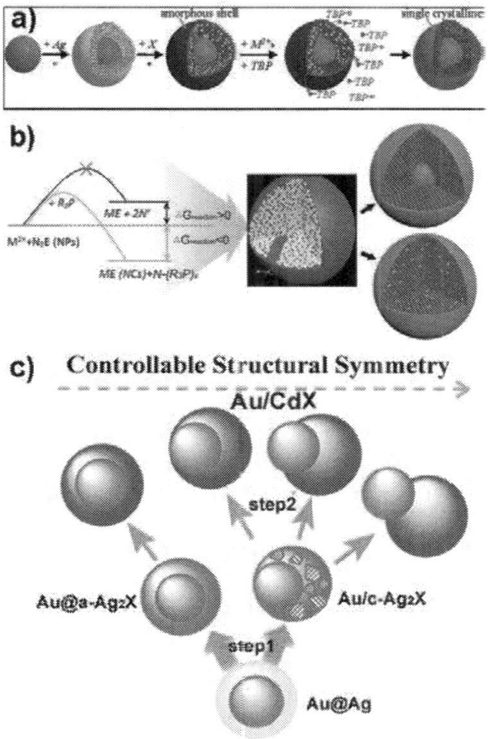

Figure 5. Schemes of large lattice mismatch-directed shape evolution and morphology control by Ouyang et al. and Zhang et al.: (A) Nonepitaxial growth process and mechanism of hybrid core–shell nanostructures with substantial lattice mismatches. From Zhang et al. [28], reprinted with permission from the AAAS. (B) The cation exchange reaction initiated by different phosphines or phosphites (R3P) and phosphine-initiated cation exchange for precisely tailoring composition and properties of metal/semiconductor nanostructures. (C) The schematic process of controllable structural symmetries in Au/CdX (X = S, Se, and Te) hybrid structures with large lattice mismatch by two steps of in situ chemical conversion. The a-Ag_2X means amorphous Ag_2X, c-Ag_2X means crystalline Ag_2X. Adapted from Gui et al. [31] and Zhao et al. [32] with permission; copyright Wiley-VCH Verlag GmbH & Co. KGaA.

First, as shown in Figure 5A, based on the Lewis acid–base reaction mechanism, where the entire nanostructure is spatially confined by an amorphous matrix, the monocrystalline growth of the semiconductor shell is fully directed by chemical thermodynamic properties of reactions within the matrix; the shell's lattice structure can be independent of that of the core NPs, thus circumventing the limitations imposed by epitaxial strategies. Starting from the core NPs, an overlayer of metal with soft Lewis acidity is grown onto the core. For all hybrid core–shell structures, we choose an Ag metal overlayer. The silver shells can be modified to form silver compound shells (Ag_2X) with an amorphous structure, providing a crucial platform for the next chemical transformation stage, ultimately leading to monocrystalline growth. It has been demonstrated that nanoscale chemical transformations, such as cation exchange, represent a versatile route for converting one crystalline solid to another [34, 35]. They show that this process can be harnessed to drive the single-crystal growth by carefully controlling the thermodynamic properties of reaction (1).

Tributylphosphine (TBP) was selected because it is a soft base and can behave as a phase-transfer agent to transport metal ions (Mn^+) to the surface of the core NPs by binding to free cations in solution. The high acid softness of Ag^+ favors the exchange process between Ag^+ in the amorphous matrix and Mn^+ in solution as long as the softness of Mn^+ is small enough to result in a positive ΔG.

Second, based on above research achievements, as shown in Figure 5B, Zhang et al. further studied the cation exchange reaction here. It has been explored that different phosphines could modulate the thermodynamic and kinetic parameters of the cation exchange reaction to synthesize complex semiconductor nanostructures [31]. Here, we take the examples of cation exchange between N^+ in amorphous N_2E (E means chalcogen) nanoparticles (NPs) and M^{2+} (such as Cd^{2+}) ions in solution, as shown in reaction (2) and Figure 1. Besides TBP, many other phosphine choices have been studied to mediate the thermodynamics and kinetics of reaction (2). Initiated by trace phosphine (R3P), the crystallization, the morphology, and the composition of metal/ME core/shell NCs have been tailored well.

The thermodynamics and kinetics of different phosphine and phosphite agents to synthesize semiconductor shell have been studied. The prerequisite to forward the cation exchange is the lone-pair electrons of P atom in phosphines or phosphites. Based on this, different π-accepting and σ-donating capabilities of them to M and N ions make reaction (2) to be exothermic (ΔG reaction < 0). Different phosphine coordinating to Ag^+ and Cd^{2+} and their coordination abilities study further prove reaction (2) is exothermic in principle. Therefore, thermodynamics and kinetics of reaction (2) could be mediated. The stronger coordination ability of R3P to N^+ than to M^{2+} ions enables in situ conversion of amorphous N_2E nanoparticles to be ME NCs. Although many kinds of phosphine or phosphite are applicable in reaction (2) thermodynamically, the steric effect derived from carbonyl ligands, such as alkyl and aryl ligands, would influence the kinetics of reaction (2), and the crystallization and composition of produced ME NCs distinctly. They took the example of cation exchange from amorphous Ag_2S nanostructure to single-crystalline CdS nanostructure. The kinetic activity order here is $PEt_3 > P(MeO)_3 > P(EtO)_3 > P(n\text{-}Bu)_3 > PPh_3 > P(n\text{-}Oct)_3 > P(PhMe)_3 > P(PhOMe)_3$. This is almost consistent with their σ donation ability order and contrary to their π back-bonding order. In particular, the steric hindrance from carbonyl ligands would influence σ donation ability and then chemical kinetics distinctly. That is the reason why $P(PhOMe)_3$ phosphines have low activities to reaction (2). These findings are concordant with Chad Tolman's classic phosphine ligands ordering in terms of their electron-donating ability and steric bulk. Besides preserving the original shape and size, phosphine-initiated cation exchange reactions show potential to precisely tune the crystallinity and composition of metal/semiconductor hybrid nanocrystals.

Third, as illustrated in Figure 5C, Zhang et al. further showed that the structural symmetry of such metal/semiconductor hybrid heterostructures can be finely tuned with controllable separation between metal and semiconductor components by taking advantage of chemical thermodynamics-directed colloidal strain tuning, from symmetric core–shell to asymmetric heterodimer gradually by in situ conversion of amorphous/crystalline Ag_2X shell to be single crystalline CdX. Typically, nanoparticle/solution interfacial energies and heterointerfacial energy (due to the lattice mismatch) can be strongly influenced by adhesion of solvent, capping ligands and precursors in colloidal phase [9]. Here, Zhang et al. took a facile two-step approach to maximize these effects to precisely control gradual shape evolution from symmetric core/shell to asymmetric heterodimer nanostructures, as shown in Figure 5C [32]. This approach is fundamentally different from generally used one-pot epitaxial overgrowth with metal nanoparticle seeds [11-25]. First, they started with concentric Au–Ag core/shell. The fact of close lattice feature between Au and Ag makes it possible to achieve precise control of core size, shell thickness, and monodispersity based on Frank-van der Merwe growth mode [9]. Moreover, the higher reactivity of silver metal nanostructures than gold enables the in situ conversion of Ag shell to silver chalcogenide (Ag_2X) without modification of the Au metal core. Depending on reaction condition (such as temperature and reactants), the crystallinity of Ag_2X can be controlled from amorphous to partial crystalline, thus leading to different lattice mismatch between Au core and Ag_2X shell. As a result, different interfacial lattice strain between Au and Ag_2X can be induced to initiate phase separation to a certain extent. Different from the long time aging-induced elemental Au diffusion in Au–Ag_2X core/shell nanocrystals via Ostwald ripening, the Au–Ag_2X here was used as intermediate precursor to carry out next step quickly to get Au/CdX heteronanocrystals. Second, they utilized cation exchange process based on the theory of hard-soft acid and base to realize the in situ conversion of Ag_2X to monocrystalline CdX [28]. The cation exchange-induced cation rearrangement and crystallography texture transformation provide further impetus to shift CdX shell gradually because of the larger strain energy between Au and monocrystalline CdX. In general, the higher crystallization of Ag_2X shell as well as the higher reaction temperature of cation exchange reaction leads to larger phase separation between Au and CdX to reduce interfacial and grain boundary energies. These tunable relocations of gold to CdX in quantum size region would enable maximum degree of tunability of their optoelectronic properties coupling.

2.3. Metal/semiconductor core/shell nanocrystals

The growth of single-crystal semiconductor-based heterostructures with modulated composition is a prerequisite for exploring fundamental nanoscale semiconductor physics [36] and can offer technological devices with optimum characteristics, including enhanced optical properties with high quantum yields [37], engineered electronic band gaps [38–40], and various solid-state optoelectronic properties [41–43]. Unintentional crystalline imperfections (such as polycrystallinity, dislocations, and other structural defects) lead to performance degradation or even premature failure of devices. For example, although the optical quality of semiconductor CdSe nanoparticles (NPs) could be improved by an overlayer of epitaxially grown CdS or ZnS, problems appear once the shell thickness becomes larger than the critical layer thickness (about two monolayers) due to the existence of strain-induced defects [38, 44, 45].

Current methods that achieve high-quality monocrystalline heterostructures are all based on epitaxial growth, as shown in Figure 1, which requires moderate lattice mismatches (<2%) between the two different materials. This lattice-matching constraint is a severe obstacle, particularly for growth of core–shell nanostructures with (quasi)spherical core NPs with highly curved surfaces that present many different crystallographic facets [46]. In addition to such lattice-matching requirements, the issues related to differences in crystal structure, bonding, and other properties have been found to inhibit epitaxial growth of dissimilar hybrid materials such as monocrystalline semiconductors on metals [47]. Attempts to use epitaxy to achieve hybrid metal core–semiconductor shell nanostructures have been unsuccessful, resulting in either polycrystalline semiconductor shells or anisotropic structures with segregation of the core and shell. This is only because as Figure 1 schemed, under large lattice mismatch (>40%), the semiconductor nanostructures would grow as small island-shaped NPs on metal core surface to decrease the surface strain energy. In this case, the Volmer—Weber (VW) mode growth would happen. Thus, based on these direct heterogeneous deposition methods, such as Figures 2–4 demonstrated, the polycrystalline semiconductor shell formed finally. As Figure 6 demonstrated, the Klimov, Talapin, and Wang groups have tried these kinds of method to prepare Au–PbS, Co–CdSe, Au–CdS, and Au–ZnS core/shell hybrid nanocrystals. Although they are highly monodispersed and could self-assemble into superlattice, the polycrystalline shell and too many defects at the interface limited their usefulness, especially in photocatalysis applications [48–53].

The photocatalysis application in the integration of semiconductor nanocrystals with noble metals uses the localized surface plasmon resonance (LSPR) effect of the metal component to enhance the light absorption, charge separation, and facilitate the absorbed light energy transfer from the metal to the semiconductor component for technologically important light-involved applications. Abundant studies have been devoted to the controllable productions of various hybrid nanocrystals. Heterostructure preparation is the basis for any application of semiconductor/noble metal hybrid nanocrystals. The crystalline imperfections of each component and the defects on the interface lead to the performance degradation or lost. Currently, high-quality monocrystalline heterostructures are usually produced by thermal decomposition epitaxial growth or through cation-exchange processes. The reported studies of monocrystalline hybrid nanocrystal synthesis are mainly based on epitaxial growth, which requires a moderate lattice mismatches between the different components. The lattice-mismatching constraint seriously limits the application of this method, particularly for growth of core/shell hybrid nanocrystals with highly curved surfaces. Therefore, it is a great challenge to synthesize the large lattice-mismatching semiconductor/metal hybrid nanocrystals with monocrystalline compounds and clear interface. Cation exchange reaction is a successful

http://dx.doi.org/10.5772/61888

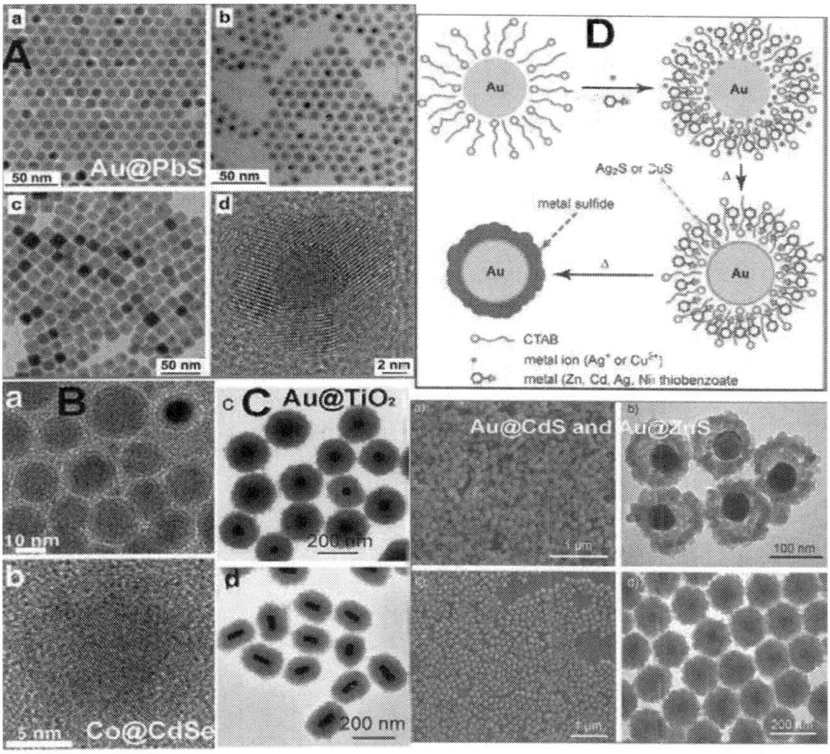

Figure 6. The metal/semiconductor core/shell NCs with polycrystalline semiconductor shell by direct heterogeneous deposition method. (A) Au–PbS core/shell NCs. (B) Co–CdSe core/shell NCs. Adapted from Lee et al. [48] and Kim et al. [49], respectively, with permission; copyright American Chemical Society. (C) Au–TiO_2 core/shell NCs. Adapted from Fang et al. [53] with permission; copyright Royal Society of Chemistry. (D) Au–CdS and Au–ZnS core/shell NCs. Adapted from Sun et al. [54] with permission; copyright Wiley-VCH Verlag GmbH & Co. KGaA.

synthetic method to preparation hybrid large lattice-mismatching heterodimer and core/shell nanocrystals with monocrystalline compounds and clear interface, which cannot be obtained by conventional epitaxial techniques [48–54]. Based on the Lewis acid–base reaction between molecular complexes and colloidal nanocrystals, the amorphous nanostructures can be transferred into monocrystalline compounds. Hence, by combining the sulfidation, selenization, or hyperoxidation of silver with cation exchange, other monocrystalline metal chalcogenide, selenide, or oxide nanostructures can been nonepitaxially grown on the large lattice-mismatching metal surface [28–33]. For the catalytic applications of semiconductor/noble metal hybrid nanocrystals, the key challenges are to obtain a clear semiconductor–metal interface and precise control over the size and shape of the heterostructures. Via cation exchange reaction, a series of semiconductor/metal hybrid nanocrystals can be obtained from concentric core–shell to nonconcentric heterodimer with precisely controlled separation and clear interface between the semiconductor and noble metal compounds. The symmetry evolution of semiconductor/metal hybrid nanocrystal has led to novel control of light absorption and photocatalytic activity, which indicates the advantage of cation exchange nonepitaxial growth and the importance of nanoscale interface control. For the phosphine-initiated cation exchange, different phosphines have been used to modulate the thermodynamic and kinetic process of the cation exchange reaction in semiconductor nanocrystal synthesis. By using different phosphines, the crystallinity, composition, morphology, and related properties of semiconductors can be precisely controlled.

Different from the reports in Figure 6, Ouyang et al. and Zhang et al. used the nonepitaxial growth scheme (Figure 5A) to get metal–semiconductor core/shell NCs with single crystalline semiconductor shell. The size of metal, semiconductor shell, and the composition of semiconductor shell could be precisely controlled (Figures 7–9) [28].

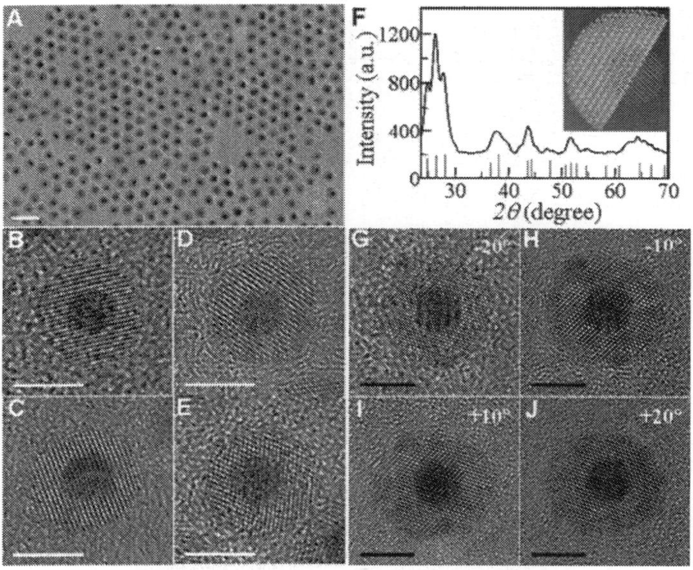

Figure 7. Au–CdS core–shell nanostructures with monocrystalline shell. (A) Typical TEM image showing uniform core–shell nanostructures. Scale bar, 20 nm. (B–E) High-resolution TEM images of core–shell nanostructures from (A). Whereas Au core NPs can manifest monocrystalline (B), single-fold twin (C), fivefold twin (D), and multiple-twin (E) lattice structures, all CdS shells are monocrystalline. The red lines highlight the lattice orientations within the Au core NPs. Scale bar, 5 nm. (F) XRD pattern of Au–CdS sample shown in panel A. Bulk Au [red solid lines, Joint Committee on Powder Diffraction Standards (JCPDS) #04-0784] and wurtzite CdS (blue solid lines, JCPDS #41-1049) are also provided for reference and comparison. (Inset) A ball-and-stick molecular model of Au–CdS, illustrating a cubic core and wurtzite shell. (G–J) Angle-dependent high-resolution TEM characterization. The sample depicted has a larger shell thickness than the one in panel A to emphasize the extremely high-quality crystallinity of the shell. The CdS shell shows perfect monocrystalline features without detectable structural defects under a different viewing angle. Scale bar, 5 nm. From Zhang et al. [28], reprinted with permission from the AAAS.

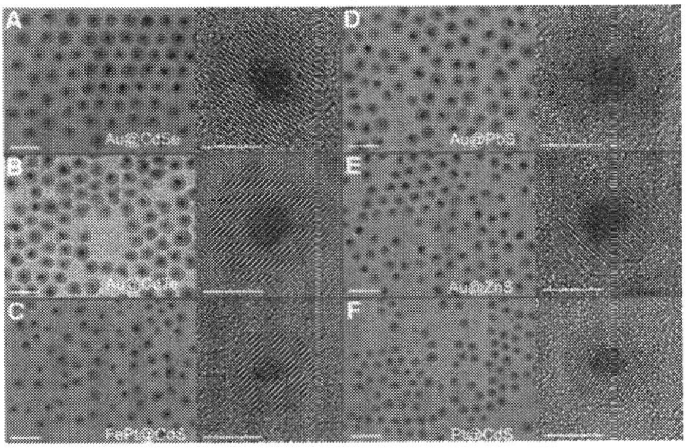

Figure 8. Large-scale (left) and high-resolution (right) TEM images of different hybrid core–shell nanostructures with various combinations of the core and shell components. All semiconductor shells show monocrystalline features. Scale bars for large-scale and high-resolution TEM images are 20 and 5 nm, respectively. (A) Au–CdSe; (B) Au–CdTe; (C) FePt–CdS; (D) Au–PbS; (E) Au–ZnS; and (F) Pt–CdS. From Zhang et al. [28], reprinted with permission from the AAAS.

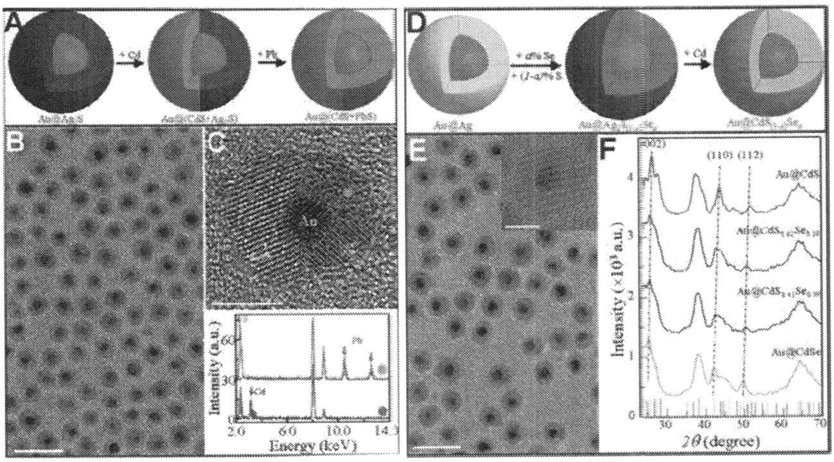

Figure 9. Growth of complex hybrid core–shell nanostructures with tailored structures and compositions of the monocrystalline shells. (A–C) Control of the monocrystalline cation species within the shell: the case of Au–(CdS+PbS). (A) Schematic of the growth procedure. (B) Large-scale TEM image. Scale bar, 20 nm. (C) (Top) High-resolution TEM image. Blue and green dashed arc curves highlight the monocrystalline CdS and PbS regimes, respectively. CdS and PbS manifest distinct lattice planes that can be assigned to (100) and (220), respectively. Scale bar, 5 nm. (Bottom) Single-particle EDS measurements in the CdS and PbS regimes. Peaks from Cd, Pb, and S elements are highlighted. (D–F) Control of the monocrystalline anion species within the shell: the case of Au–CdS1–aSea (D) Schematic growth procedure. (E) Large-scale TEM image. Scale bar, 20 nm. (Inset) High-resolution TEM image showing the monocrystalline alloy shell. Scale bar, 5 nm. (F) XRD patterns highlighting lattice evolution from CdSe to CdS with different ratio a. From Zhang et al. [28], reprinted with permission from the AAAS.

Based on the strategy in Figure 5B, Zhang et al. use air-stable PPh_3 for the first time to initiate cation exchange reaction (2). Figure 10 demonstrates the Au–CdS, Pt–CdS core/shell NCs preparation by PPh_3 initiated reaction (2). Following the nonepitaxial growth process we published before [28] after cation exchange from amorphous Ag_2S shell, Au–CdS and Pt–CdS NCs could preserve thick monocrystalline CdS shell. LRTEM and HRTEM images in Figure 10A–D confirmed their good crystallization. Especially, as shown in Figure 10D, despite of anisotropic shape of Pt nanocube, the thick CdS shell has good single-crystallinity. PPh_3 further facilitates the monocrystalline engineering to break through critical layer thickness limit of heteroepitaxy. PPh_3 is "green" choice than TBP because it is air-stable enough to enable reaction (2) under more flexibly conditions, such as higher temperature and longer time to facilitate versatile ME NCs crystallization. Moreover, besides crystallization tailoring, PPh_3 could initiate cation exchange to get more complex heterostructures with precise compositional tailoring. Figure 11A and B showed the LRTEM and HRTEM images of as-prepared Au–$CdS_{1-x}Se_x$ core/shell nanocrystals with ternary single-crystal alloys shell. The S-to-Se ratio could be tailored ($CdS_{0.58}Se_{0.42}$ and $CdS_{0.45}Se_{0.55}$, obtained by EDS elemental analysis) to engineer their band gaps. The consistent shift of their powder XRD peaks from $CdS_{1-x}Se_x$ shell (Figure 11C) and the distinct colloid color changing (insert in Figure 3D) confirmed the homogeneous composition modulation. The strong visible light absorption (550–700 nm) (Figure 11D) due to the surface plasmon resonance (SPR) and exciton coupling in as-prepared Au–$CdS_{1-x}Se_x$ NCs indicated their potential photocatalysis and photovoltaic applications [55].

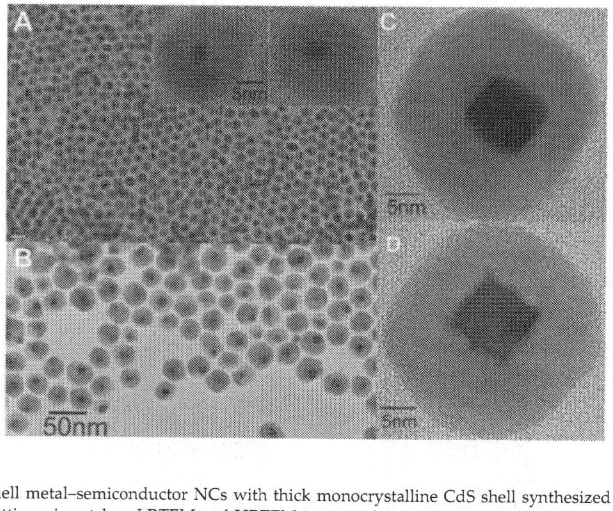

Figure 10. Core/shell metal–semiconductor NCs with thick monocrystalline CdS shell synthesized by PPh_3 initialization under large lattice mismatches: LRTEM and HRTEM images of Au–CdS NCs (A) and Pt–CdS NCs with Pt nanocube core (B-D). Adapted from Gui et al. [31] with permission; copyright Wiley-VCH Verlag GmbH & Co. KGaA)

Recently, through the strategy in Figure 5C, Zhang et al. have demonstrated evolution of relative position of Au and CdX in Au–CdX (X means S, Se, and Te) from symmetric to asymmetric configuration (Figures 12 and 13) [32].

http://dx.doi.org/10.5772/61888

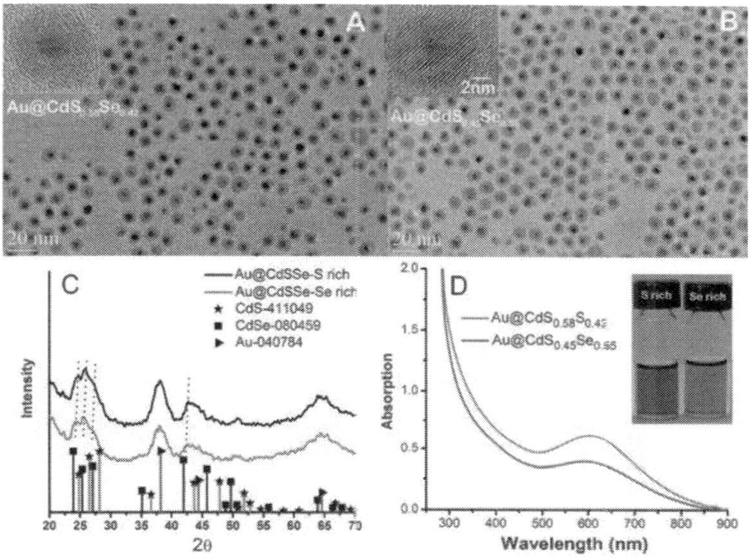

Figure 11. Au–CdS$_{1-x}$Se$_x$ NCs synthesized by PPh$_3$ initialization. (A, B) LRTEM and HRTEM images of prepared Au–CdS$_{0.58}$Se$_{0.42}$ NCs and Au–CdS$_{0.45}$Se$_{0.55}$ NCs. (C) XRD patterns comparison of them. (D) UV-Vis extinction spectra comparison of them. Inserted pictures showed their colloid color. Adapted from Gui et al. [31] with permission; copyright Wiley-VCH Verlag GmbH & Co. KGaA)

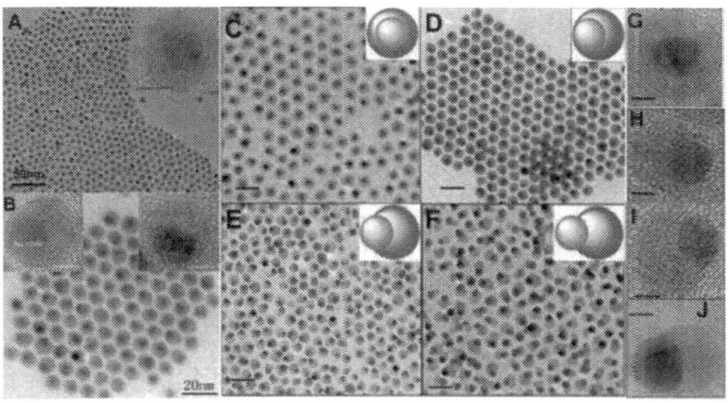

Figure 12. TEM characterizations on controllable structural symmetry from core/shell to heterodimers. (A, B) Au–Ag$_2$S with amorphous (A) and crystalline Ag$_2$S (B) shells. The inserts are HRTEM images of a-Ag$_2$S (A) and c-Ag$_2$S shell (B); scale bar, 5 nm. Red solid lines are guides for the eye, distinguishing the Au core and c-Ag$_2$S shell boundaries, respectively. (C–F) Au/CdS hybrid nanostructure with controllable structural symmetry. (C) Concentric core/shell. (D) Non-concentric core/shell. (E, F) Heterodimers. The inset diagrams highlight the phase separation-induced Au/CdS morphologies. Scale bar: 20 nm. (G–J) HRTEM images highlight the shape separations of panels C–F, respectively. Scale bar: 2.5 nm. Adapted from Zhao et al. [32] with permission; copyright Wiley-VCH Verlag GmbH & Co. KGaA.

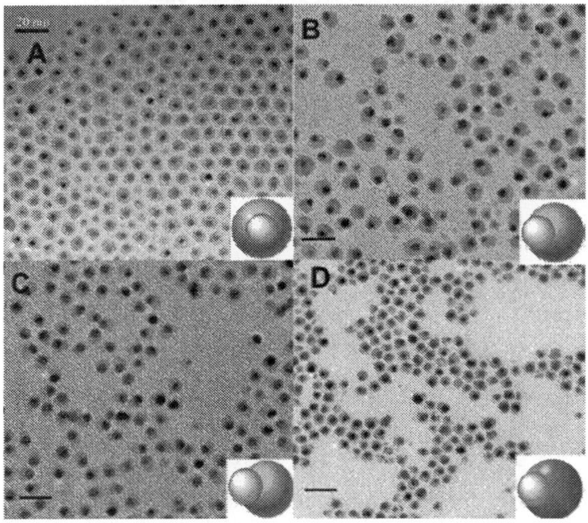

Figure 13. Large-scale TEM images of shape evolutions in Au/CdSe and Au/CdTe hybrid nanostructures with ~5 nm sized Au. (A) Concentric core/shell of Au/CdSe; (B, C) heterodimer of Au/ CdSe; (D) heterodimer of Au/CdTe. The inset diagrams highlight the phase separation-induced Au/CdSe and Au/CdTe morphologies. Scale bar: 20 nm. Adapted from Zhao et al. [32] with permission; copyright Wiley-VCH Verlag GmbH & Co. KGaA.

Based on Figure 5, the metal/semiconductor core/shell NCs can further lead to fine tuning of plasmon–exciton coupling, different hydrogen photocatalytic performance, and enhanced photovoltaic, electrical properties.

2.4. Metal/semiconductor yolk/shell nanocrystals

With a unique structure, metal-semiconductor yolk/shell nanocrystals play an important role in photocatalysis and other fields [56–63]. As photocatalysts, metal-semiconductor yolk/shell nanocrystals process some advantages, such as stronger absorption, higher catalysts selectivity, and higher quantum yield. Nowadays, different kinds of metal-semiconductor yolk/shell nanoarchitecture have been designed to enhance their photocatalysis properties. Following it are several methods for preparing metal-semiconductor yolk/shell nanocrystal.

2.4.1. Template Free for formation of metal/semiconductor yolk–shell nanocrystals

Following this method, metal/semiconductor yolk–shell nanocrystals are synthesized directly. During the process, no any nanostructure is used as template or core–shell nanocrystals forming during the process are taken as self-template [56, 64, 65]. For example, Pt–CeO_2 yolk–shell nanoparticles can be synthesized by hydrothermal method, although the mechanism may involve several factors [64]. Pt nanoparticles and CeO_2 precursors are put into the autoclave and heated at 90ºC in a period. The authors illustrate that the ratio of $CeCl_3$ concentration and

Pt colloid solution is the main factor to formation of yolk–shell nanoparticles. Only can Pt–CeO_2 core–shell nanoparticles be obtained when a larger $CeCl_3$/Pt ratio is employed. With a smaller ratio, the looser shell of Pt–CeO_2 yolk–shell nanoparticles can be obtained. The yolk–shell nanoparticles enhance the properties of photocatalysis of hydride of Pt and CeO_2.

Au–Cu_2O yolk–shell nanocrystals are synthesized by prolonging the reaction time after formation of Au–Cu_2O core–shell nanocrystals [65]. Cu_2O shell grows on the surface of Au nanoparticles to form Au–Cu_2O core–shell nanoparticles. As the reaction time increasing, Au core and Cu_2O shell spare forming yolk–shell nanostructure. The longer the reaction time, the more obvious the yolk–shell nanostructure. Furthermore, the thickness of Cu_2O shell can be tuned with Au colloid/Cu^{2+} ratio. As the authors showed in the articles, the distance between Au core and Cu_2O shell and the thickness of Cu_2O shell have a strong influence on the optical properties of Au–Cu_2O yolk–shell nanoparticles (Figure 14). It means that the method for preparation of Au–Cu_2O with tunable yolk–shell nanoparticle provides a route to tune the optical properties which is of importance for photocatalysis.

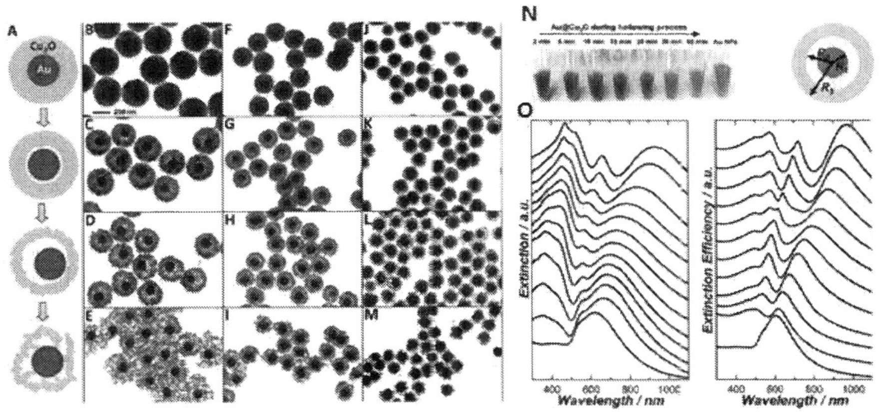

Figure 14. (A) Schematic illustration of formation of Au–Cu_2O yolk–shell nanocrystals. (B–M) TEM images of prepared Au–Cu_2O yolk–shell nanoparticles. (N) Photograph of Au–Cu_2O during hollow process. (O) Their optical properties collected by experiment and simulation. Copyright: American Chemistry Society, 2011.

Co–SiO_2 yolk–shell nanoparticles are obtained by reduction of CoO–SiO_2 core–shell nanoparticles. Park et al. coat silica on CoO nanoparticles to form CoO–SiO_2 nanoparticles then reduce CoO–SiO_2 nanoparticles with hydrogen to form Co–SiO_2 yolk–shell nanoparticles [66].

With template free method, some metal cation is introduced into hollow nanocrystals then reduced forming metal nanoparticles in the cavities of hollow structure [67, 68]. Cu–SiO_2 yolk–shell nanocrystal is synthesized by reducing Cu^{2+} cation in cavity of SiO_2 hollow particle and the size of Cu core can be tuned by multiple reduction cycles (Figure 15A–D) [67]. Besides that, yolk–shell nanocrystals can be employed because they also have cavities for reduction of metal cation. By this method, multiplied cores or cores with different materials can be synthesized. Au–SiO_2 yolk–shell with tunable core is prepared by reducing HAuCl4 in the cavity of SiO_2–SiO_2 yolk–shell nanostructure (Figure 15F–I) [68]. On account of that, metal cation can be introduced into the cavities of hollow nanoparticles or yolk–shell nanoparticles, and other kinds of metal cation can be introduced into the cavities to react with metal core forming a new metal core or alloy core. For instance, Ag–SiO_2 yolk–shell nanoparticles can be obtained by displacing Cu with Ag^+ from Cu–SiO_2 yolk–shell nanoparticles (Figure 15E) [67]. Moreover, because the cavity of yolk–shell can be taken as reactor, the synthesis way can be applied to form other yolk–shell nanostructures.

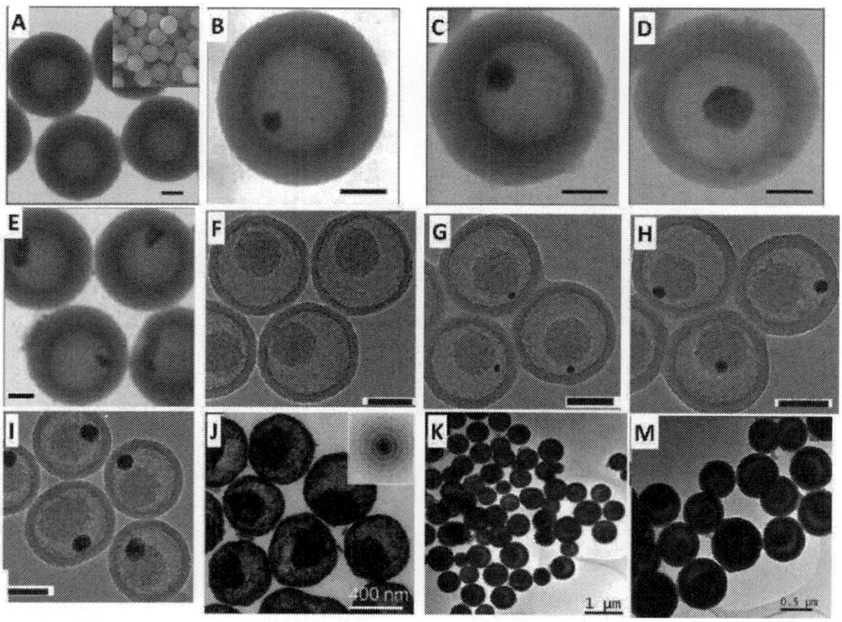

Figure 15. TEM images of various yolk–shell nanoparticles prepared with different way. (A) SiO_2 hollow nanosphere, (B–D) Cu–SiO_2 yolk–shell nanoparticles, and (E) Ag–SiO_2 yolk–shell prepared by replacing the Cu core of Cu–SiO_2 yolk–shell nanoparticles with Ag+; scale bar: 100 nm [67]. Copyright: Royal Society of Chemistry, 2004. (F) SiO_2–SiO_2 yolk–shell nanoparticles, (G-I) Au–SiO_2 yolk–shell nanoparticles with different sizes Au core; scale bar: 50 nm [68]. Copyright: Wiley-VCH, 2010. (J) Au–CeO_2 yolk–shell nanoparticles [69]. Copyright: RSC, 2012. (K-M) Pt–CeO_2 yolk–shell nanoparticles [64]. Copyright: RSC, 2011.

Although some of mechanism of template free of yolk–shell nanostructure formation is unclear, it provides the method to design and prepare metal/semiconductor yolk–shell nanocrystals.

2.4.2. Using template for metal-semiconductor yolk–shell nanocrystals

Template employed for synthesizing metal-semiconductor yolk–shell nanocrystals can be conclude into two kinds of templates: soft template and hard template.

Using soft template for metal-semiconductor yolk–shell nanocrystals: By soft template protocol, several kinds of chemical are usually employed as soft template, such as biomolecular, polymers, surfactants, or microemulsions [56, 70]. In most of case, the large molecular would form microemulsions in solution and precursors react on the boundary of microemulsions. Therefore, sizes or morphologies of yolk–shell nanoparticles could be tuned by tuning cell of microemulsions [71–73]. For metal semiconductor yolk–shell nanoparticles, metal core is reduced in the cavities of microemulsions then semiconductor shell grows on the boundary of microemulsions. For instance, Priebe and Fromm synthesize the Ag–SiO_2 yolk–shell nanoparticles in cyclohexane and Igepal CO-520 system [72]. Ag^+ is introduced into cavity of microemulsions and reduced to nanoparticles. Subsequently, silica precursor is introduced into the system and form the shell on the boundary of microemulsions (Figure 16). Besides, because Ag^+ is reduced in the cavities, the size of Ag core can be tuned by using the series of concentration of $AgNO_4$ solution.

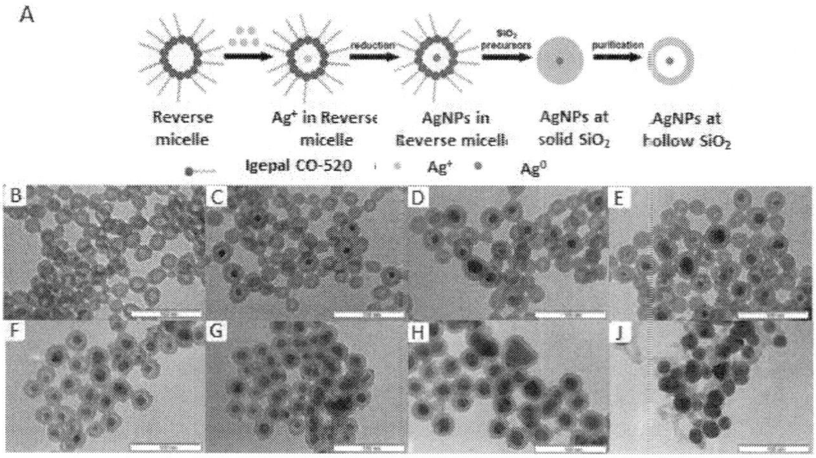

Figure 16. (A) Schematic for formation of Ag–SiO_2 yolk–shell nanoparticles and (B–I) TEM images of Ag–SiO_2 yolk–shell nanoparticles with various Ag core size; scale bar: 500 nm, Copyright: Wiley-VCH, 2014.

Metal cores can also be introduced the cavities of microemulsions and then the shell grows to form metal/semiconductor yolk–shell nanoparticles directly [73]. Hyunjoon Song and his coworkers introduce Pd nanoparticles into the system of hexanol and igepal CO-630, followed by adding tetraethyl orthosilicate (TMOS) and octadecyltrimethoxysilane (C18TMS) for shell growth (Figure 17). Their protocol provides a possible method for preparing metal/semiconductor yolk–shell nanoparticles with other metal core and semiconductor shell.

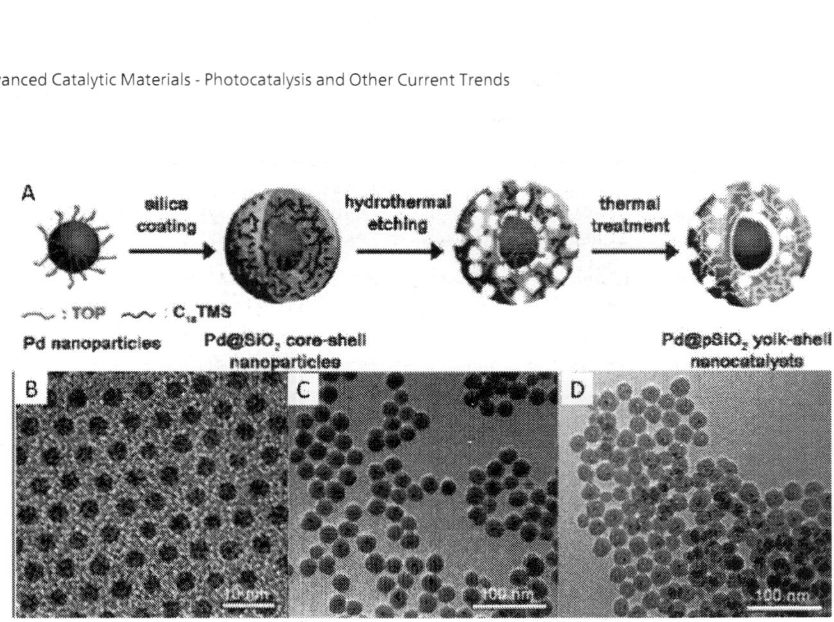

Figure 17. (A)Schematic for preparing Pb–SiO_2 yolk–shell nanoparticles and TEM images of Pb cores (B), Pb–SiO_2 core–shell nanoparticles (C), and Pb–SiO_2 yolk–shell nanoparticles(D). Copyright: ACS, 2011.

Using hard template for metal/semiconductor yolk–shell nanostructure: By hard template method, shell coating on core is synthesized as middle layer and subsequently removed or etched. The material of middle shell includes SiO_2, carbon, polymer, and so on. For example, Au–TiO_2, Au–ZrO_2, Au–SnO_2, and Au–SiO_2 yolk–shell nanoparticles are synthesized by coating semiconductor shell on Au–oxides and subsequently etching oxides layer (Figure 18) [74–77]. In most of case, SiO_2 is employed as hard template to form the middle layer and etched by NaOH, HF, or other reagents.

For example, Au–ZrO_2 yolk–shell nanoparticles can also be synthesized by etching SiO_2 layer of Au–SiO_2–ZrO_2 core–shell nanoparticles (Figure 18A) [74]. It is reported that such Au–ZrO_2 yolk–shell nanostructure is stable at high temperature and can be used as catalyst for the oxidation of CO. Taking SiO_2 as template, SnO_2 hollow nanosphere, and Au–SnO_2 yolk–shell nanoparticles can be obtained (Figure 18B). Lou et al. show their work on preparation of Au–SnO_2 yolk–shell nanoparticles, etching the SiO_2 layer with HF and they found that controlling the size of SiO_2 template cage-like and double layer shell of Au–SnO_2 can be obtained [75]. The method can also be used for synthesis of yolk–shell with other hybrid material shell [78].

3. The photocatalytic hydrogen evolution applications of metal/semiconductor hybrid nanocrystals

3.1. Mechanism of photocatalysis on semiconductors and key problems

Since Fujishima and Honda [79] discovered photoelectrochemical formation of H_2 over TiO_2 electrode, photocatalytic water splitting and H_2 evolution are an attractive solution of global

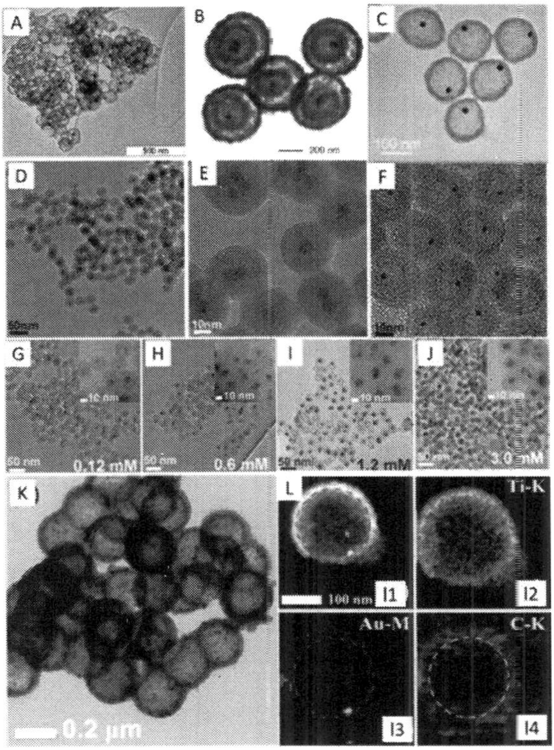

Figure 18. TEM images of (A) Au–ZrO_2 yolk–shell nanoparticles [74]. Copyright: Wiley-VCH, 2006. (B) Au–SnO_2, yolk–shell nanoparticles [75]. Copyright: Wiley-VCH, 2007. (C) Au–Zr0.5Si0.5O_2 yolk–shell nanoparticles [76]. Copyright: Springer, 2014. (D, E) Fe3O_4/Au–SiO_2 yolk–shell nanoparticles. (F) Au–SiO_2 yolk–shell nanoparticles. (G–J) Ag–SiO_2 yolk–shell nanoparticles [77]. Copyright: Royal Society of Chemistry, 2010. (K, L) TEM and HAADF-STEM images and EDS maps of Au–GO/TiO_2 yolk–shell nanoparticles [78]. Copyright: America Chemistry Society.

energy supply and related environmental issues. Numerous researchers had extensively studied water splitting using semiconductor photocatalysts since the finding [80–87]. The fundamental principle of solar water splitting for semiconductor photocatalysts is shown in Figure 19. Semiconductor materials have a band structure in which the conduction band (CB) is separated from the valence band (VB) by a band gap with a suitable width. When the energy of incident light is larger than that of a band gap, the electrons in the VB of the semiconductor photocatalyst are excited to the CB, while the holes are left in the VB. Therefore, it creates the negative-electron (e-) and the positive-hole (h+) pairs. These photogenerated electron–hole pairs may further be involved in the following possible processes: (1) successfully migrate to the surface of semiconductor, (2) be captured by the defect sites in bulk and/or on the surface region of semiconductor, and (3) recombine to release the energy in the form of heat or photon.

If the photoexcited carriers separate and migrate to the surface without recombination, adsorbed species are reduced and oxidized by the photogenerated electrons and holes to produce H_2 and O_2, respectively (Figure 19).

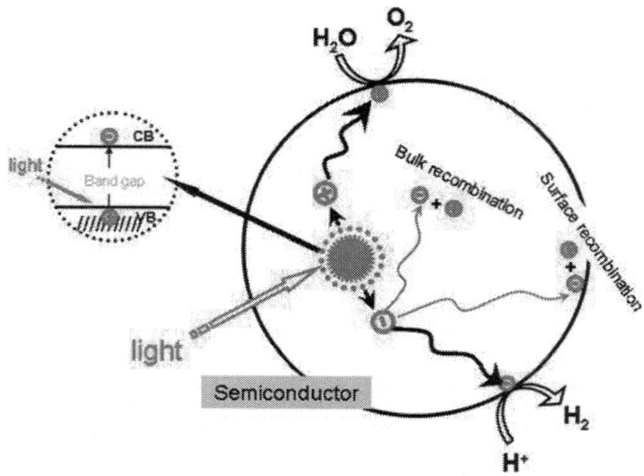

Figure 19. Process in photocatalytic water splitting [80]. Copyright: Royal Society of Chemistry, 2009.

For effective water splitting, important points in the semiconductor photocatalyst materials are their band structure, including the band gap and positions of VB and CB. The bottom level of the CB has to be more negative than the redox potential of H^+/H_2 (0 V vs. NHE), while the top level of the VB be more positive than the redox potential of O_2/H_2O (1.23 V) [80]. Therefore, a minimum band gap of 1.23 eV is required, while a much larger band gap (usually >2.0 eV) is often needed for appreciable water splitting reaction due to additional overpotential associated with each electron transfer and gas evolution step [84]. The band edge positions of some semiconductor photocatalysts are shown in Figure 20. The band structure of a semiconductor is not only determined by its own crystal phase and vacancies but can also be modified by the introduction of foreign elements into the bulk or surface of the semiconductor. Take the most studied photocatalyst TiO_2 as an example, the larger band gap (~3.2 eV) of TiO_2 limits its utilization of the solar spectrum to only the ultraviolet (UV) region (wavelength $\lambda < 400$ nm) [85]. The solar spectrum has a very small fraction of UV light (ca. 5%) in comparison with those of visible light ($400 < \lambda < 800$ nm, ca. 43%) and near-infrared (NIR) light ($800 < \lambda < 2500$ nm, ca. 52%) [83]; therefore, studies on visible-light-driven photocatalysts are more important for practical applications. There are two efficient strategies to make TiO_2 as visible-light-driven photocatalysts as shown in Figure 2B [84]: (1) narrow the band gap of TiO_2 to make it absorb visible light by introducing other elements into TiO_2 and (2) modify the TiO_2 surface with other visible light active materials (dye or quantum dot) as a light harvester to sensitize TiO_2.

http://dx.doi.org/10.5772/61888

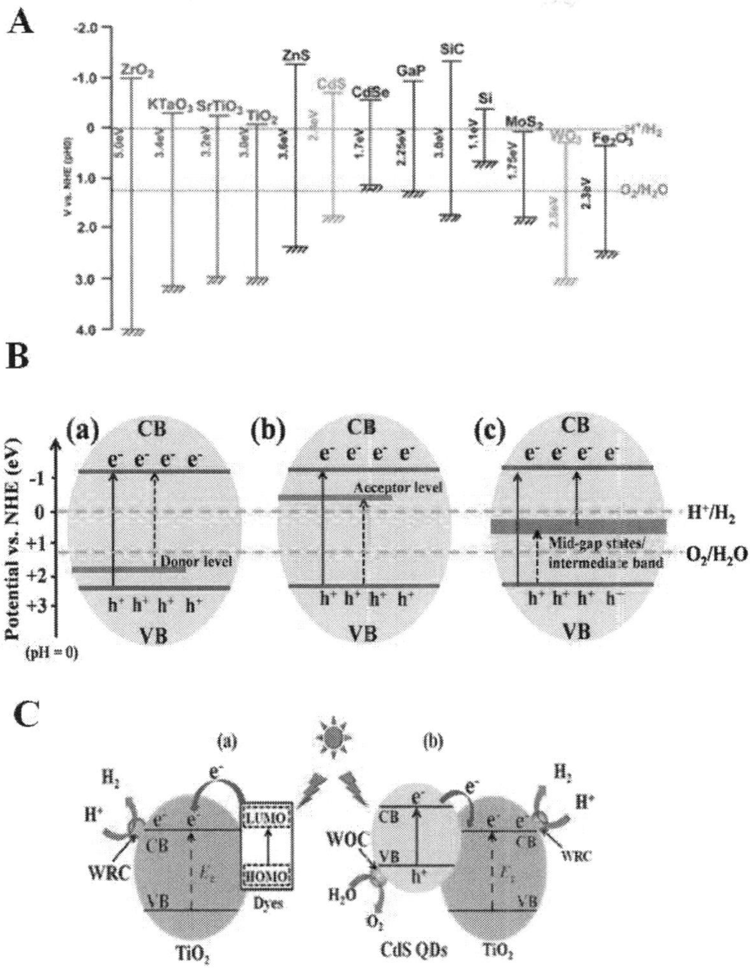

Figure 20. (A) Relationship between band structure of semiconductor and redox potential of water splitting [80]. Copyright: Royal Society of Chemistry, 2009. (B) Donor level (a), acceptor level (b), and mid-gap states (c) formed by metal ion doping. (C) The mechanism of H2 production for (a) dye-sensitized and (b) QDs-sensitized semiconductor [82]. Copyright: Royal Society of Chemistry, 2015.

After excited charges are created, efficient charge separation is next crucial factor determining the light to fuels conversion efficiency. Crystal structure, crystallinity, and particle size strongly affect the step [86]. The defects operate as trapping and recombination centers between photogenerated electrons and holes, resulting poor photocatalytic activity. Therefore, improving the crystalline quality will decrease the amount of defects and improve photocatalytic activity. If the particle size becomes small, the distance that photogenerated electrons and holes have to migrate to reaction sites on the surface becomes short, and this results in a decrease in the recombination probability. Also, the fabrication of junction structure has been recognized as an effective strategy to avoid charge recombination in semiconductors.

The surface catalytic reaction is a successive step of charge separation. The important points for this step are surface character (active sites) and quantity (surface area). Even if the photogenerated electrons and holes possess thermodynamically sufficient potentials for water splitting, they will have to recombine with each other if the active sites for redox reactions do not exist on the surface. Cocatalysts such as Pt, NiO, and RuO_2 are usually loaded to introduce active sites for H2 evolution [87], which will be discussed later.

The processes of the photocatalytic reaction on a semiconductor photocatalyst involve light absorption, charge separation, carrier migration, and surface catalytic reactions. Therefore, developing band gap engineering to narrow down the band gap of semiconductor materials for absorbing broader spectrum of solar energy and materials engineering to tune the physical properties (crystal structure, crystallinity, and particle size) for gaining efficient charge separation and migration and creating enough active sites are the key problems of improving photocatalytic hydrogen evolution efficiencies.

3.2. Metal/semiconductor heterodimer nanocrystals: the role of metal

Integration with metal is a commonly used configuration to improve the photocatalytic hydrogen evolution performance of a semiconductor. The metal may play a variety of roles in the enhancement of photocatalytic performance. In the following sections, we will focus on the role of metal as cocatalyst and the plasmonic effect of noble metals.

3.2.1. The cocatalyst role

Since the work by Kraeutler and Bard in 1978 loading Pt on the surface of TiO_2 [88], the loading of metal nanoparticles onto different semiconductor photocatalysts has been regarded as a popular strategy to improve the photocatalytic performance in photocatalytic water splitting. Besides Pt, other metal cocatalysts, including Pd, Rh, Ru, Ir, Ag, Ni, Co, etc., have been recognized as efficient cocatalysts for photocatalytic hydrogen evolution [89–96]. The processes of charge transfer between metal cocatalyst and host semiconductor photocatalyst are described in Figure 21A.

The metal cocatalysts mainly play two roles in the improvement of photocatalytic performance. One is to assist in electron–hole separation through the formation of Schottky barrier between the metal cocatalyst and the light-harvesting semiconductor. A Schottky barrier is a type of junction resulting from the intimate contact of a metal with a semiconductor (Figure 21B). The metal cocatalyst with larger work function, that is, lower Fermi energy level, should more readily trap the photogenerated electrons migrated to the surface of the host semiconductor photocatalyst. Meanwhile, the photogenerated holes stay at the host photocatalyst and

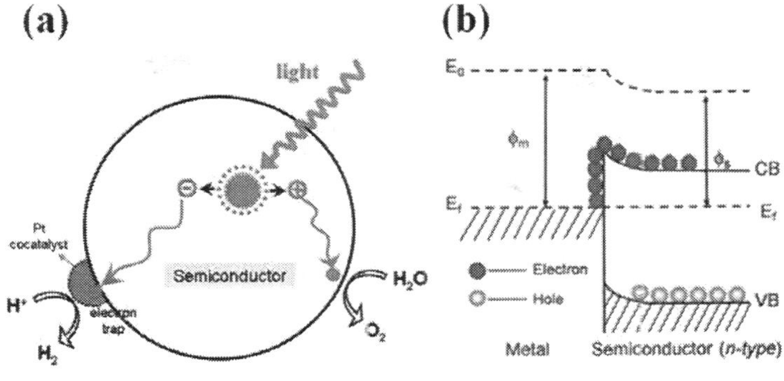

(c)

$$2e^- + 2H^+ \xrightarrow{①} 2[H_{ad}] \xrightarrow{②} H_2$$

Figure 21. (A) Processes of charge transfer between host photocatalyst and cocatalyst, taking Pt as the example of cocatalyst [86]. Copyright: America Chemistry Society, 2010 (B) Schematic of Schottky barrier [90]. Copyright: Royal Society of Chemistry, 2010. (C) Two general steps for proton reduction reaction [89]. Copyright: America Chemistry Society, 2013.

migrate to its surface. This retards the possibility of electron–hole recombination and enhances efficient separation of the photogenerated electrons and holes. It improved the overall photocatalytic activity of the water splitting because it helps to promote charge separation, which in return reduces both bulk and surface electron/hole recombination.

The other role of metal cocatalysts is to serve as the reaction sites to catalyze the proton reduction to H_2. Proton reduction course on cocatalysts goes through at least two steps: a discharge step and catalytic step (Figure 21C) [89]. Most semiconductor photocatalyst, particularly oxides, lack surface H_2 evolution sites rather than O_2 evolution sites since their CB levels are not sufficiently negative for electrons to reduce water to produce H_2 without catalytic assistance whereas their VB levels are usually positive enough for holes to oxidize water and form O_2 even in the absence of cocatalyst [92]. Metal cocatalysts could lower the activation energy or overpotential for H_2 evolution reaction on the surface of semiconductors. It also accelerates the surface chemical reaction by inhibiting the backward reaction.

Overall, the role played by the metal cocatalysts dispersed on the surface of the semiconductor photocatalysts is extremely important. Many factors can affect the capability of H_2 evolution cocatalysts in the semiconductor-based photocatalytic water splitting, such as cocatalyst loading amount and its particle size and structure [89, 94]. For example, there is a volcano-type trend between the loading amount of a given cocatalyst and the photocatalytic activity (regardless of the synthesis method, photocatalyst type, and loaded cocatalyst) [94, 95]. At the same loading amount, metal cocatalysts with smaller size and high dispersion display higher catalytic activity due to their larger specific surface area and more active sites [96].

3.2.2. The plasmonic metal case: plasmonic-metal/semiconductor heterodimer nanocrystals

The harvesting and conversion of solar energy become a renewed in recent years. Among various technologies, the direct conversion of solar to chemical energy using photocatalysts has received significant attention. Although heterogeneous photocatalysts are almost exclusively semiconductors, it has been demonstrated recently that plasmonic nanostructures of noble metals (mainly silver and gold) also show significant promise [1]. Herein, the recent progress in using plasmonic metallic nanostructures in the field of photocatalysis is reviewed. We focus on plasmon-enhanced water splitting on composite photocatalysts containing semiconductor and plasmonic-metal building blocks and recently reported plasmon-mediated photocatalytic reactions on plasmonic nanostructures of noble metals.

Plasmonic metallic nanostructures are characterized by their strong interaction with resonant photons through an excitation of surface plasmon resonance (SPR). SPR can be described as the resonant photon-induced collective oscillation of valence electrons, established when the frequency of photons matches the natural frequency of surface electrons oscillating against the restoring force of positive nuclei (Figure 22A). The resonant photon wavelength is different for different metals. For example, gold, silver, and copper nanostructures exhibit resonant behavior when interacting with ultraviolet (UV) and visible (Vis) photons. The utilization of the localized SPR (LSPR) effect of nanostructured Au, Ag, and Cu was examined for the potential photocatalysis [97–99], using Au or Ag nanoparticles (NPs) supported on ZrO_2, AgCl, or TiO_2 for the unselective degradation of organic species under visible light irradiation. As shown in Figure 22 A, LSPR is the resonant photon-induced coherent oscillation of charge at the metal-dielectric interface, established when the photon frequency matches the natural frequency of metal surface electrons oscillating against the restoring force of their positive nuclei [100]. The frequency of the surface plasmon absorption is highly dependent on the type of metal, size, shape, surrounding dielectric medium, distance between neighboring objects, and configuration their ensemble [101]. A wide range of metal/semiconductor heterostructures, including Au/TiO_2, Ag/TiO_2, Au/CdS, and Au/Fe_2O_3, have been explored to achieve enhanced photocatalytic activity [102–110].

Although the exact nature of LSPR effect on enhanced photocatalytic activity is not entirely understood, three possible enhancement mechanisms have been proposed: (1) near-field enhancement, (2) SPR-induced electron transfer from metal to semiconductor, and (3) scattering [103, 104] (Figure 22A–C). The strong SPR-induced electric field of plasmonic metal NPs can interact with the adjacent semiconductor (Figure 22B-a), this interaction may increase the rate of exciton formation and the concentration of the charge carriers generated in this part of the semiconductor [104]. If only the metal excited, the metallic plasmonic NPs can absorb resonant photons and transfer energetic photogenerated charge carrier to the semiconductor during the decay of the LSPR (Figure 22B-b). As a result of the plasmonic sensitization process, a wide band gap semiconductor could perform catalytic reduction reactions under visible light. Plasmonic structures of size larger than 50 nm are efficient in scattering the resonant photons, which increases the path length of photons in semiconductor/plasmonic metal nanostructures. Therefore, the resonant photons that are not absorbed by semiconductor photocatalysts could be scattered by the bigger plasmonic metal particles, ultimately increasing the number of

http://dx.doi.org/10.5772/61888

electron/hole pairs. The above-mentioned three mechanisms are governed by the metal/semiconductor configurations and their arrangements in the hybrid system.

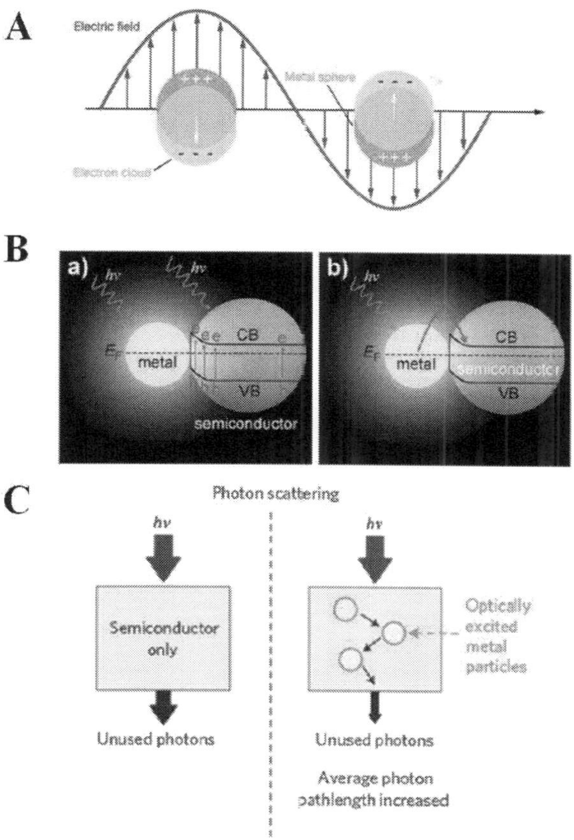

Figure 22. Schematic illustrating (A) localized surface plasmon resonance [102]. Copyright: America Chemistry Society, 2011. (B) The mechanisms for plasmon-enhanced chemical reactions with metal/semiconductor hybrid nanostructures: (a) plasmonic enhancement of light absorption; (b) hot-electron effect [103], with permission. Copyright Wiley-VCH Verlag GmbH & Co. KGaA. (C) The scattering mechanism [104]. Copyright: Royal Society of Chemistry, 2013.

In the case of SPR-mediated charge injection from metal to semiconductor, charge carriers are directly injected from excited plasmonic-metal nanostructures into the semiconductor surface. The metallic plasmonic nanoparticles essentially act as a sensitizer, absorbing resonant photons and transferring the energetic electron, formed in the process of the SPR excitation, to the nearby semiconductor (Figure 23) [1, 111]. Furthermore, the ability to tune the SPR resonance wavelength and intensity by changing the size or shape of Au or Ag nanostructures suggests that plasmon enhancement could be more dramatic when the interface between Au (Ag) and semiconductor, such as TiO_2, CdS, etc. The charge injection mechanism was functional in composite photocatalysts where the plasmonic nanoparticles and semiconductor are in direct contact with each other, allowing a rapid transfer of charge carriers. These composite systems are geometrically similar to the conventional cocatalyst/semiconductor photocatalysts that are often synthesized by an incipient wetness deposition of metal precursors and their subsequent thermal treatment on a semiconductor surface. Therefore, the defect in the heterointerface should be decreased to get more efficient charge separation and injection to be "chemically useful" energetic charge carriers for photoreduction. Otherwise, too many defects on the interface will trap the photoinduced charge.

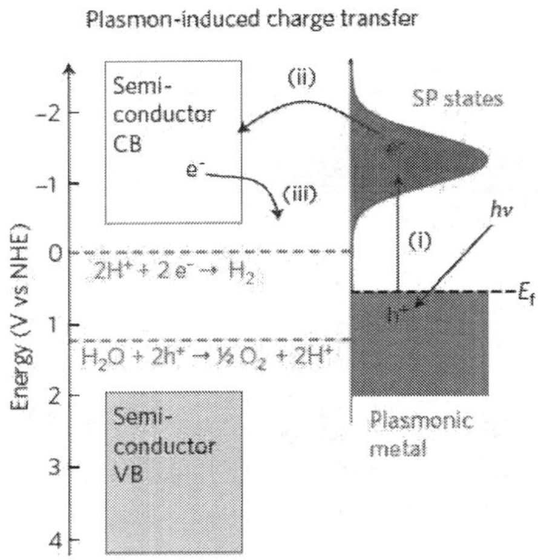

Figure 23. Mechanism of SPR-induced charge transfer with approximate energy levels on the NHE scale. Dashed red lines refer to the water-splitting redox potentials. (i) Electrons near the metal Fermi level, E_f, are excited to surface plasmon (SP) states; (ii) the electrons transfer to a nearby semiconductor particle; (iii) this activates electron-driven processes such as the hydrogen evolution half-reaction [1]. Copyright of Nature group 2011, Macmillan Publishers Limited.

3.2.3. The cocatalystic metal on semiconductor nanocrystals with well-defined crystal face exposure

Based on the photoreduction reaction of Pt^{4+} (Pt^{4+} + e− → Pt) and photooxidation reaction of Pb^{2+} ($Pb^{2+} + H_2O + h^+ \rightarrow PbO_2$), it is clearly indicated that rutile {110} and {011} facets provide reduction and oxidation sites, respectively (Figure 24Aa, c) [112–113]. The obvious separation of reduction and oxidation sites on faceted rutile crystals is attributed to photoexcited electron and hole transfer between {011} and {110} facets, which is driven by the higher electronic energy levels of {011}. Although the selective distribution of photodeposited Pt and PbO_2 particles on anatase {101} and {001} facets (Figure 24 Ab, d) is not as obvious as that on different rutile facets, anatase {101} and {001} facets can still be considered to be more reductive and more oxidative, respectively [113].

The spatial separation of reduction and the oxidation sites in TiO_2 crystals with appropriate different facets are well known now [114]. Consequently, the corresponding spatial separation of photoexcited charge carriers on different facets improves the photocatalytic activities of photocatalysts, as illustrated in Figure 24B and C. Liu et al. reported that anatase crystals with {101} and {001} facets, with selective deposition of Pt particles by a photoreduction process only on the reductive {101} facets show a much higher photocatalytic hydrogen evolution (reduction reaction) from a mixture of H_2O and methanol than do the same crystals with Pt particles on both {101} and {001} facets (Figure 24D). It is now clear that the difference in spatial distribution of cocatalysts on photocatalysts is an important but unfortunately neglected factor. The spatial separation of photoexcited electrons and holes on different facets was also seen in brookite nanosheets surrounded with four {210} and two {101} facts as reduction sites, and two {201} facets as oxidation sites. Consequently, such spatial separation produces an excellent photocatalytic activity of brookite, which is conventionally inactive [113–116].

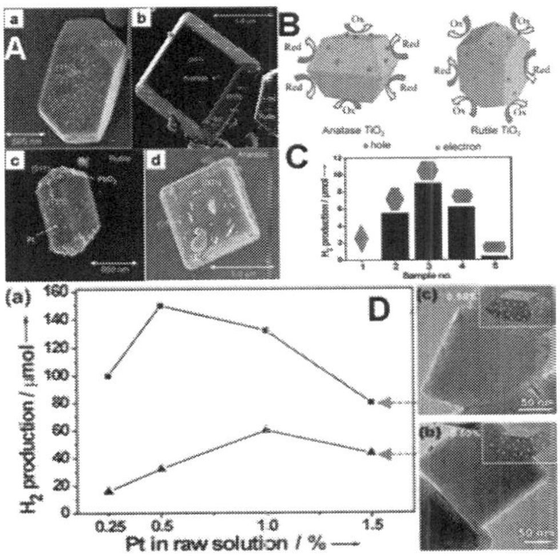

Figure 24. .(A) SEM images of (a) a rutile particle and (b) an anatase particle on which Pt fine particles were deposited by UV-irradiation in a solution of 1.0 mM H_2PtCl_6; SEM images of (c) a rutile particle and (d) an anatase particle showing PbO_2 deposits, which were loaded on the particles by UV irradiation of the Pt-deposited TiO_2 powder in a solution of 0.1 M $Pb(NO_3)_2$. Reprinted with permission from Ohno et al. [113]. Copyright 2002 Royal Society of Chemistry. (B) Schematic of the spatial separation of redox sites on anatase crystals with {101} and {001} facets and rutile TiO_2 particle with {110} and {011} facets. Reprinted with permission from Murakami et al. [115]. Copyright 2009 American Chemical Society. (C) Hydrogen production after 6 h irradiation time using TiO_2 nanocrystals (NCs) with different exposed {001} facets as photocatalysts. Reprinted with permission from Liu et al. [116]. Copyright 2013 Wiley-VCH. (D) (a) Hydrogen production amounts of Pt deposited TiO_2 NCs by means of photochemical-reduction (■) and chemical reduction routes (▲). SEM images of TiO_2 NCs with 0.5% Pt loading amount prepared by (b) chemical reduction and (c) photochemical reduction, respectively. Reprinted with permission from Liu et al. [116]. Copyright 2013 Wiley-VCH.

3.3. Metal–semiconductor core/shell nanocrystals and photocatalysis application

3.3.1. Metal core–semiconductor shell: large lattice mismatch-induced interface control and photocatalytic applications

In metal-semiconductor hetero structures, the contact interface between metal and semiconductor plays a critical role in transfer and separation of charge carriers [117–118]. Our precise symmetry control of Au–CdX can enable different contact interface between metal and semiconductor components at the nanoscale, thus offering a new dimension to control optical and electronic properties of hybrid nanostructures in a highly control manner. Figure 25A highlights evolution of plasmon–exciton coupling in Au–CdS enabled by their symmetry control. As compared with the SPR feature of a pure metal nanoparticles, optical absorption of Au–CdS is significantly broadened in the visible regime (500–700 nm), which can be attributed to the existence of plasmon–exciton coupling [119, 3]. They further performed finite-different time-domain (FDTD) simulation to evaluate optical response in Au–CdS with different structural symmetry and summarize the results in Figure 25B. The FDTD simulation reproduces symmetry dependent optical property very well. The strong SPR-induced visible light absorption of Au/CdS colloids can overlap most of the solar spectrum, and particularly important, its tunability makes it feasible to optimize photorelated processes, including solar water splitting.

It is well known that CdS or CdSe possesses conduction and valence bands at potentials appropriate for water reduction, namely, the bottoms of conduction bands locate at a more negative potential than the reduction potential of H^+ to H_2. In order to get high efficiency of H_2 evolution by sunlight irradiation, more visible light should be harvested, and then photo-excited electron/hole should be separated and migrated to the surface without recombination [120, 121]. By integrating with plasmonic Au nanoparticle in a hybrid nanostructure, its tunable optical response achieved in Figure 25A and B should make it feasible to mediate the visible light-induced electron/hole separation in CdX and electron transfer from CdX to Au surface due to the plasmon–exciton coupling [3,122]. As shown in Figure 25C and D, for comparison, photocatalysis with pure 4.5 nm CdS QDs and the CdS/Au heterodimers prepared by reported in situ deposition are also presented [123–124]. The rate of CdS QDs alone is 0.1 mmol h^{-1} g^{-1}.The concentric core/shell Au–CdS has the least H_2 evolution activity (0.009 mmol h^{-1} g^{-1}) and the heterodimer structured Au/CdS has better H_2 evolution activity (7.3 mmol h^{-1} g^{-1}) that manifest a dramatic photoactivity enhancement of 730 times. However, the CdS/Au heterodimers has less H_2 evolution activity (0.359 mmol h^{-1} g^{-1}). Figure 25C and D can be understood by the mechanism of SPR enhanced electron–hole separation and collection, as illustrated in the Figure 25E. For an asymmetric Au–CdS heterodimer, under visible light illumination, the energy band alignment between CdS and Au suggests a rapid electron transfer from conduction band of CdS to Au nanoparticle. In the meantime, concentrated electric field due to SPR of metal constituent can significantly enhance light absorption of CdS semiconductor and promote charge separation near the Au–CdS interface. Such synergetic effect enriches the electron (e-) on the Au tip for the efficient water reduction and leads to high efficient H_2 evolution activity. The bad H_2 evolution activity of the CdS/Au heterodimers mainly attributes to the weak SPR and consequent visible light harvest of smaller sized Au (2–3 nm). On the other hand, in a concentric Au–CdS core–shell nanostructure with high structure symmetry, the electrons are retained in the Au core after photo excitation and charge separation and cannot participate in photocatalytic reaction; thus, the photon absorption of CdS is suppressed. As a result, almost no H_2 evolution is observed with concentric Au–CdS.

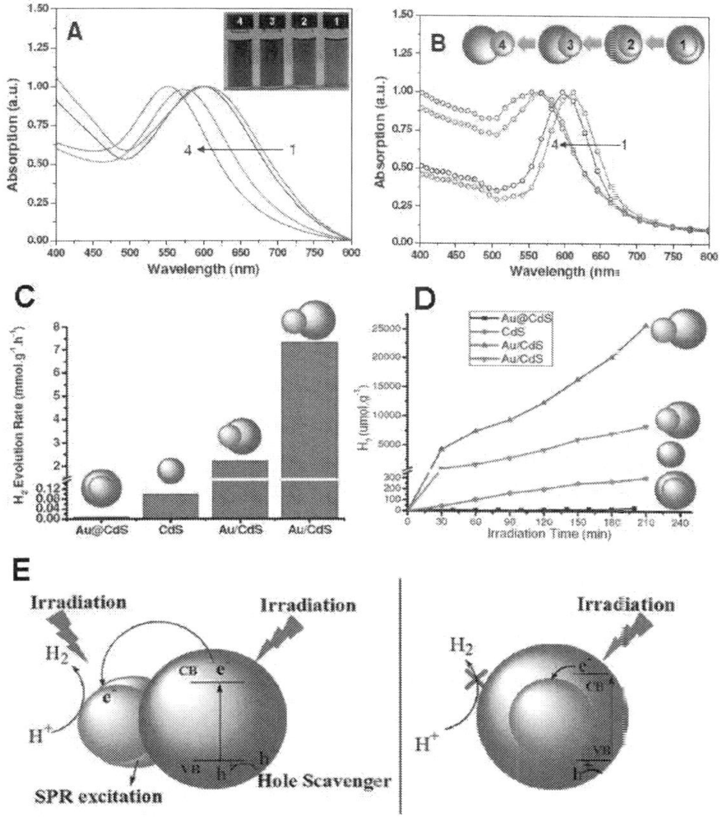

Figure 25. (A) Evolution of experimental UV-Vis absorption spectra with symmetry of Au/CdS colloid. (B) The normalized FDTD simulation of the SPR in these Au/CdS heteronanostructures. (C) Comparison of H_2 evolution activities of different Au–CdS photocatalysts. (D) Time evolution of photocatalytic generation of H_2 evolution amount versus irradiation time for them. (E) Schematic illustrations of visible light-induced electron/hole separation and transfer difference in heterodimer and core/shell Au/CdS when dispersed in aqueous solution containing sacrificial reagents. From Zhao et al. [32] with permission; copyright Wiley-VCH Verlag GmbH & Co. KGaA.

In conclusion, based on the series of metal-semiconductor hybrid nanostructures from concentric core–shell to nonconcentric heterodimer with precisely controlled separation between metal and semiconductor constituents, the gradual symmetry evolution has led to novel control of optical response (plasmon–exciton coupling) and photocatalytic activity in hybrid nanostructures, which highlight the importance of nanoscale interface control for both fundamental understanding and technology applications. Thus, it can allow us to evaluate a possible upper limit of a photocatalytic reaction and guide the design toward high efficient hybrid nanostructures. Due to the plasmon intensity difference from different sized Au NPs, the larger sized Au, such as tens of nanometers to hundreds of nanometers size, the plasmon enhancement for photocatalysis may become strong.

4. The photoreduction of CO_2 applications of metal/semiconductor hybrid nanocrystals

Photoreduction of CO_2 is another important application of metal/semiconductor photocatalyst, and it directly converses CO_2 into organic compound such methane, methanol, formalin, formic acid, and so on [125–130]. Being similar with photocatalysis for hydrogen evolution, metal/semiconductor photocatalysts for photoreduction of CO_2 need some potential to meet the photocatalysis reaction (Schemes 1–5) [125],

$$H_2O \rightarrow H_2 + 1/2\, O_2, \Delta E = 1.23V \tag{1}$$

$$CO_2 + H_2O \rightarrow HCOOH + 1/2\, O_2, \Delta E = 1.40\ V \tag{2}$$

$$CO_2 + H_2O \rightarrow HCOH + O_2, \Delta E = 1.34\ V \tag{3}$$

$$CO_2 + H_2O \rightarrow CH_3OH + 3/2\, O_2, \Delta E = 1.21\ V \tag{4}$$

$$CO_2 + H_2O \rightarrow CH4 + 2\, O_2, \Delta E = 1.06\ V \tag{5}$$

and two pathways of photoreduction of CO_2: formalin pathway and carbine pathway are proposed to explain the mechanism of photoreduction of CO_2. According to these rules, various photocatalysts have been synthesized for conversing CO_2 into fuel and modified for higher conversion efficiency. What is needed to point out here is that the reduction potential of CO_2 depends on the reaction conditions, such as pH, and as the reduction is carried out, a series of reactions occur simultaneously. Therefore, as the reaction conditions vary, the main product of photoreduction of CO_2 is different.

Nowadays, several series of semiconductor have been used for photoreduction of CO_2, such as oxides [126, 131–133], sulfides [134], phosphide [135–136], and so on. In most case, metal will be used as cocatalysts to enhance photoreduction efficiency. Habisreutinger and his coworker have concluded photoreduction of CO_2 using semiconductors (Table 1) and in most of the cases, metal would be used as cocatalyst [132]. However, as using different metal as cocatalysts, the main products of photoreduction of CO_2 are different. It hints that metal nanocrystals play an important role in the photocatalysis system. It provides impossibility for tuning the photoreduction of CO_2 by doping metal cocatalysts.

As a kind of photocatalysts, TiO_2 is one of the most popular semiconductors for photoreduction of CO_2 [130]. For example, Li and her coworkers dope Cu onto TiO_2 nanocrystals supported with mesoporous silica and enhance the photoreduction of CO_2 (Figure 26A–D) [137]. More complex structure of photocatalysts has also been synthesized for photoreduction of CO_2. Wang and his coworkers prepared the CdSe/Pt/TiO_2 and monitor decrement of CO_2 under irradiation of visible light (wavelength > 420 nm). They found that TiO_2/Pt combining with CdSe platform a higher activity for photoreduction of CO_2 (Figure 26E and F) [133].

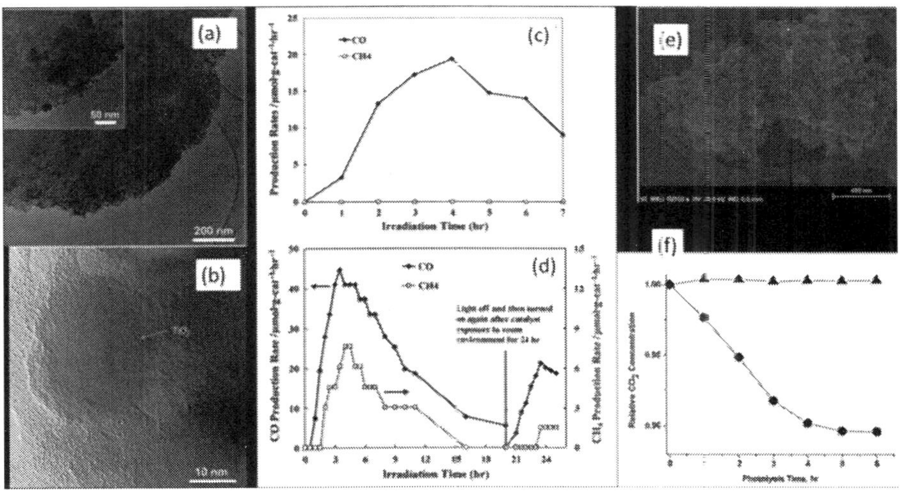

Figure 26. Cu/TiO_2 nanocrystals TEM (A) and HRTEM (B) images and its catalysis properties of photoreduction of CO_2 (C and D). Copyright: Elsevier, 2010. CdSe/Pt/TiO_2 nanocrystals and its photoreduction of CO_2 (E and F). Copyright: America Chemistry Society, 2010.

Beside TiO_2 nanocrystals photocatalyst, sulfides such as CdS nanocrystals and ZnS nanocrystals also show excellent properties for the photoreduction of CO_2. It is reported that using ZnS nanocrystals as photocatalyst, the rate of formic acid production is up to 7000 μmol h^{-1} g^{-1} [132, 134]. What's more, the shape of nanocrystals and other factors on photoreduction of CO_2 attract lots of attention and photoreduction of CO_2 (Figure 27) [138].

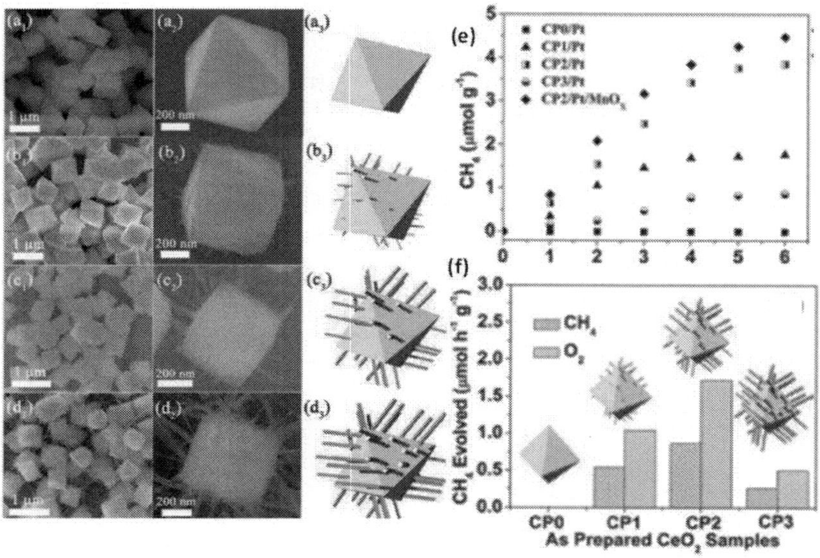

Figure 27. SEM images and schematic of CeO_2 modified with HPO_4 ions (A–D) and CH_4 evolution from photoreduction of CO_2 (E and F). Copyright: America Chemistry Society, 2015.

5. Some other recent photosynthesis applications of metal/semiconductor hybrid nanocrystals

Beside hydrogen evolution and photoreduction of CO_2 form photocatalysis, metal/semiconductor hybrid nanocrystals recently are used for photosynthesis applications. Metal or noble metal is usually employed as catalysts in organic synthesis [139]. Moreover, noble metal has plasmonic effect and induces the organic reaction under visible light and enhances the conversion efficiency [140]. Combining with semiconductor nanocrystals, metal/semiconductor hybrid nanocrystals have potential advantages in organic photosynthesis: higher conversion efficiency, higher selectivity, and lower by-products [141–143].

For instance, Au/CeO_2 is employed for oxidation of Alcohol forming aldehyde or ketone under irradiation of visible light [144]. The reaction rate depends on the surface area of Au nanoparticles, the power of irradiation and the function group of organic compound (Figure 28A–C), and a high selectivity of oxidation is also showed in such hybrid nanocrystals (Figure 28D–F). Beside, metal/semiconductor hybrid nanocrystals, such as Au/ZrO_2, can be used for other organic reaction: reduction of nitroaromatic compound, Suzuki reaction, ester reaction. and so on [140, 142–143].

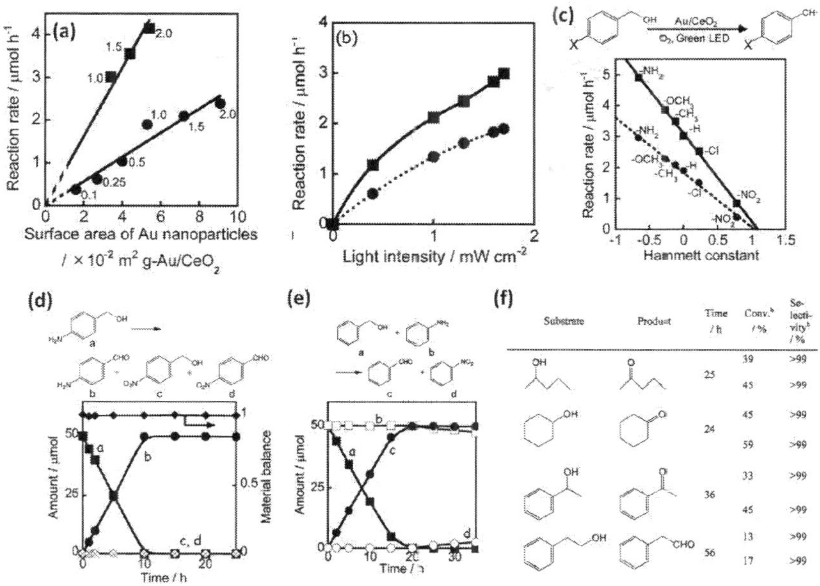

Figure 28. Reaction rate depending on surface area of Au nanoparticles (A), light intensity (B), and substituted functional groups. Copyright: America Chemistry Society, 2012.

In summary, metal/semiconductor photocatalysts have shown their potential application on hydrogen evolution, photoreduction for conversing CO_2 into fuel and organic photosynthesis. Although there is a long way before the metal/semiconductor hybrid nanocrystals being implemented in real-life application, more photocatalysts with different components, special architectures, or structures would be designed for the real-life application

6. Summary and outlook

Metal/semiconductor hybrid NCs did have advantages on integration of functionalities of noble metal and semiconductors based on well-controlled morphologies and heterointerface control. Then they could have potential applications to improve the quantum yield of photoinduced electrons or holes for photoreduction and photooxidation. This review has demonstrated the recent research efforts to synthesize metal/semiconductor hybrid NCs and to understand and control the photocatalytic applications, such as photocatalytic water splitting, CO_2 photoreduction, and photoinduced organic synthesis. First, on the long term, it can be expected that innovative design and improved synthetic capabilities in the development of elaborate metal/semiconductor NCs with defined topologies and interface control will deliver exciting opportunities in both fundamental understanding and practical exploitation of unconventional properties and functionalities stemming from properly engineered heterostructures with tailored interfaces and structural features. It can also be expected that some recent new synthesis strategy, such as the reaction between coordinated cation ions and semiconductor nanostructures (core/shell, heterodimers, doped NCs), nonepitaxial growth of metal/semiconductor core/shell, and heterodimers will speed up the precise photocatalysts synthesis and design. The new mechanism of photocatalysis, such as plasmon-enhanced photocatalysis, the synergistic cooperation of plasmonic metals, and cocatalysts based on well-defined interface will also boost each of them to an unprecedented level of catalytic performance.

Author details

Jiatao Zhang*, Muwei Ji, Jiajia Liu and Meng Xu

*Address all correspondence to: zhangjt@bit.edu.cn

Beijing Key lab of Construction Tailorable Advanced Functional Materials and Green Applications, School of materials and Engineering, Beijing Institute of Technology, Beijing, China

References

[1] Linic S., Christopher P., Ingram D. B.. Nat. Mater.. 2011; 10:911.

[2] Clavero C.. Nat. Photonics. 2014; 8:95.

[3] Zhang J. T., Tang Y., Lee K., Ouyang M.. Nature. 2010; 466:91.

[4] Jiang R. B., Li B. X., Fang C. H., Wang J. F.. Adv. Mater.. 2014; 26:5274.

[5] Yang J., Wang D., Han H., Li C.. Acc. Chem. Res.. 2013;46:1900.

[6] Xiao M., Jiang R., Wang F., Fang C., Wang J. F., Yu J. C.. J. Mater. Chem. A. 2013;1:5790.

[7] Warren S. C., Thimsen E.. Energy Environ. Sci.. 2012;5:5133.

[8] Ayers J. E. Heteroepitaxy of Semiconductors: Theory, Growth and Characterization. New York: CRC Press; 2007.

[9] Carbone L., Cozzoli P. D.. Nano Today. 2010;5:449.

[10] Casavola M., Buonsanti R., Caputo G., Cozzoli P. D.. Eur. J. Inorg. Chem.. 2008;6:837.

[11] Lee J., Hernandez P., Lee J., Govorov A. O., Kotov N. A.. Nat. Mater.. 2007;4:291.

[12] Gao J. H., Gu H. W., Xu B. Chem. Res.. 2009;42:1097.

[13] Costi R., Saunders A. E., Banin U. Angew. Chem. Int. Ed.. 2010;49:4878.

[14] Wang C., Xu C., Zeng H., Sun S. Adv. Mater.. 2009;21:3045.

[15] Jin Y. D., Gao X. H. Nat. Nanotechnol.. 2004;4:571.

[16] Yang J., Sargent E. H., Kelley S. O., Ying J. Y. Nat. Mater.. 2009;8:683.

[17] Yang J., Ying J. Y. J. Am. Chem. Soc.. 2010;132:2114.

[18] Buck M. R., Bondi J. F., Schaak R. E. Nat. Chem.. 2012;4:37.

[19] Atwater H. A., Polman A. Nat. Mater.. 2010;9:205.

[20] Gu H., Zheng R., Zhang X., Xu B. J. Am. Chem. Soc.. 2004;126:5664.

[21] Wang C., Yin H., Dai S., Sun S.H. Chem. Mater.. 2010;22:3277.

[22] Wu H., Chen O., Zhuang J., Lynch J., Montagne D. L., Nagaoka Y., Cao Y. C. J. Am. Chem. Soc.. 2011;133:14327.

[23] Figuerola A., Fiore A., Corato R. D., Falqui A., Giannini C., Micotti E., Lascialfari A., Corti M., Cingolani R., Pellegrino T., Cozzoli P. D., Manna L. J. Am. Chem. Soc.. 2008;130:1477.

[24] Cozzoli P. D., Pellegrino T., Manna L. Chem. Soc. Rev.. 2006;35:1195.

[25] Mokari T., Rothenberg E., Popov I., Costi R., Banin U. Science. 2004;304:1787.

[26] Costi R., Saunders A. E., Elmalem E., Salant A., Banin U. Nano Lett.. 2008;8:637.

[27] Du Y., Chen B., Yin Z., Liu Z., Zhang H. Small. 2014;10:4727.

[28] Zhang J. T., Tang Y., Lee K., Ouyang M. Science. 2010;327:1634.

[29] Zhang J., Tang Y., Weng L., Ouyang M. Nano lett.. 2009;9:4061.

[30] Anson C. E., Eichhöfer A., Issac I., Fenske D., Fuhr O., Sevillano P., Persau C., Stalke D., Zhang J. Angew. Chem. Int. Ed.. 2008;47:1326.

[31] Gui J., Ji M., Liu J., Xu M., Zhang J. T., Zhu H. S. Angew. Chem. Int.. 2015;54:1.

[32] Zhao Q., Ji M., Qian H., Dai B., Weng L., Gui J., Zhang J. T., Ouyang M., Zhu H. S. Adv. Mater.. 2014;26:1387.

[33] Qian H., Zhao Q., Dai B., Guo L., Zhang J., Liu J., Zhang J. T., Zhu H. S. NPG Asia Mater.. 2015;7:e152.

[34] Son D. H., Hughes S. M., Yin Y., Alivisatos A. P., Science. 2004; 306:1009.

[35] Jeong U. Y., Xia Y. N., Yin Y. D.. Chem. Phys. Lett.. 2005;416:246.

[36] Ryzhii M., Ryzhii V. Physics and Modeling of Tera- and Nano-Devices. In: World Scientific; 2008; Singapore.

[37] McBride J., Treadway J., Feldman L. C., Pennycook S. J., Rosenthal S. J. Nano Lett.. 2006;6:1496.

[38] Battaglia D., Li J. J., Wang Y., Peng X. Angew. Chem. Int. Ed.. 2003;42:5035.

[39] Kim S., Fisher B., Eisler H. J., Bawendi M. J. Am. Chem. Soc.. 2003;125:11466.

[40] Smith A. M., Nie S. Acc. Chem. Res.. 2010;43:190.

[41] Klimov V. I., Ivanov S. A., Nanda J., Achermann M., Bezel I., McGuire J. A., Piryatinski A.. Nature. 2007;441:447.

[42] Caruge J. M., Halpert J. E., Wood V., Bulovic V., Bawendi M. G. Nat. Photonics. 2008;2:247.

[43] Sargent E. H. Nat. Photonics. 2009;3:325.

[44] Chen X. B., Lou Y., Samia A. C., Burda C. Nano Lett.. 2003;3:799.

[45] Peng X. G., Schlamp M. C., Kadavanich A. V., Alivisatos A. P.. J. Am. Chem. Soc.. 1997;119:7019.

[46] Wang Z. L.. J. Phys. Chem. B.. 2000;104:1152.

[47] Palmstrøm C. J.. Annu. Rev. Mater. Sci.. 1995;25:389.

[48] Lee J. S., Shevchenko E. V., Talapin D. V.. J. Am. Chem. Soc.. 2008;130:9673.

[49] Kim H., Achermann M., Balet L. P., Hollingsworth J. A., Klimov V. I.. J. Am. Chem. Soc.. 2005;127:544.

[50] Mokari T., Sztrum C. G., Salant A., Rabani E., Banin U.. Nat. Mater.. 2005;4:855.

[51] Mokari T., Rothenberg E., Popov I., Costi R., Banin U.. Science. 2004;304:1787.

[52] Wang C., Xu C., Zeng H., Sun S.. Adv. Mater.. 2009;21:3045.

[53] Fang C., Jia H., Chang S., Ruan Q., Wang P., Chen T., Wang J.. Energy Environ. Sci.. 2014;7:3431.

[54] Sun Z., Yang Z., Zhou J., Yeung M., Ni W., Wu H., Wang J.. Angew. Chem. Int. Ed.. 2009;48:2881.

[55] Lambright S., Butaeva E., Razgoniaeva N., Hopkins T., Smith B., Perera D., Corbin J., Khon E., Thomas R., Moroz P., Mereshchenko A., Tarnovsky A., Zamkov M. ACS Nano. 2014;8:352.

[56] Priebe M., Fromm K. M.. Chemistry—A European Journal. 2015;21:3854.

[57] Güttel R., Paul M.,Schüth F.. Catal. Sci. Technol.. 2011;1:65.

[58] Li A., Zhang P., Chang X., Cai W., Wang T., Gong J.. Small. 2015;11:1892.

[59] Li G., Tang Z.. Nanoscale. 2014;6:3995.

[60] Fang Q., Xuan S., Jiang W., Gong X.. Adv. Funct. Mater.. 2011;21:1902.

[61] Jun Z., Ling X., Jiali S., Diping H., Huan. J. Surf. Coat. Technol.. 2015;271:119.

[62] Zhang Q., Lee I., Joo J. B., Zaera F., Yin Y.. Acc. Chem. Res.. 2013;46:1816.

[63] Zhao L., Peng J., Huang Q., Li C., Chen M., Sun Y., Lin Q., Zhu L., Li F.. Adv. Funct. Mater.. 2014;24:363.

[64] Zhang N., Fu X., Xu Y.J.. J. Mater. Chem.. 2011;21:8152.

[65] Zhang L., Blom D. A., Wang H.. Chem. Mater.. 2011;23:4587.

[66] Park J. C., Lee H. J., Jung H. S., Kim M.. Kim H. J., Park K. H., Song H.. ChemCatChem. 2011;3:755.

[67] Hah H. J., Um J. I., Han S. H., Koo S. M.. Chem. Commun.. 2004;:1012.

[68] Tan L., Chen D., Liu H., Tang F.. Adv. Mater.. 2010;22:4885.

[69] Fan C. M., Zhang L. F., Wang S. S., Wang D. H., Lu L. Q., Xu A. W.. Nanoscale. 2012;4:6835.

[70] Gröger H., Gyger F., Leidinger P., Zurmühl C., Feldmann C.. Adv. Mater.. 2009;21:1586.

[71] Lin Y. S., Wu S. H., Tseng C. T., Hung Y., Chang C., Mou C. Y . Chem. Commun.. 2009;:3542.

[72] Priebe M., Fromm K. M.. Part. Part. Syst. Charact.. 2014;31:645.

[73] Park J. C., Heo E., Kim A., Kim M.. Park K. H., Song H.. J. Phys. Chem. C. 2011;115:15772.

[74] Arnal P. M., Comotti M., Schuth F.. Angew. Chem. Int. Ed.. 2006;45:8224.

[75] Lou X. W., Yuan C., Archer L. A.. Small. 2007;3:261.

[76] Guan B., Wang T., Zeng S., Wang X., An D., Wang D., Cao Y., Ma D., Liu Y., Huo Q.. Nano Res.. 2014;7:246.

[77] Yeo K. M., Shin J., Lee I. S.. Chem. Commun.. 2010;46:64.

[78] Wang M., Han J., Xiong H., Guo R.. Langmuir. 2015;31:6220.

[79] Fujishima A., Honda K.. Nature. 1972;238:37.

[80] Kudo A., Miseki Y.. Chem. Soc. Rev.. 2009;38:253.

[81] Linsebigler A. L., Lu G., Yates J. T.. Chem. Rev.. 1995;95:735.

[82] Li X., Yu J., Low J., Fang Y., Xiao J., Chen X.. J. Mater. Chem. A. 2015;3:2485.

[83] Osterloh F. E.. Chem. Soc. Rev.. 2013;42:2294.

[84] Maeda K.. J. Photoch. Photobio. C. 2011;12:237.

[85] Qu Y., Duan X. F.. Chem. Soc. Rev.. 2013;42:2568.

[86] Chen X., Shen S., Guo L., Mao S. S.. Chem. Rev.. 2010;110:6503.

[87] Ma Y., Wang X., Jia Y., Chen X., Han H., Li C.. Chem. Rev.. 2014;114:9987.

[88] Kraeutler B., Bard A. J.. J. Am. Chem. Soc.. 1978;100:4317.

[89] Yang J., Wang D., Han H., Li C.. Acc. Chem. Res.. 2013;46:1900.

[90] Wang H., Zhang L., Chen Z.G., Hu J., Li S., Wang Z., Liu J., Wang X. Chem. Soc. Rev.. 2014;43:5234.

[91] Bai S., Jiang J., Zhang Q., Xiong Y. J. Chem. Soc. Rev.. 2015;44:2893.

[92] Jing L., Zhou W., Tian G., Fu H. G. Chem. Soc. Rev.. 2013;42:9509.

[93] Kochuveedu S. T., Jang Y. H., Kim D. H. Chem. Soc. Rev.. 2013;42:8467.

[94] Ran J., Zhang J., Yu J., Jaroniec M., Qiao S. Z. Chem. Soc. Rev.. 2014;43:7787.

[95] Thanminimulla C. T. K., Takata T., Hara M., Kondo J. N., Domen K. J. Catal.. 2000;196:362.

[96] Yamada Y., Miyahigashi T., Kotani H., Ohkubo K., Fukuzumi S. Energy Environ. Sci.. 2012;5:6111.

[97] Awazu K., Fujimaki M., Rockstuhl C., Tomingaga J., Murakami H., Ohiki Y., Yoshida N., Watanabe T. J. Am. Chem. Soc.. 2008;130:1676.

[98] Chen X., Zhu H., Zhao J., Zheng Z., Gao X. Angew. Chem. Int. Ed.. 2008;47:5353.

[99] Wang P., Huang B., Qin X., Zhang X., Dai Y., Wei J., Wang M. H. Angew. Chem. Int. Ed.. 2008;47:7931.

[100] SWalclawik., E. R., Zhu H. Green Chem.. 2013;15:1814.

[101] Lang X., Chen X., Zhao J. Chem. Soc. Rev.. 2014;43:473.

[102] Jones M. R., Osberg K. D., Macfarlane R. J., Langille M.R., Mirkin C. A. Chem. Rev.. 2011;111:3736.

[103] Jiang R., Li B., Fang C., Wang J. F. Adv. Mater.. 2014;26:5274.

[104] Kochuveedu S. T., Jang Y. H., Kim D. H. Chem. Soc. Rev.. 2013;42:8467.

[105] Ingram D. B., Linic S. J. Am. Chem. Soc.. 2011;133:5202.

[106] Pu Y., Wang G., Chang K., Ling Y., Lin Y., Fitzmorris B. C., Liu C., Lu X., Tong Y., Zhang J. Z., Hsu Y., Li Y. Nano Lett.. 2013;13:3817.

[107] Seh Z. W., Liu S., Low M., Zhang S., Liu Z., Mlayah A., Han M. Y. Adv. Mater.. 2012;24:2310.

[108] Pradhan S., Ghosh D., Chen S. ACS Appl. Mater. Interfaces. 2009;1:2060.

[109] Zhang Z., Zhang L., Hedhili M. N., Zhang H., Wang P. Nano Lett.. 2013;13:14.

[110] Qian K., Sweeny B. C., Johnston-Peck A. C., Niu W., Graham J. O., Duchenen J. S., Qiu J., Wang Y., Engelhard M. H., Su D., Stach E. A., Wei W. D. J. Am. Chem. Soc.. 2014;136:9842.

[111] Long R., Mao K. K., Gong M., Zhou S., Hu J. H., Zhi M., You Y., Bai S., Jiang J., Zhang Q., Wu X. J., Xiong Y. J.. Angew. Chem. Int. Ed.. 2014;53:3205.

[112] Liu G., Yang H. G., Pan J., Yang Y. Q., Lu G. Q., Cheng H. M. Chem. Rev.. 2014;114:9559.

[113] Ohno T., Sarukawa K., Matsumura M . New J. Chem.. 2002;26:1167.

[114] Yang H. G., Sun C. H., Qiao S. Z., Zou J., Liu G., Smith S.C., Cheng H. M., Lu G. Q.. Nature. 2008;453:638.

[115] Murakami N., Kurihara Y., Tsubota T., Ohno T.. J. Phys. Chem. C. 2009;113:3062.

[116] Liu C., Han, X.; Xie, S.; Kuang, Q.; Wang, X.; Jin, M.; Xie, Z.; Zheng, L.. Chem. Asian J.. 2013;8:282.

[117] Tian Y., Tasuma T. J. Am. Chem. Soc. 2005;127:7632.

[118] Silva C. G., Juárez R., Marino T., Molinari R., García H. J. Am. Chem. Soc.. 2011;133:595.

[119] Kelly K. L., Coronado E., Zhao L. L., Schatz G. C. J. Phys. Chem. B.. 2003;107:668.

[120] Maeda K., Domen K. J. Phys. Chem. C. 2007;111:7851.

[121] Hu P., Pramana S. S., Cao S., Ngaw C. K., Lin J. D., Loo S. C. J., Tan T. T. Y. Adv. Mater.. 2013;25:2567.

[122] Cao S. W., Fang J., Shahjamali M. M., Boey F. Y. C., Barber J., Loo S. C. J., Xue C. RSC Adv.. 2012;2:5513.

[123] Yan H., Yang J. H., Ma G. J., Wu G. P., Zong X., Lei Z. B., Shi J., Li C. J. Catal.. 2009;266:165.

[124] Yang J., Yan H., Wang X., Wen F., Wang Z., Fan D., Shi J., Li C. J. Catal.. 2012;290:151.

[125] Liu C., Dasgupta N. P., Yang, P.. Chem. Mater.. 2014;26:415.

[126] Schneider J., Matsuoka M., Takeuchi M., Zhang J., Horiuchi Y., Anpo M., Bahnemann D. W.. Chem. Rev.. 2014;114:9919.

[127] Ma Y., Wang X., Jia Y., Chen X., Han H., Li C.. Chem. Rev.. 2014;114:9987.

[128] Kubacka A., Fernández-García M., Colón, G.. Chem. Rev.. 2012;112:1555.

[129] Wen F., Li, C.. Acc. Chem.Res.. 2013;46:2355.

[130] Cokoja M., Bruckmeier C., Rieger B., Herrmann W. A., Kühn F. E.,. Angew. Chem. Int. Ed.. 2011;50:8510.

[131] Kang Q., Wang T., Li P.; Liu L.; Chang K.; Li M.; Ye J.. Angew. Chem. Int. Ed.. 2015;54:841.

[132] Habisreutinger S. N.; Schmidt-Mende L.; Stolarczyk J. K.. Angew. Chem. Int. Ed.. 2013;52:7372.

[133] Wang C.; Thompson R. L.; Baltrus J.; Matranga C.. J. Phy. Chem. Lett.. 2010;1:48.

[134] Johne P., Kisch H.. J. Photochem. Photobiol. A. 1997;111:223.

[135] Inoue T., Fujishima A., Konishi S., Honda, K.. Nature. 1979;277:637.

[136] Sato S., Arai T., Morikawa T., Uemura K., Suzuki T. M., Tanaka H., Kajino, T.. J. Am. Chem. Soc.. 2011;133:15240.

[137] Li Y., Wang W.N., Zhan Z., Woo M.H., Wu C.Y., Biswas P.. Appl. Catal. B. Environ.. 2010;100:386.

[138] Li P., Zhou Y., Zhao Z., Xu Q., Wang X., Xiao M., Zou Z.. J. Am. Chem. Soc.. 2015;137:9547.

[139] Wu Y., Wang D., Zhao P., Niu Z., Peng Q., Li Y.. Inorg. Chem.. 2011;50:2046.

[140] Wang F., Li C., Chen H., Jiang R., Sun L. D., Li Q., Wang J., Yu J. C., Yan C. H.. J. Am. Chem. Soc.. 2013;135:5588.

[141] Kisch H.. Angew. Chem. Int. Ed.. 2013;52:812.

[142] Xiao Q., Liu Z., Bo A., Zavahir S. Sarina S., Bottle S., Riches J. D., Zhu H.. J. Am. Chem. Soc.. 2015;137:1956.

[143] Zhu H., Ke X., Yang X., Sarina S., Li, H.. Angew. Chem. Int. Ed.. 2010;49:9657.

[144] Tanak, A., Hashimot, K., Kominam, H.. J. Am. Chem. Soc.. 2012;134:14526.

Chapter 12

Recent Advances in Visible-Light Driven Photocatalysis

Yuhua Wang, Xinlong Ma, Hao Li, Bin Liu, Huihui Li, Shu Yin and Tsugio Sato

Additional information is available at the end of the chapter

http://dx.doi.org/10.5772/61864

Abstract

Semiconductor photocatalysis has been considered a potentially promising approach for renewable energy and environmental remediation with abundant solar light. However, the currently available semiconductor materials are generally limited by either the harvesting of solar energy or insufficient charge separation ability. To overcome the serious drawbacks of narrow light-response range and low efficiency in most photocatalysts, many strategies have been developed in the past decades. This article reviews the recent advancements of visible-light-driven photocatalysts and attempts to provide a comprehensive update of some strategies to improve the efficiency, such as doping, coupling with graphene, precipitating with metal particles, crystal growth design, and heterostructuring. A brief introduction to photocatalysts is given first, followed by an explanation of the basic rules and mechanisms of photocatalysts. This chapter focuses on recent progress in exploring new strategies to design TiO_2-based photocatalysts that aim to extend the light absorption of TiO_2 from UV wavelengths into the visible region. Subsequently, some strategies are also used to endow visible-light-driven Ag_3PO_4 with high activity in photocatalytic reactions. Next, a novel approach, using long afterglow phosphor, has been used to associate a fluorescence-emitting support to continue the photocatalytic reaction after turning off the light. The last section proposes some challenges to design high efficiency of photocatalytic systems.

Keywords: Photocatalysts, TiO_2, Ag_3PO_4, graphene, long afterglow phosphor

1. Introduction

It is well known that environmental pollution is affecting human survival and development. Photocatalytic technology is considered as an efficient, stable, and environmentally friendly method in the field of environmental pollution control [1]. In general, a photocatalytic reaction includes three steps [2] (Fig. 1). First, electrons (e−) and holes (h+) are generated through photoexcitation. Then, these electrons and holes migrate to the surface of photocatalysts and subsequently react with adsorbed electron acceptors and donors, respectively. Thus, an efficient photocatalyst requires a semiconductor with a suitable band gap for harvesting light, facile separation and transportation of charge carriers, and proper valence band (VB) and conduction band (CB) edge potential for redox reaction being feasible. To date, much effort is currently focused on how to improve the light absorption, charge separation, and surface reactivity in order to achieve outstanding photocatalytic performance, such as doping, coupled with graphene, precipitated with metal particles, crystal growth design, and heterostructuring.

In the past few decades, TiO_2 nanomaterials have attracted tremendous interest in environmental pollution removal and photocatalytic hydrogen generation. However, all of them have large electronic band gaps of 3.0–3.2 eV, which means only less than 5% of the entire solar energy can be utilized. Thus, much effort has been devoted to bring about the absorption of TiO_2 into the visible-light region and improved photocatalytic activity.

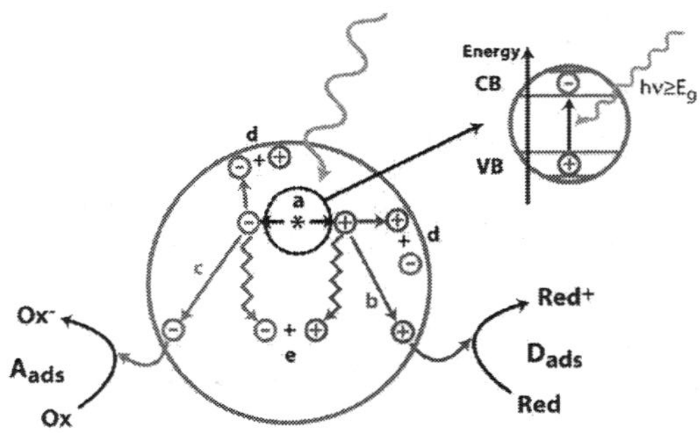

Figure 1. Schematic illustration of the reactions following light absorption by a photocatalyst. (a) Electron–hole pair formation; (b) donor oxidation by hole; (c) reduction by electron; (d) and (e) electron–hole recombination on the surface or in the bulk. Adapted with permission from ref. 2. © 1995 ACS.

Despite the fact that there are many research articles on modified TiO_2, the exploration of new active and efficient visible-light-driven photocatalysts attracts much attention [3–22]. In the effort to exploit novel photocatalyst systems working under visible light, it has been revealed that orbitals of some p-block metals with a d10 configuration [22], such as Ag 4d in Ag(I), could hybridize O 2p levels to form a new preferable hybridized VB, thus narrowing the band gap to harvest visible light.

Compared with single-phase semiconductor photocatalysts, hybrids of two or more semiconductor systems, i.e., heterostructures, seem to possess advantages in more efficiently utilizing solar light. Besides metal/semiconductor and carbon group materials/semiconductor-heterostructured photocatalysts [23–28], semiconductor/semiconductor-heterostructured photocatalysts with diverse models have been developed (Fig. 2), including type-I and type-II heterojunctions, Z-scheme, p–n heterojunctions, and homojunction band alignments [29–32].

http://dx.doi.org/10.5772/61864

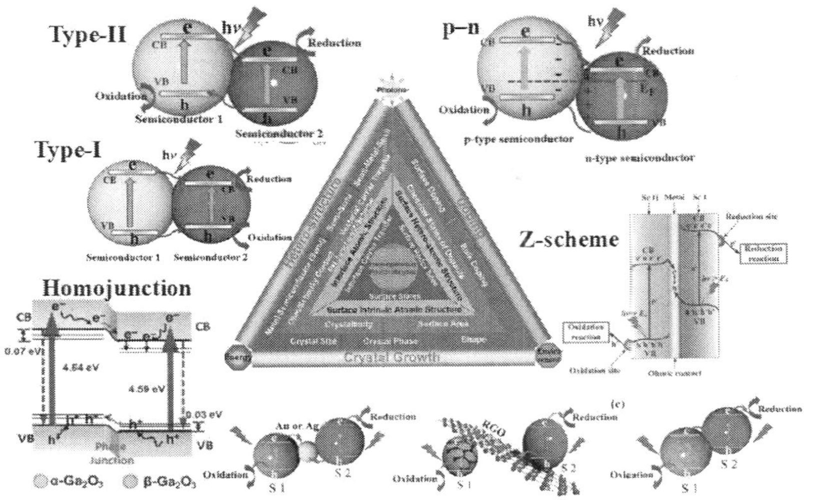

Figure 2. Schematic illustration of band alignments corresponding to five kinds of heterostructured photocatalysts, including the straddling alignment (type I), staggered alignment (type II), Z-scheme system, p–n heterojunctions, and homojunctions. Reproduced with permission from ref. 30–32 © 2011 RSC & 2011 ACS & 2012 Wiley.

2. TiO_2-based photocatalysts

In this section, strategies on the modulation of energy band structures or surface states of wide-band-gap photocatalysts, especially for TiO_2, are reviewed. Currently, many strategies have been developed to modulate the band structures of TiO_2, including doping, coupling with graphene, precipitating with metal particles, crystal growth design, and heterostructuring, which will be discussed.

2.1. Doping

Doping is an available strategy to tune the absorption band of wide-band photocatalysts. Several studies have investigated nonmetal doping. Asahi *et al.* reported nitrogen-doped TiO_2 of nonmetals, such as N [33, 34], C [35–37], S [38, 39], B [40–42], F [43–45], Br [46], I [47–50], P [51], in 2001 [33]. In our study [34], nitrogen-doped titania nanoparticles (NPs) were successfully prepared by a microwave-assisted solvothermal process in a very short time, and the prepared samples showed visible-light absorption in the range of 400–550 nm, indicating its potential applications as visible-light-induced photocatalyst. The chemical states and locations of dopants are considered to be key factors in adjusting the spectral distribution of the induced electronic states of those dopants and reconstructing favorable surface structure for photocatalysis.

Co-doping with two suitable heteroatoms can also achieve substantial synergistic effects [41, 52–55]. For example, B, N co-doped TiO_2, benefits from the B–N bonds formed, which can increase the amount of doped N on the TiO_2 surface and the promoted separation of photo-excited electron–hole pairs [41].

2.2. TiO_2-graphene composite

Graphene, a two-dimensional carbonaceous material, can be used in many applications due to its unique and remarkable properties such as high conductivity, large surface area, and good chemical stability [56–59]. Tremendous interest is devoted to fabricating numerous graphene–semiconductor composites to aid charge separation and migration and improve the performance of the photocatalysts [60–64]. Graphene works as an electron acceptor or transporter to induce electron transfer, leading to an efficient charge separation. Thus, an appropriate integration of graphene and TiO_2 could give rise to a nanocomposite that combines the desirable properties of graphene and TiO_2, e.g., the photocatalytic activity of TiO_2 can be improved. In the past few years, there were some reports about graphene–TiO_2 composites [13–17].

Recently, Liu et al. synthesized the graphene oxide–TiO_2 nanorod composites [65, 66]. Wu et al. reported the synthesis and application of graphene–TiO_2 nanorod hybrid nanostructures in microcapacitors [67]. As shown in Fig. 3, the assembling of TiO_2 nanocrystalline with exposed {001} facets on graphene sheets reported in our previous work showed a higher photocatalytic activity than the other normal TiO_2/graphene composites [68]. In another work (Fig. 4), graphene/rod-shaped TiO_2 nanocomposite was synthesized by the solvothermal method [69]. In a one-pot system, the rod-shaped TiO_2 can be homogeneously dispersed on the surface of graphene sheets by syngraphenization strategy. Owing to the combination of graphene and rod-shaped TiO_2, the graphene/rod-shaped TiO_2 nanocomposite shows a significant enhancement in the photocatalytic performance compared with that of the graphene/spherical TiO_2 nanocomposite, which can be attributed to the high electronic mobility of graphene, higher Brunauer-Emmett-Teller (BET) surface area, and rod-shaped structures of TiO_2. In our recent work [70], a series of B-doped graphene/rod-shaped TiO_2 nanocomposites were synthesized via one-step hydrothermal reaction. The photocatalytic activity of the obtained nanocomposites for the oxidative photodestruction of NOx gas showed better photocatalytic properties than pure TiO_2 and graphene/TiO_2 nanocomposites. This work provides new insight into the fabrication of TiO_2–carbon nanocomposites as high-performance photocatalysts and facilitates their application in addressing environmental protection issues.

2.3. TiO_2-based plasmonic photocatalysts

Plasmonic photocatalysis has offered a new opportunity to solve the problem of the limited efficiency of photocatalysts [71–73]. In these photocatalysts, the nanostructured plasmonic metals are often combined with a semiconductor-based material (e.g., TiO_2), and the photocatalytic activity is greatly enhanced due to the local surface plasmon resonance (LSPR) effect

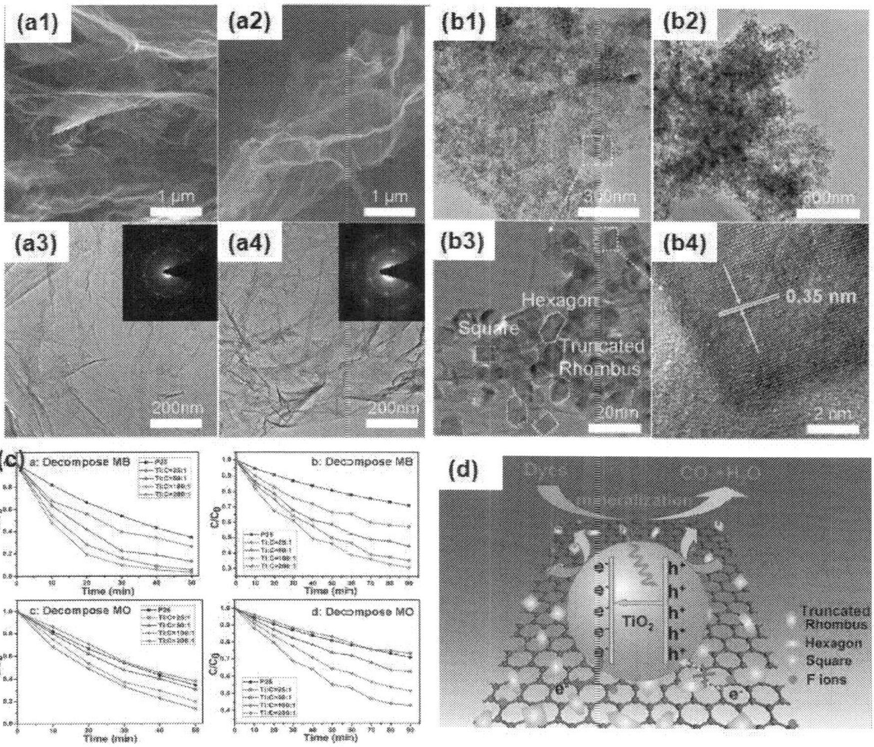

Figure 3. Scanning electron micrograph (SEM) and transmission electron micrograph (TEM) of (a1 and a3) GO and (a2 and a4) GS; (b1 and b2) TEM images of TiO_2/CS with different Ti:C ratios; (b3 and b4) high-resolution transmission electron microscopic (HRTEM) images of sample b1; (c) photocatalytic degradation of MB and MO under the irradiation of UV light and visible light over the TiO_2/GS composites; (d) a proposed schematic illustration showing the reaction mechanism for photocatalytic degradation of organic pollutants over the TiO_2/GS composites. Reproduced with permission from ref. 68 © 2012 RSC.

[74]. The LSPR effect endows the metal nanocrystals with very large absorption and scattering cross sections and local electromagnetic field enhancement in the near-field region near the surface of plasmonic metal nanocrystals, which is promising in manipulating light absorption in photocatalytic systems. The intensity of the local electromagnetic field is several orders of magnitude larger than that of the far-field incident light, and the highest charge carrier formation is observed at the semiconductor/liquid interface, which benefits the photocatalytic reactions [71–72].

Recently, noble metal nanoparticle-deposited TiO_2 has attracted significant attention. The main advantages of these noble metal NPs can be attributed to their chemically inert properties

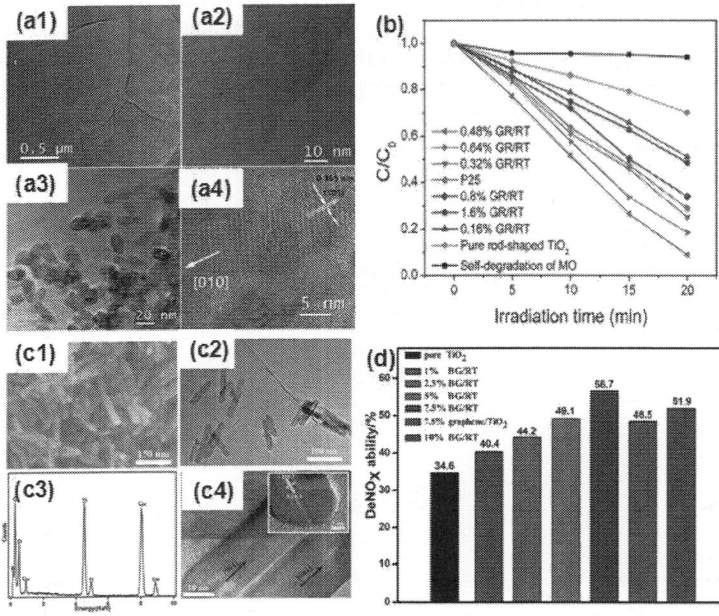

Figure 4. TEM images of (a1) graphene oxide, (a2) graphene, (a3) 0.48% GR/RT, and (a4) HRTEM image of a single rod-shaped TiO_2 nanocrystal in the composite; (b) photocatalytic degradation of MO solution over graphene/rod-shaped TiO_2 nanocomposites with various graphene contents compared with that of GR/ST; (c1) SEM image, (c2) TEM image, (c3) EDX, and (c4) HRTEM image of 7.5% BG/RT; (d) photocatalytic activity for the destruction of NO*x* gas under UV irradiation using different photocatalysts. Reproduced with permission from ref. 69 & ref. 70 © 2012 & 2014 RSC.

toward (photo) oxidation and LSPR effect on the surface [75, 76]. Meanwhile, our previous work found the plasmonic enhancement of the photocatalytic activity of semiconductor–metal nanocomposite materials [77–79]. Thus, noble metal NP-deposited TiO_2 could be an appropriate approach to improve the photocatalytic performance of TiO_2.

A variety of nanostructured Au/TiO_2 with different morphologies have been synthesized, such as highly stable mesoporous Au/TiO_2 spheres (~500 nm) [80] and mesoporous Au–TiO_2 nanocomposites using a simple spray hydrolytic method [81]. In our study, Au NPs are precipitated on the highly porous one-dimensional (1D) TiO_2 nanotubes (NTs), and the plasmonic photocatalytic properties of the material are investigated [82]. Compared with nanoparticles, there are some advantages of 1D NT structures, such as favorable recycling characteristics and the vectorial transport of photogenerated charge carriers [83, 84], which have great potential for superior photocatalytic performance. The Au NPs/TiO_2 NTs were synthesized by emulsion electrospinning followed by deposition–precipitation (DP) method. The results in Fig. 5 show that the modified porous TiO_2 NTs with the presence of Au NPs increased photocatalytic destruction of methylene blue (MB) solution under visible-light irradiation. Furthermore, the migration of Au NPs from the rutile phase to the interface of rutile/anatase was found when the calcination temperature changed from 250 °C to 350 °C. The optimal photocatalytic activity was obtained in the sample $Au_3(DP_{350})/TiO_2$, due to the plasmon activation of the Au NPs followed by consecutive electron transfer that induced efficient charge separation. Therefore, such a highly porous Au/TiO_2 heterojunction structure provides a new pathway for the design and fabrication of other energy- and environment-related applications.

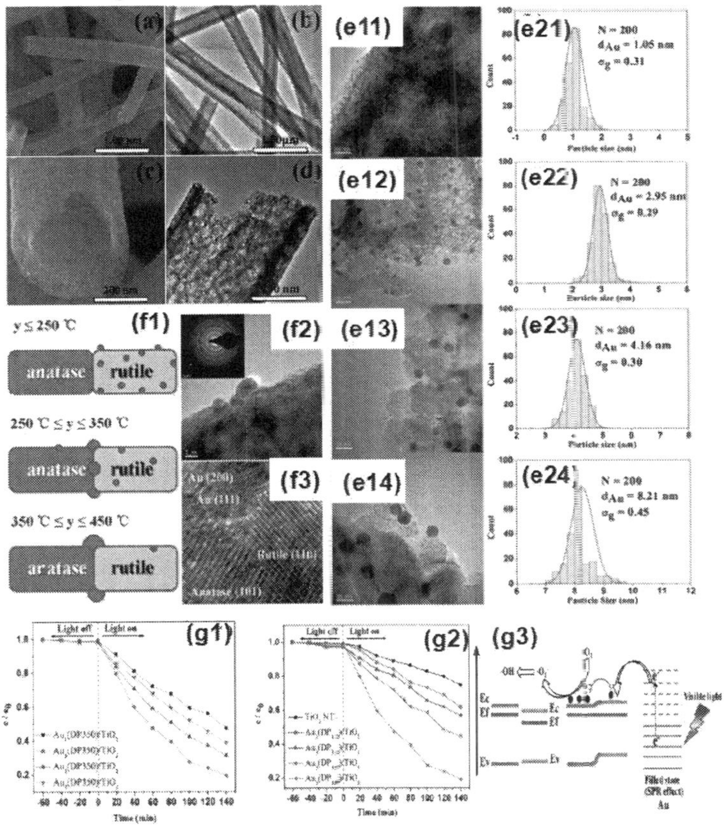

Figure 5. (a) SEM and (b) TEM images of as-prepared TiO_2 NTs and (c) SEM and (d) TEM images of the representative sample $Au_3(DP_{350})/TiO_2$ NTs; TEM images of $Au_3(DP_y)/TiO_2$ (y = 150 (e11), 250 (e12), 350 (e13), and 450 °C (e14), respectively) and the dispersion of Au co-catalysts (e21), (e22), (e23), and (e24), respectively; (f1) effect of calcination temperature on the location and size of the Au NPs on TiO_2 NTs; (f2 and f3) TEM and HRTEM images of $Au_3(DP_{350})/TiO_2$. Variation of normalized C/C_0 of MB concentration as a function of visible-light irradiation time for (g1) $Aux(DP_{350})/TiO_y$ and (g2) $Au_3(DP_y)/TiO_2$ (y = 150, 250, 350, 450 °C, respectively); (g3) schematic diagram for the possible mechanism for photocatalytic degradation of MB over $Au_3(DP_{350})/TiO_2$ under visible-light irradiation. Reproduced with permission from ref. 82 © 2013 RSC.

2.4. TiO_2-based heterostructure photocatalysts

The heterojunctions provide a facile way to enable the effective separation of photoexcited electron–hole pairs, thus to enhance the photocatalytic performance. Table 1 lists recent prominent heterostructured photocatalysts as well as their photocatalytic application.

	Heterostructured photocatalysts
Type I	CdS/ZnS,[85] Bi2S3/CdS,[86] $V_2O_5/BiVO_4$[30]
Type II	CdS/TiO_2,[87–89] $SrTiO_3/TiO_2$,[90, 91] Fe_2O_3/TiO_2,[92] ZnO/CdS,[93] $AgIn5S8/TiO_2$,[94] Ag_3VO_4/TiO_2,[95] $ZnFe_2O_4/TiO_2$[96]
p–n	$CuFe_2O_4/TiO_2$,[97] CuO/ZnO,[98] MoS2/CdS,[99, X] Ag_2O/TiO_2,[101] $CuInSe_2/TiO_2$,[102] TiO_2/ZnO,[103] $ZnFe_2O_4/TiO_2$,[104] NiO/ZnO[105]
Homojunction	Anatase/rutile TiO_2,[106] α/β-Ga2O3,[31] p–n Cu2O,[107] α/γ-Bi_2O_3,[108] Co-doped TiO_2/TiO_2,[109] W-doped $BiVO_4$,[110] Pt/n-Si/n+-Si/Ag[111]
Z-scheme	CdS/Au/TiO1.96C0.04,[112] $CdS/Au/TiO_2$,[113] ZnO/CdS,[114] CdS/Au/ZnO,[115] CuO/TiO_2,[116] $CaFe_2O_4/WO_4$[117]

Table 1. Recent reports of diverse heterostructured photocatalysts.

According to different electronic energy levels and band gaps of photocatalysts, five kinds of semiconductor heterojunctions have been reported: straddling alignment (type I), staggered alignment (type II), Z-scheme system, homojunctions, and p–n heterojunctions [118, 119]. In those work, the dynamics of electron and hole were studied, including the band gap, the electron affinity, and the work function of different semiconductor heterojunctions.

3. Visible-light-driven Ag_3PO_4 photocatalysts

A breakthrough was made in finding a novel semiconductor material, Ag_3PO_4, as an active visible-light-induced photocatalyst [4]. Ag_3PO_4 demonstrates extremely high capability for O_2 evolution from H_2O and organic dye decomposition under visible-light irradiation [120]. More importantly, this novel photocatalyst can achieve a quantum efficiency up to 90% at wavelengths longer than 420 nm, which is clearly higher than that reported previously by using semiconductor photocatalysis.

So far, various methods have been proposed to further enhance the photocatalytic activity of Ag_3PO_4 under visible-light irradiation. One approach is the synthesis of Ag_3PO_4 with various morphologies. This is because photocatalytic reactions are typically surface-based processes; thus, the photocatalytic efficiency is closely related to the morphology and microstructure of a photocatalyst [83]. Recently, some new morphologies of Ag_3PO_4 have been developed [120–126]. For example, Bi et al. fabricated the single-crystalline Ag_3PO_4 rhombic dodecahedrons and cubes, and they found that both of these samples exhibited higher photocatalytic activity than the microsized spherical Ag_3PO_4 particles [120]. Liang et al. synthesized hierarchical Ag_3PO_4 porous microcubes with enhanced photocatalytic property [123]. Wang et al. reported the synthesis of Ag_3PO_4 tetrapod microcrystals, and they demonstrated that Ag_3PO_4 tetrapod showed higher photocatalytic activity than the microsized spherical Ag_3PO_4 particles [126].

Another approach is to couple Ag_3PO_4 with other semiconductors, carbon materials, or noble metals to improve the photocatalytic activity, such as Ag_3PO_4/TiO_2 [127, 128], Ag_3PO_4/AgX (X = Cl, Br, I) [129], $Ag_3PO_4/BiOCl$ [130], Ag_3PO_4/Fe_3O4 [131], Ag_3PO_4/SnO_2 [132], Ag_3PO_4/carbon quantum dots [133], Ag_3PO_4/reduced graphite oxide sheets [134], and Ag_3PO_4/Ag composites [135–137].

3.1. Crystal growth design

Manipulating the crystal structure will result in controlling the percentage of exposed facets on crystal surfaces and thus can lead to a dramatic change in reactivity, which has been widely investigated in sensing [138], electronics [139], magnetic memory devices [140], and catalysis [141]. It is widely accepted in catalysis that a higher surface energy leads to a more reactive surface. Therefore, the control of the exposed facets is one of the most available and efficient methods to obtain more active surface [142], which has been investigated recently to promote photocatalytic activity [83].

Herein, we controllably prepared Ag_3PO_4 crystals with various new morphologies (including branched, tetrapod, nanorod-shaped, and triangular-prism-shaped Ag_3PO_4 crystals) via a facile and efficient synthesis process in the solvent mixture of N,N dimethylformamide (DMF) and H_2O at room temperature (Fig. 6). The results indicate that the branched Ag_3PO_4 sample shows highly enhanced photocatalytic activity compared with other as-prepared Ag_3PO_4 samples, and the BET-specific surface area makes a greater contribution to the enhanced photocatalytic activity of as-prepared Ag_3PO_4 crystals [143].

3.2. Ag_3PO_4–graphene composite

The photocorrosion phenomenon of Ag_3PO_4 reduces the photocatalytic activity during the process. Up to now, some solutions have been put forward to improve its stability and photocatalytic properties under visible-light irradiation. Apart from its unique electronic properties [58], graphene has several other excellent attributes, such as the large theoretical specific surface area [57] and good chemical stability [59]. The large specific surface area of graphene facilitates the attachment of inorganic nanomaterials [144, 145]. Thus, the combination of graphene and Ag_3PO_4 could be a good choice to construct a stable and efficient photocatalyst composite.

In one of our works, Ag_3PO_4/reduced graphite oxide (RGO) nanocomposites were synthesized to enhance the visible-light photocatalytic activity and the stability of Ag_3PO_4. The results show that the graphene content obviously affects the photocatalytic activity of Ag_3PO_4/RGO nanocomposites [134]. As shown in Fig. 7, among a series of Ag_3PO_4/RGO, the Ag_3PO_4/2.1 wt % RGO shows the best photocatalytic activity despite the degradation of MB or Methyl Orange (MO) solution. In addition, the Ag_3PO_4/RGO is more stable than pure Ag_3PO_4 since the RGO

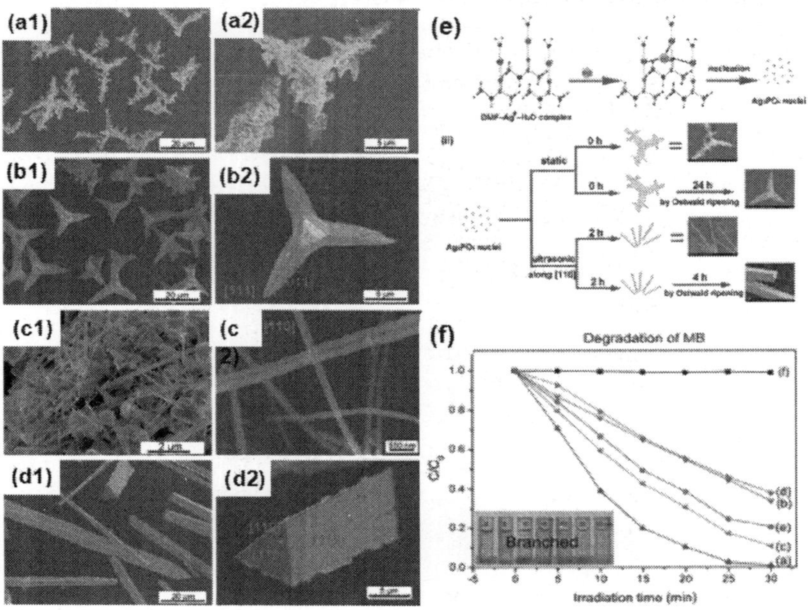

Figure 6. SEM images of (a) branched, (b) tetrapod, (c) nanorod-shaped, and (d) triangular-prism-shaped Ag_3PO_4 crystals; (e) schematic illustration of the possible formation mechanism of Ag_3PO_4 crystals with four typical morphologies prepared under static and ultrasonic conditions; (f) variation of MB solution concentration against illumination time in the presence of branched, tetrapod, nanorod-shaped, triangular-prism-shaped, and irregular spherical Ag_3PO_4 products. Reproduced with permission from ref. 143 © 2013 RSC.

can be used as protective coatings that inhibit the photocorrosion of Ag_3PO_4. Thus, the Ag_3PO_4/RGO nanocomposites with excellent photocatalytic performance and enhanced stability can find promising applications in addressing environmental protection issues.

3.3. Ag_3PO_4-based heterostructure photocatalysts

Many narrow-band-gap metal oxides or chalcogenides have been coupled with Ag_3PO_4 photocatalyst to enhance its photocatalytic activity and/or improve its stability. In Yao's work, Ag_3PO_4/TiO_2 has been synthesized via the in situ deposition of Ag_3PO_4 nanoparticles onto the TiO_2 surface, which facilitates electron–hole separation, thereby leading to enhanced photocatalytic activity [128].

In our work, the phenomenon of "self-corrosion" was first observed in the simple physical mixed Ag_3PO_4/TiO_2 compounds (Fig. 8). It is found that both self-corrosion and photocorrosion in Ag_3PO_4/TiO_2 compounds alter the chemical environment of Ag. The corrosion degree, however, is different due to a slight difference in the chemical environment of Ag. Furthermore, it is the strong adsorption capacity that determines the photocatalytic activities of

http://dx.doi.org/10.5772/61864

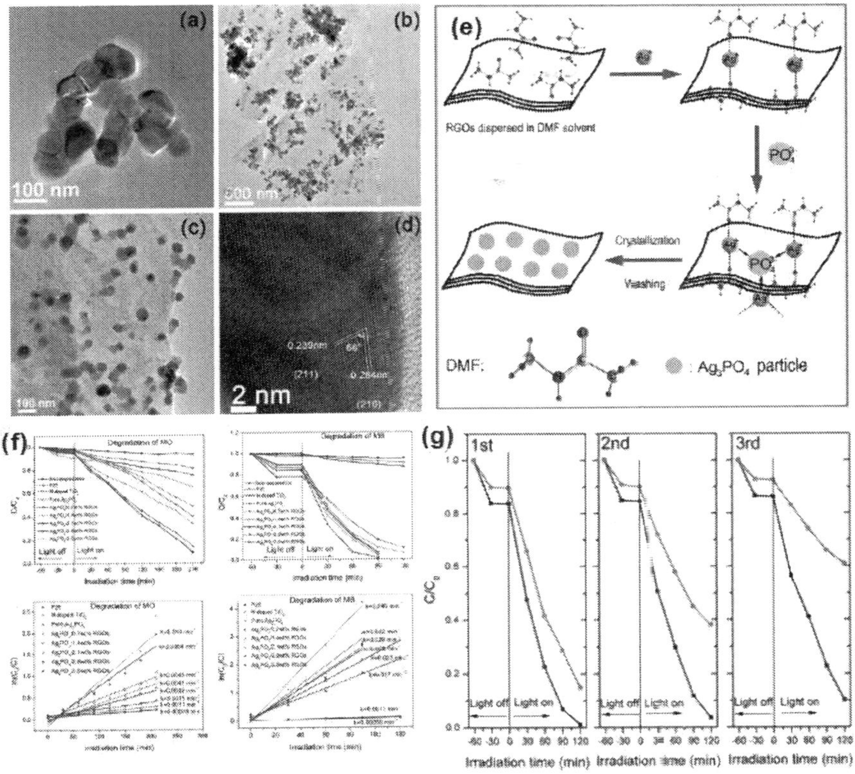

Figure 7. (a) TEM image of pure Ag_3PO_4 nanoparticles; (b and c) TEM images of Ag_3PO_4/2.1 wt% RGO nanocomposite; (d) HRTEM image of a single Ag_3PO_4 nanoparticle in the Ag_3PO_4/2.1 wt% RGO nanocomposite; (e) scheme of synthetic procedure for the Ag_3PO_4/RGO nanocomposite; (f) variation of MO and MB concentration against irradiation time using Ag_3PO_4/RGO nanocomposites with various RGO contents under visible-light irradiation and plots of $\ln(C_0/C)$ versus irradiation time representing the fit using a pseudo-first-order reaction rate; (g) repeated photocatalytic degradation of MB solution under visible-light irradiation. Reproduced with permission from ref. 134 © 2013 Elsevier B.V.

Ag_3PO_4/TiO_2compounds under UV light irradiation, which is almost independent of self-corrosion. In contrast, it is the amount of visible-light response of Ag_3PO_4 in Ag_3PO_4/TiO_2 compounds that mainly determines the photocatalytic activities under visible-light irradiation, which is highly relevant to self-corrosion [146].

Besides Ag_3PO_4/TiO_2 composite heterostructures, AgX/Ag_3PO_4 (X = Cl, Br, I) heterocrystals have also attracted much attention due to their excellent photocatalytic activity [129]. Bi and coworkers have reported that the AgX/Ag_3PO_4 (X = Cl, Br, I) heterocrystals embodied some advantages compared to the single Ag_3PO_4, and it is a more promising and fascinating visible-

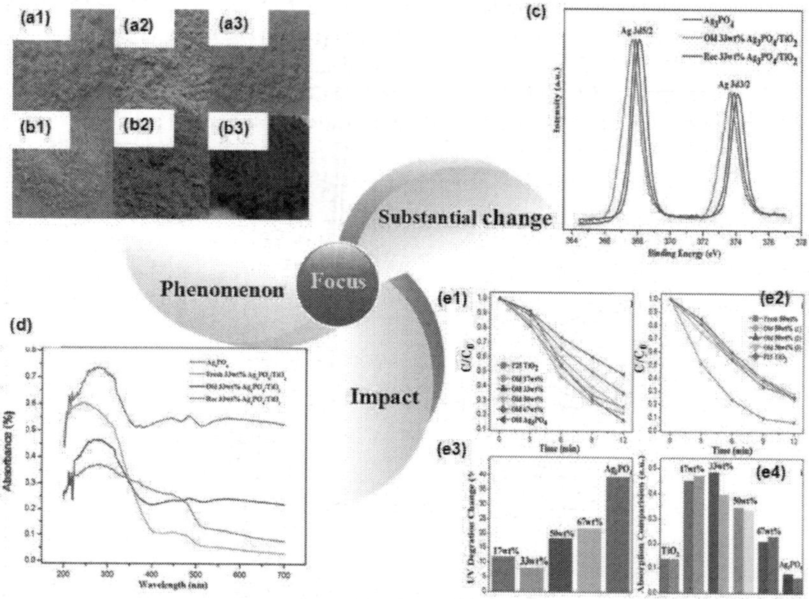

Figure 8. Images of fresh samples of (a1 and a3) P25 TiO_2, bare Ag_3PO_4, and Ag_3PO_4/TiO_2 compounds and their counterparts (b1 and b3) kept in dark for 5 days; (c) XPS spectra of Ag 3d; (d) UV–Vis absorption spectra for related samples; (e1) effects of "self-corrosion" on the photocatalytic activity under UV light irradiation, (e2) maintained time, (e3) decreased degradation percentage, (e4) adsorption capacity. Reproduced with permission from ref. 146 © 2014 Elsevier B.V.

light-driven photocatalyst than pure Ag_3PO_4 [129]. In their work, the $AgBr/Ag_3PO_4$ hybrid displayed much higher photocatalytic activity than single AgBr or Ag_3PO_4, as well as high stability under visible-light irradiation.

4. Long afterglow phosphor-assisted photocatalysts

Long afterglow phosphor-assisted novel photocatalysts become promising functional materials due to their effective utilization of solar light in practical applications of environmental purification. In this section, the recent development on TiO_2-based and Ag_3PO_4-based fluorescence photocatalyst composites with full-time active photocatalytic properties is reviewed.

The long afterglow phosphor ($CaAl_2O_4$:(Eu, Nd)) has a high luminescent brightness around 440 nm of wavelength, long afterglow time, good chemical stability, and low toxicity. The luminescent brightness around 440 nm can excite the visible-light-responsive nitrogen-doped titania ($TiO_{2-x}N_y$). Therefore, $TiO_{2-x}N_y$ photocatalyst was expected to possess a novel photocatalytic property after coupling with $CaAl_2O_4$:(Eu, Nd).

http://dx.doi.org/10.5772/61864

In our research [147–149], the degradation of continuous NO gas was achieved by $TiO_{2-x}N_y$ surface-immobilized $CaAl_2O_4$:(Eu, Nd) microparticles (Fig. 9). The results show a persistent deNOx ability of the $CaAl_2O_4$:(Eu, Nd)/$TiO_{2-x}N_y$ composite. In another work [149], the persistent fluorescence-assisted photocatalysts of Ag_3PO_4 composites have been synthesized to expand the application of Ag_3PO_4 on dye decomposition day and night. $Sr_4Al_{14}O_{25}$:(Eu,Dy)/Ag_3PO_4 composites exhibited excellent photocatalytic activities for the Rhodamine B (RhB) decomposition reaction in the dark without additional light sources (Fig. 9c).

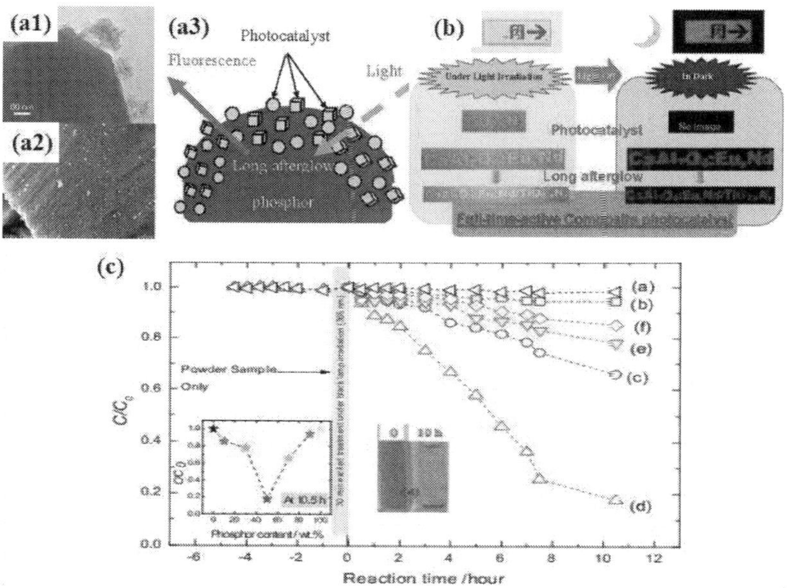

Figure 9. (a) Composition and (b) the mechanism of the persistent fluorescence-assisted photocatalysts; (c) variation of RhB solution concentration against illumination time using $Sr_4Al_{14}O_{25}$:(Eu,Dy)/Ag_3PO_4 composites. Reproduced with permission from ref. 148 © 2013 Elsevier B.V. & ref. 149 © 2013 RSC.

5. Challenges and perspectives

Photocatalysis appears to be a promising avenue to solve environmental and energy issues in the future. A variety of strategies, such as doping, coupling with graphene, precipitating with metal particles, crystal growth design, heterostructuring, were explored to enhance the efficiencies of photocatalytic activities. Besides modified TiO_2 and Ag_3PO_4, other visible-light-driven photocatalysts, including CdS, $BiVO_4$, Bi_2WO_6, and g-C_3N_4, even fluorescence-assisted photocatalyst composites, have attracted increased attention [147–156].

Although great progresses have been achieved, some challenges still exist to design high efficiency of photocatalytic systems. First, fundamental studies are essential to tackle the bottleneck problems in the field, including improved charge separation and transfer, promoted optical absorption, optimized band-gap position, lowered cost, and toxicity. Second, faceted photocatalysts remain a challenge and the development of surfactant-free synthesis routes is highly desirable, since most synthesis strategies involve the use of morphology-controlling agents that must be eventually removed in order to obtain clean facets. Finally, photostability of photocatalyst is and will continue to be a main challenge for practical applications. Therefore, new material design and innovative strategies for improving the efficiency and increasing the visible-light absorption of photocatalysts will be the key challenge and opportunity in this field.

Acknowledgements

This work is supported by the National Nature Science Young Foundation of China (no. 10904057), National Natural Science Foundation of China (51402139), the International Sci. & Tech. Cooperation Foundation of Gansu Provincial, China (Grant No. 1304WCGA177), and Specialized Research Fund for the Doctoral Program of Higher Education (No. 20120211130003).

Author details

Yuhua Wang[1,2*], Xinlong Ma[1,2], Hao Li[1,2], Bin Liu[1,2], Huihui Li[1,2], Shu Yin[3] and Tsugio Sato[3]

*Address all correspondence to: wyh@lzu.edu.cn

1 Key Laboratory for Special Function Materials and Structure Design of the Ministry of Education, Lanzhou University, Lanzhou, P. R. China

2 Department of Materials Science, School of Physical Science and Technology, Lanzhou University, Lanzhou, P. R. China

3 Institute of Multidisciplinary Research for Advanced Materials, Tohoku University, Aoba-ku, Sendai, Japan

References

[1] A. Kudo and Y. Miseki, Chem. Soc. Rev., 2009, 38, 253–278.

[2] M. Hoffmann, S. Martin, W. Choi and D. Bahnemann, Chem. Rev., 1995, 95, 69–96.

[3] Y. Hosogi, Y. Shimodaira, H. Kato, H. Kobayashi and A. Kudo, Chem. Mater., 2008, 20, 1299–1307.

[4] Z. Yi, J. Ye, T. Kako, S. Ouyang, H. Yang, W. Luo and R. Withers, Nat. Mater., 2010, 9, 559–564.

[5] L. Zhou, W. Wang, H. Xu, S. Sun and M. Shang, Chem. Eur. J., 2009, 15, 1776–1782.

[6] H. Cheng, B. Huang, X. Zhang and Y. Dai, Phys. Chem. Chem. Phys., 2010, 12, 15468–15475.

[7] J. Tang, Z. Zou and J. Ye, Angew. Chem. Int. Ed., 2004, 43, 4463–4466.

[8] F. Amano, A. Yamakata, M. Osawa and B. Ohtani, J. Am. Chem. Soc., 2008, 130, 17650–17651.

[9] L. Zhang, Y. Wang, H. Cheng, W Yao and Y. Zhu, Adv. Mater., 2009, 21, 1286–1290.

[10] L. Zhang, H. Wang, Z. Chen, P. Wong and J. Liu, Appl. Catal. B:Environ., 2011, 106, 1–13.

[11] S. Yao, J. Wei, B. Huang, S. Feng, X. Zhang et al., J. Solid State Chem., 2009, 182, 236–239.

[12] A. Kudo, K. Omori and H. Kato, J. Am. Chem. Soc., 1999, 121, 11459–11467.

[13] G. Xi and J. Ye, Chem. Commun., 2010, 46, 1893–1895.

[14] R. Li, F. Zhang, D. Wang, J. Yang, M. Li, J. Zhu et al., Nat. Commun., 2013, 4, 1432.

[15] Y. Liu, B. Huang, Y. Dai, X. Zhang, X. Qin et al., Catal. Commun., 2009, 11, 210–213.

[16] Y. Liu, Z. Wang, B. Huang, X. Zhang, X. Qin et al., J. Colloid Interface Sci., 2010, 348, 211–215.

[17] W. Yao, X. Xu, H. Wang, J. Zhou, X. Yang et al., Appl. Catal. B:Environ., 2004, 52, 109–116.

[18] W. Wei, Y. Dai and B. Huang, J. Phys. Chem. C, 2009, 113, 5658–5663.

[19] H. Cheng, B. Huang, K. Yang, X. Y. Qin, Z. Wang et al., Chem. Phys. Chem., 2010, 11, 2167–2173.

[20] Y. Liu, Z. Wang, B. Huang, K. Yang, X. Zhang et al., Appl. Surf. Sci., 2010, 257, 172–175.

[21] W. Wang, B. Huang, X. Ma, Z. Wang, X. Qin et al., Chem. Eur. J., 2013, 19, 14777–14780.

[22] H. Cheng, B. Huang and Y. Dai, Nanoscale, 2014, 6, 2009–2026.

[23] W. Tu, Y. Zhou and Z. Zou, Adv. Funct. Mater., 2013, 23, 4996–5008.

[24] H. Chen, L. Shao, Q. Li, J. Wang and Chem. Soc. Rev., 2013, 42, 2679–2724.

[25] H. Tada, T. Kiyonaga and S. Naya, Chem. Soc. Rev., 2009, 38, 1849–1858.

[26] H. Tang, C. Hessel, J. Wang, H. Zhao and D. Wang, Chem. Soc. Rev., 2014, 43, 4281–4299.

[27] Q. Xiang, J. Yu and M. Jaroniec, Chem. Soc. Rev., 2012, 41, 782–796.

[28] K. Manga, Y. Zhou, Y. Yan and K. Loh, Adv. Funct. Mater., 2009, 19, 3638–3643.

[29] H. Li, Y. Zhou, W. Tu, J. Ye and Z. Zou, Adv. Funct. Mater. 2015, 25, 998–1013.

[30] J. Su, X. Zou, G. Li, Y. Wang, L. Zhou and J. Chen, J. Phys. Chem. C, 2011, 115, 8064–8071.

[31] X. Wang, Q. Xu, M. Li, H. Han and C. Li, Angew. Chem. Int. Ed., 2012, 51, 13089–13092.

[32] G. Liu, J. Yu, G. Lu and H. Cheng, Chem. Commun., 2011, 47, 6763–6783.

[33] R. Asahi, T. Morikawa, T. Ohwaki, K. Aoki and Y. Taga, Science, 2001, 293, 269.

[34] B. Liu, Y. Wang, Y. Huang, P. Dong and S. Yin, J. Austr. Ceram. Soc., 2012, 48, 249–252.

[35] S. Khan, M. Al-Shahry and W. Ingler, Science, 2002, 297, 2243.

[36] C. Di Valentin, G. Pacchioni and A. Selloni, Chem. Mater., 2005, 17, 6656–6665.

[37] S. Sakthivel and H. Kisch, Angew. Chem., Int. Ed., 2003, 42, 4908–4911.

[38] T. Umebayashi, T. Yamaki, H. Itoh and K. Asai, Appl. Phys. Lett., 2002, 81, 454–456.

[39] W. Ho, J. Yu and S. Lee, J. Solid State Chem., 2006, 179, 1171–1176.

[40] W. Zhao, W. Ma, C. Chen, J. Zhao and Z. Shuai, J. Am. Chem. Soc., 2004, 126, 4782–4783.

[41] G. Liu, Y. Zhao, C. Sun and H. Cheng, Angew. Chem., Int. Ed., 2008, 47, 4516–4520.

[42] S. In, A. Orlov, R. Berg, F. Garcia and R. Lambert, J. Am. Chem. Soc., 2007, 129, 13790–13791.

[43] J. Yu, J. Yu, W. Ho, Z. Jiang and L. Zhang, Chem. Mater., 2002, 14, 3808–3816.

[44] D. Li, H. Haneda and N. Labhsetwar, Chem. Phys. Lett., 2005, 401, 579–584.

[45] C. Di Valentin, E. Finazzi, M. Paganini and E. Giamello, Chem. Mater., 2008, 20, 3706–3714.

[46] H. Luo, T. Takata, Y. Lee, J. Zhao, K. Domen and Y. Yan, Chem. Mater., 2004, 16, 846–849.

[47] X. Hong, Z. Wang, W. Cai, F. Lu and J. Zhang, Chem. Mater., 2005, 17, 1548–1552.

[48] G. Liu, Z. Chen, C. Dong, G. Lu and H. Cheng, J. Phys. Chem. B, 2006, 110, 20823–20828.

[49] W. Su, Y. Zhang, Z. Li, L. Wu, X. Wang, J. Li and X. Fu, Langmuir, 2008, 24, 3422–3428.

[50] S. Tojo, T. Tachikawa, M. Fujitsuka and T. Majima, J. Phys. Chem. C, 2008, 112, 14948–14954.

[51] L. Lin, W. Lin, Y. Zhu, B. Zhao and Y. Xie, Chem. Lett., 2005, 34, 284–285.

[52] G. Liu, L. Wang, H. Yang, H. Cheng and G. Lu, J. Mater. Chem., 2010, 20, 831–843.

[53] G. Liu, C. Sun, L. Cheng, G. Lu and H. Cheng, J. Phys. Chem. C, 2009, 113, 12317–12324.

[54] J. Graciani, A. Nambu, J. Evans and J. Sanz, J. Am. Chem. Soc., 2008, 130, 12056–12063.

[55] T. Ikeda, T. Nomoto, K. Eda, A. Kudo and H. Onishi, J. Phys. Chem. C, 2008, 112, 1167–1173.

[56] V. Singh, D. Joung, L. Zhai, S. Khondaker and S. Seal, Prog. Mater Sci., 2011, 56, 1178–1271.

[57] H. Chae, D. Pérez, J. Kim, Y. Go, M. Eddaoudi et al., O. Yaghi, Nature, 2004, 427, 523–527.

[58] A. Geim and K. Novoselov, Nat. Mater., 2007, 6, 183–191.

[59] K. Loh, Q. Bao, P. Ang, J. Yang and J. Mater. Chem., 2010, 20, 2277–2289.

[60] Y. Liang, H. Wang, H. Casalongue, Z. Chen and H. Dai, Nano Res., 2010, 3, 701–705.

[61] Y. Zhang, Z. Tang, X. Fu and Y. Xu, ACS Nano., 2010, 4, 7303–7314.

[62] Y. Wang, R. Shi, J. Lin and Y. Zhu, Appl. Catal. B, 2010, 100, 179–183.

[63] X. Meng, D. Geng, J. Liu, R. Li and X. Sun, Nanotechnology, 2011, 22, 165602–165612.

[64] P. Wang, Y. Zhai, D. Wang and S. Dong, Nanoscale, 2011, 3, 1640–1645.

[65] J. Liu, H. Bai, Y. Wang, Z. Liu, X. Zhang and D. Sun, Adv. Funct. Mater., 2010, 20, 4175–4181

[66] J. Liu, L. Liu, H. Bai, Y. Wang and D. Sun, Appl. Catal. B:Environ., 2011, 106, 76–82.

[67] C. Wu, X. Huang, L. Xie, X. Wu, J. Yu and P. Jiang, J. Mater. Chem., 2011, 21, 17729–17736.

[68] B. Liu, Y. Huang, Y. Wen, L. Du, W. Zeng et al., J. Mater. Chem., 2012, 22, 7484–7491.

[69] P. Dong, Y. Wang, L. Guo, B. Liu, S. Xin et al., Nanoscale, 2012, 4, 4641–4649.

[70] H. Li, B. Liu, Y. Wang, Y. Shi, X. Ma et al., RSC Advance, 2014, 4, 37992–37997.

[71] H. Cheng, K. Fuku, Y. Kuwahara, K. Yamashita, J. Mater. Chem. A, 2015, 3, 5244–5258.

[72] X. Liu and M. T. Swihart, Chem. Soc. Rev., 2014, 43, 3908–3920.

[73] G. Baffou and R. Quidant, Chem. Soc. Rev., 2014, 43, 3898–3907.

[74] C. Clavero, Nat. Photonics, 2014, 8, 95–103.

[75] X. Chen, Z. Zheng, X. Ke, E. Jaatinen, T. Xie et al., Green Chem., 2010, 12, 414–419.

[76] S. Sun, W. Wang, L. Zhang, M. Shang and L. Wang, Catal. Commun., 2009, 11, 290–293.

[77] T. Takahashi, A. Kudo, S. Kuwabata, A. Ishikawa, H. Ishihara et al., J. Phys. Chem. C, 2013, 117, 2511–2520.

[78] X. Zhou, G. Liu, J. Yu and W. Fan, J. Mater. Chem., 2012, 22, 21337–21354.

[79] S. Kochuveedu, D. Kim and D. Kim, J. Phys. Chem. C, 2012, 116, 2500–2506.

[80] C. Li, Y. Zheng and T. Wang, J. Mater. Chem., 2012, 22, 13216–13222.

[81] M. Zhou, J. Zhang, B. Cheng and H. Yu, Int. J. Photoenergy, 2012, 2012, 532843.

[82] Y. Wen, B. Liu, W. Zeng and Y. Wang, Nanoscale, 2013, 5, 9739–9746.

[83] H. Tong, S. Ouyang, Y. Bi, N. Umezawa, J. Ye et al., Adv. Mater., 2012, 24, 229–251.

[84] X. Zhang, V. Thavasi, S. G. Mhaisalkar and S. Ramakrishna, Nanoscale, 2012, 4, 1707–1716.

[85] L. Huang, X. Wang, J. Yang, G. Liu, J. Han and C. Li, J. Phys. Chem. C, 2013, 117, 11584–11591.

[86] Z. Fang, Y. Liu, Y. Fan, Y. Ni, X. Wei et al., J. Phys. Chem. C, 2011, 115, 13968–13976.

[87] J. Jang, S. Choi, H. Kim and J. Lee, J. Phys. Chem. C, 2008, 112, 17200–17205.

[88] C. Li, J. Yuan, B. Han, L. Jiang and W. Shangguan, Int. J. Hydrogen. Energ., 2010, 35, 7073–7079.

[89] L. Sang, H. Tan, X. Zhang, Y. Wu, C. Ma and C. Burda, J. Phys. Chem. C, 2012, 116, 18633–18640.

[90] J. Ng, S. Xu, X. Zhang, H. Yang and D. Sun, Adv. Funct. Mater., 2010, 20, 4287–4294.

[91] H. Bai, Z. Liu and D. Sun, J. Am. Ceram. Soc., 2013, 96, 942–949.

[92] K. DeKrafft, C. Wang and W. Lin, Adv. Mater., 2012, 24, 2014–2018.

[93] X. Wang, G. Liu, G. Lu and H. Cheng, Int. J. Hydrogen. Energ., 2010, 35, 8199–8205.

[94] K. B. Chai, T. Peng, J. Mao and L. Zan, ACS Catal., 2013, 3, 170–177.

[95] J. Wang, H. Ruan, W. Li, D. Li, Y. Hu et al., J. Phys. Chem. C, 2012, 116, 13935–13943.

[96] X. Li, Y. Hou, Q. Zhao and G. Chen, Langmuir, 2011, 27, 3113–3120.

[97] A. Kezzim, N. Nasrallah, A. Abdi and M. Trari, Energ. Convers. Manag., 2011, 52, 2800–2806.

[98] Q. Simon, D. Barreca, A. Gasparotto, C. Maccato, T. Montini et al., J. Mater. Chem., 2012, 22, 11739–11747.

[99] Y. Liu, Y. X. Yu and W. D. Zhang, J. Phys. Chem. C, 2013, 117, 12949–12957.

[100] X. Zong, H. Yan, G. Wu, G. Ma, F. Wen et al., J. Am. Chem. Soc., 2008, 130, 7176–7177.

[101] D. Sarkar, C. Ghosh, S. Mukherjee and K. K. Chattopadhyay. ACS Appl. Mater. Interfaces, 2013, 5, 331–337.

[102] Y. Liao, H. Zhang, J. Zhang, L. Jia, F. Bai et al., ACS Appl. Mater. Interfaces, 2013, 5, 11022–11028.

[103] L. Lin, Y. Yang, L. Men, X. Wang, D. He et al., Nanoscale, 2013, 5, 588–593.

[104] Y. Hou, X. Li, Q. Zhao, X. Quan and G. Chen, Adv. Funct. Mater., 2010, 20, 2165–2174.

[105] Z. Zhang, C. Shao, X. Li, C. Wang, M. Zhang et al., ACS Appl. Mater. Interfaces, 2010, 2, 2915–2923.

[106] J. Yang, W. Liao and J. Wu, ACS Appl. Mater. Interfaces, 2013, 5, 7425–7431.

[107] L. Liau, Y. Lin and Y. Peng, J. Phys. Chem. C, 2013, 117 26426–26431.

[108] Y. Sun, W. Wang, L. Zhang and Z. Zhang, Chem. Eng. J., 2012, 211, 161–167.

[109] S. Chen, W. Liu, S. Zhang and Y. Chen, J. Sol–Gel Sci. Technol., 2010, 54, 258–267.

[110] F. Abdi, L. Han, A. Smets, M. Zeman, B. Dam and R. Krol, Nat. Commun., 2013, 4, 2195.

[111] Y. Qu, L. Liao, R. Cheng, Y. Wang, Y. Lin et al., Nano Lett., 2010, 10, 1941–1949.

[112] H. Yun, H. Lee, N. Kim, D. Lee, S. Yu and J. Yi, ACS Nano, 2011, 5, 4084–4090.

[113] H. Tada, T. Mitsui, T. Kiyonaga, T. Akita and K. Tanaka, Nat. Mater., 2006, 5, 782–786.

[114] X. Wang, L. Yin, G. Liu, L. Wang, R. Saito et al., Energy Environ. Sci., 2011, 4, 3976–3979.

[115] Z. Yu, Y. Xie, G. Liu, G. Lu, X. Ma and H. Cheng, J. Mater. Chem. A, 2013, 1, 2773–2776.

[116] S. Qin, F. Xin, Y. Liu, X. Yin and W. Ma, J. Colloid. Interf. Sci., 2011, 356, 257–261.

[117] M. Miyauchi, Y. Nukui, D. Atarashi and E. Sakai, ACS Appl. Mater. Interfaces, 2013, 5, 9770–9776.

[118] Y. Wang, Q. Wang, X. Zhan, F. Wang, M. Safdar and J. He, Nanoscale, 2013, 5, 8326–8339.

[119] T. Teranishi and M. Sakamoto, J. Phys. Chem. Lett., 2013, 4, 2867–2873.

[120] Y. Bi, S. Ouyang, N. Umezawa, J. Cao and J. Ye, J. Am. Chem. Soc., 2011, 133, 6490–6492.

[121] Y. Bi, H. Hu, Z. Jiao, H. Yu, G. Lu et al., Phys. Chem. Chem. Phys., 2012, 14, 14486–14488.

[122] W. Jun, F. Teng and M. Chen, CrystEngComm, 2013, 15, 39–42.

[123] Q. Liang, W. Ma, Y. Shi, Z. Li and X. Yang, CrystEngComm, 2012, 14, 2966–2973.

[124] Y. Bi, H. Hu, S. Ouyang, G. Lu, J. Cao and J. Ye, Chem. Commun., 2012, 48, 3748–3750.

[125] J. Liu, C. Luo, J. Wang, X. Yang and X. Zhong, CrystEngComm, 2012, 14, 8714–8721.

[126] H. Wang, L. He, L. Wang, P. Hu, L. Guo et al., CrystEngComm, 2012, 14, 8342–8344.

[127] S. Rawal, S. Sung and W. Lee, Catal. Commun., 2011, 17, 131–135.

[128] W. Yao, B. Zhang, C. Huang, C. Ma, X. Song and Q. Xu, J. Mater. Chem., 2012, 22, 4050–4055.

[129] Y. Bi, S. Ouyang, J. Cao and J. Ye, Phys. Chem. Chem. Phys., 2011, 13, 10071–10075.

[130] B. Cao, P. Dong, S. Cao and Y. Wang, J. Am. Ceram. Soc., 2013, 96, 544–548.

[131] G. Li and L. Mao, RSC Adv., 2012, 2, 5108–5111.

[132] L. Zhang, H. Zhang and H. Huang, New J. Chem., 2012, 36, 1541–1544.

[133] H. Zhang, H. Huang, H. Ming and L. Zhang, J. Mater. Chem., 2012, 22, 10501–10506.

[134] P. Dong, Y. Wang, B. Cao, S. Xin, L. Guo et al., Appl. Catal. B: Environ., 2013, 132–133, 45–53.

[135] Y. Liu, L. Fang, H. Lu, L. Liu, H. Wang and C. Hu, Catal. Commun., 2011, 17, 200–204.

[136] Y. Bi, H. Hu, S. Ouyang, Z. Jiao, G. Lu and J. Ye, J. Mater. Chem., 2012, 22, 14847–14850.

[137] Y. Liu, L. Fang, H. Lu, Y. Li, C. Hu et al., Appl. Catal. B: Environ., 2012, 115-116, 245–252.

http://dx.doi.org/10.5772/61864

[138] C. Haynes, A. McFarland and R Duyne, Anal. Chem., 2005, 77, 338A–346A.

[139] B. Wiley, Z. Wang, J. Wei, Y. Yin, D. Cobden and Y. Xia, Nano Lett., 2006, 6, 2273–2278.

[140] M. Aslam, R. Bhobe, N. Alem, S. Donthu and V. Dravid, J. Appl. Phys., 2005, 98, 1–8.

[141] G. Somorjai and D. Blakely, Nature, 1975, 258, 580–583.

[142] D. Martin, N. Umezawa, X. Chen, J. Ye and J. Tang, Energy Environ. Sci., 2013, 6, 3380–3386.

[143] P. Dong, Y. Wang, H. Li, H. Li, X. Ma and L. Han, J. Mater. Chem. A, 2013, 1, 4651–4656.

[144] E. Bekyarova, M. Itkis, P. Ramesh, C. Berger, M. Sprinkle et al., J. Am. Chem. Soc., 2009, 131, 1336–1337.

[145] G. Williams, B. Seger and P.V. Kamat, ACS Nano, 2008, 2, 1487–1491.

[146] X. Ma, H. Li, Y. Wang, H. Li, B. Liu, et al., Appl. Catal. B: Environ., 2014, 158–159, 314–320.

[147] H. Li, S. Yin, Y. Wang, T. Sato, J. Catal., 2012, 286, 273–278.

[148] H. Li, S. Yin, Y. Wang, T. Sato, Appl. Catal. B:Environ., 2013, 132–133, 487–492.

[149] H. Li, S. Yin, Y. Wang, T. Sekino, S. Lee and T. Sato, J. Mater. Chem. A, 2013, 1, 1123–11261.

[150] J. Savio, A. Stephen, D. Martin, Z. Guo and J. Tang, Energy Environ. Sci., 2015, 8, 731–759.

[151] R. Li, H. Han, F. Zhang, D. Wang and C. Li, Energy Environ. Sci., 2014, 7, 1369–1376.

[152] T. Kim and K. Choi, Science, 2014, 343, 990.

[153] T. Saison, N. Chemin, C. Chanéac, O. Durupthy, V. Ruaux et al., *J. Phys. Chem. C*, 2011, 115, 5657–5666.

[154] Y. Huang, Z. Ai, W. Ho, M. Chen and S. Lee, *J. Phys. Chem. C*, 2010, 114, 6342–6349.

[155] Z. Zhao, Y. Sun and F. Dong, Nanoscale, 2015, 7, 15–37.

[156] C. Li, S. Wang, T. Wang, Y. Wei, P. Zhang and J. Gong, Small, 2014, 10, 2783–2790.

Chapter 13

First-Transition Metal Oxocomplex–Surface-Modified Titanium(IV) Oxide for Solar Environmental Purification

Hiroaki Tada and Qiliang Jin

Additional information is available at the end of the chapter

http://dx.doi.org/10.5772/62008

Abstract

The ongoing global energy and environmental issues warrant the development of environmental catalysts for decomposing pollutants in water and air by utilizing solar energy named as "solar environmental catalysts." This chapter describes the recent studies on a novel class of solar environmental catalysts consisting of TiO_2 and molecular-scale first-row transition metal oxide clusters (or metal oxocomplexes) on the surface (MOs/TiO_2). The TiO_2 surface modification with the oxocomplexes by the chemisorption–calcination cycle (CCC) technique presents a novel band engineering for fine-tuning the band energy. The unique physicochemical and electronic properties of MOs/TiO_2 give rise to the outstanding photocatalytic activity for the decomposition of organic pollutants. The combination with the rapidly growing technique for preparation of TiO_2 nanostructures allows us to expect further improvement of the performances and the wide application to the solar chemical transformation for producing useful substances.

Keywords: Visible-light photocatalyst, Titanium(IV) oxide, Surface modification, Metal oxide cluster, Metal oxocomplex, Band energy turning, Solar environmental purification

1. Introduction

Environmental pollution is beyond limits, and the development of environmental catalysts is a critical subject for scientists and engineers all over the world. Here, the catalysts for purifying polluted water and air by utilizing the solar energy are termed as "solar environmental catalyst." The TiO_2 photocatalyst possesses great potential as the "solar environmental catalyst" owing to its strong oxidation ability, high physicochemical stability, abundance in nature, and nontoxicity [1,2]. Scheme 1 shows the fundamental reaction mechanism on the TiO_2-photocatalyzed decomposition of organic pollutants with the characteristic time for each process [3]. UV-light absorption by TiO_2 causes the excitation of electrons in the valence band (VB) to the conduction band (CB) in the order of femtoseconds. Most of the photogenerated charge carriers are lost by the recombination within ~100 ns. The charge carriers surviving the recombination are trapped at the TiO_2 surface to induce the redox reactions. In general, the CB electrons reduce oxygen (O_2), whereas the VB holes oxidize organic pollutants. The VB holes have a highly positive potential (+2.67 V versus standard hydrogen electrode (SHE) at pH 7) to oxidize most organic compounds. Conversely, the driving force for one-electron O_2 reduction (standard potential, $E^0(O_2/O_2^-) = -0.284$ V versus SHE) by the CB electrons (−0.53 V versus SHE at pH 7) is small. Consequently, the O_2 reduction reaction (ORR) is much slower (~ms) than the oxidation process (~100 ns). TiO_2 usually takes crystal forms of rutile and anatase. The flatband potential of anatase is ~0.2 V, which is more negative than that of rutile, and anatase has a higher UV-light activity for the oxidation of organic compounds as compared with rutile [4]. This fact also points to the importance of the ORR in TiO_2 photocatalytic reactions. Also, the coupling of anatase and rutile TiO_2 can further increase the UV-light activity because of the enhancement of charge separation due to the interfacial electron transfer [5].

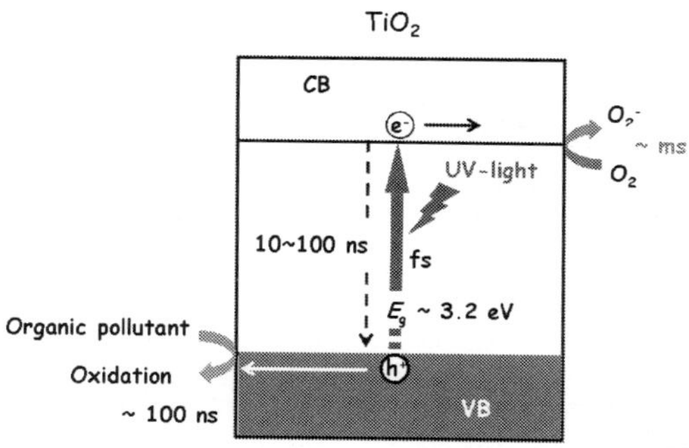

Sheme 1. The basic mechanism on the TiO_2 photocatalytic reaction (the surface trapping processes for the CB electrons and VB holes are abbreviated) with the characteristic time for each step shown.

Recently, the visible-light activation of TiO_2 by its surface modification with metal oxide nanoparticles (NPs) or oxocomplexes has been developed [6,7]. This approach has a major advantage over the conventional doping [8–14], in that visible-light activation can be achieved by a simple procedure without the introduction of the impurity/vacancy levels into the bulk TiO_2. To date, the impregnation method has been mainly used for the surface modification with metal oxide NPs, including chromium oxides [15], iron oxides [16–18], and copper oxides [19]. Unfortunately, the surface modification by the impregnation method is effective for rutile but less effective for anatase.

This chapter deals with our recent studies on the surface modification of anatase TiO_2 with the first (3d) transition metal oxocomplexes (MOCs) by the chemisorption–calcination cycle technique (MOCs/TiO_2) [20] and the characterization and photocatalytic activities for the degradation of organic pollutants. We show that some MOCs/TiO_2 fulfill the requirements for the "solar environmental catalyst."

2. Design for solar environmental catalysts

The requirements for the highly active TiO_2-based "solar environmental catalyst" are described below. As shown in Figure 1, the excitation from the VB electrons to the CB needs UV-light irradiation, occupying only 3% of the incident sunlight. From a viewpoint of the effective use of the sunlight, the response to the visible-light occupying 45% of the solar energy (Figure 1) should be particularly imparted to anatase TiO_2 and anatase–rutile-mixed TiO_2 (Requirement 1). However, even if TiO_2 can be endowed with the visible-light activity, it is usually much smaller than the UV-light activity. Therefore, the inherent excellent UV-light activity of TiO_2 should be compatible with the visible-light activity (Requirement 2). For high levels of visible and UV-light activities to be achieved, the enhancement of the ORR is crucial because it is usually the rate-determining step in the TiO_2-photocatalyzed reactions [21] (Requirement 3).

3. Catalyst preparation

3.1. Chemisorption–calcination cycle technique

The adsorption mechanism of metal acetylacetonates (acac) on TiO_2 depends on the kind of complexes. As an example, Fe oxocomplex formation by the CCC technique is represented in Scheme 2. In the first step, $Fe(acac)_3$ is chemisorbed on the TiO_2 surface via the ligand exchange between the acac ligand and the surface Ti-OH group from the nonaqueous solution (Eq. 1) [22]:

$$Fe(acac)_3 + l(Ti_s\text{-}OH) \rightarrow Fe(acac)_{3-l}(O\text{-}Ti_s)_l + lAcacH \ (l \approx 1) \quad (1)$$

where the subscript s denotes the surface atom.

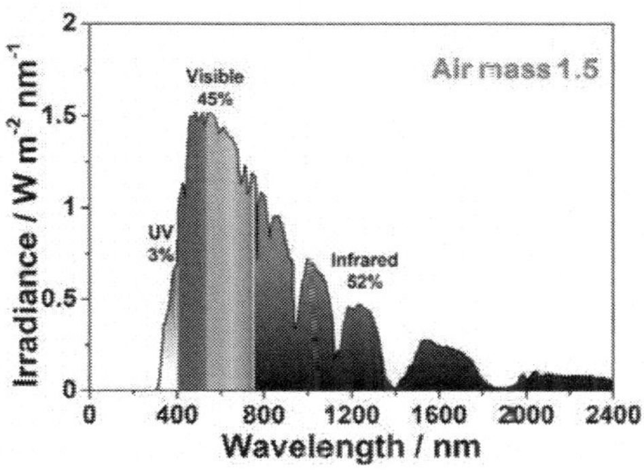

Figure 1. Solar spectrum (AM 1.5).

Conversely, $[Sn(acac)_2]Cl_2$ is adsorbed on TiO_2 via the ion exchange between H^+ and $[Sn(acac)_2]^{2+}$ ion (Eq. 2) [23]. In each case, the adsorption apparently obeys the Langmuir model. The saturated adsorption amount and the adsorption constant for the adsorption of various metal acetylacetonates on TiO_2 at 298 K are summarized in Table 1. The adsorption constants range from 10^2 to 10^4, indicating that they are strongly adsorbed on the TiO_2 surface by chemical bonds. Exceptionally, $Cr(acac)_3$ is not adsorbed because of its large ligand-field stabilization energy ($1.2\Delta_0$).

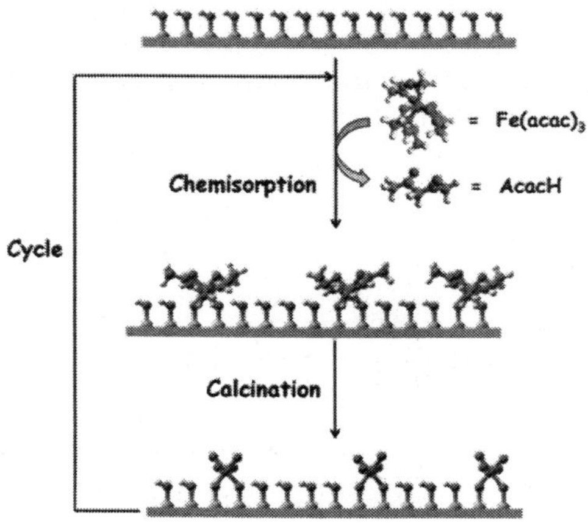

Sheme 2. Fe oxocomplex (Fe_2O_3) formation on the TiO_2 surface by the CCC technique.

TiO_2 (crystal form, specific surface area)	Precursor	Solvent	Saturated adsorption amount / ions nm^{-2}	Adsorption equilibrium constant / 10^3 mol^{-1} dm^3	Mechanism	Oxidation number of metal ion in oxide cluster *
ST-01 (A, 308 $m^2 g^{-1}$)	$VO(acac)_2$	ethanol	0.18	1.67	ligand exchange	V^{5+} (V_2O_5)
P-25 (A-R, 50 $m^2 g^{-1}$)	$Cr(acac)_3$	ethanol	0	0	—	—
P-25 (A-R, 50 $m^2 g^{-1}$)	$Mn(acac)_3$	ethanol	1.65	0.17	ligand exchange	Mn^{3+} (Mn_2O_3)
P-25 (A-R, 50 $m^2 g^{-1}$)	$Fe(acac)_3$	ethanol : n-hexane 3:17 v/v	0.46	2.06	ligand exchange	Fe^{3+} (Fe_2O_3)
ST-01 (A, 308 $m^2 g^{-1}$)			0.35	1.52		Fe^{3+} (Fe_2O_3)
ST-01 (A, 308 $m^2 g^{-1}$)	$Co(acac)_2(H_2O)_2$	methanol	0.18	1.23	surface Ti-OH coordination	Co^{3+} (Co_2O_3)
P-25 (A-R, 50 $m^2 g^{-1}$)	$Ni(acac)_2(H_2O)_2$	ethanol : n-hexane 3:17 v/v	0.56	0.74	ligand exchange	Ni^{2+} (NiO)
P-25 (A-R, 50 $m^2 g^{-1}$)	$Cu(acac)_2$	acetonitrile	0.54	1.34	surface Ti-OH coordination	Cu^{2+} (CuO)
A-100 (A, 8.1 $m^2 g^{-1}$)	$[Sn(acac)_2]Cl_2$	ethanol	0.29	17.0	ion-exchange	Sn^{4+} (SnO_2)
MT-700B (R, 12.5 $m^2 g^{-1}$)			0.10	46.0		Sn^{4+} (SnO_2)

Table 1. Adsorption properties of 3d metal acetylacetonates on TiO_2 at 298 K.

http://dx.doi.org/10.5772/62008

$$\left[\mathrm{Sn(acac)_2}\right]\mathrm{Cl_2} + 2\left(\mathrm{Ti_s\text{-}OH}\right) \rightarrow \mathrm{Sn(acac)_2}\left(\mathrm{O\text{-}Ti_s}\right)_2 + 2\mathrm{HCl} \quad (2)$$

In the second step, the oxidation of the acac ligands by heating the samples in air at 773 K yields iron oxides on the TiO_2 surface Further, these procedures are repeated to control the Fe-loading amount. Chemical analysis confirmed that all the Fe was confirmed to be present only on the TiO_2 surface. The Fe-loading amount is expressed by the number of Fe ions per unit TiO_2 surface area (Γ/ions nm^{-2}).

3.2. Control of loading amount

A feature of the CCC technique is precise control of the loading amount of metal oxides. As an example, the manner in which the Fe-loading amount is controlled in the iron oxide/TiO_2 system is described. Figure 2A shows the relation between Γ and initial complex concentration: black, $Fe(acac)_3$ and blue, $Mn(acac)_3$. In each case, the Γ gradually increases with an increase in the initial concentration. Figure 2B shows plots of Γ versus CCC cycle number (N): black, $Fe(acac)_3]_0$ = 0.65 mmol dm^{-3} (black) and blue, $Mn(acac)_3]_0$ = 8 mmol dm^{-3}. The Γ further increases in proportional to N in both the systems. The linear Γ–N relation is also observed in the other metal oxide/TiO_2 systems. In this manner, the loading amount of metal oxides can be controlled in a wide range using the precursor complex concentration and the cycle number.

4. Structure of surface metal oxocomplexes

Another feature of the CCC technique is the formation of molecular-scale metal oxide species on TiO_2. Figure 3 shows transmission electron micrographs (TEMs) of iron oxide/TiO_2 with varying Γ. No particulate deposits are observed on the TiO_2 surface at $\Gamma < 1$ ions nm^{-2}. This fact suggests that iron oxides exist as molecular-scale iron oxide species on the TiO_2 surface.

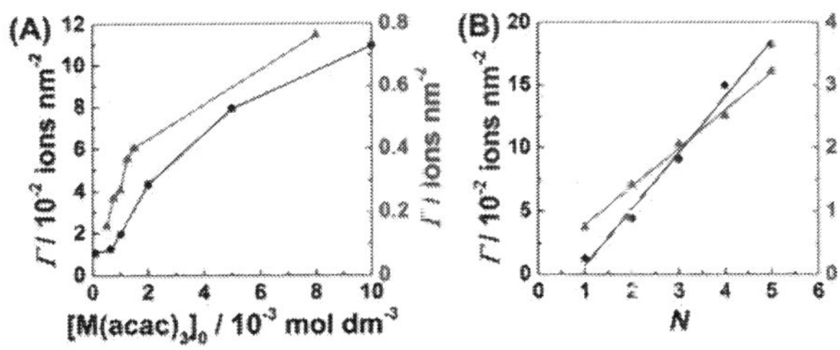

Figure 2. (A) Plots of Fe-loading amount (Γ/ions nm^{-2}) versus initial concentration of the complex ($[M(acac)_3]_0$): M = Fe (black) and M = Mn (blue). (B) Plots of Γ versus cycle number (N) at $Fe(acac)_3]_0$ = 0.65 mmol dm^{-3} (black) and $Mn(acac)_3]_0$ = 8 mmol dm^{-3} (blue).

To obtain the detailed structural information, Fe K-edge X-ray absorption fine-structure spectra were measured for the iron oxide/TiO_2 samples with varying Γ [24]. Figure 4A shows

X-ray absorption near-edge structure (XANES) spectra for the iron oxide/TiO_2 samples, and authentic Fe, Fe_3O_4, and α-Fe_2O_3 for comparison. The absorption edge of the iron oxide/TiO_2 sample is in agreement with that of α-Fe_2O_3, indicating the oxidation number of the iron to be +3 (Table 1). Figure 4B shows the Fourier transforms of the k^3-weighted X-ray absorption fine structure (EXAFS) for the iron oxide/TiO_2 samples. The peaks around 1.6 Å and 2.8 Å arise from the scattering from the nearest neighbor O and Fe, respectively. It is worth noting that the latter peak becomes very weak at $\Gamma \leq 0.5$. Evidently, the iron oxides exist as a mononuclear Fe oxocomplex on TiO_2 designated as Fe_2O_3/TiO_2 below.

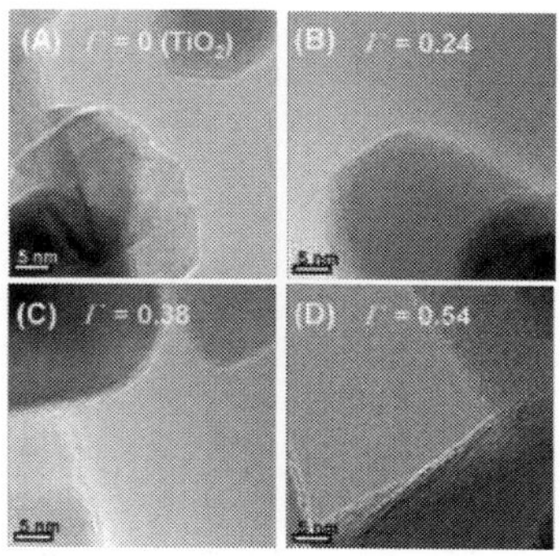

Figure 3. TEM images of Fe_2O_3/TiO_2 with varying Γ: (A) $\Gamma = 0$ (P-25); (B) $\Gamma = 0.23$, (C) $\Gamma = 0.38$, (D) $\Gamma = 0.54$.

In this manner, MOCs are formed on the TiO_2 surface in a highly dispersed state by the CCC technique, whereas the conventional impregnation method usually yields metal oxide NPs. As illustrated in Scheme 2, the molecular size of $Fe(acac)_3$ (~0.5 nm^2 $complex^{-1}$) is much larger than the reciprocal number density of the Ti-OH groups on the TiO_2 surface (~0.1 nm^2 $group^{-1}$) [25]. In the first step of the CCC process, $Fe(acac)_3$ complexes are chemisorbed isolatedly each other owing to the bulky acac-ligands. Also, the strong bond between the complexes and the TiO_2 surface suppresses the aggregation of the oxocomplexes during the second step. Consequently, mononuclear MOCs can be formed on the TiO_2 surface.

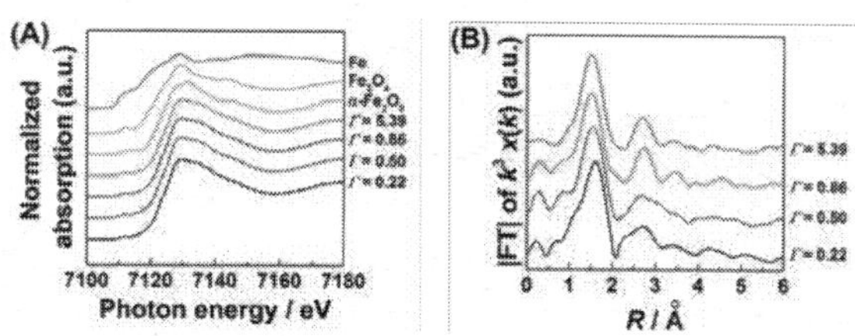

Figure 4. XANES and EXAFS spectra. (A) XANES spectra for Fe, Fe_3O_4, α-Fe_2O_3, and Fe_2O_3/TiO_2 with varying Γ. (B) Fourier transforms of the k^3-weighted EXAFS spectra for Fe_2O_3/TiO_2. The figures were taken from Ref. 24.

5. Characteristics of metal oxocomplex–surface-modified TiO_2

5.1. Optical property

The optical property is a fundamental factor affecting the photocatalytic activity. Figure 5 shows UV–visible absorption spectra for Fe oxocomplex–surface-modified mesoporous TiO_2 nanocrystalline films (Fe_2O_3/mp-TiO_2) with varying Γ. Impregnation samples usually have absorption approximately 470 nm in addition to absorption at 410 nm [16,17,26,27]. The former and latter absorptions can be attributed to the d–d transition and electronic transition from Fe^{3+} levels to the CB of TiO_2, respectively [28]. A strong d–d absorption is also observed for a heavy-loading CCC sample (Fe_2O_3(Γ = 5.5)/mp-TiO_2). In contrast, the absorption spectra of Fe_2O_3($\Gamma \leq 2.1$)/mp-TiO_2 apparently show a marked bandgap narrowing from 3.3 to 2.85 eV with an increase in Γ, whereas the d–d transition absorption is very weak [24]. Similar spectra were previously observed for TiO_2 doped with Cr [12] and N [13] prepared by ion implantation and magnetron sputtering. The weak d–d transition absorption is a common feature for the CCC samples including Co_2O_3/TiO_2 [29], NiO/TiO_2 [30,31], and CuO/TiO_2 [32]. Clearly, the unique optical properties of the CCC samples originate from the highly dispersed MOCs on the TiO_2 surface.

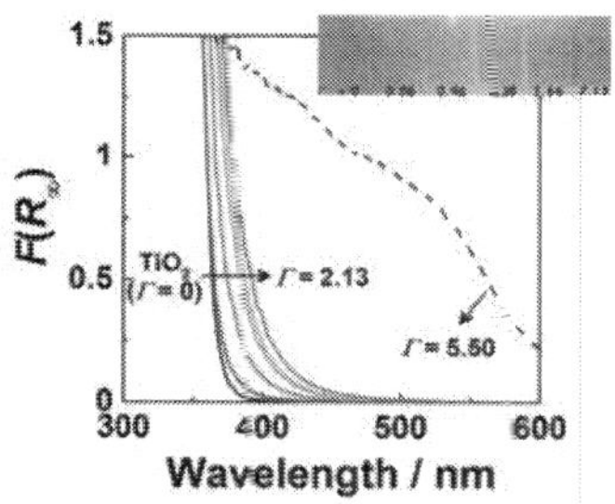

Figure 5. UV–Vis absorption spectra of Fe_2O_3/mp-TiO_2 prepared by the CCC technique.

5.2. Fine-tuning of band energy

The VB maximum determines the oxidation ability of the holes, and thus is a key factor for the decomposition of organic pollutants by semiconductor photocatalysts. The VB maximum level can be estimated from the VB-X-ray photoelectron spectroscopy (XPS) [24]. Since the VB maximum of TiO_2 is almost independent of its crystal form and size, the VB maximum of MOC-modified TiO_2 can be compared with respect to that of unmodified TiO_2. Figure 6 shows the VB-XPS spectra for Fe_2O_3/TiO_2 with varying Γ. The emission from the O 2*p*–VB extends from 2 to 9 eV. As a result of the surface modification, the top of VB rises, whereas the bottom remains invariant. Figure 6B compares the energy shift in the VB maximum level with respect to that of unmodified TiO_2 (ΔE_{VBM}) as a function of Γ for the Fe_2O_3/TiO_2, Co_2O_3/TiO_2, NiO/TiO_2, and SnO_2/TiO_2 systems. Interestingly, the ΔE_{VBM} for the MOC-modified TiO_2, except SnO_2/TiO_2, goes up with an increase in Γ, which means that the oxidation ability of the VB holes can be tuned by Γ. This is the most unique and important feature of the TiO_2 modification with MOCs using the CCC technique.

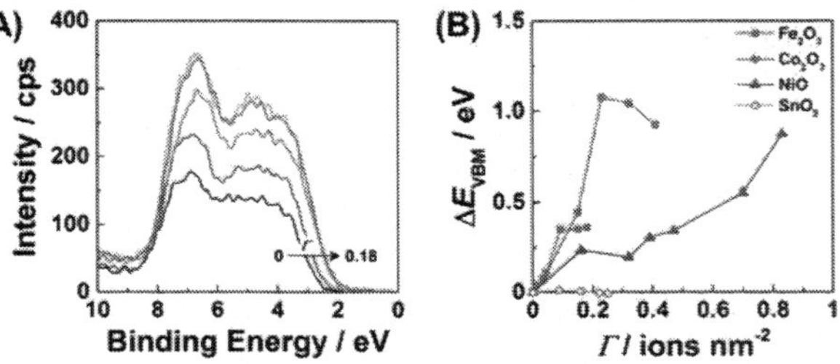

Figure 6. (A) VB-XPS spectra for Fe_2O_3/TiO_2 with varying Γ. (B) Energy shift in the VB maximum level (ΔE_{VBM}) as a function Γ for various MOCs/TiO_2. The figure (A) was taken from ref. 35.

These results were further simulated by the density functional theory (DFT) calculations [7]. In the DFT-optimized structure for a model Fe_2O_3 cluster-adsorbed TiO_2 system, plural Fe–O–Ti interfacial bonds were observed. The PEDOS (projected electronic density of states) plots showed that states from the adsorbed Fe_2O_3 clusters lie above the VB of TiO_2, that is, the iron oxide-derived states make it to the top of the VB. This changes the nature of the VB edge that moves the top of the VB to higher energy. The offsets between the TiO_2 VB edge and the iron oxide states around the VB are ~0.3 eV for Fe_2O_3-modified TiO_2, which is comparable with the experimental value. The effective mixing between the surface Fe_2O_3 levels and O 2*p* through the Ti_s–O–Fe interfacial bond is considered as yielding a d-band overlapping the VB of TiO_2. Thus, the excitation of Fe_2O_3/TiO_2 by the visible light with wavelength below 500 nm can induce the interfacial electron transfer from the surface d-band to the CB of TiO_2.

Conversely, the information about empty levels can be obtained by photoluminescence spectroscopy. TiO_2(ST-01) has a broad emission band centered at 538 nm (E_1) [22]. The E_1 signal intensity remarkably weakens with heating ST-01 at 773 K for 1 h in air. This PL band is assignable to the emission from the surface oxygen vacancy levels of anatase TiO_2 [33]. On modifying ST-01 with the Fe oxocomplexes, the intensity further decreases to disappear at $\Gamma > 0.044$ ions nm^{-2}, while a new emission appears at 423 nm (E_2). The E_2 signal can be attributed to the emissions from extrinsic levels. These findings strongly suggest that the excited electrons in the CB of TiO_2 are transferred to the empty surface Fe oxocomplex levels with the energy distributed around 0.27 V below the CB of TiO_2.

5.3. Electrocatalytic activity for oxygen reduction reaction

As stated above, the ORR is frequently the key process in the TiO_2-photocatalyzed reactions as well as low-temperature polymer electrolyte membrane fuel cells (PEMFCs) [34,35]. Figure 7 shows dark current (*I*)–potential (*E*) curves of the mp-TiO_2 film-coated F-doped tin oxide (FTO) electrodes (mp-TiO_2/FTO) with and without the surface modification by Fe oxocomplexes. In the absence of O_2, a small current due to water reduction is observed regardless of the surface modification. In the presence of O_2, the current for ORR markedly increases with the surface modification (Fe_2O_3/mp-TiO_2/FTO), whereas it remains weak without O_2. In this manner, the surface Fe oxocomplex has an electrocatalytic activity for the ORR, and a similar ORR-promoting effect is also observed for the NiO/TiO_2 [30] and Co_2O_3/TiO_2 [29] systems. This is also the unique feature of the MOC/TiO_2 systems.

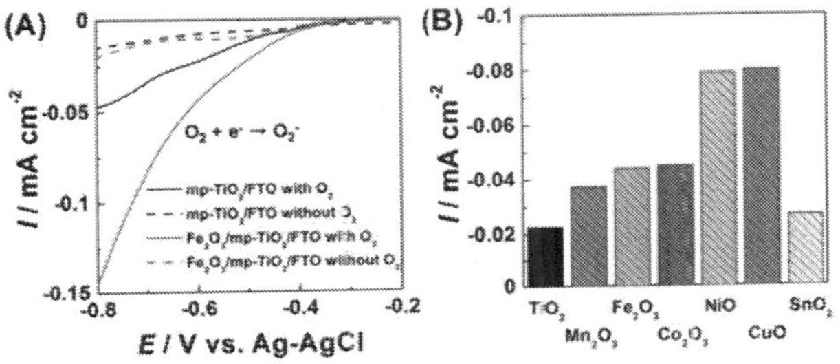

Figure 7. (A) Dark current (I)–potential (E) curves for the Fe_2O_3/mp-TiO_2/FTO electrodes. (B) Comparison of the electrocatalytic activity of the MOC/mp-TiO_2/FTO electrodes for the ORR.

6. Photocatalytic activity

Acetaldehyde is a toxic volatile organic compound (VOC), while 2-naphthol is widely used as the starting material of azo dyes. Both of them are optically transparent in the visible region, and then, acetaldehyde and 2-naphthol were used as model air and water pollutants, respectively. The relative photocatalytic activities of various MOCs/TiO_2 were evaluated with respect to that of highly active commercial TiO_2 particles with a crystal form of anatase (ST-01, Ishihara Sangyo Co.). The photocatalytic degradation of 2-naphthol and acetaldehyde apparently follows the first-order rate law, and the rate constants for irradiation of UV light ($330 < \lambda < 400$ nm, k_{UV}) and visible light ($\lambda > 400$ nm, k_{vis}) were used as the indicators for the photocatalytic activities. Figure 8 shows the k_{vis} and k_{UV} of Fe_2O_3/TiO_2 for the degradations of 2-naphthol (A) and acetaldehyde (B) as a function of Γ. Surprisingly, the surface modification of TiO_2 by the Fe-oxocomplex gives rise to a high level of visible-light activity and a concomitant increase in the UV-light activity of anatase TiO_2 [24]. Each plot exhibits a volcano-shaped curve, which is a general feature in the activity-Γ plots for the MOC/TiO_2 systems. By using an atomic layer deposition technique, Libera et al. have recently prepared Fe(III) oxospecies–surface-modified TiO_2 showing a reactivity for the decoloration of methylene blue under visible-light irradiation [18].

Particulate systems have high photocatalytic activity due to the large surface area, but needs the troublesome separation of the particles from purified water. Oppositely, in supported films, the photocatalytic activity is generally much lower due to the smaller surface area,

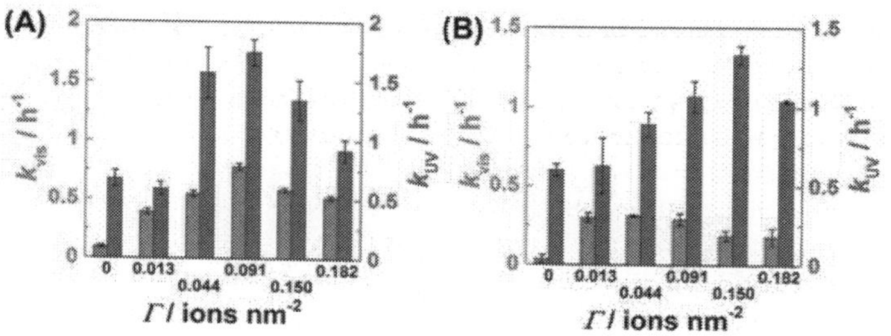

Figure 8. (A) UV-light activity (k_{UV}, blue) and visible-light activity (k_{vis}, red) of Fe_2O_3/TiO_2(ST-01) for the liquid-phase decomposition of 2-NAP as a function of Γ. (B) UV-light activity (k_{UV}, blue) and visible-light activity (k_{vis}, red) of Fe_2O_3/TiO_2(ST-01) for the gas-phase decomposition of CH_3CHO as a function of Γ.

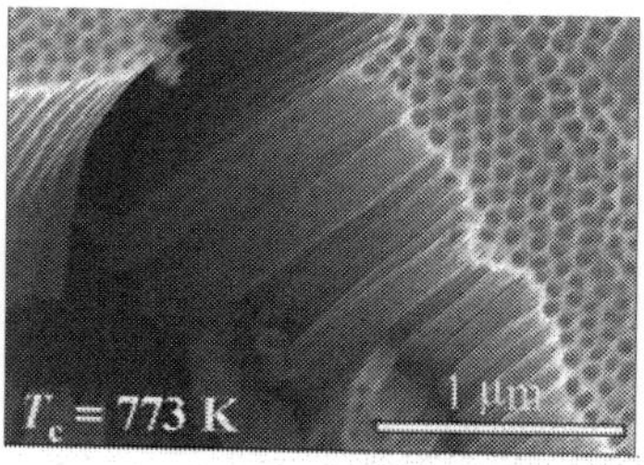

Figure 9. TiO_2 NTA parepared by a two-step anodization of Ti plate (first anodization 40 V, 0.5 h/second anodization 40 V-1 h/heating temperature, T_c = 773 K). The figure was taken from Ref. 35.

while the separation process is unnecessary. TiO_2 nanotube array (NTA) has the advantages of the particulate and film systems is promising. Figure 9 shows TEM image for TiO_2 NTA prepared by two-step anodization. The application of the CCC technique to the TiO_2 NTA led to a high visible-light activity for 2-naphthol degradation comparable with that of the particulate system [36].

Figure 10 compares the relative visible-light activity (k_{vis}) and UV-light activity (k_{UV}) of 3d MOCs/TiO_2(ST-01) with respect to the activities of unmodified TiO_2(ST-01) (k_{vis}^0 and k_{UV}^0) for the 2-naphthol degradation under the same conditions. Each Γ shows the optimum value for visible-light activity in each MOC/TiO_2 system. Among MOCs, the surface modification by Fe_2O_3 [24], Co_2O_3 [25], and NiO [30] is effective in the visible-light activation. Particularly, the Co_2O_3/TiO_2 system exhibits a very high level of visible-light activity [29]. The activity is on the order of $Co_2O_3 > Fe_2O_3 > NiO > CuO > V_2O_5 \approx Mn_2O_3 > SnO_2 \approx$ unmodified TiO_2. However, the surface modification with Fe_2O_3, NiO, and Co_2O_3 by the CCC technique can endow anatase TiO_2 with high levels of visible-light activity, with the high UV-light activity further increased (Fe_2O_3) or maintained (Co_2O_3, NiO). Although the effect of the surface modification by SnO_2 was small for anatase, a significant increase in the UV-light activity was induced for rutile [37]. Interestingly, Boppana and Lobo have recently reported that loading of SnO_x clusters on $ZnGa_2O_4$ by the impregnation method causes visible-light activity for the decomposition of p-cresol [38]. Besides metal oxides, the surface modification of TiO_2 with halogeno complexes of rhodium(III) and platinum(IV) on the TiO_2 surface is known to induce visible-light activity [39,40].

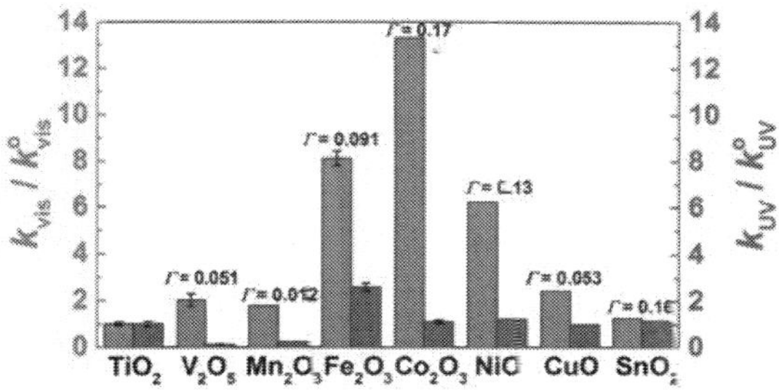

Figure 10. Comparison of the visible-light activities (k_{vis}) and UV-light activities (k_{UV}) of MOCs/TiO_2(ST-01) with respect to those of unmodified TiO_2 for the 2-naphthol degradation under the same conditions.

The degradation of formic acid was further carried out in the aqueous phase with Co_2O_3/TiO_2 at 298 K under visible-light irradiation. The Co_2O_3 surface modification greatly enhanced the decomposition of formic acid to CO_2. The visible-light activity reached a maximum at Γ = 0.17 with the conversion to CO_2 reaching ~100% within 5 h [29] (Eq. 3).

$$HCOOH+1/2O_2 \xrightarrow[h\nu(\lambda>400nm)]{Co_2O_3/TiO_2} CO_2+H_2O \quad (3)$$

Also, prolonging irradiation decomposed 2-naphthol to CO_2, but the conversion was only ~6% at 96 h. The decomposition of 2-naphthol to CO_2 would proceed stepwise via oxidative cleavage of the naphthalene ring.

On the basis of the energy band diagram, the action mechanism of MOCs in the TiO_2 photocatalysis can be explained. In the nanoscale Fe_2O_3–TiO_2 coupling system, Fe_2O_3 NP with a band gap of 2.2 eV is excited by the visible-light irradiation. However, the potential of the CB electrons is more positive than the TiO_2 CB minimum of TiO_2 (−0.48 V) and the standard redox potential of O_2 ($E^0(O_2/O_2^-)$ = −0.284 V). Thus, the electron transfer from the CB electrons of Fe_2O_3 to neither TiO_2 nor O_2 can occur. Consequently, nano-coupling does not show visible-light activity [41].

Scheme 3 illustrates the surface modification effects of the Fe oxocomplex on the TiO_2 photocatalytic decomposition of organic pollutants. In this case, the surface modification raises the VB maximum with the CB minimum unchanged, due to the effective electronic coupling through the Fe–O–Ti interfacial bonds (Effect 1). The resulting decrease in the band gap shifts the light absorption to the visible region (Effect 2). The visible-light absorption triggers electronic excitation from the highest-energy oxocomplex-derived VB states to the empty CB

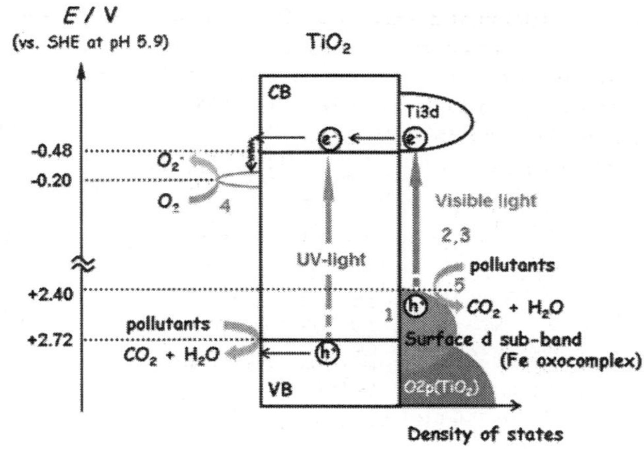

Sheme 3. Surface modification effects of the Fe oxocomplex on the TiO_2-photocatalyzed decomposition of organic pollutants. The levels around −0.2 V show the vacant Fe oxocomplex.

of TiO_2 in order to generate charge carriers. This surface-to-bulk interfacial electron transfer enhances charge separation (Effect 3). The surface modification permits the electron transfer from the CB of TiO_2 to shallow vacant surface oxocomplex levels, which distribute around ca. −0.2 V [22]. The formation of O_2^- radicals was confirmed by chemiluminescence photometry in the Cu^{2+}-grafted TiO_2 system under visible-light irradiation [42]. In this cathodic process, the electrons efficiently reduce adsorbed O_2 with the aid of the electrocatalytic activity of the surface-adsorbed oxocomplex (Effect 4). This effect should also contribute to the increase in the UV-light activity. In the anodic process, the holes generated in the VB could take part in the oxidation process without diffusion (Effect 5) [15]. Consequently, Fe_2O_3/TiO_2 as well as NiO/TiO_2 and Co_2O_3/TiO_2 satisfy the three requirements of the "solar environmental catalyst."

7. Conclusions and future prospect

The surface of TiO_2 can be modified by oxocomplexes of the first transition metals (MOCs/TiO_2) with the loading amount precisely controlled by using the CCC technique. Among the MOCs/TiO_2, Fe_2O_3-, Co_2O_3- and NiO-surface-modified TiO_2 possess unique physicochemical properties such as strong visible-light absorption and the excellent reduction ability of O_2. Spectroscopic experiments and first-principles DFT simulation have revealed that the surface modification with the MOCs raises the VB maximum of TiO_2 due to the formation of plural metal–O–Ti interfacial bonds. Surface-to-bulk and/or bulk-to-surface interfacial electron transfer induced by visible-light absorption enhances charge separation. This novel coupling system consisting of MOCs and TiO_2 would be promising as the "solar environmental catalyst."

The standard potentials of multiple-electron ORRs ($E^0(O_2/H_2O_2)$ = +0.695 V and $E^0(O_2/H_2O)$ = +1.229 V versus SHE) are much more positive than that of one-electron ORR. Therefore, the hybridization of appropriate electrocatalysts for the multiple-electron ORR can impart visible-light activity to many metal oxide semiconductors with E_g < 3 eV. The effectiveness of this approach has recently been verified in the Pt NP-WO_3 (E_g = 2.7 eV) [43] and Cu(acac)$_2$-$BiVO_4$ (E_g = 2.4 eV) hybrid systems [44], where Pt NP and O_2-bridged Cu complex work as excellent electrocatalysts for multiple ORRs, respectively.

As a future subject, we further suggest the importance of the effective use of the infrared ray occupying 52% of the solar energy for the catalytic reactions (Figure 1). For example, Co_2O_3/TiO_2 exhibits high levels of photocatalytic and thermocatalytic activities [29], whereas Mn_2O_3/TiO_2 exhibits a high thermocatalytic activity for the oxidation of organic compounds [45]. Further, MOCs/TiO_2 with the VB maximum level (or the oxidizing ability of the VB holes) fine-tuned by the loading amount may open up the application of MOCs/TiO_2 to "green" and selective chemical transformations [46–48].

Acknowledgements

A series of studies on metal oxocomplex–surface-modified TiO_2 have been performed in collaboration with Dr. Michael Nolan and Dr. Anna Iwaszuk (Tyndall National Institute, University College Cork). The authors are sincerely grateful for their very useful DFT simulations. Also, the authors acknowledge Dr. M. Fujishima (Kinki University) for a helpful discussion, and T. Hattori, S. Okuoka, and Y. Sumida (Nippon Shokubai Co.) for EXAFS measurements and a valuable discussion. Ishihara Sangyo Co. gifted us with ST-01, and K. Fujiwara aided us in the collection and arrangement of the materials. H.T. acknowledges the support from the Ministry of Education, Science, Sport, and Culture, Japan, through a Grant-in-Aid for Scientific Research (C) No. 24550239, No. 15K05654, MEXT-Supported Program for the Strategic Research Foundation at Private Universities, and Nippon Sheet Glass Foundation for Materials Science and Engineering, and by Sumitomo Foundation.

Author details

Hiroaki Tada[1*] and Qiliang Jin[2]

*Address all correspondence to: h-tada@apch.kindai.ac.jp

1 Department of Applied Chemistry, School of Science and Engineering, Kinki University, Kowakae, Higashi-Osaka, Osaka, Japan

2 R&D Placement Business Division, WORLD INTEC CO., Sakae, Naka-ku, Nagoya, Aichi, Japan

References

[1] Fujishima A, Zhang X, Tryk D A: TiO_2 photocatalysis and related surface phenomena. Surf. Sci. Rep. 2008; 63: 515–582. DOI: j.surfrep.2008.10.001

[2] Hashimoto K, Irie H, Fujishima A: TiO_2 photocatalysis: A historical overview and future prospects. Jpn. J. Appl. Phys. 2005; 44: 8269–8285.

[3] Hoffmann M R, Martin S T, Choi W, Bahnemann D W: Environmental applications of semiconductor photocatalysis. Chem. Rev. 1995; 95: 69–96. DOI: 10.1021/cr00033a004

[4] A. Y. Ahmed, T. A. Kandiel, T. Oekermann, D. Bahnemann: Photocatalytic activities of different well-defined single crystal TiO_2 surfaces: Anatase versus rutile. J. Phys. Chem. Lett. 2011; 2: 2461–2465. DOI: 10.1021/jz201156b

[5] Kawahara T, Konishi Y, Tada H, Tohge N, Nishii J, Ito S: A patterned TiO_2(anatase)/TiO_2(rutile) bilayer-type photocatalyst: Effect of the anatase/rutile junction on the photocatalytic activity. Angew. Chem. Int. Ed. 2002; 41: 2811–2813. DOI: 10.1002/1521-3757(20020802)114:15<2935:: AID-ANGE2935>3.0.CO;2-6

[6] Kisch H: Semiconductor photocatalysis–mechanistic and synthetic aspects. Angew. Chem. Int. Ed. 2013; 52: 812–847. DOI: 10.1002/anie.201201200

[7] Nolan M, Iwaszuk A, Tada H: Molecular metal oxide cluster-surface modified titanium(IV) dioxide photocatalysts. Aust. J. Chem. 2012; 65: 624–632.

[8] Asahi R, Morikawa T, Ohwaki T, Aoki K, Taga Y: Visible-light photocatalysis in nitrogen-doped titanium oxides. Science 2001; 293: 269–271. DOI: 10.1126/science.1061051

[9] Khan S U M, Al-Shahry M, Ingler Jr. W B: Efficient photochemical water splitting by a chemically modified n-TiO_2. Science 2002; 297: 2243–2245. DOI: 10.1126/science.1075035

[10] Serpone N: Is the band gap of pristine TiO_2 narrowed by anion- and cation-doping of titanium dioxide in second-generation photocatalysts? J. Phys. Chem. B 2006; 110: 24287–24293. DOI: 10.1021/jp065659r

[11] Zhang H, Chen G, Bahnemann D W: Photoelectrocatalytic materials for environmental applications. J. Mater. Chem. 2009; 19: 5089–5121. DOI: 10.1039/B821991E

[12] Anpo M, Takeuchi M: The design and development of highly reactive titanium oxide photocatalysts operating under visible light irradiation. J. Catal. 2003; 216: 505–516. DOI: 10.1016/S0021-9517(02)00104-5

[13] Kitano M, Funatsu K, Matsuoka M, Ueshima M, Anpo M: Preparation of nitrogen-substituted TiO_2 thin film photocatalysts by the radio frequency magnetron sputtering deposition method and their photocatalytic reactivity under visible light irradiation. J. Phys. Chem. B 2006; 110: 25266–25272. DOI: 10.1021/jp064893e

[14] Liu G, Wang L, Yang H G, Cheng H-M, Lu G Q: Titania-based photocatalysts-crystal growth, doping and heterostructuring. J. Mater. Chem. 2010; 20: 831–843. DOI: 10.1039/B909930A

[15] Irie H, Shibanuma T, Kamiya K, Miura S, Yokoyama T, Hashimoto K: Characterization of Cr(III)-grafted TiO_2 for photocatalytic reaction under visible light. Appl. Catal. B. 2010; 96: 142–148. DOI: 10.1016/j.apcatb.2010.02.011

[16] Murakami N, Chiyoya T, Tsubota T, Ohno T: Switching redox site of photocatalytic reaction on titanium(IV) oxide particles modified with transition-metal ion controlled by irradiation wavelength. Appl. Catal. A. 2008; 348: 148–152. DOI: 10.1016/j.apcata.2008.06.040

[17] Yu H, Irie H, Shimodaira Y, Hosogi Y, Kuroda Y, Miyauchi M, Hashimoto K: J. Phys. Chem. C 2010; 114: 16481–16487. DOI: 10.1021/jp1071956

[18] Libera J A, Elam J W, Sather N F, Rajh T M, Dimitrijevic N M: Iron(III)-oxo centers on TiO_2 for visible-light photocatalysis. Chem. Mater. 2010; 22: 409–413. DOI: 10.1021/cm902825c

[19] Irie H, Miura S, Kamiya K, Hashimoto K: Efficient visible light-sensitive photocatalysts: Grafting Cu(II) ions onto TiO_2 and WO_3 photocatalysts. Chem. Phys. Lett. 2008; 457: 202–205. DOI: 10.1016/j.cplett.2008.04.006

[20] Tada H: Encyclopedia of Surface and Colloid Science. Ed. Hubbard A T, New York: Marcel Dekker, 2002. ISBN: 0-8247-0633-1

[21] Wang C-M, Heller A, Gerischer H: Palladium catalysis of O_2 reduction by electrons accumulated on TiO_2 particles during photoassisted oxidation of organic compounds. J. Am. Chem. Soc. 1992; 114: 5230–5234. DOI: 10.1021/ja00039a039

[22] Jin Q, Fujishima M, Tada H: Visible-light-active iron oxide-modified anatase titanium(IV) dioxide. J. Phys. Chem C 2011; 115: 6478–6483. DOI: 10.1021/jp201131t

[23] Fujishima M, Jin Q, Yamamoto H, Tada H, Nolan M: Tin oxide-surface modified anatase titanium(IV) dioxide with enhanced UV-light photocatalytic activity. Phys. Chem. Chem. Phys. 2012; 14: 705–711. DOI: 10.1039/C1CP22708D

[24] Tada H, Jin Q, Nishijima H, Yamamoto H, Fujishima M, Okuoka S-i, Hattori T, Sumida Y, Kobayashi H: Titanium(IV) dioxide surface-modified with iron oxide as a visible light photocatalyst. Angew. Chem. Int. Ed. 2011; 50: 3501–3505. DOI: 10.1002/anie.201007869

[25] Nosaka Y, Nosaka A. Nyumon: Hikarisyokubai (Introduction to Photocatalyst). Tokyo: Tokyo-Tosho; 2004, 68 p. ISBN: 4-489-00684-5 C3043

[26] Navio J A, Colon G, Litter M I, Bianco G N: Synthesis, characterization and photocatalytic properties of iron-doped titania semiconductors prepared from TiO_2 and

iron(III) acetylacetonate. J. Mol. Catal. A: Chem. 1996; 106: 267–276. DOI: 10.1016/1381-1169(95)00264-2

[27] Wang C-Y, Bahnemann D W, Dohrmann J K: A novel preparation of iron-doped TiO_2 nanoparticles with enhanced photocatalytic activity. Chem. Commun. 2000; 1539–1540. DOI: 10.1039/B002988M

[28] Serpone N, Lawless D, Disdier J, Hermann J-M: Spectroscopic, photoconductivity, and photocatalytic studies of TiO_2 colloids: Naked and with the lattice doped with Cr^{3+}, Fe^{3+}, and V^{5+} cations. Langmuir 1994; 10: 643–652. DOI: 10.1021/la00015a010

[29] Jin Q, Yamamoto H, Yamamoto K, Fujishima M, Tada H: Simultaneous induction of high level thermal and visible-light catalytic activities to titanium(IV) oxide by surface modification with cobalt(III) oxide clusters. Phys. Chem. Chem. Phys. 2013; 15: 20313–20319. DOI: 10.1039/C3CP54304H

[30] Jin Q, Ikeda T, Fujishima M, Tada H: Nickel(II) oxide surface-modified titanium(IV) dioxide as a visible-light-active photocatalyst. Chem. Commun. 2011; 47: 8814–8816. DOI: 10.1039/C1CC13096J

[31] Iwaszuk A, Nolan M, Jin Q, Fujishima M, Tada H: Origin of the visible-light response of nickel(II) oxide cluster surface modified titanium(IV) dioxide. J. Phys. Chem. C 2013; 117: 2709–2718. DOI: 10.1021/jp306793r

[32] Jin Q, Fujishima M, Iwaszuk A, Nolan M, Tada H: Loading effect in copper(II) oxide cluster-surface-modified titanium(IV) oxide on visible- and UV-light activities. J. Phys. Chem. C 2013; 117: 23848–23857. DOI: 10.1021/jp4085525

[33] Shi J, Chen J, Feng Z, Chen T, Lian Y, Wang X, Li C: Photoluminescence characteristics of TiO_2 and their relationship to the photoassisted reaction of water/methanol mixture. J. Phys. Chem. 2007; 111: 693–699. DOI: 10.1021/jp065744z

[34] Gewirth A A, Thorum M S: Electroreduction of dioxygen for fuel-cell applications: Materials and challenges. Inorg. Chem. 2010; 49: 3557–3566. DOI: 10.1021/ic9022486

[35] Fermández J L, Walsh D A, Bard A J: Thermodynamic guidelines for the design of bimetallic catalysts for oxygen electroreduction and rapid screening by scanning electrochemical microscopy. M-co (M: Pd, Ag, Au). J. Am. Chem. Soc. 2005; 127: 357–365. DOI: 10.1021/ja0449729

[36] Muramatsu Y, Jin Q, Fujishima M, Tada H: Visible-light-activation of TiO_2 nanotube array by the molecular iron oxide surface modification. Appl. Catal. B: Environ. 2012; 119–120: 74–80. DOI: 10.1016/j.apcatb.2012.02.012

[37] Jin Q, Fujishima M, Nolan M, Iwaszuk A, Tada H: Photocatalytic activities of tin(IV) oxide surface-modified titanium(IV) dioxide show a strong sensitivity to the TiO_2 crystal form. J. Phys. Chem. C 2012; 116: 12621–12626. DOI: 10.1021/jp302493f

[38] Boppana V B R, Lobo R F: SnOx-$ZnGa_2O_4$ photocatalysts with enhanced visible light activity. ACS Catal. 2011; 1: 923–928. DOI: 10.1021/cs200137h

[39] Kisch H: Semiconductor photocatalysis-mechanistic and synthetic aspects. Angew. Chem. Int. Ed. 2013; 52: 812–847. DOI: 10.1002/anie.201201200

[40] Kitano S, Murakami N, Ohno T, Mitani Y, Nosaka Y, Asakura H, Teramura K, Tanakia T,

[41] Tada H, Hashimoto K, Kominami H: Bifunctionality of Rh^{3+} modifier on TiO_2 and working mechanism of Rh^{3+}/TiO_2 photocatalyst under irradiation of visible light. J. Phys. Chem. C 2013; 117: 11008–11016. DOI: 10.1021/jp311801e

[42] Lin X, Li D-Z, Wu Q-P, Fu X-Z, Wang X-X: Chem. Photocatalytic activity and mechanism of heterojunction thin films. J. Chin. Univ. 2005; 26: 727–730.

[43] Nosaka Y, Takahashi S, Sakamoto H, Nosaka A: Reaction mechanism of Cu(II)-grafted visible-light responsive TiO_2 and WO_3 photocatalysts studied by means of ESR spectroscopy and chemiluminescence photometry. J. Phys. Chem. C 2011; 115: 21283–21290. DOI: 10.1021/jp2070634

[44] Abe R, Takami H, Murakami N, Ohtani B: Pristine simple oxides as visible light driven photocatalysts: Highly efficient decomposition of organic compounds over platinum-loaded tungsten oxide. J. Am. Chem. Soc. 2008; 130: 7780–7781. DOI: 10.1021/ja800835q

[45] Naya S, Niwa T, Negishi R, Kobayashi H, Tada H: Multi-electron oxygen reduction by a hybrid visible-light-photocatalyst consisting of metal-oxide semiconductor and self-assembled biomimetic complex. Angew. Chem. Int. Ed. 2014; 53: 13894–13897. DOI: 10.1002/anie.201408352

[46] Jin Q, Arimoto H, Fujishima M, Tada H: Manganese oxide-surface modified titanium(IV) dioxide as environmental catalyst. Catalysts 2013; 3: 444–454. DOI: 10.3390/catal3020444

[47] Ide Y, Hattori H, Ogo S, Sadakane M, Sano T: Highly efficient and selective sunlight-induced photocatalytic oxidation of cyclohexane on an eco-catalyst under a CO_2 atmosphere. Green Chem. 2012; 14: 1264–1267. DOI: 10.1039/C2GC16594E

[48] Higasihmoto S, Shirai R, Osano Y, Azuma M, Ohue H, Sakata Y, Kobayashi H: Influence of metal ions on the photocatalytic activity: Selective oxidation of benzyl alcohol on iron (III) ion-modified TiO_2 using visible light. J. Catal. 2014; 311: 137–143. DOI: 10.1016/j.jca.2013.11.013

[49] Tang J, Grampp G, Liu Y, Wang, B-X, Tao F-F, Wang L-J, Liang X-Z, Xiao H-Q, Shen, Y-M: Visible light mediated cyclization of tertiary anilines with maleimides using nickel(II) oxide surface-modified titanium dioxide catalyst. J. Org. Chem. 2015; 80: 2724–2732. DOI: 10.1021/jo502901h

Section 4

Recent Progress in Catalytic Materials for Electrocatalytic Applications

Chapter 14

Electrocatalytic Applications of Graphene–Metal Oxide Nanohybrid Materials

Arnab Halder, Minwei Zhang and Qijin Chi

Additional information is available at the end of the chapter

http://dx.doi.org/10.5772/61808

Abstract

Development of state-of-the-art electrocatalysts using commercially available precursors with low cost is an essential step in the advancement of next-generation electrochemical energy storage/conversion systems. In this regard, noble metal-free and graphene-supported nanocomposites are of particular interest. Graphene-based nanocomposite is an excellent candidate as energy-device and sensor-related electrode materials, largely due to their high electrical conductivity, large specific surface area, high-speed electron/heat mobility, and reasonably good mechanical strength. Among many types of graphene-based composite materials, graphene–metal oxide nanohybrids hold great promise toward engineering efficient electrocatalysts and have attracted increasing interest in both scientific communities and industrial partners around the world. The goal of this chapter is primarily set on an overview of cutting-edge developments in graphene–metal oxide nanohybrid materials, with the recently reported results from worldwide research groups. This chapter is presented first with an introduction, followed by synthetic methods and structural characterization of nanocomposites, an emphasis on their applications in energy and sensor-related fields, and finally completed with brief conclusions and outlook.

Keywords: Graphene, metal oxides, graphene–metal oxide nanohybrids, electrocatalysis, energy storage and conversion, sensors, electrochemistry

1. Introduction

Graphene, a single atomic thick layer of graphite with closely packed conjugated and hexagonally connected carbon atoms, has attracted tremendous attention since its discovery in 2004 because of its large specific surface area, high-speed electron mobility, good mechanical strength, high electric and thermal conductivity, room temperature quantum hall effect, good optical transparency, and tunable band gap [1, 2]. This material has a theoretical surface area of 2630 m^2 g^{-1}, a mobility of 200,000 cm^2 V^{-1} s^{-1} at a carrier density of ~10^{12} cm^{-2}, and the highest electrical conductivity of 106 s cm^{-1} at room temperature [3, 4]. Also, graphene has a Young's modulus of ~1 TPa, breaking strength of 42 N m^{-1} [5], and thermal conductivity of 5000 W m^{-1} K^{-1} [6]. For all these versatile properties and obvious advantages, a rapid development of graphene-based materials has been witnessed in the fields of chemistry, physics, biology, and many other interdisciplinary fields such as nanotechnology and nanomedicine.

Electrocatalysis is a special type of catalysis that speeds up the rate of an electrochemical reaction occurring on electrode surfaces or at liquid/solid interfaces. Various kinds of electrochemical reactions are involved for different electrocatalytic applications in energy and sensor-related fields. Therefore, design and fabrication of advanced electrocatalysts with outstanding performance and low cost are of great significance for the commercialization of electrocatalytic-system-based energy devices. Metal oxide nanostructures have been identified as one of the most important electrocatalytic materials due to their several advantages. *Firstly*, they have exceptional electrical, optical, and molecular properties. *Secondly*, there is further possibility to insert more functional groups on the surface for the immobilization of other biological catalysts. *Thirdly*, metal oxides have higher alkaline corrosion resistance compared to other materials in electrochemical environmental due to the stabilization of the higher oxidation state of the transition metals. *Finally*, their unique crystalline structures benefit in preventing the agglomeration of metal oxide nanostructures with their size retained [7]. Besides electrocatalytic activity and cost, the stability and durability of a catalyst are very critical issues for practical applications. In addition, a good catalyst support is needed, which should have a large surface area for catalyst dispersion, excellent electronic conductivity, and high electrochemical stability in different electrolytes. In recent years, various catalyst support materials have been proposed and studied. In this regard, graphene nanosheets have shown promising characteristics for wide applications as 2D support materials for different electrocatalysts. In the last decade, intensive efforts have been devoted to functionalize graphene-based nanomaterials and to explore their applications in sensors [8],[9], electrochemical energy storage [10, 11], electronics, optoelectronics [12], and others. Graphene-based nanocomposites provide a new option as electrode materials in the field of electrocatalytic applications due to their high electrical conductivity, high surface area, and richness of functional groups for further modification. The rapid development of low cost and facile preparation methods of graphene-based nanocomposites has promoted their practical industrial applications. The catalytic activity of the graphene-supported catalysts can be improved, due to enhanced electronic communication (e.g., charge transfer) between catalysts and support. Furthermore, arising from their synergistic effects of graphene sheets and functionalized components, the nanocomposites can offer novel physicochemical properties and consequently improve electrochemical performances. As a result, graphene-based nanocomposites have thus been regarded as one of the most promising hybrid materials that can drive the development of more efficient next-generation energy devices as well as applied in the fabrication of electrochemical and gas sensors. This chapter is focused on one of such nanocomposites, made of chemically exfoliated graphene and metal oxides. We overview and discuss their synthesis methods, structural features, electrocatalytic applications, and future perspectives.

2. Synthesis of graphene–metal-oxide-based nanohybrid composites

2.1. Introduction

In terms of electrocatalysis, graphene-supported metal oxides nanohybrid materials have exhibited promising applications because of the following well-recognized advantages. *Firstly,* the large surface area and 2D flexibility of graphene nanosheets can offer sufficient space to accommodate different nanomaterials and also prevent their agglomeration. *Secondly,* due to the good superficial characteristics of graphene, solid-air contact efficiency increases and simultaneously the amount of oxygen adsorption also increases. *Thirdly,* the electrical conductivity of graphene promotes the electron transfer rate on the surface. *Finally,* the structural defects of graphene also provide more active sites for further modification with different functional groups to promote the selective electrocatalysis.

Incorporation of inorganic nanomaterial onto the surface of graphene has attracted tremendous attention for the development of new-generation catalytic materials [13]. These novel nanostructures show superior electrocatalytic activity, selectivity, and long-term stability, which can serve as promising electrode material for different electrochemical reactions. Different metal oxides are promising in this regard for the replacement of noble-metal-based electrocatalysts. But normally metal oxides are poorly conductive and often suffer from dissolution and aggregation during the electrochemical reactions [14, 15]. It was proposed that anchoring metal oxide nanostructures onto graphene surface by suitable synthetic procedure could solve the problems. Indeed, some graphene-supported metal oxide nanocomposites have displayed remarkable improvements in electrocatalytic activity and stability toward some crucial electrochemical reactions in amperometric sensors and energy storage and conversion.

2.2. Major synthetic methods

The increasing interest in graphene-based metal oxides nanohybrid materials for different electrocatalytic applications has led to a variety of new processes proposed for the synthesis of nanocomposite materials. Some recently established methods are discussed in this section.

2.2.1. Solvothermal and hydrothermal synthetic approaches

Solvothermal and hydrothermal synthetic approaches are one of the most common synthetic strategies for the development of different graphene-based nanohybrid electrocatalysts. Song et al. [16] reported a new hydrothermal approach for the preparation of CuO/GO nanocomposite materials, which used cupric acetate and graphene oxide as precursors. The reaction was conducted for 10 h under different temperatures (120, 150, and 180°C). The final product was washed by deionized water and dried for the electrocatalytic applications. Dong et al. [17] introduced another hydrothermal approach for the preparation of Graphene/Co_3O_4 nanowire composite. They used presynthesized 3D graphene and $CoCl_2.6H_2O$ as starting materials. This reaction was proceeded in autoclave at 120°C for 16 h. Dai et al. [18] also used a hydrothermal approach for the synthesis of covalent hybrid of spinel manganese–cobalt oxide and graphene.

They employed a two-step procedure to synthesize $MnCo_2O_4$–graphene oxide nanohybrids. In the first nucleation step, $Co(OAc)_2$ and $Mn(OAc)_2$ were mixed at 80°C with mildly oxidized graphene oxide in an ethanol/water NH_4OH solution. In the second step, hydrothermal treatment was done at 150°C to achieve the nitrogen-doped graphene. And the final material showed an excellent electrocatalytic activity for oxygen reduction reaction (ORR). Dai et al. [19] also used cobalt acetate and GO as a precursor for the hydrothermal synthesis of Co_3O_4/ graphene hybrid bifunctional catalyst for ORR and water oxidation or oxygen evolution reaction (OER).

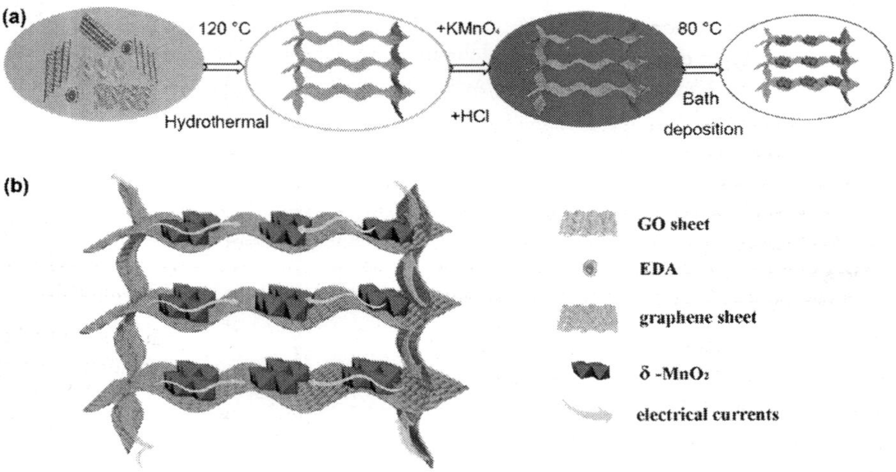

Figure 1. Schematic illustration of the hydrothermal synthesis of 3D graphene/MnO_2 (a), and schematic illustration of electrons transfer on the 3D hybrid material (b). (Reproduced with permission from ref. 44 Copyright 2014 American Chemical Society)

Wang et al. [20] reported a hydrothermal approach for the synthesis of CoO/rGO nanocomposite using GO, Co $(Ac)_2.4H_2O$, $Co(NH_2)_2$ as a precursor (190°C, 2 h). Mullen et al. [21] successfully synthesized 3D nitrogen-doped graphene aerogel-supported Fe_3O_4 nanoparticles by hydrothermal approach for the efficient electrocatalysis of ORR. Graphene oxide, iron acetate, and polypyrrole were hydrothermally assembled at 180°C for 12 h to form a 3D graphene-based hydrogel. The hydrogel was further dehydrated and annealed at 600°C for 3 h under nitrogen atmosphere.

Hydrothermal methods for the synthesis of graphene-based hybrid materials can be carried out at different temperature ranges, up to 190°C [22]. Figure 1 shows an example for the preparation of 3D graphene–MnO_2 composites. There are several advantages of using hydrothermal process. *Firstly*, the strong electrolyte water possesses a high diffusion coefficient and dielectric constant under hydrothermal reaction conditions, which helps to remove the oxygen-containing groups via dehydration and also accelerates heterolytic bond cleavage.

Secondly, this is a green reduction approach and does not add any further impurities to the final product. *Thirdly,* the degree of reduction and the properties of the hybrid material can be tuned by adjusting temperature and pressure of the reaction. And *finally,* the process could be easily implemented in an industry scale with a relatively simple setup (an autoclave) and low cost. A possible disadvantage of the method could be the consumption of significant amount of energy.

2.2.2. Chemical reduction approaches

Chemical reduction is another method among the most common approaches for the preparation of different graphene-based nanohybrid electrocatalysts. Different research groups have used different types of reducing agents for the specific synthesis purposes. For example, Qiao et al. [23] reported a simple chemical reduction method for the preparation of CuO/N-rGO nanohybrid using NaOH in the mixture of $CuCl_2$ and N-doped reduced graphene oxide. Guo et al. [24] used diethylene glycol (DEG) as a reducing agent for the preparation of Cu_2O/RGO composite material. Yang et al. [25] prepared PDDA-G/Fe_3O_4 nanohybrid material using ammonia solution as a reducing agent. Khezrian et al. [26] also used ammonia solution as a reducing agent for the synthesis of Fe_3O_4 magnetic nanoparticles/RGO hybrid nanosheets. Wang et al. [27] used citric acid for the preparation of MnO_2/GO. Ruoff et al. [28] used hydrazine as a reducing agent for the synthesis of RGO/tin oxide (TiO_2) nanocomposite.

The major advantage of this approach is that one can tune the degree of reduction and other properties by using specific reducing agents. And also for most of cases the reactions are very energy-efficient due to low temperatures and slow time. An obvious drawback is the need of purifying the final product from different reducing agents, which is in some cases quite challenging.

2.2.3. MW-assisted synthetic approaches

Microwave (MW)-assisted synthesis is a simple and popular technique for the fast production of nanomaterials with small particle dimensions, uniform particle size distribution, and high purity. It is a uniform heating procedure compared to the other conventional heating systems. Moreover, microwave can facilitate the nucleation of nanoparticles and shorten the synthesis time. There are some excellent examples of using this approach. For instance, Peng et al. [29] synthesized CuO/SG hybrid materials by MW-assisted method using graphene oxide and cupric acetate as a precursor (Fig. 2). Ruoff et al. [30] also reported MW-assisted method for the synthesis of RGO/Fe_2O_3. Ferric chloride and graphene oxide were used as precursors for this method. And the as-synthesized nanocomposite was used as a high-performance anode material for lithium ion batteries. Wang et al. [31] synthesized highly dispersed titanium dioxide (TiO_2) nanoclusters on RGO in a toluene–water system by MW-irradiation-assisted method. The main advantages for the MW-assisted synthetic approaches include rapid reaction time, possibility for scale-up production and impurity-free final nanohybrid product. The relatively high cost needed for experimental setups could be a major drawback.

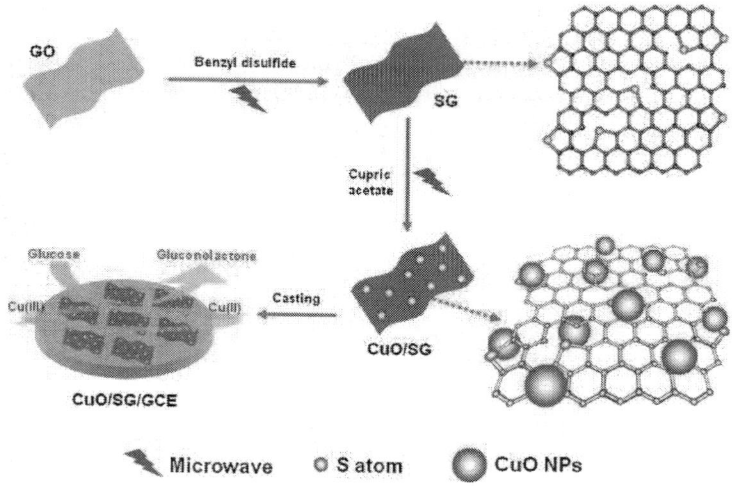

Figure 2. Schematic illustration of the MW-assisted synthesis of CuO nanoparticle supported on S-doped graphene/SG and CuO/SG on glassy carbon electrode for glucose sensing. (Reproduced with permission from ref. 29 Copyright Elsevier 2015)

2.2.4. Electrochemical synthetic approaches

The electrochemical synthetic method is an efficient technique for transforming electronic states by regulating the external power source to change the Fermi energy level of the electrode material surface [32]. Kong et al. [33] successfully prepared CuO nano needle/graphene/carbon nanofiber modified electrodes by electrochemical synthetic approaches for nonenzymatic glucose sensing in saliva. Duan et al. [34] reported a novel electrochemical approach to deposit MnO_2 nanowires on graphene for the sensor applications. The MnO_2 nanowires were anodically electrodeposited onto a graphene paper electrode using a CV technique in the potential range from 1.4 V to 1.5 V with a scan rate of 250 mV s^{-1} (Fig. 3) by cyclic voltametry.

The electrochemical synthetic approach is a fast, controllable, and green technique. By this procedure, one can achieve impurity-free nanohybrid material using little power consumption. However, this approach normally yields solid nanocomposite products, which is difficult for further processing. And also it would be difficult for large-scale production.

2.2.5. Other synthetic approaches

There are several other techniques that have also been explored for the synthesis of different nanocomposites, such as electrospinning, template-based synthesis, light- or radiation-induced methods, etc. However, although among them each technique has its specific advantages for specific systems; to date, few have been widely used.

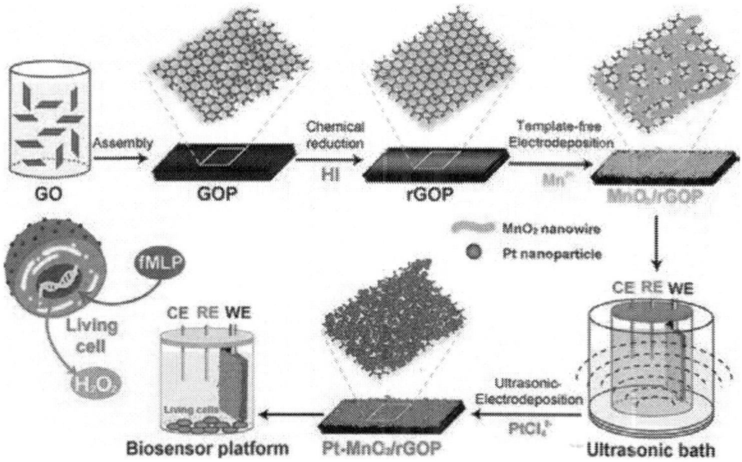

Figure 3. Schematic illustration of electrodeposition-assisted synthesis of MnO_2 nanoparticle supported on reduced graphene oxide paper electrode for biosensor applications. (Reproduced with permission from ref. 34 Copyright 2012 WILEY-VCH Verlag GmbH & Co. KGaA, Weinheim)

In summary, a number of methods have been developed for the synthesis of metal oxide/graphene composites, among which hydrothermal procedures have been most extensively used in particular. However, each method has its advantages as well as disadvantages. Table 1 summarizes the preparation methods and their major applications of most studied metal oxide/graphene nanocomposites.

Metal oxides	Nanohybrid electrocatalyst	Preparation method	Precursors	Applications	Ref
Copper oxide	RGO-Cu_2O	Heat treatment and sonication	GO, $CuSO_4.5H_2O$, $[C_{16}MMIm]Br$	Nonenzymatic amperometric glucose sensing	[35]
	CuNiO-graphene	Hydrothermal synthesis	$CuCl_2$, $NiCl_2$ Graphene	Nonenzymatic glucose sensors	[36]
	Cu_2O–rGO	Physical adsorption, in situ reduction and one pot synthesis	GO, $CuSO_4.5H_2O$, PDDA	Nonenzymatic H_2O_2 sensing	[37]
	CuO/ rGO/ CNF	Electrodeposition	GO, $CuCl_2$, carbon nanofiber	Nonenzymatic sensing of glucose in saliva	[33]

Metal oxides	Nanohybrid electrocatalyst	Preparation method	Precursors	Applications	Ref
	CuO/graphene	Heat treatment and annealing	Copper nitrate, GO	Nonenzymatic glucose sensors	[38]
	CuO/GO	Hydrothermal synthesis	GO, Cupric acetate, DMF	Nonenzymatic glucose sensors	[16]
	Cu_2O/GNs	Ultrasonication, stirring	GO, $CuCl_2$	Nonenzymatic H_2O_2 and glucose sensing	[39]
	CuO/SG	microwave-assisted solvothermal method	GO, Benzyl disulfide, Cupric acetate	Nonenzymatic glucose sensors	[29]
	CuO/N-rGO	Chemical reduction	N-rGO, $CuCl_2$	Direct methanol fuel cells	[23]
	Cu_2O/RGO	Chemical reduction	GO, Cupric acetate, DEG	Alkaline ORR	[24]
Cobalt oxide	Co_3O_4 /graphene	Ultrasonication and pyrolyzation	GO, cobalt phthalocyanine	Anode Materials for Li- Ion Batteries	[40]
	3D graphene/ Co_3O_4	Hydrothermal synthesis	Graphene, $CoCl_2.6H_2O$	Supercapacitors and enzymeless glucose sensors	[17]
	$MnCo_2O_4$/N-rGO	Hydrothermal method	GO, $Co(OAc)_2$, $Mn(OAc)_2$, NH_4OH	Alkaline ORR	[18]
	Co_3O_4/rmGO	Hydrothermal synthesis	GO, $Co(OAc)_2$, NH_4OH	Alkaline ORR	[19]
	CoO/CG	Aerosolization/ high-temperature	GO, $CoCl_2$	Enzymeless glucose sensors	[41]
	CoO/rGO	Hydrothermal synthesis	GO, Co $(Ac)_2$. $4H_2O$, $CO(NH_2)_2$	Electrochemical nonenzymatic sensor	[20]
Iron oxide	RG-O/Fe_2O_3	Microwave irradiation	GO, $FeCl_3$, N_2H_4	Anode Materials for Li- Ion Batteries	[30]
	RGO/Fe_3O_4	Chemical coprecipitation method	GO, $FeCl_3.6H_2O$, $FeCl_2.4H_2O$, NH_4OH	Nonenzymatic H_2O_2 sensing	[42]
	PDDA-G/Fe_3O_4	Chemical reduction	GO, $FeCl_3.6H_2O$, $FeSO_4$, PDDA	Amperometric H_2O_2 sensing	[25]
	Fe_3O_4/r-GO	Chemical reduction	GO, $FeCl_3.6H_2O$, $FeCl_2.4H_2O$	Eletrochemical NADH sensors	[26]
	3D Fe_3O_4/ N-GAs	Hydrothermal/ heat treatment	GO, iron(III) acetate, polypyrrole	Alkaline ORR	[43]

Metal oxides	Nanohybrid electrocatalyst	Preparation method	Precursors	Applications	Ref
Manganese oxide	MnO_2 /rGOP	Electrodeposition	GO, $MnSO_4$	Nonenzymatic H_2O_2 sensing	[34]
	GO/MnO_2	Chemical reduction	GO, $KMnO_4$	Nonenzymatic H_2O_2 sensing	[27]
	GO/MnO_2	Solution process	GO, MnO_2	Voltammetric sensor for hydroquinone and catechol	[44]
	PtAu–MnO_2/ GP	Electrodeposition	GO, $MnSO_4$	Nonenzymatic glucose sensors	[45]
	MnO_2/3D rGO	Chemical reduction	GO, $KMnO_4$	Electrochemical energy storage	[46]
	Mn_3O_4/GO	Hydrothermal method	GO, N_2H_4, $KMnO_4$	Alkaline ORR	[47]
	GO-MnO_2	Electrodeposition	GO, MnO_2	Alkaline ORR	[48]
Nickel oxide	NiONPs/GO	Electrodeposition	GO, $Ni(NO_3)_2.6H_2O$	Supercapacitor and enzymeless glucose sensors	[49]
	NA/NiONF-rGO	Electrospinning	GO, $C_4H_6O_4Ni.4H_2O$	Nonenzymatic glucose sensor	[50]
	NiO-GR	Electrodeposition	GO, $NiSO_4$	Nonenzymatic glucose sensor	[51]
Tin oxide	RG-O/SnO_2	Chemical reduction	GO, $SnCl_4.6H_2O$	Anode Materials for Li- Ion Batteries	[28]
	SnO_2-rGO	Hydrothermal method	GO, $SnCl_4.6H_2O$	NO_2 sensors	[52]
Titanium dioxide	RGO–TDN	Microwave synthesis	GO, titanium (IV) isopropoxide	Glucose sensors	[31]
	TiO_2-graphene	Hydrothermal method	GO, titanium (IV) isopropoxide	Adenine and Guanine sensors	[53]
	GN/TiO2	Template-based synthesis, Calcination	GO, titanium (IV) isopropoxide, CTAB	Determination of trace colorants in foods	[54]
	TiO_2@TiO_xN_y/TiN-GS	Hydrothermal/heat treatment	GO, TiO_2	Anode Materials for Li- Ion Batteries	[55]
	TiO_2-FGS	Chemical reduction	$TiCl_3$, CNT	Li- Ion Batteries	[56]
	TiO_2-FGS	Heat treatment	GO, sodium dodecyl sulfate (SDS), $TiCl_3$	Alkaline ORR	[57]

Metal oxides	Nanohybrid electrocatalyst	Preparation method	Precursors	Applications	Ref
Vanadium oxide	Graphene/ V_2O_5	Chemical reduction	GO, vanadyl tri isobutoxide	High-performance cathodes in Li-Ion batteries	[58]
	V_2O_5/GO	Chemical reduction, freeze drying	GO, HVO_3	Li-ion batteries	[59]
	3D VO_2 NR/ rGO	Hydrothermal/heat treatment	V_2O_5, GO	High-capacity supercapacitor electrodes	[60]
Zinc Oxide	Graphene/ZnO	Hydrothermal method	$ZnCl_2$, Graphene	Supercapacitors and electrochemical sensors	[61]
	rGO quantum dots/ZnO	Electrospinning	GO, ZnO	Intracellular H_2O_2 sensors	[62]

Table 1. Summary of the main synthetic methods for preparation of metal oxide–graphene nanocomposites and their major applications.

3. Structural characterization and physicochemical properties of graphene–metal oxide nanocomposites

As the synthesized graphene–metal oxide nanohybrid materials have tremendous effects on electrocatalytic applications, structural characterization is of paramount importance for understanding the correlation between their nanostructures and catalytic activity. The fundamental knowledge about the structural features of a nanocomposite is also essential for the building up of catalysts with optimal electrocatalytic activity. These structural features also offer helpful clues for further modification of the catalysts. In this section, we briefly summarize characterization of the structural features of different graphene–metal oxide nanocomposites by various instrumental techniques. These widely used techniques include scanning electron microscopy (SEM), transmission electron microscopy (TEM), atomic force microscopy (AFM), Raman spectroscopy, Fourier transform infrared spectroscopy (FTIR), X-ray photoelectron spectroscopy (XPS), energy dispersive X-ray spectroscopy (EDX), and thermogravimetric analysis (TGA).

3.1. Microscopic imaging of chemically exfoliated graphene and composites

The atomic structural features of a material affect its electronic, chemical, and catalytic properties. Various microscopic imaging techniques such as SEM, TEM, and AFM are proven as very essential tools to characterize nanomaterials. Their high-resolution capability enables to provide detailed information regarding shape, size, chemical composition, and phase of

http://dx.doi.org/10.5772/61808

nanomaterials. As a typical example, it has been well-shown that particle size, morphology, and exposed active facets have significant impacts on the catalytic efficiency of metal or metal oxide nanomaterials toward ORR electrocatalysis.

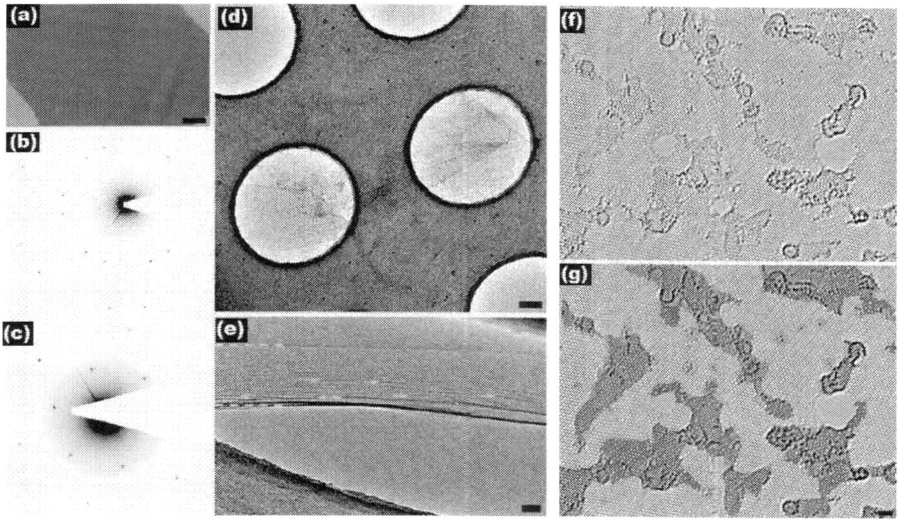

Figure 4. (a) Optical micrograph of graphene oxide sheets on the Quantifoil (QF) TEM grid. (b) Electron diffraction pattern of a bilayer area, displaying the stacking structure of the sheets. (c) Diffraction pattern from a single layer. (d) TEM image of RGO sheets on the QF grid. (e) TEM image with the sample tilted to 60°. The region between the horizontal dashed lines is a single layer (region above is a double layer, below is vacuum). Arrows indicate horizontal dark lines where the RGO sheet appears parallel to the beam, indicating local deformations up to 30°. Scale bars are 10 μm (a), 200 nm (d), and 10 nm (e). (f) Atomic resolution, aberration-corrected TEM image of a single layer reduced graphene oxide membrane. (g) Highlight with color in different areas. The light-gray color indicates the defect-free crystalline graphene area. Dark-gray shaded regions show the contaminated area. Blue regions indicate the disordered single-layer carbon networks, or extended topological defects due to the remnants of the oxidation–reduction process. Red areas indicate individual ad-atoms or substitutions. Green areas highlight isolated topological defects, that is, single bond rotations or dislocation cores. Holes and their edge reconstructions are marked in yellow color. Scale bar in (f) and (g) is 1 nm. (Reproduced with permission from ref. 65. Copyright ACS 2010)

SEM is mainly used to get the overall morphology of nanocomposites at large scales from several microns to 500 nm. However, to get more detailed and descriptive view of metal oxide nanocomposites and the crystal lattices or defects on graphene sheets, TEM and high-resolution TEM (HRTEM) are the most appropriate tools [63, 64]. The working principles of HRTEM are based on the interference of different transmitted and diffracted electron beams for building an image that can show the variation in phase. This is quite different from other traditional microscopic techniques, where sample image is derived from the variation of beam amplitudes due to the absorption of specimens Compared to mechanically exfoliated graphene, the chemically synthesized graphene contains notable structural defects, but they could hardly be detected by normal spectroscopic and microscopic characterizations. TEM is one of the leading methods for atomic scale imaging of graphene-based materials. For example, Fig. 4a shows an optical micrograph of RGO-coated Quantifoil (QF) TEM grid with the coverage visible as grayish spots [65]. TEM imaging analysis of the sample (Fig. 4d) showed that only ~1% of the holes of the whole grid were covered with sheets. Diffraction analysis was done to find holes covered by single-layered graphene sheets, which exhibited only one hexagonal pattern (Fig. 4c). These hexagonal patterns indicate the presence of a long-range hexagonal order orientation in the graphene sheets. Figure 4b indicates the parallel-beam diffraction pattern from a bilayer area, where two hexagonal patterns can be clearly detected. Then, the sample imaging was done under a high tilt (60°, Fig. 4e), which showed a high level of roughness, much more than that in mechanically exfoliated graphene sheets. The horizontal dark lines in Fig. 4e (arrows) was explained as representing the graphene layer in parallel to the electron beam. This image interprets that the wrinkles might be due to solution processing and drying, to stress, or to form defects in RGO. The high-resolution imaging of single-layered graphene sheets showed the actual atomic structure of the RGO layers, as represented in Figs. 4f and 4g. Various regions of the image are highlighted by colors in Fig. 4g. It is clearly observed that the largest part of the layer is formed of clean well-crystallized graphene areas where the hexagonal lattice is clearly seen (light-gray color in Fig. 4f). The visible well-crystallized areas are around 3–6 nm in the domain size and they cover ~60% of the whole plane. However, it is impossible to determine the exact structure of graphene with the adsorbed contamination, which covers ~30% of the total area. As most of the contaminants prefer to stick on the defects, the part of the defective areas is most likely underestimated. In spite of the presence of such a significant number of topological defects, the long-range oriented order of RGO is maintained. But such defects were normally not seen in any mechanically exfoliated graphene samples. Because the RGO and mechanically cleaved graphene samples were prepared from the same graphite source, it can be ruled out that the defects were from the starting material. These observations suggest that the high density of topological defects in these samples were introduced during graphene exfoliation by strong chemical oxidation and reduction. Although HRTEM can magnify the micro view into 1 nm, there is still limitation at around 0.2 nm. And, as electron beam heating can destroy small nanoparticles, the possibility of particle melting should be considered.

Song et al. [66] reported an alternative important strategy using scanning transmission electron microscopy (STEM), which combines the advantages of SEM and TEM and has extensively used in characterization of morphology and crystal structures of nanomaterials. Scanning probe microscopy techniques such as AFM and scanning tunneling microscopy (STM) also play an important role in the structural characterization of material surfaces with atomic resolution. Although the theoretical thickness of monolayer graphene is approximately 0.34 nm, the detected thickness by AFM analysis of a single layer of chemically synthesized GO is around 0.6–1.2 nm [67]. Although AFM can partially provide information regarding the number of layers in graphene, it is always better to associate it with Raman and XPS measurements for a complete chemical information. In addition, the STEM technique is also very helpful to obtain the lattice structure, surface morphology, particle size, and distribution of graphene-based materials surface at atomic resolution [68],[69],[70]. Figure 5a shows an overview of blue-shaded and uniformly folded graphene nanosheets taken by STEM at 200

http://dx.doi.org/10.5772/61808

kV with a high-angle annular dark field (HAADF) detector. Figures 5b and 5c show the 3 and 28 layers of graphene edges corresponding to the two arrows pointed area in Fig. 5a [66]. Figure 5d displays AFM images of chemically synthesized RGO nanosheets with ~0.6 nm thickness and with lateral dimensions in a few hundred nanometers to one micrometer [67]. And Figs. 5e and 5f show the AFM images of self-assembled RGO nanosheets with a 0.9 nm thickness [71].

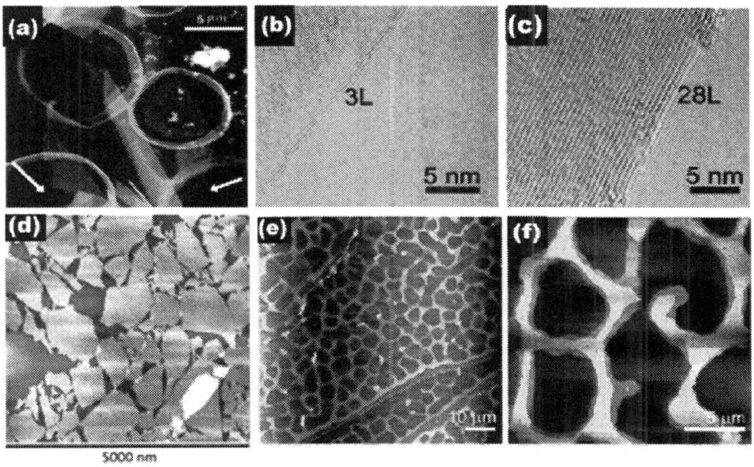

Figure 5. (a) HAADF-STEM image showing an overview of graphene flakes supported by a holey Formvar film covered with Cu grids. The arrows indicate areas where graphene is freely suspended on the holey film. (b) High-resolution TEM images of a three-layered graphene edge. (c) High-resolution TEM images of a 28-layered graphene sheet edge. (d) An AFM image of RGO nanosheet. (e,f) AFM images as an example of the self-assembled patterns of RGO nanosheets forming a bilayer nanofilm on mica surfaces. Scanned areas of AFM images are 80 μm × 80 μm (c) and 20 μm × 20 μm. (Combiningly reproduced with permission from ref. 66 Copyright Elsevier 2010, ref. 67 copyright 2013 WILEY-VCH Verlag GmbH & Co. KGaA, Weinheim, and ref. 71 copyright 2012 WILEY-VCH Verlag GmbH & Co. KGaA, Weinheim)

Figure 6 shows the morphology and structure of the 3D graphene/Co_3O_4 nanowire nanohybrid materials by SEM and TEM imaging. Figure 6a shows the SEM images of 3D porous structured graphene. Figure 6b-6d shows the uniform coverage of Co_3O_4 nanowire on 3D graphene skeleton. The high-resolution SEM image shows that the diameter of Co_3O_4 nanowire on 3D graphene is around 200–300 nm and the length is around several micrometers (Fig. 6d). And the TEM image (Fig. 6e) shows that the Co_3O_4 nanowires are composed of numerous nanoparticles.

3.2. Spectroscopic characterization

FTIR spectroscopy has been widely used to characterize different functional groups with specific chemical bonds (such as hydroxyl, carbonyl, carboxylic, and epoxy), which absorb

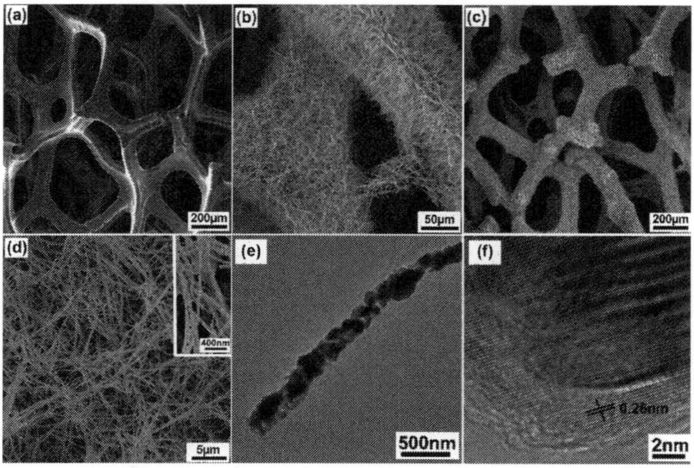

Figure 6. SEM images of (a) 3D graphene structure, (b) 3D graphene/Co_3O_4 nanowire nanohybrid (c, d) Low- and high-magnification SEM images of 3D graphene/Co_3O_4 nanowire nanohybrid material. Inset panel d shows an enlarged image. (e, f) Low- and high-resolution TEM images of Co_3O_4 nanowire grown on the 3D graphene surface. (Reproduced with permission from ref. 17 Copyright © 2012 American Chemical Society)

light energy (4000 cm^{-1} to 400 cm^{-1}) and exhibit a frequency corresponding to the fundamental vibrations [72]. Raman spectroscopy deals with the frequencies of Raman-scattered monochromatic light. As a supplementary to FTIR, Raman spectroscopy can provide sufficient information for characterization of graphene-based materials. Vibrations of different groups in polar/nonpolar molecules can be efficiently detected by these two methods. From the Raman spectra of graphene, three typical peaks of the G band at around 1580 cm^{-1}, the D band at around 1350 cm^{-1}, and the 2D band at 2700 cm^{-1} are often observed. The G band is an indicator of the stacking structures; the D is generally associated with the order/disorder of the material; the 2D-band is sensitive to the layer number of graphene sheets [2]. The ratio of the intensity of the two bands (D/G) is used for determining the number of layers in a particular graphene sample and its overall stacking behavior [2]. From Fig. 7a [16], two prominent peaks for the D and G bands are observed in the range of 1000–2000 cm^{-1}. And the intensity ratio of the D and G band indicates the density of structural defects on the graphene surface. In Fig. 7a, the I_D/I_G ratio of CuO/GO composite is 1.03 (S3), 0.96 (S2), and 0.87 (S1), which is much higher than the calculated value of GO (0.77). These ratios clearly indicate that the CuO modification introduced additional defects into the GO structure [16].

Figure 7b displays the FT-IR spectra of GO and different metal oxide nanocomposites. GO shows some typical peaks at 3420 and 1712 cm^{-1}, respectively, for the stretching vibrations of O–H and C=O. And also the bending vibration peak at 1408 cm^{-1} for O–H, at 1223 cm^{-1} for C–OH, the stretching peak at 1052 cm^{-1} for C–O, and the vibration peak at 1633 cm^{-1} for C=C [73]. For the spectra of Mn_3O_4–GNS, the vibration peaks at 610 cm^{-1} and 491 cm^{-1} are an indi-

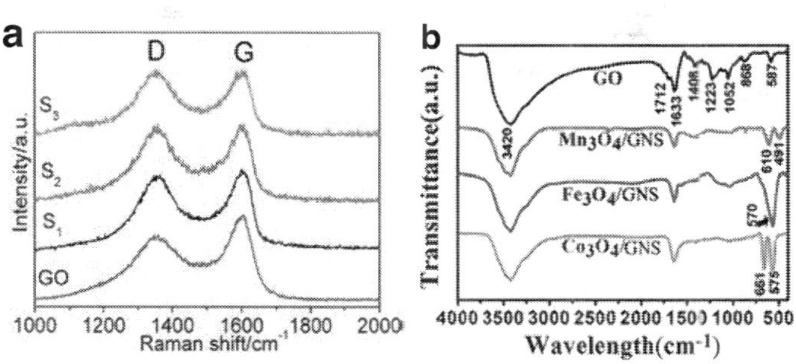

Figure 7. (a) Raman spectra of GO and CuO/GO composites and (b) FTIR spectra of GO, and Mn_3O_4–GNS, Fe_3O_4–GNS, and Co_3O_4–GNS nanocomposites. (Reproduced with permission from ref. 16. Copyright ACS 2013; from ref. 74 Copyright RSC 2012)

cation of the stretching modes of the Mn–O. In the spectrum of Fe_3O_4–GNS, the peak at 570 cm^{-1} is assigned to the vibration of the Fe–O bonds. In addition, for the spectrum of Co_3O_4–GNS, the peaks at 611 cm^{-1} and 575 cm^{-1} correspond to Co–O bonds [74].

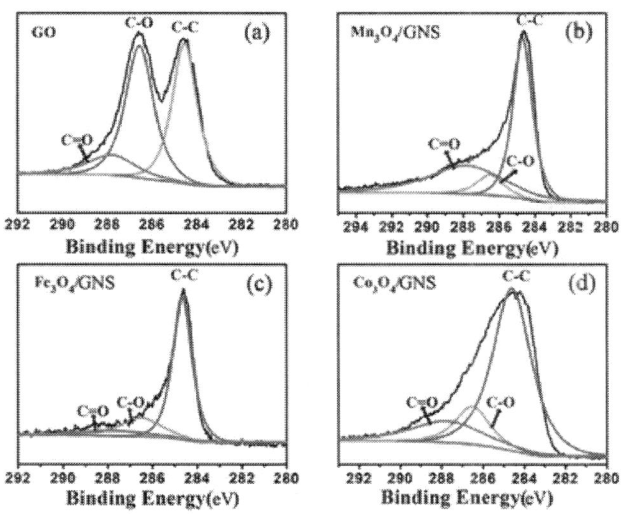

Figure 8. C 1s XPS spectra of (a) GO, (b) Mn_3O_4–GNS composite, (c) Fe_3O_4–GNS composite, and (d) Co_3O_4–GNS composite. The blue, pink, and green curves denote C–C, C–O, and C=O spectra, respectively. (Reproduced with permission from ref. 74 Copyright RSC 2012)

XPS is a powerful tool for the investigation of chemical composition, elemental states, and the nature of heteroatom functionalized or doped-graphene-based nanohybrid materials. For example, the presence of different metal or different heteroatom on functionalized graphene can be reliably identified by XPS. Zhang et al. [74] synthesized different inorganic–organic hybrid nanocomposite materials based on reduced graphene oxide and three different metal oxides. Figure 8a shows the C 1s spectrum of GO with two strong peaks of C–C, C–O, and a relatively weak peak of C=O species. After the formation of the composites, the peaks of C–O and C=O are significantly weakened due to the removal of oxygen-containing functional groups on GO [75]. Also, by comparing Fig. 8(b)–(d), Fe_3O_4–GNS displays the lowest C–O and C=O intensity, indicating that Fe^{2+} is the most efficient ion for the reduction of GO. In addition, C 1s, O 1s, and M 2p peaks existing in the wide scan spectra of different metal nanocomposites with graphene clearly indicate the combination of graphene nanosheets and different metal oxide NPs.

Other spectroscopic techniques are also available for the characterization of nanocomposites. UV-vis spectroscopic analysis often shows some typical absorption peaks for graphene and graphene oxide at around 268, 230, and 300 nm, which is relevant to the π–π^* transitions of aromatic C=C bonds, and the n-π^* transitions of C=O bonds, respectively [76]. TGA technique has been extensively used in characterizing the thermal stability and loading amount of different metal nanoparticles in graphene-based nanocomposites [77, 79]. Energy dispersive X-ray (EDX) spectroscopy is very helpful for qualitative and quantitative analysis of the element distribution on graphene-based nanocomposite materials. EDX is more effective for determination of the locations of different metals in nanocomposites when combined with TEM or SEM [80, 81].

3.3. Main physicochemical properties

The physicochemical properties of different graphene–metal oxide nanohybrid material can be characterized by a series of electrochemical methods and instruments, such as cyclic voltammetry (CV) measurements, linear sweep voltammetry analysis (LSV), and electrochemical impedance spectroscopy (EIS). In some cases, rotating-disk electrodes (RDE) and rotating ring-disk electrodes (RRDE) are needed. Each method has its specific advantages for studying electrocatalytic performance of the hybrid materials. For the different characterizations, one method is normally not sufficient but several methods involved. In this section, we focus on discussing some electrochemical characterization techniques and their specialties and limitations.

3.3.1. *Cyclic Voltammetry (CV)*

CV is arguably the most common and straightforward method to determine the electrocatalytic activity of a nanocomposite material. CV is normally measured in a typical electrolyte solution at room temperature. An electrochemical cell consists of three electrodes, i.e., a working electrode loaded with the catalyst, a reference electrode, and a Pt wire as a counter electrode [82].

As shown in Fig. 9, the electrochemical kinetics of CuO/GO/GCEs compared with CuO/GCEs was studied by CV systematically. The CV was performed in 0.1 M NaOH electrolyte solution with a scan rate of 100 mV/s. Figure 9a shows a peak at +0.67 V versus Ag/AgCl from all the three CuO/GCEs electrodes, which corresponds to the Cu(II)/Cu(III) redox couple. The three

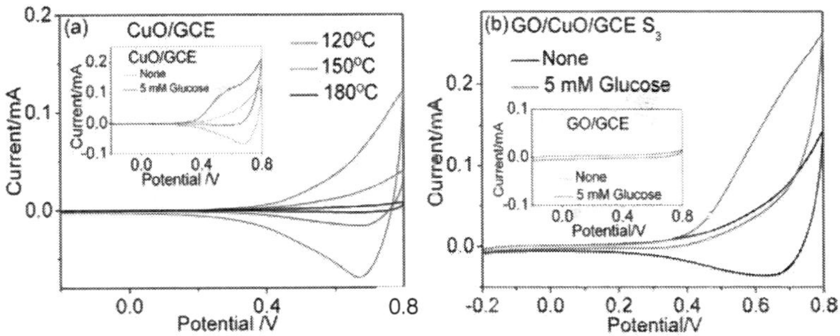

Figure 9. CV curves of (a) CuO NPs synthesized at 120, 150, and 180°C. The inset is the CuO NPs synthesized at 120°C in 0.1 M NaOH before (black trace) and after (red trace) the injection of 5 mM glucose. (b) CuO/GO composite S3 and GO sheet (inset) in 0.1 M NaOH before (black trace) and after (red trace) the injection of 5 mM glucose. (Reproduced with permission from ref. 16. Copyright ACS 2013)

different types of CuO NP generated different peak currents. CuO NPs obtained at 120°C displayed the most efficient electron transfer. The inset in Fig. 9a represents the CVs of before and after the addition of 5 mM glucose to the electrolye solutions for CuO/GCE based on 120°C CuO NPs. The Cu (II)/Cu(III) redox couple is the important factor for nonenzymatic glucose detection. Figure 9b shows the CV response of the CuO/GO/GCE before and after the addition of 5 mM glucose to the electrolye solutions, and the inset shows the CV curve of the GO/GCE electrode as a reference. As GO is electroinactive, there is no electrocatalytic oxidation of glucose. In contrast, CuO/GO/GCE showed high electrocatalytic activity, and the peak current significantly increases. Thus, the cyclic voltammetry offers a convenient way to study the electrocatalytic oxidation process of glucose.

Figure 10. (a) CVs of ORR on different electrodes in N_2- and O_2-saturated 0.1 M KOH with a scan rate of 10 mV s^{-1}. (b) CV curves of $MnCo_2O_4$/N-rmGO, $MnCo_2O_4$ + N-rmGO mixture, Co_3O_4/N-rmGO, and N-rmGO on GCEs in O_2-saturated (solid line) or N_2-saturated (dash line) 1 M KOH. The peak position of Pt/C is displayed as a dashed line for comparison. (Reproduced with permission from ref. 19 Copyright Nature 2011; from ref. 18 Copyright ACS 2012)

As shown in Fig. 10a, the commercial Pt/C displays typical CV responses of ORR in N_2 and O_2-saturated 0.1 M KOH. The black dash line shows CV response in the N_2-saturated 0.1 M KOH within a potential window from 0.36 to 1.1 V. It is clearly seen that there is no sign of the typical ORR peak of Pt/C at +0.9 V [19]. However, a distinct performance of ORR (black solid line) is clearly seen in the case of O_2-saturated 0.1 M KOH. The clear peak at ~0.9 V of Pt/C indicates that the Pt/C exhibits excellent ORR activity with its standard onset potential (~1.0 V) and peak potential (~0.9 V). Figure 10b compares the electrocatalytic ORR performance of $MnCo_2O_4$/N-rmGO, N-rmGO mixture, Co_3O_4/N-rmGO, and N-rmGO in an N_2- and O_2-saturated 0.1 M KOH solution, respectively. However, these four electrodes showed different ORR activity in O_2 saturated electrolyte solutions. The ORR activity of $MnCo_2O_4$/N-rmGO (red solid line) has a more positive peak potential and higher peak current density (0.88 V, 0.5 mA cm^{-2}) than those of the N-rmGO mixture (~0.84 V, 0.38 mA cm^{-2}), Co_3O_4/N-rmGO (~0.86 V, 0.44 mA cm^{-2}), and N-rmGO (~0.82 V, 0.29 mA cm^{-2}). Therefore, $MnCo_2O_4$/N-rmGO material is more promising for ORR application, with a similar performance to that obtained at commercially available Pt/C [18].

3.3.2. *Linear Sweep Voltammetry analysis (LSV)*

LSV analysis is a vital method for evaluating the ORR activity of synthesized nanocomposites combining a rotating-disk electrode (RDE) or rotating ring-disk electrode (RRDE) [83, 86]. Similar to CV, the LSV analysis is also done in an O_2-saturated 0.1 M KOH electrolyte solution at different rotation rates of electrode under room temperature. For LSV, the electrolytic bath contains three electrodes: an RDE or RRDE quantitatively coated with the synthesized catalyst, an Ag|AgCl/KCl (saturated) reference electrode, and a Pt wire as an auxiliary electrode. During electrochemical measurement, the working electrode steadily rotates at the required rotation rate, and the current density changes in the potential range corresponding to the CV for one linear sweep are recorded. LSV technique has been extensively used in studying the electrocatalytic ORR kinetics and mechanism.

3.3.3. *Amperometric technique (I-t)*

Amperometric technique is very crucial for electrocatalytic sensing applications. For amperometric sensors, current is produced proportional to the concentration of the analyte to be detected.

Figure 11a compares the I–t curves of CuO/GCE and three CuO/GO/GCEs (S1–S3) electrode sensing glucose. Figures 11a and 11b show that the sensors produced an excellent amperometric current with a short response time. Figures 11c and 11d show the amperometric responses of CuO/GO/GCEs S3 sensor toward the electrocatalytic oxidation of glucose in human serum and the corresponding calibration curve [16]. The sensitivity, lower detection limit, and linear range can thus be calculated from these amperometric data.

3.3.4. *Electrochemical Impedance Spectroscopy (EIS)*

Electrochemical impedance spectroscopy (EIS) has become a popular and effective technique in recent decades for the determination of double-layer capacitance, characterization of

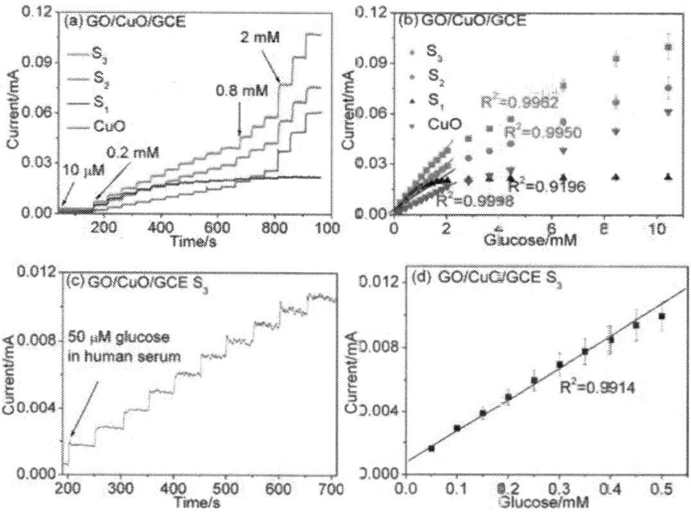

Figure 11. Amperometric responses: (a) CuO (120 °C) and three different RGO-CuO (S1, S2, and S3); (b) the corresponding calibration curves of (a); (c) S3 to successive additions of human serum contained 50 μM glucose; (d) corresponding calibration curve of (c). (Reproduced with permission from ref. 16. Copyright ACS 2013)

electrode processes, and identification of complex interfaces. EIS records the response from an electrochemical system by the stimulation of an imposed periodic small amplitude AC signal. EIS measurements are normally done at various AC frequencies, and then, the EIS can be measured by the changes of the ratio between the AC potential and current signal with the corresponding sinusoid frequencies (ω). EIS analysis has extensively been performed for ORR to explore the kinetic process of the reaction with some essential information, including the interface and structure of the ORR electrode materials, properties of electric double layer, and diffusion of oxygen. The kinetic electron transfer process for ORR is explained by an EIS plot with a semicircle and a linear portion corresponding to a charge transfer and mass transportation procedure at the high-frequency region and the low-frequency region, respectively.

4. Applications of graphene–metal oxide nanohybrids as electrocatalysts

Graphene–metal-oxide-based nanohybrid materials have a wide range of applications from electrochemical sensing to the ORR electrocatalysis, due to their superior properties and low cost [87, 88]. The new types of materials could open up a new window for superior electrocatalytic activity as well as selectivity and durability, which can act as promising electrode materials for various electrochemical reactions [89, 90, 91]. In this section, we present some examples for the applications of graphene–metal oxide nanohybrids for different kinds of electrocatalytic reactions.

4.1. Metal-oxide-decorated 2D graphene structures and electrocatalysis

If metal oxide nanoparticles are randomly loaded on graphene sheets but with little control in size and structure [92], then it can result in make the poor interaction between nanoparticles and graphene sheet [93]. Recently, the methods have been continuously improved, to some degree with controlling their locations and amount. In this context, a number of metal oxides are synthesized and loaded onto graphene matrix used as nonenzymatic sensor materials [16, 29, 31, 35, 45, 94, 95]. For example, Sun and coworkers [96] made a presynthesized monodisperse Co/CoO core/shell nanoparticles on the graphene surface (Fig. 12). In Co/CoO core/shell nanoparticles, Co core size and thickness can be tuned by controlling the oxidation conditions. In their work, they demonstrated the significance of Co/CoO size and graphene support for the tuning of electrocatalysts for efficient ORR with a selective 4e process (Fig. 12b,c).

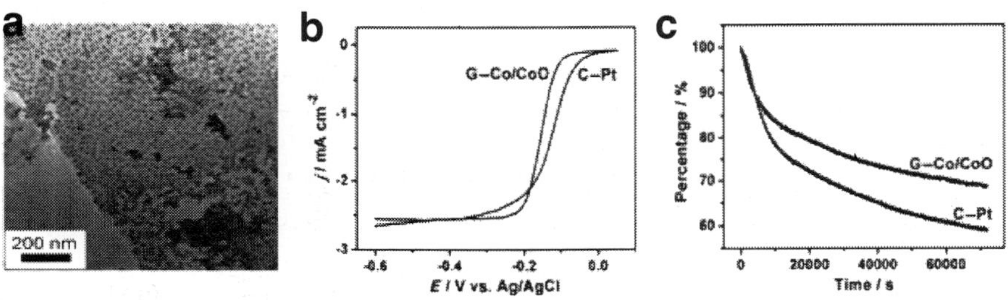

Figure 12. (a) A TEM image of the Co/CoO core/shell NPs deposited on graphene surface. (b) ORR polarization curves of the G–Co/CoO NPs and commercial C–Pt catalyst. Scan rate: 10 mV s^{-1} and rotation rate 400 rpm. (c) The chronoamperometric responses for the ORR on the G–Co/CoO NPs and commercial C–Pt catalyst at −0.3 V. Rotation rate: 200 rpm. The measurements were performed in O_2-saturated KOH (0.1 M) solution. (Reproduced with permission from ref. 96 Copyright 2012 WILEY-VCH Verlag GmbH & Co. KGaA, Weinheim)

The optimized Co/CoO–graphene electrocatalyst exhibited reasonably high activity and better stability than the commercial Pt/C catalyst. It is also evident that the ORR activity of cobalt oxide–graphene hybrid electrocatalysts is significantly enhanced with the Co content and its coupling with N-doped graphene [97]. Cobalt oxides/graphene nanohybrid materials can also serve as the ORR catalysts used in Li–O_2 batteries. Graphene sheets with Co_3O_4 nanofiber immobilized on both sides can act as a bifunctional catalyst for the ORR[98]. This excellent electrochemical performance relies on the facile electron transport and fast O_2 diffusion between the porous Co_3O_4 nanofiber networks and the ultrathin graphene layer. Manganese-oxides-based graphene nanocomposites have also been used as a stable and low-cost cathode electrocatalysts for fuel cells and Li–air batteries [99, 101]. It is clearly evident that electrocatalytic performance of metal oxide–graphene nanocomposites is closely associated with morphology and size of metal oxide nanoparticles and metal oxide–graphene electronic couplings [102]. Qiao and coworkers showed a mesoporous structure of Mn_3O_4/graphene hybrid nanomaterials with good ORR activity, excellent stability, and high selectivity [103]. Kim and coworkers also showed the ORR mechanism of a system with a lower loading (19.2%)

of Mn_3O_4 nanoparticles on graphene sheets is comparable to that of the Pt/C electrode with a 4e transfer, whereas the composite with a higher Mn_3O_4 content (52.5%) undergoes a conventional 2e process [104]. Also, other graphene-supported metal oxides, e.g., Fe_3O_4 [105], Fe_2O_3 [106], Cu_2O [107, 108], and Ru_2O [109] were also studied as electrocatalysts for ORR in fuel cells and Li–air batteries.

4.2. Metal-oxide-decorated 3D graphene structures and electrocatalysis

3D structured graphene (e.g., graphene foam) is an ideal candidate for noble metal oxide catalyst support for electrocatalytic applications because of its high electron conductivity, large surface area, sufficient porosity, and thermal stability [110]. 3D graphene can offer high surface areas for higher loading of metal oxide nanoparticles, which can show enhanced electrocatalytic activity. For example, Dong and coworkers [17] synthesized a 3D graphene–cobalt oxide nanohybrid material for high-performance supercapacitor and enzymeless glucose detection. Figure 13 shows the electrocatalytic oxidation of glucose in low-concentration alkaline solutions. Figure 13a shows the CV curves of the 3D graphene/cobalt oxide composite electrode obtained at different scan rates. Two pairs of redox peaks (I/ II and III/ IV) are observed from the CV. The redox peak currents increase with increasing scan rate proportionally, which implies a surface-controlled electrochemical process. Figure 13b shows that the oxidation current at peak III (at ~0.58 V) started increasing with introduction of glucose, but the other peak remained almost constant. Figure 13c shows the amperometric responses of the graphene/Co_3O_4 composite electrode to glucose with various concentrations. The calibration curve of the amperometric response was plotted in Fig. 13d. This composite material showed an excellent sensitivity of 3.39 mA mM^{-1} cm^{-2}, a relatively narrow linear range (up to 80 μM) and sub-100 nM lower detection limit (LOD).

Feng and Mullen have studied the controllable structural assembly of Fe_3O_4 nanoparticles on 3D N-doped graphene aerogel support [21]. Figure 14 (a-d) shows interconnected macroporous graphene hybrid network uniformly decorated with Fe_3O_4 nanoparticles. The Fe_3O_4/N–graphene aerogel network displayed excellent electrocatalytic activity for the ORR in alkaline electrolytes with a high current density, low ring current and H_2O_2 yield, being a four electron transfer number, and high stability (Fig. 14e,f). The electrocatalytic ORR has an onset potential of +0.16 V (vs Ag/AgCl) and a high current density of 1.46 mA cm^{-2}, which is well-comparable with the performance by commercial Pt/C.

4.3. Heteroatom doped-graphene-materials and their electrocatalysis

Nitrogen and sulfur are the mostly used elements for doping graphene[111, 114]. Specific doping on graphene could lead to remarkable increase in charge carrier concentration, specific surface area, and enhanced capacitance retention. The N- or S-doped graphene materials show new exciting properties compared to pristine graphene. For instance, the spin density and charge distribution of carbon atoms are modulated by the neighboring nitrogen dopants, which induce the "activation region" on the graphene surface. This kind of activated region can directly participate in electrocatalytic reactions such as ORR [115, 116] or anchor metal nanoparticles with specific catalytic activity desired [117].

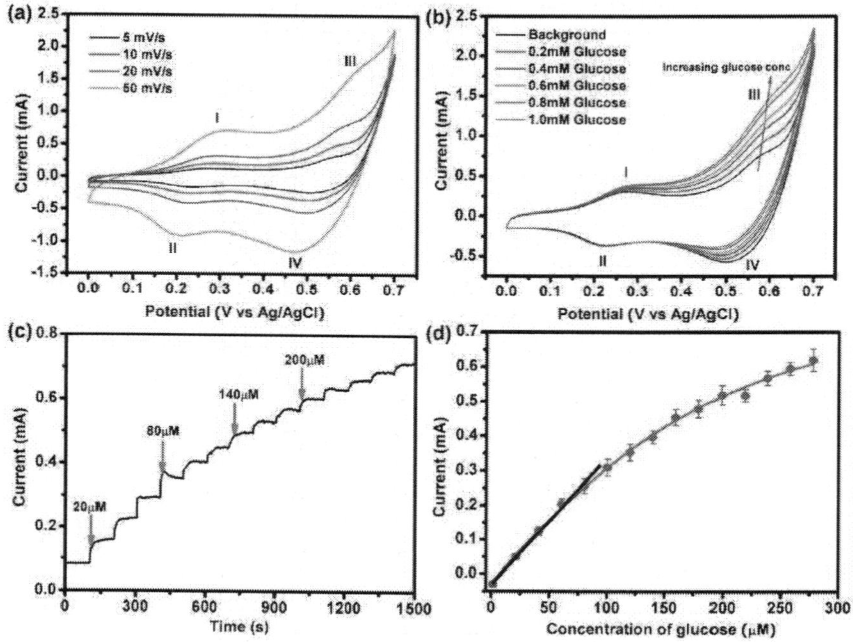

Figure 13. Electrochemical sensing of glucose in 0.1 M NaOH solution using the 3D graphene/Co_3O_4 composite electrode. (a) CV curves measured at different scan rates (5, 10, 20, and 50 mV/s). (b) CV curves in the presence of different concentrations of glucose (0, 0.2, 0.4, 0.6, 0.8, and 1 mM), at the scan rate of 20 mV/s. (c) Amperometric data of the composite electrode (potential 0.58 V) upon addition of glucose to increasing concentrations. (d) Average dose response curve (amperometric current response *vs* glucose concentration) obtained from three different sensors, with a linear fitting at lower concentration range and an exponential fitting at higher concentration range. The error bars indicate the standard deviations. (Reproduced with permission from ref. 17. Copyright ACS 2012)

Sheng et al. [118] reported a facile catalyst-free method for the synthesis of N-doped graphene via thermal annealing graphene oxide with melamine for the electrocatalytic application in ORR. Figure 15 compares cyclic voltammograms (CVs) for the electrochemical reduction of O_2 at a bare glassy carbon electrode (GCE), graphene/GCE, and NG/GCE in O_2-saturated 0.1 M KOH solutions. The onset potential of ORR at the NG/GCE occurs at 0.1 V, which is about 0.1 V more positive than that of graphene/GCE. The electrocatalytic process of NG/GCE is a one-step four-electron pathway for ORR, which is almost twice as large as that for pristine graphene in the current density. Sulfur-doped graphene was successfully prepared using GO and benzyl disulfide as precursors under high temperature, and the as-prepared temperature-dependant S-Doped graphene was tested as a metal-free cathode catalyst for oxygen reduction. All the results further confirmed that the S-doped graphene is a promising material with high catalytic activity for ORR [119]. N and S co-doped graphene was also developed and used for ORR recently; compared to the single element doped graphene, co-doped graphene displayed even more efficient electrocatalysis toward ORR [120].

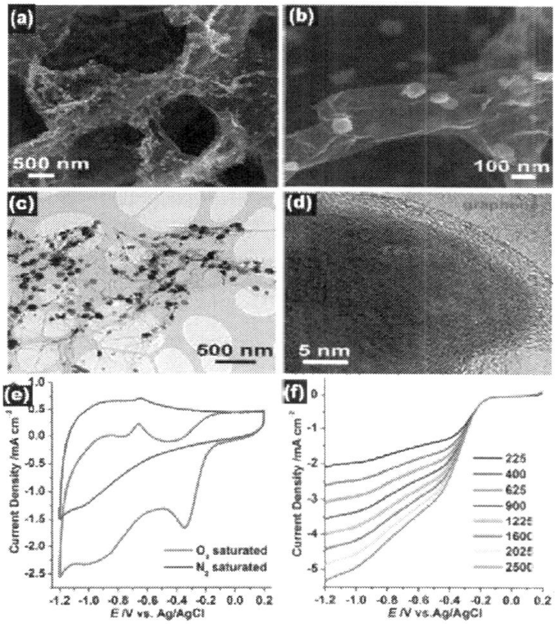

Figure 14. (a,b) Typical SEM images of Fe_3O_4/N-Gas, revealing the 3D macroporous structure and uniform distribution of Fe_3O_4 NPs in the GAs. The red rings in (d) indicate Fe_3O_4NPs encapsulated in thin graphene layers. Representative (c) TEM and (d) HRTEM images of Fe_3O_4/N-Gas, revealing an Fe_3O_4 NP wrapped by graphene layers. (e) CVs of Fe_3O_4/N-GAs in N_2- and O_2-staturated 0.1 M aqueous KOH electrolyte solution at a scan rate of 100 mV s^{-1}. (f) LSVs of Fe_3O_4/N-GAs in O_2-saturated 0.1 M KOH at a scan rate of 10 mV s^{-1} at different RDE rotation rates (in rpm). (Reproduced with permission from ref. 21. Copyright ACS 2012)

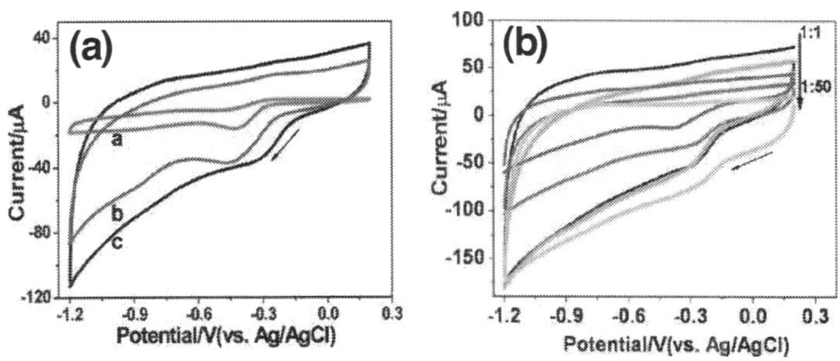

Figure 15. (a) Cyclic voltammograms (CVs) for ORR obtained at a bare GCE (a), graphene/GCE (b), and NG5/GCE (N % = 7.1%) (c) in O_2-saturated 0.1 M KOH aqueous solution. (B) CVs for ORR at NGs, synthesized with different mass ratio of GO and melamine (1:1, 1:2, 1:5, 1:10, 1:50) at 800°C, modified GCE in O_2-saturated 0.1 M KOH aqueous solution. Scan rate: 100 mV/s. (Reproduced with permission from ref. 118 Copyright ACS 2011)

5. Conclusions and outlook

The rapid development of graphene-based nanohybrid electrocatalysts for energy conversion, storage, and various electrochemical sensors has been driven by their unique structural features, novel physicochemical properties, high stability, and low cost. The immobilization of transition metal oxides on graphene by various methods via the use of interactions between the structural defects and functional groups on graphene's surface, contributed to the improved catalytic activity and stability of graphene-based transition metal oxide nanohybrid catalysts. In this chapter, we have discussed the recent development of graphene-supported transition metal oxide nanohybrid materials for their sensor and energy applications, including their synthesis, structural characterizations, key properties, and major applications. The control in the morphology and dimension of the graphene-based nanocomposite is of crucial importance for their electrocatalytic activities. As a consequence, multivalent transition metal oxides with special structural properties exhibit more efficiency for the electrocatalytic reactions than amorphous nanocomposites under similar experimental conditions. Electrocatalytic performance of graphene-based transition metal nanohybrid materials can be characterized by different electrochemical techniques in detail. Stability and durability are among the most important factors for promising electrocatalytic applications. The flexibility and large surface area of graphene sheets could prevent particles from agglomerationand facilitate accommodation of large amount of particles. Monotransition metal oxides anchored on graphene have already shown good electrocatalytic performance with long-term stability. However, the strategy of synthesizing bimetallic transition metal oxides is very promising for further elevating the electrocatalytic activity. Also, covalently bonded bimetallic oxides on graphene can offer better activity and longer durability than the physical mixture of two types of metallic nanoparticles. Therefore, the rational design of cationic substitution and covalent coupling with graphene supports can instruct the construction of advanced electrocatalysts for sensor- and energy-related applications. Thus, the advancement of sophisticated structure-controlled methods and processes for the *in situ* synthesis of graphene-based transition metal oxide nanocomposites is crucial for the development of next-generation electrocatalysts. It should be noted, however, that the fundamental mechanisms behind the electrocatalytic performance of graphene-based nanocomposites are far from being fully understood. Moreover, the electronic interactions between graphene and nanoparticles and the synergistic effects are needed to be explored. These remained challenges will motivate strongly many ongoing research and further development of this field.

Author details

Arnab Halder, Minwei Zhang and Qijin Chi*

*Address all correspondence to: cq@kemi.dtu.dk

Department of Chemistry, Technical University of Denmark, Kongens Lyngby, Denmark

References

[1] A.K. Geim and K.S. Novoselov, "The Rise of Graphene," *Nat. Mater.*, vol. 6, no. 3, pp. 183–191, Mar. 2007.

[2] D. R. Dreyer, S. Park, C. W. Bielawski, and R. S. Ruoff, "The Chemistry of Graphene Oxide," *Chem. Soc. Rev.*, vol. 39, no. 1, pp. 228–40, Jan. 2010.

[3] Y. Zhu, S. Murali, W. Cai, X. Li, J. W. Suk, J. R. Potts, and R. S. Ruoff, "Graphene and Graphene Oxide: Synthesis, Properties, and Applications," *Adv. Mater.*, vol. 22, no. 35, pp. 3906–3924, 2010.

[4] X. Huang, Z. Yin, S. Wu, X. Qi, Q. He, Q. Zhang, Q. Yan, F. Boey, and H. Zhang, "Graphene-Based Materials: Synthesis, Characterization, Properties, and Applications," *Small*, vol. 7, no. 14, pp. 1876–1902, 2011.

[5] C. Lee, X. Wei, J. W. Kysar, and J. Hone, "Measurement of the Elastic Properties and Intrinsic Strength of Monolayer Graphene," *Science*, vol. 321, no. 5887, pp. 385–388, Jul. 2008.

[6] A. A. Balandin, S. Ghosh, W. Bao, I. Calizo, D. Teweldebrhan, F. Miao, and C. N. Lau, "Superior Thermal Conductivity of Single-Layer Graphene," *Nano Lett.*, vol. 8, no. 3, pp. 902–907, Mar. 2008.

[7] J. A. Farmer and C. T. Campbell, "Ceria Maintains Smaller Metal Catalyst Particles by Strong Metal-Support Bonding," *Science*, vol. 329, no. 5994, pp. 933–936, Aug. 2010.

[8] R. S. Dey, H. A. Hjuler, and Q. Chi, "Approaching the Theoretical Capacitance of Graphene through Copper Foam Integrated Three-Dimensional Graphene Networks," *J. Mater. Chem. A*, vol. 3, no. 12, pp. 6324–6329, 2015.

[9] R. S. Dey and C. R. Raj, "Development of an Amperometric Cholesterol Biosensor Based on Graphene-Pt Nanoparticle Hybrid Material," *J. Phys. Chem. C*, vol. 114, no. 49, pp. 21427–21433, 2010.

[10] S. Giri, D. Ghosh, and C. K. Das, "Growth of Vertically Aligned Tunable Polyaniline on Graphene/ZrO 2 Nanocomposites for Supercapacitor Energy-Storage Application," *Adv. Funct. Mater.*, pp. 1312–1324, Oct. 2013.

[11] B. Anasori, M. Beidaghi, and Y. Gogotsi, "Graphene - Transition metal oxide hybrid materials," *Mater. Today*, vol. 17, no. 5, pp. 253–254, 2014.

[12] S. Gan, L. Zhong, C. Engelbrekt, J. Zhang, D. Han, J. Ulstrup, Q. Chi, and L. Niu, "Graphene Controlled H- and J-Stacking of Perylene Dyes into Highly Stable Supramolecular Nanostructures for Enhanced Photocurrent Generation," *Nanoscale*, vol. 6, no. 18, p. 10516 - 10523, 2014.

[13] W. Il Park, C.-H. Lee, J. M. Lee, N.-J. Kim, and G.-C. Yi, "Inorganic Nanostructures Grown on Graphene Layers," *Nanoscale*, vol. 3, no. 9, pp. 3522–3533, 2011.

[14] Y. Wang, L. Cheng, F. Li, H. Xiong, and Y. Xia, "High Electrocatalytic Performance of Mn_3O_4/Mesoporous Carbon Composite for Oxygen Reduction in Alkaline Solutions," *Chem. Mater.*, vol. 19, no. 8, pp. 2095–2101, 2007.

[15] F. Cheng, Y. Su, J. Liang, Z. Tao, and J. Chen, "MnO_2-Based Nanostructures as Catalysts for Electrochemical Oxygen Reduction in Alkaline Media," *Chem. Mater.*, vol. 22, no. 3, pp. 898–905, 2010.

[16] J. Song, L. Xu, C. Zhou, R. Xing, Q. Dai, D. Liu, and H. Song, "Synthesis of Graphene Oxide Based CuO Nanoparticles Composite Electrode for Highly Enhanced Nonenzymatic Glucose Detection," *ACS Appl. Mater. Interf.*, vol. 5, no. 24, pp. 12928–12934, Dec. 2013.

[17] X.-C. Dong, H. Xu, X.-W. Wang, Y.-X. Huang, M. B. Chan-Park, H. Zhang, L.-H. Wang, W. Huang, and P. Chen, "3D Graphene-Cobalt Oxide Electrode for High-Performance Supercapacitor and Enzymeless Glucose Detection," *ACS Nano*, vol. 6, no. 4, pp. 3206–3213, Apr. 2012.

[18] Y. Liang, H. Wang, J. Zhou, Y. Li, J. Wang, T. Regier, and H. Dai, "Covalent Hybrid of Spinel Manganese-Cobalt Oxide and Graphene as Advanced Oxygen Reduction Electrocatalysts," *J. Am. Chem. Soc.*, vol. 134, no. 7, pp. 3517–3523, 2012.

[19] Y. Liang, Y. Li, H. Wang, J. Zhou, J. Wang, T. Regier, and H. Dai, "Co_3O_4 Nanocrystals on Graphene as a Synergistic Catalyst for Oxygen Reduction Reaction," *Nat. Mater.*, vol. 10, no. 10, pp. 780–786, 2011.

[20] M. Wang, J. Huang, M. Wang, D. Zhang, and J. Chen, "Electrochemical Nonenzymatic Sensor Based on CoO Decorated Reduced Graphene Oxide for the Simultaneous Determination of Carbofuran and Carbaryl in Fruits and Vegetables," *Food Chem.*, vol. 151, pp. 191–197, 2014.

[21] Z.-S. Wu, S. Yang, Y. Sun, K. Parvez, X. Feng, and K. Müllen, "3D Nitrogen-Doped Graphene Aerogel-Supported Fe_3O_4 Nanoparticles as Efficient Electrocatalysts for the Oxygen Reduction Reaction," *J. Am. Chem. Soc.*, vol. 134, no. 22, pp. 9082–9085, Jun. 2012.

[22] Y. Zhou, Q. Bao, L. A. L. Tang, Y. Zhong, and K. P. Loh, "Hydrothermal Dehydration for the 'Green' Reduction of Exfoliated Graphene Oxide to Graphene and Demonstration of Tunable Optical Limiting Properties," *Chem. Mater.*, vol. 21, no. 13, pp. 2950–2956, Jul. 2009.

[23] R. Zhou, Y. Zheng, D. Hulicova-Jurcakova, and S. Z. Qiao, "Enhanced Electrochemical Catalytic Activity by Copper Oxide Grown on Nitrogen-Doped Reduced Graphene Oxide," *J. Mater. Chem. A*, vol. 1, no. 42, p. 13179 - 13184, 2013.

[24] X.-Y. Yan, X.-L. Tong, Y.-F. Zhang, X.-D. Han, Y.-Y. Wang, G.-Q. Jin, Y. Qin, and X.-Y. Guo, "Cuprous Oxide Nanoparticles Dispersed on Reduced Graphene Oxide as an

Efficient Electrocatalyst for Oxygen Reduction Reaction," *Chem. Commun.*, vol. 48, no. 13, p. 1892 - 1895, 2012.

[25] X. Liu, H. Zhu, and X. Yang, "An Amperometric Hydrogen Peroxide Chemical Sensor Based on Graphene-Fe $_3$O $_4$ Multilayer Films Modified ITO Electrode," *Talanta*, vol. 87, no. 1, pp. 243–248, 2011.

[26] H. Teymourian, A. Salimi, and S. Khezrian, "Fe_3O_4 Magnetic Nanoparticles/Reduced Graphene Oxide Nanosheets as a Novel Electrochemical and Bioeletrochemical Sensing Platform," *Biosens. Bioelectron.*, vol. 49, pp. 1–8, Nov. 2013.

[27] L. Li, Z. Du, S. Liu, Q. Hao, Y. Wang, Q. Li, and T. Wang, "A Novel Nonenzymatic Hydrogen Peroxide Sensor Based on MnO_2/Graphene Oxide Nanocomposite," *Talanta*, vol. 82, no. 5, pp. 1637–1641, 2010.

[28] X. Zhu, Y. Zhu, S. Murali, M. D. Stoller, and R. S. Ruoff, "Reduced Graphene Oxide/Tin Oxide Composite as an Enhanced Anode Material for Lithium Ion Batteries Prepared by Homogenous Coprecipitation," *J. Power Sources*, vol. 196, no. 15, pp. 6473–6477, 2011.

[29] Y. Tian, Y. Liu, W. Wang, X. Zhang, and W. Peng, "CuO Nanoparticles on Sulfur-Doped Graphene for Nonenzymatic Glucose Sensing," *Electrochim. Acta*, vol. 156, pp. 244–251, 2015.

[30] X. Zhu, Y. Zhu, S. Murali, M. D. Stoller, and R. S. Ruoff, "Nanostructured Reduced Graphene Oxide/Fe_2O_3 Composite as a High-Performance Anode Material for Lithium Ion Batteries," *ACS Nano*, vol. 5, no. 4, pp. 3333–3338, 2011.

[31] Z. Luo, X. Ma, D. Yang, L. Yuwen, X. Zhu, L. Weng, and L. Wang, "Synthesis of Highly Dispersed Titanium Dioxide Nanoclusters on Reduced Graphene Oxide for Increased Glucose Sensing," *Carbon*, vol. 57, pp. 470–476, 2013.

[32] H.-L. Guo, X.-F. Wang, Q.-Y. Qian, F.-B. Wang, and X.-H. Xia, "A Green Approach to the Synthesis of Graphene Nanosheets," *ACS Nano*, vol. 3, no. 9, pp. 2653–2659, Sep. 2009.

[33] D. Ye, G. Liang, H. Li, J. Luo, S. Zhang, H. Chen, and J. Kong, "A Novel Nonenzymatic Sensor Based on CuO Nanoneedle/Graphene/Carbon Nanofiber Modified Electrode for Probing Glucose in Saliva," *Talanta*, vol. 116, pp. 223–230, 2013.

[34] F. Xiao, Y. Li, X. Zan, K. Liao, R. Xu, and H. Duan, "Growth of Metal-Metal Oxide Nanostructures on Freestanding Graphene Paper for Flexible Biosensors," *Adv. Funct. Mater.*, vol. 22, no. 12, pp. 2487–2494, Jun. 2012.

[35] D. L. Zhou, J. J. Feng, L. Y. Cai, Q. X. Fang, J. R. Chen, and A. J. Wang, "Facile Synthesis of Monodisperse Porous Cu_2O Nanospheres on Reduced Graphene Oxide for Non-Enzymatic Amperometric Glucose Sensing," *Electrochim. Acta*, vol. 115, pp. 103–108, 2014.

[36] X. Zhang, Q. Liao, S. Liu, W. Xu, Y. Liu, and Y. Zhang, "CuNiO Nanoparticles Assembled on Graphene as an Effective Platform for Enzyme-Free Glucose Sensing," *Anal. Chim. Acta*, vol. 858, pp. 49–54, 2015.

[37] F. Xu, M. Deng, G. Li, S. Chen, and L. Wang, "Electrochemical Behavior of Cuprous Oxide-Reduced Graphene Oxide Nanocomposites and their Application in Nonenzymatic Hydrogen Peroxide Sensing," *Electrochim. Acta*, vol. 88, pp. 59–65, 2013.

[38] C. L. Sun, W. L. Cheng, T. K. Hsu, C. W. Chang, J. L. Chang, and J. M. Zen, "Ultrasensitive and Highly Stable Nonenzymatic Glucose Sensor by a CuO/Graphene-Modified Screen-Printed Carbon Electrode Integrated with Flow-Injection Analysis," *Electrochem. commun.*, vol. 30, pp. 91–94, 2013.

[39] M. Liu, R. Liu, and W. Chen, "Graphene Wrapped Cu_2O Nanocubes: Non-Enzymatic Electrochemical Sensors for the Detection of Glucose and Hydrogen Peroxide with Enhanced Stability," *Biosens. Bioelectron.*, vol. 45, pp. 206–212, Jul. 2013.

[40] S. Yang, G. Cui, S. Pang, Q. Cao, U. Kolb, X. Feng, J. Maier, and K. Müllen, "Fabrication of Cobalt and Cobalt Oxide/Graphene Composites: Towards High-Performance Anode Materials for Lithium Ion Batteries," *ChemSusChem*, vol. 3, no. 2, pp. 236–239, 2010.

[41] S. Ci, S. Mao, T. Huang, Z. Wen, D. A. Steeber, and J. Chen, "Enzymeless Glucose Detection Based on CoO/Graphene Microsphere Hybrids," *Electroanalysis*, vol. 26, no. 6, pp. 1326–1334, 2014.

[42] Y. Ye, T. Kong, X. Yu, Y. Wu, K. Zhang, and X. Wang, "Enhanced Nonenzymatic Hydrogen Peroxide Sensing with Reduced Graphene Oxide/Ferroferric Oxide Nanocomposites," *Talanta*, vol. 89, pp. 417–421, 2012.

[43] Z. Wu, S. Yang, Y. Sun, K. Parvez, and X. Feng, "3D Nitrogen-Doped Graphene Aerogel-Supported Fe_3O_4 Nanoparticles as E ffi cient Electrocatalysts for the Oxygen Reduction Reaction," *J. Am. Chem. Soc.*, vol. 134, no. 22, pp. 9082–9085, 2012.

[44] T. Gan, J. Sun, K. Huang, L. Song, and Y. Li, "A Graphene Oxide-Mesoporous MnO_2 Nanocomposite Modified Glassy Carbon Electrode as a Novel and Efficient Voltammetric Sensor for Simultaneous Determination of Hydroquinone and Catechol," *Sensors Actuators B Chem.*, vol. 177, pp. 412–418, 2013.

[45] F. Xiao, Y. Li, H. Gao, S. Ge, and H. Duan, "Growth of Coral-like PtAu-MnO_2 Binary Nanocomposites on Free-Standing Graphene Paper for Flexible Nonenzymatic Glucose Sensors," *Biosens. Bioelectron.*, vol. 41, pp. 417–423, Mar. 2013.

[46] X. Zhu, P. Zhang, S. Xu, X. Yan, and Q. Xue, "Free-Standing Three-Dimensional Graphene/Manganese Oxide Hybrids As Binder-Free Electrode Materials for Energy Storage Applications," *ACS Appl. Mater. Interfaces*, vol. 6, no. 14, pp. 11665–11674, 2014.

[47] Q. Tang, L. Jiang, J. Liu, S. Wang, and G. Sun, "Effect of Surface Manganese Valence of Manganese Oxides on the Activity of the Oxygen Reduction Reaction in Alkaline Media," *ACS Catal.*, vol. 4, no. 2, pp. 457–463, 2014.

[48] W. J. Basirun, M. Sookhakian, S. Baradaran, Z. Endut, M. R. Mahmoudian, M. Ebadi, R. Yousefi, H. Ghadimi, and S. Ahmed, "Graphene Oxide Electrocatalyst on MnO_2 Air Cathode as an Efficient Electron Pump for Enhanced Oxygen Reduction in Alkaline Solution," *Sci. Rep.*, vol. 5, p. 9108-9115, 2015.

[49] B. Yuan, C. Xu, D. Deng, Y. Xing, L. Liu, H. Pang, and D. Zhang, "Graphene Oxide/Nickel Oxide Modified Glassy Carbon Electrode for Supercapacitor and Nonenzymatic Glucose Sensor," *Electrochim. Acta*, vol. 88, pp. 708–712, 2013.

[50] Y. Zhang, Y. Wang, J. Jia, and J. Wang, "Nonenzymatic Glucose Sensor Based on Graphene Oxide and Electrospun NiO Nanofibers," *Sensors Actuators B Chem.*, vol. 171–172, pp. 580–587, Aug. 2012.

[51] X. Zhu, Q. Jiao, C. Zhang, X. Zuo, X. Xiao, Y. Liang, and J. Nan, "Amperometric Nonenzymatic Determination of Glucose Based on a Glassy Carbon Electrode Modified with Nickel(II) Oxides and Graphene," *Microchim. Acta*, vol. 180, no. 5–6, pp. 477–483, 2013.

[52] H. Zhang, J. Feng, T. Fei, S. Liu, and T. Zhang, "SnO_2 Nanoparticles-Reduced Graphene Oxide Nanocomposites for NO_2 Sensing at Low Operating Temperature," *Sensors Actuators, B Chem.*, vol. 190, no. 2, pp 472–478, 2014.

[53] Y. Fan, K.-J. Huang, D.-J. Niu, C.-P. Yang, and Q.-S. Jing, "TiO_2-Graphene Nanocomposite for Electrochemical Sensing of Adenine and Guanine," *Electrochim. Acta*, vol. 56, no. 12, pp. 4685–4690, Apr. 2011.

[54] T. Gan, J. Sun, W. Meng, L. Song, and Y. Zhang, "Electrochemical Sensor Based on Graphene and Mesoporous TiO_2 for the Simultaneous Determination of Trace Colourants in Food," *Food Chem.*, vol. 141, no. 4, pp. 3731–3737, 2013.

[55] A. Titanium, O. Titanium, and H. C. Performance, "Synthesis of Size-Tunable Anatase TiO_2 Nanospindles and Their Assembly into Anatase@Titanium Oxynitride/Titanium Nitride–Graphene Nanocomposites for Rechargeable Lithium Ion Batteries with High Cycling Performance," *ACS Nano*, vol. 4, no. 11, pp. 6515–6526, 2010.

[56] D. Wang, D. Choi, J. Li, Z. Yang, Z. Nie, R. Kou, D. Hu, C. Wang, L. V Saraf, J. Zhang, I. a Aksay, and J. Liu, "Self-Assembled TiO_2 – Graphene Hybrid Nanostructures for Enhanced Li-Ion Insertion," *ACS Nano*, vol. 3, no. 4, pp. 907–914, 2009.

[57] K. Tiido, N. Alexeyeva, M. Couillard, C. Bock, B. R. MacDougall, and K. Tammeveski, "Graphene–TiO_2 Composite Supported Pt Electrocatalyst for Oxygen Reduction Reaction," *Electrochim. Acta*, vol. 107, pp. 509–517, Sep. 2013

[58] Z. Li, H. Zhang, Q. Liu, Y. Liu, L. Stanciu, and J. Xie, "Hierarchical Nanocomposites of Vanadium Oxide Thin Film Anchored on Graphene as High-Performance Cathodes in Li-Ion Batteries," *ACS Appl. Mater. Interfaces*, vol. 6, pp. 18894-18900, 2014

[59] Q. Liu, Z. Li, Y. Liu, H. Zhang, Y. Ren, C. Sun, W. Lu, Y. Zhou, L. Stanciu, E. a Stach, and J. Xie, "Graphene-Modified Nanostructured Vanadium Pentoxide Hybrids with Extraordinary Electrochemical Performance for Li-Ion Batteries," *Nat. Commun.*, vol. 6, pp. 1–10, 2015.

[60] G. Ye, Y. Gong, K. Keyshar, E. a. M. Husain, G. Brunetto, S. Yang, R. Vajtai, and P. M. Ajayan, "3D Reduced Graphene Oxide Coated V_2 O_5 Nanoribbon Scaffolds for High-Capacity Supercapacitor Electrodes," *Part. Part. Syst. Charact.*vol. 32, pp. 817–821, 2015.

[61] X. Dong, Y. Cao, J. Wang, M. B. Chan-Park, L. Wang, W. Huang, and P. Chen, "Hybrid Structure of Zinc Oxide Nanorods and Three Dimensional Graphene Foam for Supercapacitor and Electrochemical Sensor Applications," *RSC Adv.*, vol. 2, no. 10, pp. 4364–4369, 2012.

[62] C. Yang, L. Hu, H.-Y. Zhu, Y. Ling, J.-H. Tao, and C.-X. Xu, "rGO Quantum Dots/ZnO Hybrid Nanofibers Fabricated Using Electrospun Polymer Templates and Applications in Drug Screening Involving an Intracellular H_2O_2 Sensor," *J. Mater. Chem. B*, vol. 3, pp. 2651–2659, 2015.

[63] M. Sawangphruk, P. Srimuk, P. Chiochan, A. Krittayavathananon, S. Luanwuthi, and J. Limtrakul, "High-Performance Supercapacitor of Manganese Oxide/Reduced Graphene Oxide Nanocomposite Coated on Flexible Carbon Fiber Paper," *Carbon*, vol. 60, pp. 109–116, 2013.

[64] Y. Liu, S. Shrestha, and W. E. Mustain, "Synthesis of Nanosize Tungsten Oxide and Its Evaluation as an Electrocatalyst Support for Oxygen Reduction in Acid Media," *ACS Catal.*, vol. 2, no. 3, pp. 456–463, 2012.

[65] C. Gómez-Navarro, J. C. Meyer, R. S. Sundaram, A. Chuvilin, S. Kurasch, M. Burghard, K. Kern, and U. Kaiser, "Atomic Structure of Reduced Graphene Oxide," *Nano Lett.*, vol. 10, no. 4, pp. 1144–1148, 2010.

[66] F. Q. Song, Z. Y. Li, Z. W. Wang, L. He, M. Han, and G. H. Wang, "Free-Standing Graphene by Scanning Transmission Electron Microscopy," *Ultramicroscopy*, vol. 110, no. 12, pp. 1460–1464, 2010.

[67] N. Zhu, S. Han, S. Gan, J. Ulstrup, and Q. Chi, "Graphene Paper Doped with Chemically Compatible Prussian Blue Nanoparticles as Nanohybrid Electrocatalyst," *Adv. Funct. Mater.*, vol. 23, no. 42, pp. 5297–5306, Nov. 2013.

[68] J. C. Koepke, J. D. Wood, D. Estrada, Z.-Y. Ong, K. T. He, E. Pop, and J. W. Lyding, "Atomic-Scale Evidence for Potential Barriers and Strong Carrier Scattering at Gra-

phene Grain Boundaries: A Scanning Tunneling Microscopy Study," *ACS Nano*, vol. 7, no. 1, pp. 75–86, 2013.

[69] E. Y. Polyakova (Stolyarova), K. T. Rim, D. Eom, K. Douglass, R. L. Opila, T. F. Heinz, A. V Teplyakov, and G. W. Flynn, "Scanning Tunneling Microscopy and X-ray Photoelectron Spectroscopy Studies of Graphene Films Prepared by Sonication-Assisted Dispersion," *ACS Nano*, vol. 5, no. 8, pp. 6102–6108, 2011.

[70] G. Dong and J. W. M. Frenken, "Kinetics of Graphene Formation on Rh(111) Investigated by In Situ Scanning Tunneling Microscopy," *ACS Nano*, vol. 7, no. 8, pp. 7028–7033, 2013.

[71] S. Gan, L. Zhong, T. Wu, D. Han, J. Zhang, J. Ulstrup, Q. Chi, and L. Niu, "Spontaneous and Fast Growth of Large-Area Graphene Nanofilms Facilitated by Oil/Water Interfaces.," *Adv. Mater.*, vol. 24, no. 29, pp. 3958–64, Aug. 2012.

[72] T. Ando, M. Ishii, M. Kamo, and Y. Sato, "Thermal Hydrogenation of Diamond Surfaces Studied by Diffuse Reflectance Fourier-Transform Infrared Temperature-Programmed Desorption and Laser Raman Spectroscopy," *J. Chem. Soc. Faraday Trans.*, vol. 89, no. 11, pp. 1783–1789, 1993.

[73] Y. Xu, H. Bai, G. Lu, C. Li, and G. Shi, "Flexible Graphene Films via the Filtration of Water-Soluble Noncovalent Functionalized Graphene Sheets," *J. Am. Chem. Soc.*, vol. 130, no. 18, pp. 5856–5857, May 2008.

[74] W. Zhang, F. Liu, Q. Li, Q. Shou, J. Cheng, L. Zhang, B. J. Nelson, and X. Zhang, "Transition Metal Oxide and Graphene Nanocomposites for High-Performance Electrochemical Capacitors," *Phys. Chem. Chem. Phys.*, vol. 14, no. 47, pp. 16331–16337, 2012.

[75] Z.-J. Fan, W. Kai, J. Yan, T. Wei, L.-J. Zhi, J. Feng, Y. Ren, L.-P. Song, and F. Wei, "Facile Synthesis of Graphene Nanosheets via Fe Reduction of Exfoliated Graphite Oxide," *ACS Nano*, vol. 5, no. 1, pp. 191–198, Jan. 2011.

[76] Z. Sun, Z. Yan, J. Yao, E. Beitler, Y. Zhu, and J. M. Tour, "Growth of Graphene from Solid Carbon Sources," *Nature*, vol. 468, no. 7323, pp. 549–552, Nov. 2010.

[77] J. Shen, X. Li, N. Li, and M. Ye, "Facile Synthesis of $NiCo_2O_4$-Reduced Graphene Oxide Nanocomposites with Improved Electrochemical Poperties," *Electrochim. Acta*, vol. 141, pp. 126–133, 2014.

[78] M. Sun, Y. Dong, G. Zhang, J. Qu, and J. Li, "α-Fe_2O_3 Spherical Nanocrystals Supported on CNTs as Efficient Non-noble Electrocatalysts for the Oxygen Reduction Reaction," *J. Mater. Chem. A*, vol. 2, no. 33, pp. 13635–13640, 2014.

[79] Q. Tang, Z. Shan, L. Wang, and X. Qin, "MoO_2–Graphene Nanocomposite as Anode Material for Lithium-Ion Batteries," *Electrochim. Acta*, vol. 79, pp. 148–153, 2012.

[80] G. Xi, J. Ye, Q. Ma, N. Su, H. Bai, and C. Wang, "In Situ Growth of Metal Particles on 3D Urchin-like WO3 Nanostructures," *J. Am. Chem. Soc.*, vol. 134, no. 15, pp. 6508–6511, 2012.

[81] H. Chen, Y. Li, F. Zhang, G. Zhang, and X. Fan, "Graphene Supported Au-Pd Bimetallic Nanoparticles with Core-Shell Structures and Superior Peroxidase-like Activities," *J. Mater. Chem.*, vol. 21, no. 44, pp. 17658–17661, 2011.

[82] M. Sun, H. Liu, Y. Liu, J. Qu, and J. Li, "Graphene-Based Transition Metal Oxide Nanocomposites for the Oxygen Reduction Reaction," *Nanoscale*, vol. 7, no. 4, pp. 1250–1269, 2015.

[83] J. Chlistunoff, "RRDE and Voltammetric Study of ORR on Pyrolyzed Fe/Polyaniline Catalyst. On the Origins of Variable Tafel Slopes," *J. Phys. Chem. C*, vol. 115, no. 14, pp. 6496–6507, 2011.

[84] L. Cui, G. Lv, Z. Dou, and X. He, "Fabrication of Iron Phthalocyanine/Graphene Micro/Nanocomposite by Solvothermally Assisted π–π Assembling Method and its Application for Oxygen Reduction Reaction," *Electrochim. Acta*, vol. 106, pp. 272–278, 2013.

[85] C.-C. Chou, C.-H. Liu, and B.-H. Chen, "Effects of Reduction Temperature and pH Value of Polyol Process on Reduced Graphene Oxide Supported Pt Electrocatalysts for Oxygen Reduction Reaction," *Energy*, vol. 70, pp. 231–238, 2014.

[86] C. Du, Y. Sun, T. Shen, G. Yin, and J. Zhang, "7 - Applications of {RDE} and {RRDE} Methods in Oxygen Reduction Reaction," in *Rotating Electrode Methods and Oxygen Reduction Electrocatalysts*, W. X. Y. Zhang, Ed. Amsterdam: Elsevier, 2014, pp. 231–277.

[87] W. Il Park, C.-H. Lee, J. M. Lee, N.-J. Kim, and G.-C. Yi, "Inorganic Nanostructures Grown on Graphene Layers," *Nanoscale*, vol. 3, no. 9, pp. 3522–3533, 2011.

[88] L. De Rogatis, M. Cargnello, V. Gombac, B. Lorenzut, T. Montini, and P. Fornasiero, "Embedded Phases: A Way to Active and Stable Catalysts," *ChemSusChem*, vol. 3, no. 1, pp. 24–42, 2010.

[89] S. Cui, S. Mao, G. Lu, and J. Chen, "Graphene Coupled with Nanocrystals: Opportunities and Challenges for Energy and Sensing Applications," *J. Phys. Chem. Lett.*, vol. 4, no. 15, pp. 2441–2454, 2013.

[90] W. Sun, X. Wang, X. Sun, Y. Deng, J. Liu, B. Lei, and Z. Sun, "Simultaneous Electrochemical Determination of Guanosine and Adenosine with Graphene-ZrO_2 Nanocomposite Modified Carbon Ionic Liquid Electrode," *Biosens. Bioelectron.*, vol. 44, no. 1, pp. 146–151, 2013.

[91] C.-J. Cai, M.-W. Xu, S.-J. Bao, C. Lei, and D.-Z. Jia, "A Facile Route for Constructing a Graphene-Chitosan-ZrO_2 Composite for Direct Electron Transfer and Glucose Sensing," *RSC Adv.*, vol. 2, no. 21, p. 8172-8179, 2012.

[92] S. Guo, S. Zhang, and S. Sun, "Tuning Nanoparticle Catalysis for the Oxygen Reduction Reaction," *Angew. Chemie Int. Ed.*, vol. 52, no. 33, pp. 8526–8544, 2013.

[93] J. Xiao, Q. Kuang, S. Yang, F. Xiao, S. Wang, and L. Guo, "Surface Structure Dependent Electrocatalytic Activity of Co_3O_4 Anchored on Graphene Sheets toward Oxygen Reduction Reaction," *Sci. Rep.*, vol. 3:2300, 2013.

[94] Y. Zhang, F. Xu, Y. Sun, Y. Shi, Z. Wen, and Z. Li, "Assembly of $Ni(OH)_2$ Nanoplates on Reduced Graphene Oxide: A Two Dimensional Nanocomposite for Enzyme-Free Glucose Sensing," *J. Mater. Chem.*, vol. 21, no. 42, p. 16949-16954, 2011.

[95] Z. Wang, Y. Hu, W. Yang, M. Zhou, and X. Hu, "Facile One-Step Microwave-Assisted Route Towards Ni Nanospheres/Reduced Graphene Oxide Hybrids for Non-Enzymatic Glucose Sensing," *Sensors*, vol. 12, no. 4, pp. 4860–4869, 2012.

[96] S. Guo, S. Zhang, L. Wu, and S. Sun, "Co/CoO Nanoparticles Assembled on Graphene for Electrochemical Reduction of Oxygen," *Angew. Chemie – Int. Ed.*, vol. 51, no. 47, pp. 11770–11773, 2012.

[97] Q. He, Q. Li, S. Khene, X. Ren, F. E. López-Suárez, D. Lozano-Castelló, A. Bueno-López, and G. Wu, "High-Loading Cobalt Oxide Coupled with Nitrogen-Doped Graphene for Oxygen Reduction in Anion-Exchange-Membrane Alkaline Fuel Cells," *J. Phys. Chem. C*, vol. 117, no. 17, pp. 8697–8707, 2013.

[98] W.-H. Ryu, T.-H. Yoon, S. H. Song, S. Jeon, Y.-J. Park, and I.-D. Kim, "Bifunctional Composite Catalysts Using Co_3O_4 Nanofibers Immobilized on Nonoxidized Graphene Nanoflakes for High-Capacity and Long-Cycle Li–O_2 Batteries," *Nano Lett.*, vol. 13, no. 9, pp. 4190–4197, 2013.

[99] J. Feng, Y. Liang, H. Wang, Y. Li, B. Zhang, J. Zhou, J. Wang, T. Regier, and H. Dai, "Engineering Manganese Oxide/Nanocarbon Hybrid Materials for Oxygen Reduction Electrocatalysis," *Nano Res.*, vol. 5, no. 10, pp. 718–725, 2012.

[100] T. N. Lambert, D. J. Davis, W. Lu, S. J. Limmer, P. G. Kotula, A. Thuli, M. Hungate, G. Ruan, Z. Jin, and J. M. Tour, "Graphene-Ni-[small alpha]-MnO_2 and -Cu-[small alpha]-MnO_2 Nanowire Blends as Highly Active Non-Precious Metal Catalysts for the Oxygen Reduction Reaction," *Chem. Commun.*, vol. 48, no. 64, pp. 7931–7933, 2012.

[101] Y. Cao, Z. Wei, J. He, J. Zang, Q. Zhang, M. Zheng, and Q. Dong, "[small alpha]-MnO_2 Nanorods Grown In Situ on Graphene as Catalysts for Li-O_2 Batteries with Excellent Electrochemical Performance," *Energy Environ. Sci.*, vol. 5, no. 12, pp. 9765–9768, 2012.

[102] J. Zhang, C. Guo, L. Zhang, and C. M. Li, "Direct Growth of Flower-Like Manganese Oxide on Reduced Graphene Oxide Towards Efficient Oxygen Reduction Reaction," *Chem. Commun.*, vol. 49, no. 56, pp. 6334–6336, 2013.

[103] J. Duan, Y. Zheng, S. Chen, Y. Tang, M. Jaroniec, and S. Qiao, "Mesoporous Hybrid Material Composed of Mn_3O_4 Nanoparticles on Nitrogen-Doped Graphene for High-

ly Efficient Oxygen Reduction Reaction," *Chem. Commun.*, vol. 49, no. 70, pp. 7705–7707, 2013.

[104] J.-S. Lee, T. Lee, H.-K. Song, J. Cho, and B.-S. Kim, "Ionic Liquid Modified Graphene Nanosheets Anchoring Manganese Oxide Nanoparticles as Efficient Electrocatalysts for Zn-air Batteries," *Energy Environ. Sci.*, vol. 4, no. 10, pp. 4148–4154, 2011.

[105] J. Xiao, G. Xu, S.-G. Sun, and S. Yang, "MFe_2O_4 and MFe@Oxide Core–Shell Nanoparticles Anchored on N-Doped Graphene Sheets for Synergistically Enhancing Lithium Storage Performance and Electrocatalytic Activity for Oxygen Reduction Reactions," *Part. Part. Syst. Charact.*, vol. 30, no. 10, pp. 893–904, 2013.

[106] W. Zhang, Y. Zeng, C. Xu, H. Tan, W. Liu, J. Zhu, N. Xiao, H. H. Hng, J. Ma, H. E. Hoster, R. Yazami, and Q. Yan, "Fe_2O_3 Nanocluster-Decorated Graphene as O_2 Electrode for High Energy Li-O_2 Batteries," *RSC Adv.*, vol. 2, no. 22, pp. 8508–8514, 2012.

[107] R. Zhou, Y. Zheng, D. Hulicova-Jurcakova, and S. Z. Qiao, "Enhanced Electrochemical Catalytic Activity by Copper Oxide Grown on Nitrogen-Doped Reduced Graphene Oxide," *J. Mater. Chem. A*, vol. 1, no. 42, pp. 13179–13185, 2013.

[108] X.-Y. Yan, X.-L. Tong, Y.-F. Zhang, X.-D. Han, Y.-Y. Wang, G.-Q. Jin, Y. Qin, and X.-Y. Guo, "Cuprous Oxide Nanoparticles Dispersed on Reduced Graphene Oxide as an Efficient Electrocatalyst for Oxygen Reduction Reaction," *Chem. Commun.*, vol. 48, no. 13, pp. 1892–1894, 2012.

[109] H.-G. Jung, Y. S. Jeong, J.-B. Park, Y.-K. Sun, B. Scrosati, and Y. J. Lee, "Ruthenium-Based Electrocatalysts Supported on Reduced Graphene Oxide for Lithium-Air Batteries," *ACS Nano*, vol. 7, no. 4, pp. 3532–3539, 2013.

[110] Y. Li, L. Zhang, Z. Hu, and J. C. Yu, "Synthesis of 3D Structured Graphene as a High Performance Catalyst Support for Methanol Electro-Oxidation," *Nanoscale*, vol. 7, no. 25, pp. 10896–10902, 2015.

[111] D. Deng, X. Pan, L. Yu, Y. Cui, Y. Jiang, J. Qi, W.-X. Li, Q. Fu, X. Ma, Q. Xue, G. Sun, and X. Bao, "Toward N-Doped Graphene via Solvothermal Synthesis," *Chem. Mater.*, vol. 23, no. 5, pp. 1188–1193, Mar. 2011.

[112] X. Feng, Y. Zhang, J. Zhou, Y. Li, S. Chen, L. Zhang, Y. Ma, L. Wang, and X. Yan, "Three-Dimensional Nitrogen-Doped Graphene as an Ultrasensitive Electrochemical Sensor for the Detection of Dopamine," *Nanoscale*, vol. 7, no. 6, pp. 2427–2432, 2015.

[113] H. M. Jeong, J. W. Lee, W. H. Shin, Y. J. Choi, H. J. Shin, J. K. Kang, and J. W. Choi, "Nitrogen-Doped Graphene for High-Performance Ultracapacitors and the Importance of Nitrogen-Doped Sites at Basal Planes," *Nano Lett.*, vol. 11, no. 6, pp. 2472–2477, Jun. 2011.

[114] H. Wang, T. Maiyalagan, and X. Wang, "Review on Recent Progress in Nitrogen-Doped Graphene: Synthesis, Characterization, and Its Potential Applications," *ACS Catal.*, vol. 2, no. 5, pp. 781–794, May 2012.

[115] S. Yang, L. Zhi, K. Tang, X. Feng, J. Maier, and K. Müllen, "Efficient Synthesis of Heteroatom (N or S)-Doped Graphene Based on Ultrathin Graphene Oxide-Porous Silica Sheets for Oxygen Reduction Reactions," *Adv. Funct. Mater.*, vol. 22, no. 17, pp. 3634–3640, 2012.

[116] J. Li, X. Li, P. Zhao, D. Y. Lei, W. Li, J. Bai, Z. Ren, and X. Xu, "Searching for Magnetism in Pyrrolic N-Doped Graphene Synthesized via Hydrothermal Reaction," *Carbon*, vol. 84, pp. 460–468, Apr. 2015.

[117] W. Zhou, J. Zhou, Y. Zhou, J. Lu, K. Zhou, L. Yang, Z. Tang, L. Li, and S. Chen, "N-Doped Carbon-Wrapped Cobalt Nanoparticles on N-Doped Graphene Nanosheets for High-Efficiency Hydrogen Production," *Chem. Mater.*, vol. 27, no. 6, pp. 2026–2032, Mar. 2015.

[118] Z.-H. Sheng, L. Shao, J.-J. Chen, W.-J. Bao, F.-B. Wang, and X.-H. Xia, "Catalyst-Free Synthesis of Nitrogen-Doped Graphene via Thermal Annealing Graphite Oxide with Melamine and Its Excellent Electrocatalysis," *ACS Nano*, vol. 5, no. 6, pp. 4350–4358, Jun. 2011.

[119] Z. Yang, Z. Yao, G. Li, G. Fang, H. Nie, Z. Liu, X. Zhou, X. Chen, and S. Huang, "Sulfur-Doped Graphene as an Efficient Metal-free Cathode Catalyst for Oxygen Reduction," *ACS Nano*, vol. 6, no. 1, pp. 205–211, Jan. 2012.

[120] S. Bag, B. Mondal, A. K. Das, and C. R. Raj, "Nitrogen and Sulfur Dual-Doped Reduced Graphene Oxide: Synergistic Effect of Dopants Towards Oxygen Reduction Reaction," *Electrochim. Acta*, vol. 163, no. 0, pp. 16–23, 2015.

Chapter 15

Ag-Cu Nanoalloys as Oxygen Reduction Electrocatalysts in Alkaline Media for Advanced Energy Conversion and Storage

Adnan Qaseem, Fuyi Chen and Nan Zhang

Additional information is available at the end of the chapter

http://dx.doi.org/10.5772/62050

Abstract

Silver-based nanoalloys owing to their cost, performance and stability are an attractive electrocatalyst system for oxygen reduction reaction (ORR) in the alkaline fuel cells and metal air batteries. A systematic computational and experimental approach has been adopted to investigate their performance for ORR in alkaline environment. Firstly, genetic algorithm (GA) based calculations have been performed to look for the stable compositions and structures of these nanoalloys. Later, density functional theory (DFT) is employed to simulate the working of those stable nanoalloys in actual working conditions. Finally, the most promising nanoalloys have been synthesized by physical and chemical routes to confirm their performance in real-life conditions. It has been found that the alloying of silver with copper enhances the catalytic performance of Ag nanoparticles. The enhancement in performance can be related to the modification of the electronic and physical structure of Ag due to copper doping. The superior performance of Ag–Cu nanocatalysts in alkaline fuel cells and metal air batteries along with their modest cost and long-term stability make them a promising candidate for deployment as a catalyst for ORR in alkaline media.

Keywords: Electrocatalyst, silver nanoalloys, metal air batteries, oxygen reduction reaction, rechargeable zinc air battery

1. Introduction

Oxygen reduction reaction (ORR) plays a vital role in the working of fuel cells and metal-air batteries. Both of these technologies utilize oxygen from the air to generate electrical energy. The ORR mechanism in acidic environment is accompanied with the formation of water.

$$O_2 + 4H^+ + 4e^- \rightarrow 2H_2O \tag{1}$$

For alkaline medium, hydroxyl formation takes place as

$$O_2 + 2H_2O + 4e^- \rightarrow 4OH^- \tag{2}$$

The current Li-ion battery technology is unable to offer the solutions for the long-range electric vehicles and energy storage grids. It is also postulated that the Li-ion battery system may reach its theoretical limit which will still be short of the demand for the long-range EVs. Metal air batteries such as Li-air (11,000 Wh/kg) and Zn–air systems (1,084 Wh/kg) offer much higher energy capacities [1–3]. A great deal of contemporary research is directed toward the realization of these high energy systems. Apart from being dense in energy, metal air batteries and fuel cells are green source of energy. The product of their working is free from toxic and harmful waste gases which damage the environment. These systems provide a valuable opportunity to cut the greenhouse gas emissions on a tremendous scale [4,5].

Electrocatalyst plays a crucial role in the working of metal air batteries and fuel cells. The ORR and OER mechanism are strongly related to the functionality of the electrocatalyst. Pt and Pt-based alloy catalysts are widely used for ORR, but prohibitive cost and catalytic poisoning are major drawbacks associated with Pt. Therefore, it is highly desirable to explore novel Pt-free cost-effective catalysts. Different non-platinum catalysts have been considered as a replacement of Pt. Silver being about 50 times cheaper than platinum is an attractive choice for catalyst in alkaline fuel cells. The pourbaix diagram reveals the superior stability of Ag over platinum in alkaline environment [6,7].

The ORR is accompanied with the formation of various adsorbed intermediates such as O, OH and OOH. Norkosov et al. evaluated the effect of potential on the free energy of various intermediates on Pt (111) by DFT calculations [8]. At high potential, the adsorbed oxygen was found to be stable. The ORR reaction was found to proceed only by lowering the potential, hence giving rise to overpotential. Bond energies of oxygen and hydroxyl on different metals were also calculated by DFT. The rate of ORR is limited by the removal of O, OH for metals which bind oxygen strongly where as in case of metals with poor oxygen binding, the rate is limited by the weak bonding of the adsorbed species. A volcano plot as shown in Figure 1 is obtained as a result of DFT calculations performed on various systems. Although platinum sits near the top of volcano plot developed by DFT calculations, alloying of metals can result in new materials with adsorption energies for the intermediates that are different from the constituent pure metals. Therefore, new generation of superior electrocatalysts can be developed by alloying of metals to yield optimum binding of adsorbates onto alloy surface.

Tremendous amount of research has been performed in the past decade to enhance the activity of ORR catalyst. A myriad of bimetallic and multi metallic alloy compositions have been developed in a variety of structures such as core shell, skin alloys, thin films, ordered intermetallics and solid solutions [9–12]. Skin alloys have been widely popular because of their

http://dx.doi.org/10.5772/62050

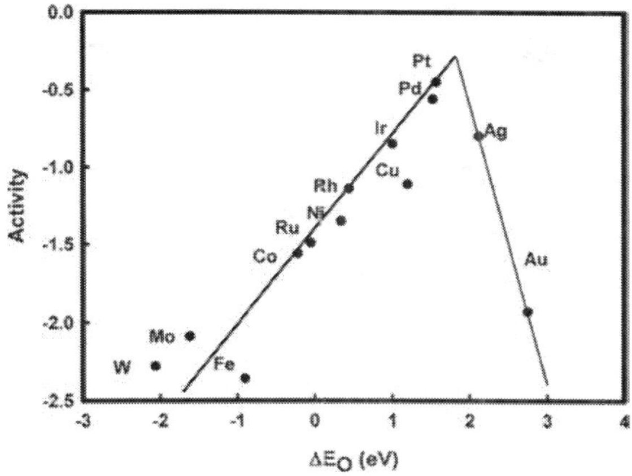

Figure 1. Variation of activity with oxygen adsorption energy for metals. Reproduced with permission from ref. 8. Copyright 2004, American Chemical Society.

superior activity as compared to the bulk alloys. This is accompanied with the reduction of cost of precious metal cost such as Pt in skin alloys by using a monolayer thick platinum on top of a non-precious metal/alloy core. As the skin is of Pt, therefore, stability in the corrosive chemical environment is still maintained in these alloys. The enhancement in the activity of these skin alloys is attributed to the geometric and ligand effect of the subsurface atoms on the skin. Recently, these structural form alloys have been reported to improve the ORR performance of the silver-based alloys. Some possible modifications of the surface electronic structure by the ligand mechanism are revealed in Figure 2.

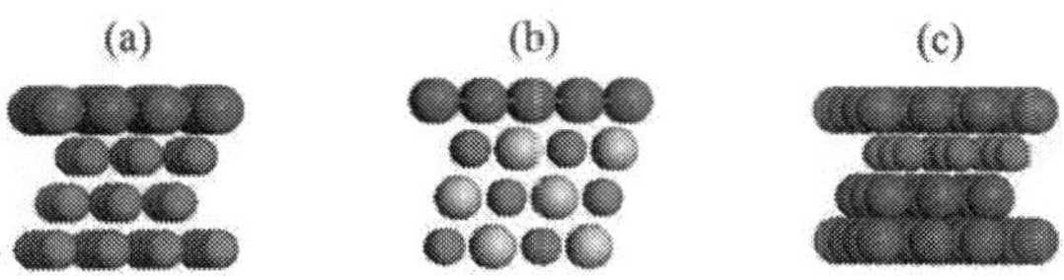

Figure 2. Modification of the surface geometric and electronic structure by subsurface configuration (a) skin on pure metal, (b) skin on alloy/compound, (c) subsurface ligand.

The rate-limiting step for ORR in case of Pt catalyst is the removal of adsorbed OH species. It is well established that the decrease in binding energy by 0.1–0.2 eV will optimize the performance of the Pt-based catalyst [13,14]. By using the Pt skin on a transition metal core, this reduction in binding energy can be achieved. This is attributed to the modification of the electronic structure of the surface atoms by the core. In case of transition metals such as Ti, Co, Ni, and Ru, charge transfers from the core to the dband of the surface atoms of platinum [15]. This occupancy of dband lowers the dband centre of the platinum atoms which in turns decreases the adsorption of the OH on the Pt skin. As a result of this mechanism, the activity of these alloy systems has been reported to be much superior than that of commercial Pt/C catalyst. The skin of platinum atoms can be either in compression or tension. This is because of the subsurface structure effect on the skin alloy. The compression of the Pt skin increases the overlap of the d orbitals which consequently increases the dband width. The result of this perturbation of the structure is the lowering of the dband center of the surface atoms [16, 17]. According to the dband theory, the lower the dband center of the surface, the lower its reactivity and vice versa. By the combination of geometric and ligand effects, the dband center of the surface can be tuned to the desired value so as to achieve the optimum adsorption of adsorbates such as O, OH and OOH on the surface of the metal. This is in accordance with the Sabatier principle which implies that for the catalytically induced chemical reactions catalyst for the reaction should have neither strong nor weak adsorption for the reaction species [18].

From Figure 1, it is evident that weak binding of oxygen onto silver is the cause of its shift from the volcano peak and weak ORR activity.

The oxygen reduction reaction for alkaline can occur by a direct four electron transfer method as in Eq. 2 or by indirect 2e transfers [19–21] as:

$$O_2 + H_2O + 2e^- \rightarrow HO_2^- + OH^- \quad (3)$$

$$H_2O + 2e^- + HO_2^- \rightarrow 3OH^- \quad (4)$$

In the indirect mechanism, second step, i.e., Eq. (4), is the rate-determining step. The oxygen reduction reaction by indirect mechanism seriously limits the performance of the cell. This is because of the fact that if the reaction 4 does not occur by direct mechanism, then the total electrons transferred during ORR reduce to two only. As the result, this lowers the total output voltage and energy density. Also the peroxide formed in Eq. (3) can undergo catalyst-induced conversion to O_2 and OH^- by Eq. (5).

$$HO_2^- \rightarrow O_2 + 2OH^- \quad (5)$$

Reaction 5 limits the catalytic activity of the catalyst and hence it is desirable that the ORR proceeds by direct four electron transfer mechanism. The effect of pH on the ORR mechanism has been studied by Blizanac et al. [22]. In case of ORR on Ag(111) in alkaline medium, four electron transfer was found to be the dominant mechanism at all overpotentials, but in case of low pH, i.e., in acidic solutions, 2 electron pathway was favored at low overpotentials. It was observed that the ORR on Ag (111) by 4 electron pathway could take place only at high overpotentials in the low pH electrolytes.

Owing to the stability and efficient ORR mechanism of silver-based catalysts in alkaline media, various research groups have focused on the synthesis of Ag-based catalysts. Holewinski et al. have investigated the effect of alloying on silver for the ORR performance [23]. Density functional theory (DFT) based calculations were performed on Ag alloy slabs with an Ag skin on top to investigate the effect of alloy core on the ORR performance. The 4 electrons transfer has been proposed by the following mechanism:

$$Ag + O_2 + e^{-1} \overset{H-OH}{\rightarrow} Ag\cdots OOH - OH^{-1} \tag{6}$$

$$Ag\cdots OOH + e^{-1} \overset{H-OH}{\rightarrow} Ag\cdots O + OH^{-1} \tag{7}$$

$$Ag\cdots O + e^{-1} \overset{H-OH}{\rightarrow} Ag\cdots OH + OH^{-1} \tag{8}$$

$$Ag\cdots OH + e^{-1} \overset{H-OH}{\rightarrow} Ag + OH^{-1} \tag{9}$$

A reaction coordinate diagram was developed for different alloys of silver on the basis of DFT calculation for the above mentioned reactions. From the theoretical calculations it was concluded that in case of silver-based catalysts, the rate-determining step for ORR is the initial adsorption of oxygen on the catalyst surface to form OOH adsorbate. In order to confirm the theoretical findings, the group synthesized Ag–Co alloys. After chemical etching of the alloy an Ag-skin with an Ag-Co core was developed. The resulting silver alloy showed phenomenal improvement in the ORR as compared to the unalloyed catalyst. Functionality tests on the Ag-skin /Ag-Co alloy core yielded an improvement in the area specific activity by a factor of 6 @ 0.8 V_{RHE}. This significant enhancement was attributed to the perturbation of the electronic structure of the surface silver atoms which resulted in the lowering of the activation energy barriers for the ORR.

Composites of silver/graphene oxide and silver/graphene oxide/carbon were developed to investigate their ORR performance in alkaline environment [24]. The composites were facially synthesized by the reduction of $AgNO_3$ with graphene oxide with or without the presence of Vulcan XC-72 carbon black. The average particle size of Ag/CO/C composites (ca d = 12.9 nm) was found to be almost twice of Ag/CO composite (ca d = 6.9). The composites were electrochemically characterized which revealed the superior performance of Ag/CO/C for ORR as compared to Ag/CO composite. Rotating disc electrode (RDE) analysis revealed that the onset potential and the half wave potential shift positively for Ag/CO/C as compared to Ag/CO composite. This enhancement in ORR performance of Ag/CO/C composites as compared to Ag/CO composites was attributed to the 3D composite support which not only improves the electrical conductivity but also facilitated the mass transport in the catalyst layer. A similar beneficial effect of catalyst support was observed in the case of $Ag/Mn_3O_4/C$ catalysts [25]. The catalyst performance for ORR in alkaline media was found to be superior to the simple Ag/C catalyst. This improvement was ascribed to the Mn_3O_4 support which perturbed the electronic structure of the silver particles. Charge was transferred to Mn_3O_4 support from Ag which was manifested by the lowering of the binding energy of the Ag 3d electrons in XPS measurements. This was accompanied with the rise of the d-band center of Ag in the Ag/Mn_3O_4/C catalyst as compared to the Ag/C catalyst. This was attributed to the tensile strain which results in less overlap of d orbitals and a corresponding rise of the dband center which in turn favors the kinetics of ORR by O—O bond breakage. Figure 3 reveals the oxygen reduction polarization curves for Ag/C, Ag/Mn_3O_4/C and Pt/C at 1600 rpm in O_2-saturated 0.1 M KOH. The onset potential for Ag/Mn_3O_4/C was 0.92 V (vs. RHE) which is close to the onset potential for Pt/C catalyst. The limiting current incase of Ag/Mn_3O_4/C (c.a. I_l = 5.5 mA/cm^2) was also found to be very close to the limiting current of Pt/C (c.a. I_l = 5.62 mA/cm^2). Moreover, the electron transfer mechanism was found to proceed by four electrons from the Koutecky–Levich plots.

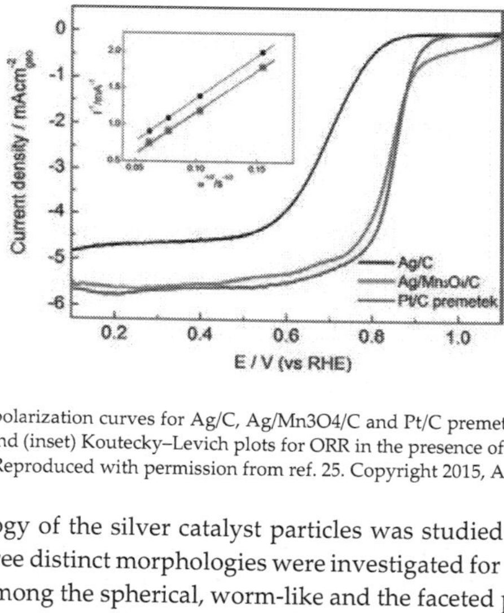

Figure 3. Oxygen reduction polarization curves for Ag/C, Ag/Mn3O4/C and Pt/C premetek at 1,600 rpm in O_2-saturated 0.1 M KOH at 10 mV s^{-1}, and (inset) Koutecky–Levich plots for ORR in the presence of Ag/C, Ag/Mn3O4/C and Pt/C premetek at 0.32 V vs. RHE. Reproduced with permission from ref. 25. Copyright 2015, American Chemical Society.

The effect of morphology of the silver catalyst particles was studied by Ohyama et al. [26]. Silver particles with three distinct morphologies were investigated for their ORR performance in alkaline medium. Among the spherical, worm-like and the faceted particles, the maximum specific activity was observed in the worm-like particles with subsurface oxygen at surface defects. This was followed by the multifaceted particles with surface $AgCO_3$ layer and defects. The smooth and spherical particles had the least specific activity of the three types. This increase in activity is justified by the large number of defects on these irregular-shaped particles which increase the reactivity of the silver catalyst toward the O_2 during ORR.

2. Computational work

The computational work involves initially the search for the structure with global minimum energy by genetic algorithm. The structure with minimum energy is further optimized by density functional theory (DFT) calculations. The most optimum structures are employed for simulations of the ORR reactions by density functional theory calculations.

2.1. Structural optimization

The first step toward computational modeling of the nanoalloys is to search for the most stable geometry at absolute zero. This requires the modeling of the potential energy surface for the multi-element alloy. Gupta potential was used for the atomistic modeling of the nanoalloy [27, 28]. It is a semi-empirical method for the approximation of the potential energy surface. This method is based on the second moment approximation to the tight binding theory (SMATB). Total energy E for the nanoalloy A_xB_{N-x} is written as a sum of an attractive term E_j^b and a repulsive term E_j^r.

$$E = \sum_j (E_j^b + E_j^r) \tag{10}$$

where E_j^b and E_j^r are further defined as:

$$E_j^b = \sqrt{\sum_{i \neq j}^{N} \xi^2 e^{-2q(r_{ij}/r_0 - 1)}} \tag{11}$$

and

$$E_j^r = \sum_{i \neq j}^{N} A e^{-p\left(\frac{r_{ij}}{r_o} - 1\right)} \tag{12}$$

N is the number of atoms, r_{ij} is the distance between atoms i and j in the cluster and r_o is the nearest-neighbor distance. The parameters A, r_o, ξ, p and q for the pure species are fitted to several bulk experimental values, such as the cohesive energy, the lattice parameter and the elastic constants. The heteroatom interactions are fitted to the solubility energy of an impurity A into a B bulk. The Gupta potential parameters used for the Ag–Cu system are listed in Table 1 [29].

Parameters	$A(eV)$	ξ (eV)	p	q	$r_o(A^o)$
Ag–Ag	0.1031	1.1895	10.85	3.18	2.8291
Ag–Cu	0.0980	1.2274	10.70	2.805	2.72405
Cu–Cu	0.0894	1.2799	10.55	2.43	2.556

Table 1. Gupta potential parameters for Ag–Cu system

From the modeling of the potential energy surface the next step forward is to search for the global minimum (GM) by optimization. This is performed by the help of Birmingham cluster genetic algorithm (GA) [30]. From the initially randomly generated cluster population, the algorithm looks for the most stable cluster structure by mutation and cross over. For each generation, parent clusters are chosen with a probability depending on their fitness and offsprings are developed from parents by a cross-over scheme which is followed by a mutation step on the offspring to bring diversity in population. The whole population is ranked by fitness and the less fit, i.e. high-energy, clusters are replaced with more stable structures. The whole process is repeated till a predefined convergence criterion is met.

The 13 atom Ag–Cu bimetallic cluster was chosen for further geometric optimization by Dmol^3 module available in the materials studio software package [31,32]. The 13 atom cluster is a magic size owing to special stability and relative abundance in case of Ag–Cu alloy clusters [33,34]. Spin-polarized DFT calculations are performed in real space within the framework of DFT-based semi core pseudo potentials (DSPPs) with the double numerical plus polarization (DNP) function. Grid integration is performed with a global cutoff of 5.0 °A. Self-consistent field procedures are performed with a convergence criterion of 10^{-6} Hartree on the total energy and the electron density. The Perdew et al. generalized gradient approximation (PBE/GGA) is used for the exchange–correlation functional during the geometry optimization for the Ag cluster. The ascending order of stability for the pure 13 atom Ag cluster is icosahedron (Ih) with binding energy E_b = –18.682 eV, decahedron (Dh) with E_b = –18.731 eV and cuboctahedron (COh) with E_b = –18.958 eV, where cuboctahedron (COh) structure was found to be the most stable configuration for the 13 atom Ag cluster. In case of the single Cu surface-doped $Ag_{12}Cu$ cluster, the increasing order of stability is also icosahedron (Ih) with E_b = –19.18727 eV, decahedron (Dh) with E_b = –19.40207 eV and cuboctahedron (COh) with E_b = –19.55135 eV. So for both pure Ag_{13} and $Ag_{12}Cu_{surface}$ clusters, the most stable structural form was found to be cuboctahedron (COh).

2.2. ORR on 13 atom Ag-Cu clusters

Ma et al. performed first principle calculations for the ORR process in alkaline media on the 13 atom pure Ag and Cu doped Ag clusters [35]. 13 atom Ag-Cu nanoalloy clusters have been previously identified as a potential candidate for ORR catalyst [36]. Pure Ag_{13} and $Ag_{12}Cu_{surface}$ clusters with cuboctahedron (COh) symmetry were used for these calculations. The doping of copper significantly improves the ORR process. The ORR reaction was observed to occur by the efficient four electron transfer mechanism. Pure silver is a poor catalyst for the ORR because of its weak adsorption of oxygen. Doping of silver with copper atom on the surface improves the binding of the intermediates such as O, OH and OOH on the nanoalloy cluster. This optimum binding is critical for the efficient ORR. Binding energies of different adsorbates are described in Table 2.

From Table 2 it is evident that the binding energy of every adsorbate is more negative on the $Ag_{12}Cu_{surface}$ cluster as compared to pure Ag_{13} cluster. This stronger binding facilitates the electron transfer reactions in the ORR process. A schematic of the ORR mechanism is provided in Figure 4.

ORR pathway	$^{\bullet}O_2$	$^{\bullet}O_2^-$	$^{\bullet}OOH$	$^{\bullet}OOH^-$	$^{\bullet}OH + ^{\bullet}OH$	$^{\bullet}OH$
Ag_{13}(eV)	−0.216	−0.259	−1.144	−0.980	−4.594	−2.958
$Ag_{12}Cu$ (eV)	−1.019	−1.154	−1.697	−1.304	−5.405	−3.312

Table 2. Binding energy of different adsorbates on nanoclusters

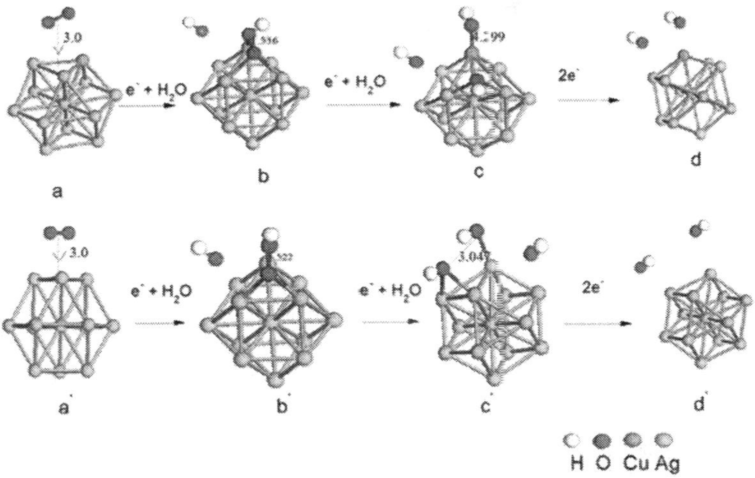

Figure 4. ORR pathway on Ag_{13}(a,b,c,d) and $Ag_{12}Cu$ (a', b', c',d'). Reprinted with permission from ref. 35. Copyright 2014, Springer.

The improvement of the ORR performance of the copper doped 13 atom nanocluster can be attributed to the modification of the dband. The dband center of the Ag_{13} cluster is at −3.078 eV. Doping with a copper atom at surface raises the dband center to −1.507eV. This can be explained by the dband theory of the reactivity of metal surfaces. According to that theory, the deeper the dband from the Fermi level, the lower is the surface reactivity. Alloying with copper raises the dband of the nanoalloy as compared to the pure metal cluster, which in turn raises the binding ability of the intermediates onto the nanoalloy cluster. The stronger binding of the intermediates is the reason behind the better ORR performance of $Ag_{12}Cu$ cluster as compared to the Ag_{13} cluster.

2.3. ORR on 38 atom Ag–Cu clusters

The 38 atom $Ag_{38-x}Cu_x$ cluster has been studied for ORR because of its relative stability by another group [37]. Truncated octahedron (TO) $Ag_{32}Cu_6$ alloy cluster was observed to perform as a better catalyst for ORR as compared to the TO $Ag_{32}Cu_6$ core-shell clusters. With the use of Gupta potential-based potential energy surface and genetic algorithm (GA) search for global minimum, polyicosahedron (PIh) $Ag_{32}Cu_6$ core-shell structure is found to be the most stable structure for $Ag_{32}Cu_6$ nanoalloy by Zhang et al.[38]. The stability of polyicosahedron (PIh) $Ag_{32}Cu_6$ core-shell structure exceeds that of truncated octahedron (TO) $Ag_{32}Cu_6$ core shell by 0.564 eV. For both pure Ag_{38} and Cu_{38} clusters truncated octahedron (TO) was the most stable geometry. Figure 5 shows that the minimum energy structure of the pure 38 atom silver took 35 iterations while more than 70 iterations were required for the $Ag_{32}Cu_6$ nanoalloy cluster.

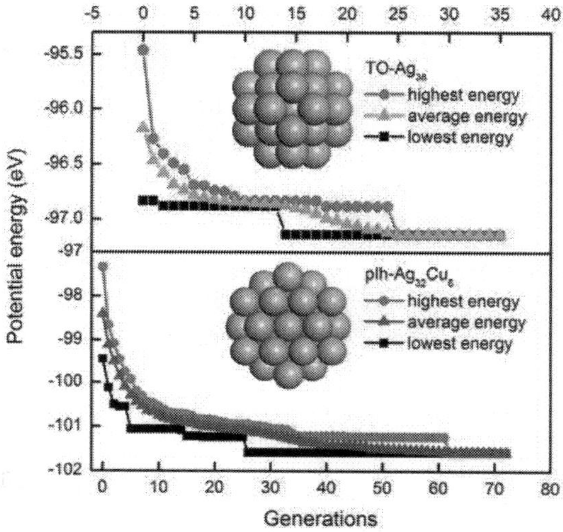

Figure 5. Generations to reach global minimum by genetic algorithm. Adapted from ref. 38.

The polyicosahedron (PIh) $Ag_{32}Cu_6$ core-shell structure was further investigated for ORR because of its stability. ORR was found to proceed more favorably by the dissociation mechanism as compared to the associative mechanism by 0.1 eV. Hence, dissociative ORR mechanism involving the scission of molecular oxygen to atomic form, i.e., $O_2 \rightarrow O + O$, was considered for computational analysis. Four non-equivalent sites were identified on the $Ag_{32}Cu_6$ core-shell structure as B1 to B4. The ORR was then followed with the bond fracture and subsequent adsorption of atomic oxygen at hollow sites marked H1, H2 and H3 as shown in Figure 6.

Of the four adsorption configurations, we notice that the adsorption energy on B4 site has a highest value of –0.149 eV, and also the maximum value of 1.209 eV for the dissociation barrier, and an energy release of 0.259 eV, dissociating to H2 and H3 sites. The B1 site, which has similar adsorption energy to B4 site, –0.146 eV, further dissociates to two H2 sites with barrier of 0.993 eV and exothermicity of 0.259 eV. The O_2 on B2 and B3 sites have bond-cleavage barriers of 0.715 and 1.134 eV and energy release of 1.088 and 0.368 eV, respectively. It is clear that the most favorable pathway for O_2 dissociation is B2 sites with a minimum value of activation energy barrier.

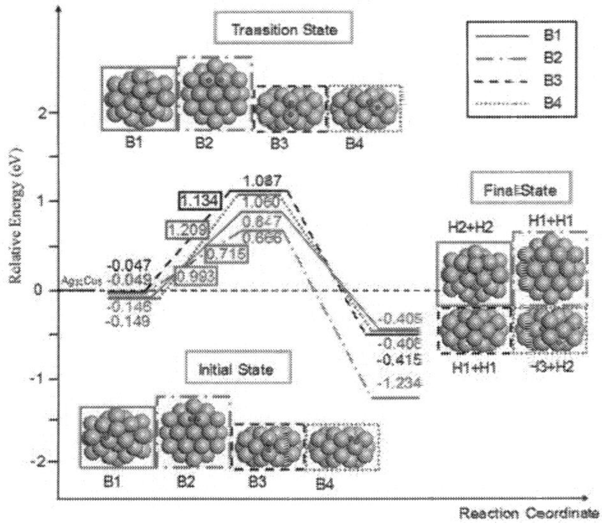

Figure 6. Reaction coordinate diagram. Adapted from ref 38.

The interaction strength of atoms and molecules with metal surface is defined by the d-band center of the metal. In order to explain further that B2 site is most favorable to display a good catalytic behavior, the electronic structure of these four adsorption configurations was addressed and the position of the d-band center relative to the Fermi energy for these different sites was calculated as shown in Figure 7. The $d$band center of B2 site is –3.395 eV, which is closest to the Fermi level. By having the dband center closest to the Fermi level as compared to the other adsorption sites on the polyicosahedron (PIh) $Ag_{32}Cu_6$ core-shell structure, B2 site is the most conducive for ORR as it enhances the otherwise weak affinity of silver alloys for the reaction intermediates.

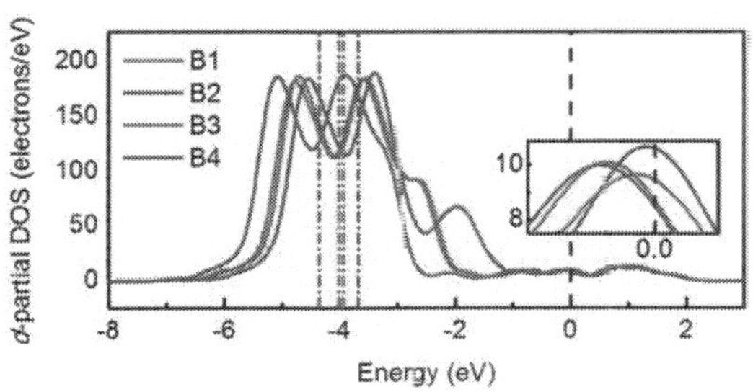

Figure 7. Partial density of states (PDOS). Adapted from ref. 38

Moreover, frontier orbital theory describes that the states of a metal which are involved in electron transfer with the adsorbates are closest to the Fermi level [39]. Therefore, the density of states (DOS) at the Fermi level is an indicator of the chemical activity. We notice that the density of states at the Fermi energy level is maximal for B2 site. This further endorses the superior activity of the B2 site for ORR on $Ag_{32}Cu_6$ core-shell nanoalloy.

3. Synthesis of Ag–Cu nanoalloy

Bimetallic Ag–Cu nanoalloy particles were developed by both physical and chemical routes.

3.1. Pulse laser deposition

Pulse laser deposition (PLD) is widely used in the field because the approach is a feasible way to control the thickness of catalyst layer [40]. As the utilization of catalyst in the fuel cell is determined mainly by the surface area of catalyst with electrolyte, the reduction of the thickness of catalytic layer can lead to improvement of the catalyst utilization and reduction of the fuel cell cost [41]. Moreover, compared to chemical preparation techniques such as electrochemical deposition, chemical vapor deposition, reduction of salts and facile hydrothermal method, PLD method owns high repeatability and stability in process, making it to be a suitable route to obtain electrocatalyst with film state [42–44].

Ag–Cu alloyed catalyst was developed by PLD method in a vacuum chamber [45]. The target of Ag–Cu alloy with atomic ratio of 50:50 was irradiated with a nanosecond Q-switched Nd:YAG laser beam (EKSPLA, Lithuania). The wavelength was set to be 266 nm, and the pulse duration was ranging from 3 to 6 ns. The laser beam diameter was around 1 mm, with an energy density of 200 mJ/pulse. Both target and substrate (nickel foam) rotated at a speed of 5 rpm during deposition, and target was irradiated for 2 min at 10 Hz to clear away the oxide on the surface before deposition. The laser was operated at the frequency of 10 Hz. The deposition time is set as 90 min. The as-prepared product is $Ag_{50}Cu_{50}$ catalyst.

Figure 8 shows series of TEM analysis on $Ag_{50}Cu_{50}$ catalyst. According to Figure 8a, plenty of nanoparticles are distributed in a continuous film. The tiny nanoparticles with size under 5 nm dominate the film. Magnifying the blue rectangle area, the obtained HRTEM is shown in Figure 8b. It can be seen that they display two different states: few are amorphous, and the left are with crystallized state.

HAADF result shown in Figure 8c displays that contrast of the particles is brighter than the gap area between particles, demonstrating a higher atomic number Z for nanoparticles. The lower Z corresponding gap area then is attributed from Cu element. This is because Z of Cu ($Z = 29$) is smaller than Ag ($Z = 47$). Combining the amorphous state in gap area observed in Figure 8b, we can draw that $Ag_{50}Cu_{50}$ catalyst actually is Ag–Cu alloyed nanoparticles embedded in amorphous Cu film.

Electrochemical characterizations have been carried out on PLD synthesized $Ag_{50}Cu_{50}$ catalyst. Figure 9a shows RDE polarization curves of $Ag_{50}Cu_{50}$ catalyst with rotation rate 1,600 rpm in

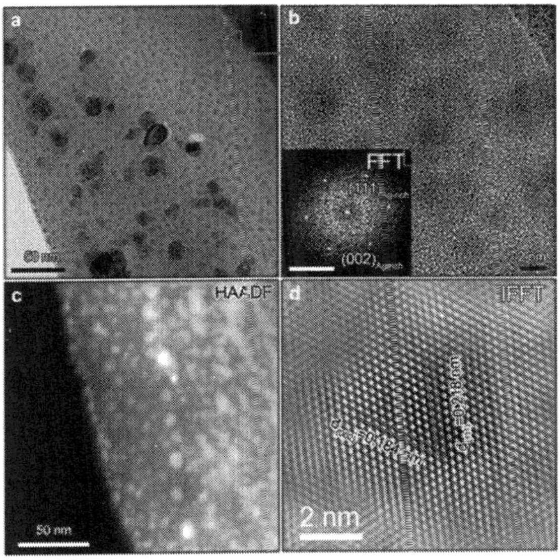

Figure 8. TEM and HAADF characterization of $Ag_{50}Cu_{50}$ catalysts. (a) bright field image, (b) HRTEM, (c) HAADF result and (d) IFFT image. Adapted from ref. 45.

N_2 and O_2 saturated 0.1 M KOH solutions. It can be seen that there is reduction in current density in O_2-saturated KOH solution, while that in N_2-saturated solution is flat. This shows that the catalyst indeed works on O_2. Figure 9b shows a set of RDE curves with rotation rates of 400, 800, 1,200 and 1,600 rpm. The Koutecky–Levich plots were then obtained from the limiting current density, as shown in Figure 9c. The plots show the inverse current density J^{-1} as a function of $\omega^{-1/2}$. From these plots the number of electrons transferred during ORR was found to be 3.76, 3.87, 3.85 and 3.97 when the potential was 0.5, 0.4, 0.3 and 0.2 V, respectively. Hence, four electrons route was found to be dominant for ORR in case of synthesized $Ag_{50}Cu_{50}$ catalyst. Finally, 9d gives a comparison of the performance of Ag, $Ag_{50}Cu_{50}$ and Pt/C(20 wt%).

The catalytic layer was used to assemble a zinc-air battery and results showed open-circuit voltage (OCV) of the cell was around 1.48 V close to the theoretical value, and the maximum power density is 67 mW cm^{-2} at 100 mA cm^{-2}. The resulting rechargeable zinc-air battery exhibits low charge–discharge voltage polarization of 1.1 V at 20 mAcm^{-2} and high durability over 100 cycles in natural air.

Wu et al. deposited Ag–Cu nanoalloys on nickel foam by pulse laser deposition. Several Ag–Cu alloys with Ag/Cu atom ratios of 90:10, 50:50 and 25:75 are used as the target material with the high-purity nickel foam (99.97%) as substrate [46]. Ag–Cu nanoalloys were used as the catalyst layer of the air cathode for a single zinc-air battery, and found to exhibit good

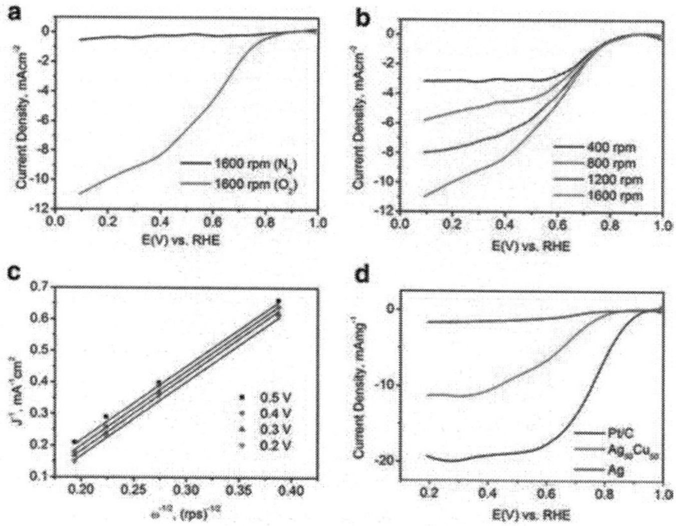

Figure 9. Electrochemical characterization of $Ag_{50}Cu_{50}$ catalyst. (a) The RDE curves of $Ag_{50}Cu_{50}$ catalyst in O_2- and N_2-saturated 0.1 M KOH solution; (b) the RDE curves at the rotation rates of 400, 800, 1,200 and 1,600 rpm; (c) the Koutecky–Levich plot of $Ag_{50}Cu_{50}$ catalyst; and (d) the ORR mass activity for Ag, $Ag_{50}Cu_{50}$ and Pt/C (20 wt%) catalysts. Adapted from ref. 45.

bifunctional catalytic performance. The effect of the Ag/Cu atom ratio on the average electron transfer numbers of the ORR was systematically investigated. This carbon-free binder-free bimetallic catalyst layer was found to possess both ORR and OER catalytic activity in the rechargeable zinc-air battery. Figure 10(a) shows representative microscopic images of the $Ag_{50}Cu_{50}$ alloys used in this work. The Ag–Cu nanoparticles were uniformly distributed in the substrate, and the electron diffraction pattern revealed the single phase and polycrystalline structure of the Ag–Cu alloy. Figure10 (b) shows a HRTEM image of more than 100 nanoparticles in the substrates. It is clear that the nanoparticles have an average size of 2.58 nm with a narrow size distribution between 1 and 5 nm. Figure 10(c,d) indicate that the nanoparticles are enriched with Ag atoms, with few doped copper atoms in them, and the films are enriched with Cu atoms.

Hence, it can be concluded that the nanocatalyst has crystalline Ag-enriched nanoparticles embedded in an amorphous Cu-enriched matrix. These copper-doped silver nanoparticles with composition $Ag_{50}Cu_{50}$ were observed to have superior catalytic performance for ORR as compared to pure silver as shown in Figure 11. The ORR was found to proceed via four electron transfer mechanism. It is for the first time the Ag-based electrocatalysts in amorphous films were created from the vapor phase under far-from-equilibrium condition by pulse laser ablation, previous works demonstrated that the face-centered cubic Ag–Cu solid solutions or completely amorphous Ag–Cu metal glasses were formed by rapid quenching from the liquid or vapor phase as corrosion-resistant non-equilibrium alloys and metastable phases.

http://dx.doi.org/10.5772/62050

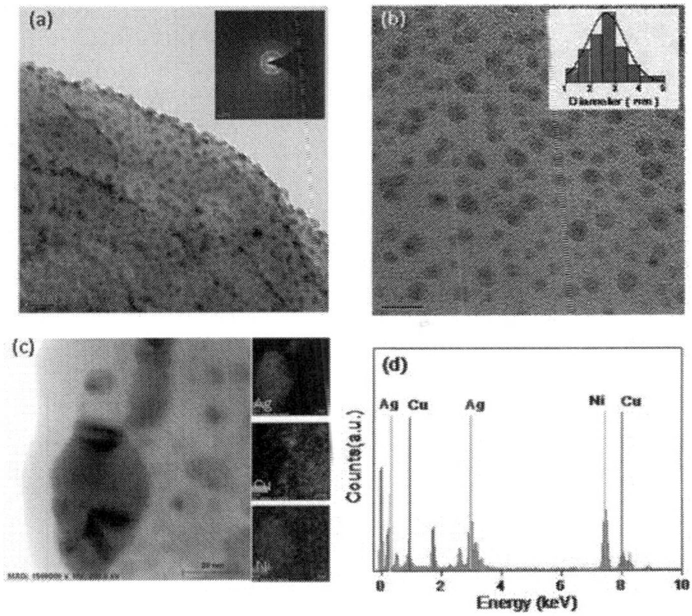

Figure 10. (a)TEM bright-field images and SAED patterns (inset) of the $Ag_{50}Cu_{50}$ film prepared by PLD. (b) HRTEM images and the particle size distribution (inlet) of Ag–Cu nanoparticles in the film. (c) TEM element mapping for Ag, Cu and Ni on a Ag–Cu nanoparticle deposited on nickel grid. (d) EDS of the $Ag_{50}Cu_{50}$ film on nickel grid. Reprinted with permission from ref. 46. Copyright 2015, American Chemical Society.

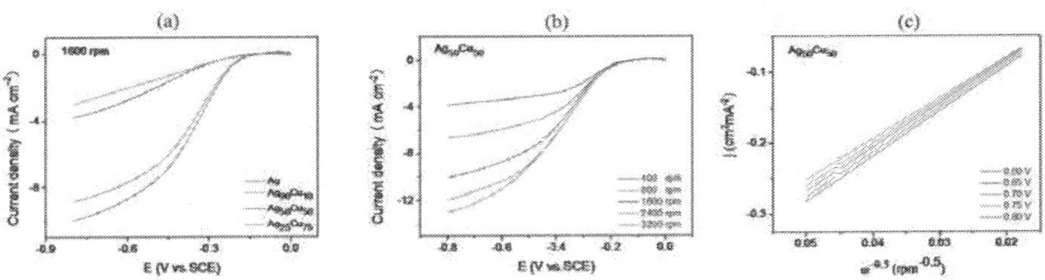

Figure 11. (a) RDE polarization curve of different compositions in O_2-saturated 0.1 M KOH solution. (b) RDE polarization curve for $Ag_{50}Cu_{50}$ same environment (c) Koutecky–Levich plots for (b). Reprinted with permission from ref. 46. Copyright 2015, American Chemical Society.

Zinc-air batteries were assembled using $Ag_{90}Cu_{10}$ and $Ag_{50}Cu_{50}$ catalysts in the air cathode, and the discharge performance is shown in Figure 12(a). The cell voltage decreases nearly with increasing current density, demonstrating that the cell performance shows a strong dependence on the resistance of the battery. For $Ag_{50}Cu_{50}$ and $Ag_{90}Cu_{10}$ catalysts, the open-circuit voltages of the single cell are about 1.42 V and 1.44 V, the peak power densities of the zinc-air batteries are 86.3 mW.cm^{-2} and 82.1 mW.cm$^{-2,}$ and the current densities at a voltage of 1 V are 60 mA cm^{-2} and 50 mA cm^{-2}, which is higher than Ag/C, N-doped CNTs and silver-molybdate catalysts [47,48]. Figure 12 (b) records the change of cell voltage with time at a current density of 20 mA cm^{-2}. It is clear that $Ag_{90}Cu_{10}$ catalysts have a higher initial discharge voltage of 1.15 V than $Ag_{50}Cu_{50}$ catalysts (1.0V); however, after 30 h discharging, the discharge voltage of $Ag_{90}Cu_{10}$ catalysts gradually reduces to 1.11 V, while $Ag_{50}Cu_{50}$ catalysts gradually increase to 1.18 V. The discharge curve of the $Ag_{50}Cu_{50}$ catalysts gradually rises to a stable cell voltage and decrease by 16% as compared with the open-circuit potential, but for $Ag_{90}Cu_{10}$ catalysts, the curve gradually reduces to a stable voltage after about 30 h of discharge, showing a decrease of 20%. It can be concluded that the $Ag_{50}Cu_{50}$ catalyst has higher discharge voltage stability, being more stable than the $Ag_{90}Cu_{10}$ catalysts for applications in zinc-air batteries.

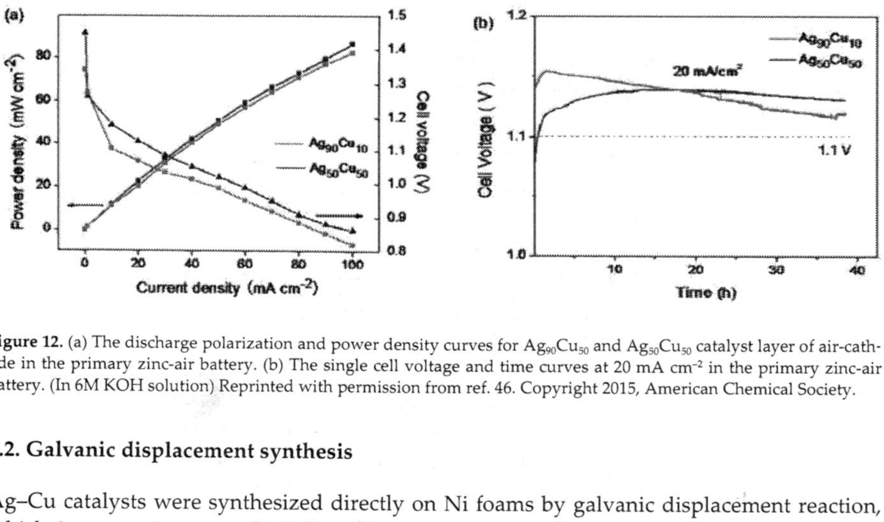

Figure 12. (a) The discharge polarization and power density curves for $Ag_{90}Cu_{50}$ and $Ag_{50}Cu_{50}$ catalyst layer of air-cathode in the primary zinc-air battery. (b) The single cell voltage and time curves at 20 mA cm^{-2} in the primary zinc-air battery. (In 6M KOH solution) Reprinted with permission from ref. 46. Copyright 2015, American Chemical Society.

3.2. Galvanic displacement synthesis

Ag–Cu catalysts were synthesized directly on Ni foams by galvanic displacement reaction, which is an environment-friendly and straightforward process [42]. In this method, the catalysts were directly grown on Ni foams, thereby freeing the catalytic layer from carbon and binder. The driving mechanism to grow the various catalysts in this work is the large difference of the redox potentials of Ni^{2+}/Ni (–0.25 V vs. SHE), Cu^{2+}/Cu (0.34 V vs. SHE) and Ag+/Ag (0.799 V vs. SHE). The galvanic displacement reaction can be described by the following equations:

$$Ni(s) + Cu^{+2}(aq) \rightarrow Cu(s) + Ni^{+2}(aq) \tag{13}$$

$$Ni(s) + 2Ag^{+}(aq) \rightarrow 2Ag(s) + Ni^{+2}(aq) \tag{14}$$

$$Cu(s) + 2Ag^{+}(aq) \rightarrow 2Ag(s) + Cu^{+2}(aq) \tag{15}$$

Figure 13 shows the typical SEM images of (a) Ni foam, (b) Cu nanoparticles, (c) AgCu-10 and (d) Ag catalysts. Figure 13(a) shows the pure Ni foam smooth surface. Immersion into $CuSO_4$ solution for 3 h makes the surface of the foam rough (Figure 13(b)) along with coverage of octahedral copper nanoparticles (Figure 13(b) inset). The SEM of AgCu-10 catalyst prepared by two-step galvanic displacement reaction is shown in Figure 13(c). The catalyst has dendritic morphology. The dendrites shown in Figure 13(c) have a dense and uniform distribution, and the shape is complete. On the other hand, similarly dendritic morphologies are obtained for pure Ag catalyst (Figure 13(d)) prepared by directly immersing the as-prepared Ni foam into $AgNO_3$ solution. The dendrites of Ag catalyst are thinly distributed on the Ni foam compared to that in Figure 13(c). The difference of the dendrites between AgCu-10 and Ag catalysts can be because of the different sacrificial templates. For the AgCu-10 catalyst, the Ni foam was already covered by octahedral copper nanoparticles while the nickel surface was free from copper in case of pure Ag particles.

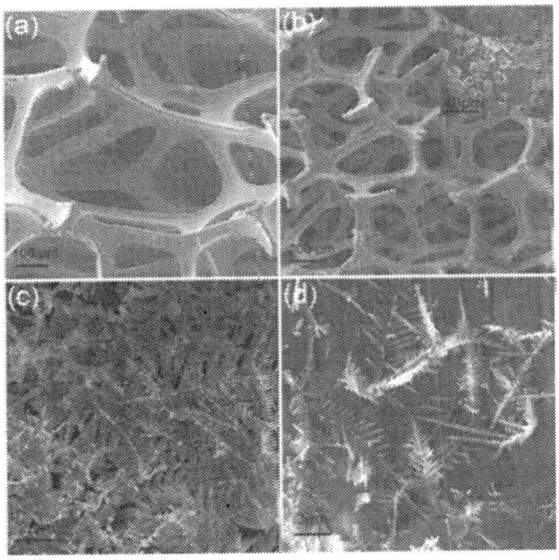

Figure 13. The FE-SEM images of (a) Ni foam. (b) Cu nanoparticles supported on Ni foam. The inset in (b) shows the high-magnification image of Cu nanoparticles. (c) Ag–Cu dendrites in AgCu-10 catalyst and (d) Ag dendrites supported on Ni foam. Reprinted with permission from ref. 42. Copyright 2015, Elsevier.

As shown in Figure 13(c), the one-dimensional dendrites prefer to form in a relatively high $AgNO_3$ concentration (10 mM). The AgCu-10 bimetallic catalyst possessed a hierarchical structure characteristic and highly rough surface which provided more catalytic active sites so it showed higher catalytic current during the RDE polarization measurements. The SEM elemental mapping analysis for AgCu-10 catalyst shown in Figure 14(a–d) clearly prove that Ag and Cu are uniformly distributed. HRTEM and SAED images are shown in Figure 14 (e, f). The clear lattice fringes are observed for AgCu-10 catalyst. The SAED pattern shows a typical polycrystalline structure with the sharp diffraction rings for AgCu-10 catalyst. The diffraction spots for Ag (111), (2 0 0), (2 2 0), (311) facets and Cu (111), (2 2 0) facets are observed. A fast

Fourier transform (FFT) diffraction pattern as shown in Figure 14 (g) is obtained on the area marked by the red rectangle in Figure 14 (e), exhibiting the two phases characteristic of AgCu-10 catalyst. The FFT image also reveals that Ag and Cu crystallites are in an epitaxial relationship relative to each other in parallel orientation. The grown orientation schematic is shown in Figure 14 (h).

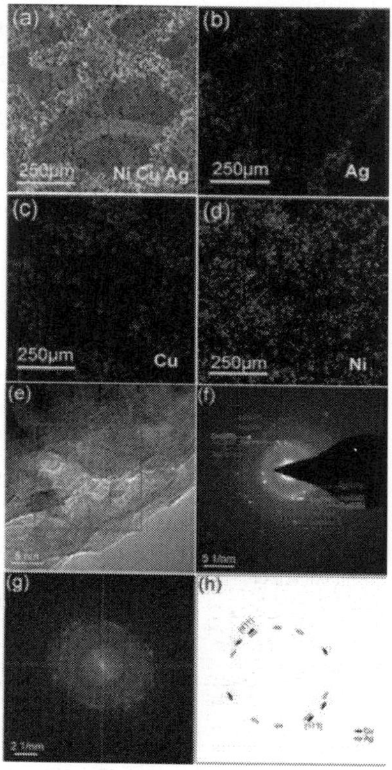

Figure 14. EDS elemental map of the AgCu-10 catalyst: (a) Overlay image. (b) Ag map. (c) Cu map. (d) Ni map. The TEM characteristic for AgCu-10 catalyst: (e) HRTEM image. (f) SAED pattern. (g) Fast Fourier transform (FFT) image corresponding to (e) and schematic drawing with index of reflections (h). Reprinted with permission from ref. 42. Copyright 2015, Elsevier.

Rotating disc measurements were performed on the catalysts made by galvanic displacement and comparison was made with pure silver. AgCu-10 performed efficiently as compared to the pure silver particles in terms of both onset potential and the limiting current. The Koutecky–Levich plots revealed the four electron transfer mechanism during ORR. Rechargeable zinc-air battery was also fabricated by as-prepared AgCu-10 catalyst-based air cathode, 6 M KOH solution with 0.2 M zinc acetate (zinc acetate was dissolved in KOH to form zincate to

http://dx.doi.org/10.5772/62050

ensure reversible Zn electrochemical reactions at the anode) and pure zinc plate anode. A charge–discharge cycle experiment was performed with a short cycle period of 20 min and a long cycle period of 4 h at 20 mA cm^{-2}. At the short cycle period, the initial charge and discharge potentials of AgCu-10 based rechargeable zinc-air battery are 2.04 and 1.1 V, respectively. The round-trip efficiency corresponding to the first cycle is 53.9%. There is almost no apparent fluctuation for the charge and discharge potentials of the rechargeable zinc-air battery through all the cyclic process. The round-trip efficiency after 100 cycles is 53.08% compared to the initial 53.9% with a little decline of 0.82%. A further cycle performance study with a long cycle period of 4 h was carried out with the same rechargeable zinc-air battery after replacing the zinc anode and the electrolyte. The rechargeable zinc-air battery also shows high cycling stability at the long cycle period. The increase in charge and discharge potentials difference from the first to tenth cycle is as little as 0.06 V, which is comparable to the tri-electrode rechargeable zinc-air battery [49]. The cycling stability obtained on AgCu-10 catalyst based zinc-air battery is certainly appealing and significant for the large-scale application of metal-air batteries and fuel cells.

3.3. Electro-deposition of Ag–Cu nanoalloys

Ag–Cu catalysts were synthesized by the electrodeposition method under a potential of −0.4 V for a period of 50 s by using the conventional three-electrode cell system [50]. For synthesis, $AgNO_3$, $Cu(NO_3)_2\,3H_2O$ and 3 mM sodium citrate ($Na_3C_6H_5O_7$) are mixed by deionized water with the formula of Ag_xCu_{100-x} (labelled as $Ag_{25}Cu_{75}$, $Ag_{50}Cu_{50}$ and $Ag_{75}Cu_{25}$ for $x = 25$, 50 and 75 mM, respectively).

Figure 15(a) describes the TEM and SEM (the inset) images of the bimetallic Ag–Cu catalyst. The nanoplatelets have diameters of 40–50 nm. The HRTEM analysis in Figure 15(b) clearly shows lattice fringes, indicating good crystallinity. The lattice spacing obtained from the HRTEM image is 0.239 nm. The particle demonstrates a single crystal pattern in the SAED of Figure15(c), indicating that Ag–Cu catalysts grow larger by the oriented attachment from small Ag–Cu nanoparticles. The cell constant of the single crystal is 0.3986 nm, which is between the standard cell parameter of FCC-Ag (a = 0.4086 nm) and FCC-Cu (a = 0.3615 nm), suggesting that the Cu atoms are partially alloyed with the Ag atoms.

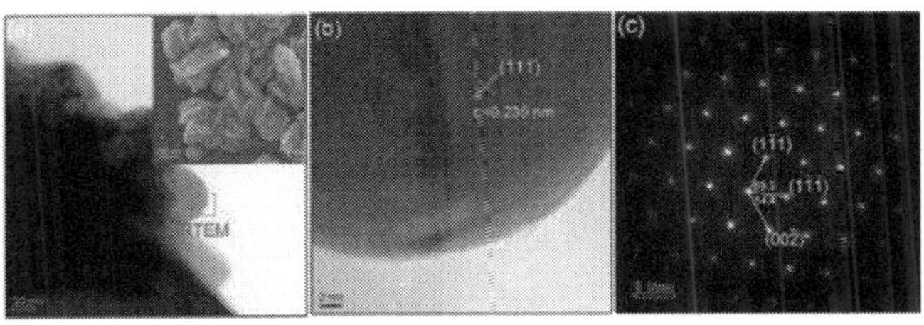

Figure 15. Images of the Ag–Cu catalyst (a) TEM (inset SEM). (b) HRTEM. (c) SAED. Reprinted with permission from ref. 50. Copyright 2015, John Wiley & Sons.

The EDX spectrum of the Ag–Cu nanoalloy in Figure 16(a) exhibits that the Ag–Cu deposits contain both Ag and Cu elements and the nominal atomic composition are 1.5:1, 5:1 and 10:1 for the $Ag_{25}Cu_{75}$, $Ag_{50}Cu_{50}$ and $Ag_{75}Cu_{25}$ samples, respectively.

The survey spectrum of XPS for the Ag–Cu catalyst is shown in Figure 16(b), which shows clear Ag and Cu peaks. The binding energies of the $Ag3d_{3/2}$ and $Ag3d_{5/2}$ orbits observed from the high-resolution spectrum (Figure 16(b) inset) are 374.4 and 368.4 eV. This result indicates that the Ag atoms are zero-valent (Ag^0) metals [51,52]. The binding energies of the $Cu2p_{1/2}$ and $Cu2p_{3/2}$ orbits are 952.6 and 932.7 eV (Figure 16(f)). An analysis of the Auger electron spectrum for Cu LMM (Figure 16(f) inset) is also conducted to determine the specific valence of Cu; an apparent kinetic energy peak at 918.6 eV is seen, corresponding to the zero-valent Cu[53]. Thus, the Ag and Cu atoms exist as elemental metallic substances in Ag–Cu catalyst.

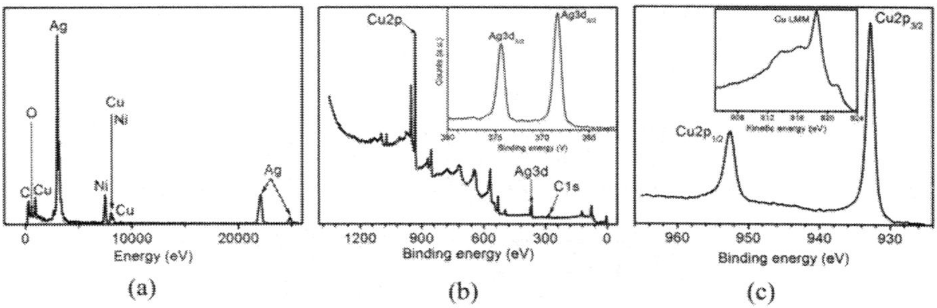

Figure 16. (a) EDX. (b) XPS survey spectrum, inset: high-solution spectrum of Ag3d. (c) High-solution spectrum of Cu2p, inset: Cu LMM. Reprinted with permission from ref. 50. Copyright 2015, John Wiley & Sons.

As shown in Figure 17(a), a reduction current peak is observed from the CV curves of the Ag–Cu catalyst in O_2-saturated 0.1 M KOH at −0.3 V (vs. SCE) but not in N_2-saturated solution, indicating that Ag–Cu catalyst has catalytic activity for ORR. To study the catalysis kinetics of Ag–Cu catalyst for ORR, an RDE experiment is performed in O_2-saturated 0.1 M KOH solution (Figure 17(b)). The number (n) of electrons transferred on the Ag–Cu catalyst during ORR, which determines the catalytic efficiency, is calculated by the Koutecky–Levich plots (see inset). The result with n = 3.8 and 3.7 at −0.7 and −0.8 V, respectively, indicates that the ORR catalyzed by the Ag–Cu catalyst occurs through a four-electron pathway, which is more efficient than a two-electron pathway.

The performance of primary battery fabricated from $Ag_{50}Cu_{50}$-based air cathode was evaluated as shown in Figure 18(a). The open-circuit voltage (OCV) and maximum power density are 1.49 V and 87 $mWcm^{-2}$, respectively, which have significant improvements [54,55]. After the primary zinc–air battery undergoes 10 discharging cycles, the OCV and power density decrease slightly. This result may be attributed to the polarization of the zinc anode caused by the zincate produced during the discharge process. Replacing the zinc anode and electrolyte revitalizes the battery performance. The battery has no obvious voltage loss compared with the first cycle, suggesting that the $Ag_{50}Cu_{50}$-based air cathode is stable in alkaline solution and

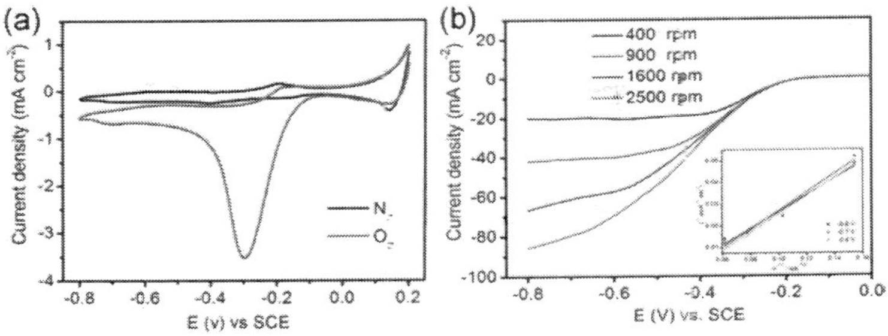

Figure 17. (a) CV curves for Ag–Cu catalyst in O_2- and N_2-saturated 0.1 M KOH solutions at a scan rate of 10 mV s^{-1}. (b) RDE polarization curves at different rotation rates in O_2-saturated 0.1 M KOH solution at a scan rate of 10 mV s^{-1}, inset: Koutecky–Levich plots. Reprinted with permission from ref. 50. Copyright 2015, John Wiley & Sons.

can be a potential candidate for rechargeable zinc-air batteries. Further analysis of discharging performance of this primary zinc-air battery is conducted at different constant current densities as in Figure 18(b). The cell voltages are stable during the whole discharge period of 20 h, except for a slight drop of approximately 0.05 V at the current density of 50 mA cm^{-2}. This voltage drop can be attributed to the electrode polarization under high current density.

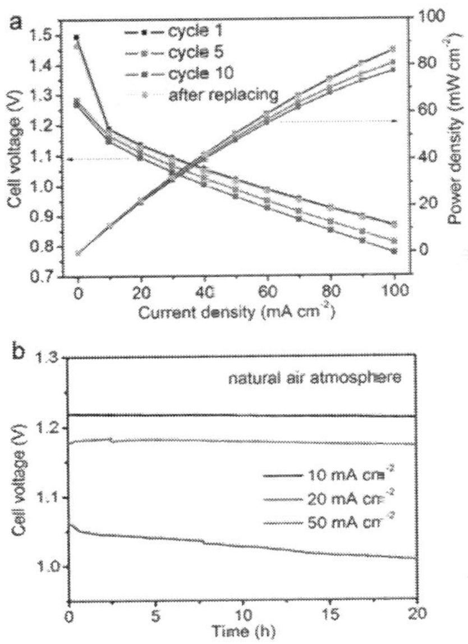

Figure 18. Performance of the primary zinc-air battery fabricated with an $Ag_{50}Cu_{50}$ catalyst-based air cathode: a) Cell voltage and power density polarization curves of the battery over 10 cycles and the polarization curve after replacing Zn anode and electrolyte; b) discharge voltage curves at different current densities. Reprinted with permission from ref. 50. Copyright 2015, John Wiley & Sons.

Figure 19(a) shows the charge and discharge polarization curves of the secondary zinc-air battery. An abrupt polarization occurs when the current densities increase from 0 to 10 mA cm^{-2} because of the activation polarization and anode polarization [56]. However, once the zinc-air battery begins its function, the polarization exhibits a steady increase with varying current densities from 10 to 100 mA cm^{-2}. The charge–discharge voltage gap (i.e., the overpotential) at 20 mA cm^{-2} is 0.9 V, which is lower than that of Co_3O_4-based rechargeable zinc-air batteries [57,58]. The cycle performance with different cycle periods is shown in Figure 19(b). As seen in the bottom of Figure 19(b), the secondary zinc-air battery undergoes 100 charge and discharge cycles at 20 mA cm^{-2} with 20 min per step. The difference between the charge and discharge potentials is 0.9 V, and the overpotential shows no apparent fluctuation through all 100 cycles. The round-trip efficiency is up to 56.4%, which is a considerable improvement. A more violent charge and discharge cycle experiment is conducted with a cycle period of 4 h for the same rechargeable zinc-air battery after replacing the zinc anode and electrolyte. As shown in the top of Figure 19(b), the charge and discharge voltages are still stable even with the long cycle period; this result is comparable to the result of the tri-electrode rechargeable zinc-air battery.

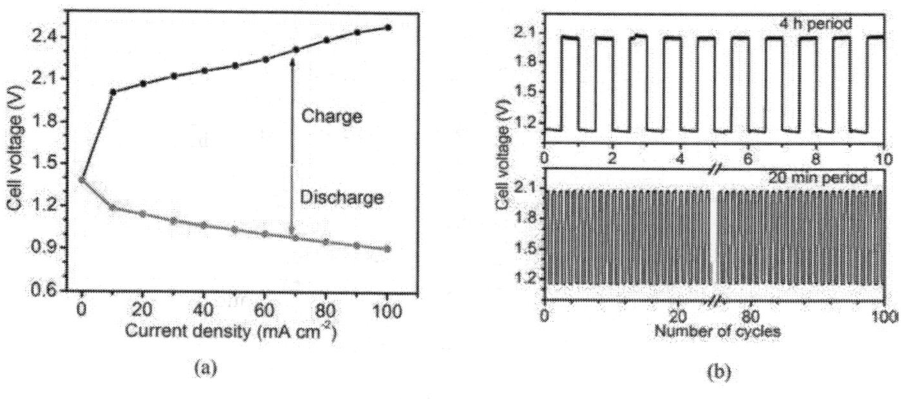

Figure 19. (a) Charge–discharge polarization curves for the rechargeable zinc air battery. b) Cycle performance for the rechargeable zinc-air battery at 20 mA cm–2 with a 20 min cycle period for 100 cycles and a 4 h cycle period for 40 h. Reprinted with permission from ref. 50. Copyright 2015, John Wiley & Sons.

4. Conclusions

Silver copper Ag–Cu nanoalloy particles have been investigated for prospective application as an electrocatalyst for oxygen reduction reaction in alkaline fuel cells and metal air battery systems. A holistic approach has been adopted incorporating density functional theory simulations along with synthesis of potential candidate compositions of Ag–Cu nanoalloys. Following conclusions can be drawn from this work:

http://dx.doi.org/10.5772/62050

1. Alloying of silver with copper significantly enhances the catalytic activity of silver for ORR. The enhancement of this catalytic activity is attributed to the modification of the surface electronic structure of the Ag–Cu nanoalloy catalyst.

2. Our DFT-based calculations confirm the rise of the d-band center of Ag–Cu nanoalloy catalyst toward Fermi level which in turn increases the adsorption strength of ORR intermediates on the catalyst surface during oxygen reduction. It has been well established that the optimum binding reaction intermediates such as O, OH, OOH are crucial for the fast kinetics of ORR.

3. The doping of a single Cu atom on the surface of 13 atom silver cluster improves the adsorption strength of ORR intermediates on the catalyst surface. We report that in case of 38 atom nanoalloy clusters, polyicosahedron (PIh) $Ag_{32}Cu_6$ core-shell structure is found to be the most stable structure by genetic algorithm calculations. The most active catalytic site on the $Ag_{32}Cu_6$ nanoalloy was also identified by our DFT calculations.

4. Ag–Cu nanoalloys have been successfully synthesized on Ni foams by pulse laser deposition as well as by chemical routes such as galvanic displacement and electrodeposition. Characterization of these nanoalloys by electron microscopy techniques (TEM, HRTEM, SEM) endorse the alloying of Ag and Cu atoms at the nanoscale.

5. Electrochemical characterization by rotating disc electrode methods (RDE) reveals the improvement of the ORR performance of the Ag–Cu nanoalloys due to the synergistic role of silver and copper in the Ag–Cu nanoalloys. Out of various compositions produced by PLD $Ag_{50}Cu_{50}$ was observed to be the most active toward ORR in alkaline media. A twofold increase in limiting current density along with more positive onset potential was recorded in case of AgCu-10 nanoalloy compared to pure silver. We report four electron transfer pathway to be dominant during ORR in alkaline conditions on the surface of Ag–Cu nanoalloys.

6. Ni foam is reported here to be a suitable support for Ag–Cu nanoalloy catalysts which enables the elimination of binder-based carbon supports.

5. Future outlook

Ag–Cu nanoalloy catalysts offer an attractive alternative for the otherwise costly Pt-based alloys. The performance of these Ag–Cu nanoalloy catalysts was found to be reasonable while the stability in alkaline media is much superior compared to the Pt-based alloy catalysts. The realization of high energy density metal air batteries and alkaline fuel cells for stationary and mobile applications demand swift ORR kinetics at cathode which can be achieved by highly active silver-based nanoalloy electrocatalysts. Very few Ag-based Ag/transition metal nanoalloy electrocatalysts have been reported for ORR. Recently, the number of publications on silver-based nanoalloy electrocatalyst for ORR in alkaline media has been on the rise. DFT-based calculations show that the surface electronic structure of the nanoalloy catalyst is sensitive to doping. This opens the window for carrying out investigations on the ternary Ag–

Cu–M catalyst systems with different morphologies to further improve the catalytic activity. Moreover, the effect of support such as CNT, MWCNT, RGO is still to be investigated for the Ag–Cu nanoalloys. These advance catalyst supports have been previously reported to be highly beneficial for Ag-based systems. The highest amount of catalytic activity in Ag-based systems is reported in the Ag composites owing to the ensemble effect. Incorporation of Ag–Cu nanoalloys into composite structures can provide a big leap forward toward improving the catalytic activity of these nanoalloys. At present, the activity of silver nanoalloys in alkaline media is limited because of weak adsorption of ORR intermediates. Highly active silver nanoalloy compositions can be identified by DFT calculations which when combined with active supports can rival or even surpass commercial Pt/C in alkaline media. A universal approach incorporating first principle simulations and experimental analysis is pivotal for the realization of highly active and stable silver-based nanoalloy electrocatalysts for ORR in alkaline media.

Acknowledgements

This study was supported by the National Natural Science Foundation of China (grant nos. 51271148 and 50971100), the Aeronautic Science Foundation Program of China (grant no. 2012ZF53073), the Science and Technology Innovation Fund of Western Metal Materials (grant no. XBCL-2-11) and the Doctoral Fund of Ministry of Education of China (grant no. 20136102110013).

Author details

Adnan Qaseem, Fuyi Chen* and Nan Zhang

*Address all correspondence to: fuyichen@nwpu.edu.cn

State Key Laboratory of Solidification Processing, Northwestern Polytechnical University, Xian, China

References

[1] Nobuyuki Imanishi and Osamu Yamamoto. Rechargeable lithium–air batteries: characteristics and prospects. Materials Today. 2014;17(1):24–30. DOI: 10.1016/j.mattod.2013.12.004

[2] Md. Arafat Rahman, Xiaojian Wang and Cuie Wen. A review of high energy density lithium–air battery technology. Journal of Applied Electrochemistry. 2014;44(1):5–22. DOI: 10.1007/s10800-013-0620-8

[3] Nuria Garcia-Araez and Petr Novák. Critical aspects in the development of lithium–air batteries. Journal of Solid State Electrochemistry. July 2013; 17 (7):1793–1807. DOI: 10.1007/s10008-013-1999-1

[4] Mark K. Debe. Electrocatalyst approaches and challenges for automotive fuel cells. NATURE. 2012;486(7401):43–51. DOI: 10.1038/nature11115

[5] İbrahim Dinçer and Calin Zamfirescu. Hydrogen and fuel cell systems. In: Sustainable Energy Systems and Applications. New York: Springer US 2012. pp. 519–632. DOI: 10.1007/978-0-387-95861-3_13

[6] Marcel Pourbaix. Atlas of Electrochemical Equilibria in Aqueous Solutions. 2nd ed. US: National Association of Corrosion; 1974. 644 p. DOI: 0915567989

[7] Edward McCafferty. Thermodynamics of Corrosion: Pourbaix Diagrams. In: Edward McCafferty, editor. Introduction to Corrosion Science. Springer New York; 2010. pp. 95–117. DOI: 10.1007/978-1-4419-0455-3_6

[8] J. K. Nørskov, J. Rossmeisl, A. Logadottir and L. Lindqvist. Origin of the overpotential for oxygen reduction at a fuel-cell cathode. The Journal of Physical Chemistry B. 2004;108 (46):17886–17892. DOI: DOI: 10.1021/jp047349j

[9] Daojian Cheng, Xiangguo Qiua and Haiyan Yua. Enhancing oxygen reduction reaction activity of Pt-shelled catalysts via subsurface alloying. Physical Chemistry Chemical Physics. 2014;16(38): 20377-20381. DOI: 10.1039/c4cp02863e

[10] Bongjin Simon Mun, Guofeng Wang, Philip N. Ross, Christopher A. Lucas, and Nenad M. Marković. Improved Oxygen Reduction Activity on $Pt_3Ni(111)$ via Increased Surface Site Availability. Science. 2007;315 (5811):493–497. DOI: 10.1126/science.1135941

[11] Gongwei Wang, Bing Huang, Li Xiao, Zhandong Ren, Hao Chen, Deli Wang, Héctor D. Abruña, Juntao Lu, and Lin Zhuang. Pt Skin on AuCu Intermetallic Substrate: A Strategy to Maximize Pt Utilization for Fuel Cells. Journal of the American Chemical Society (JACS). 2014;136(27):9643–9649. DOI: 10.1021/ja503315s

[12] O A Petrii. Electrosynthesis of nanostructures and nanomaterials. Russian Chemical Reviews. 2015;84(2):159–193. DOI: 10.1070/RCR4438

[13] J. Greeley, I. E. L. Stephens, A. S. Bondarenko, T. P. Johansson H. A. Hansen, T. F. Jaramillo, J. Rossmeisl, I. Chorkendorff and J. K. Nørskov. Alloys of platinum and early transition metals as oxygen reduction electrocatalysts Nature Chemistry. 2009;1:552–556. DOI: 10.1038/nchem.367

[14] Hongliang Xin, Adam Holewinski, and Suljo Linic. Predictive Structure- Reactivity Models for Rapid Screening of Pt-Based Multimetallic Electrocatalysts for the Oxygen Reduction Reaction. ACS Catalysis. 2012;2:12–16. DOI: 10.1021/cs200462f

[15] Mitsuru Wakisaka, Satoshi Mitsui, Yoshikazu Hirose, Katsura Kawashima,Hiroyuki Uchida, and Masahiro Watanabe. Electronic structures of Pt-Co and Pt-Ru alloys for

CO-tolerant anode catalysts in polymer electrolyte fuel cells studied by EC-XPS. The Journal of Physical Chemistry B. 2006;110(46):23489–23496. DOI: 10.1021/jp0653510

[16] B. Hammer and J.K. Nørskov. Theoretical surface science and catalysis—calculations and concepts. In: Bruce C. Gates, Helmut Knozinger, editors. Advances in Catalysis. 45th ed. Academic Press; 2000. pp. 71–129. DOI: 10.1016/S0360-0564(02)45013-4

[17] Hongliang Xin, Adam Holewinski, Neil Schweitzer, Eranda Nikolla and Suljo Linic. Electronic structure engineering in heterogeneous catalysis: identifying novel alloy catalysts based on rapid screening for materials with desired electronic properties. Topics in Catalysis. 2012;55 (5-6):376–390. DOI: 10.1007/s11244-012-9794-2

[18] Paul Sabatier. Hydrogénations et déshydrogénations par catalyse. Berichte der deutschen chemischen Gesellschaft. 1911;44(3):1984–2001. DOI: 10.1002/cber.19110440303

[19] Ernest Yeager. Dioxygen electrocatalysis: mechanisms in relation to catalyst structure. Journal of Molecular Catalysis. 1986; 38(1–2):5–25. DOI:10.1016/0304-5102 (86)87045-6

[20] F. Bidault, D.J.L. Brett, P.H. Middleton and N.P. Brandon. Review of gas diffusion cathodes for alkaline fuel cells. Journal of Power Sources. 2009;187(1):39–48. DOI: 10.1016/j.jpowsour.2008.10.106

[21] Vladimir Neburchilov, Haijiang Wang, Jonathan J. Martin and Wei Qu. A review on air cathodes for zinc–air fuel cells. Journal of Power Sources. 2010;195(5):1271–1291. DOI: 10.1016/j.jpowsour.2009.08.100

[22] B.B. Blizanac, P.N. Ross and N.M. Markovic. Oxygen electroreduction on Ag(1 1 1): The pH effect. Electrochimica Acta. 2007;52 (6):2264–2271. DOI: doi:10.1016/j.electacta.2006.06.047

[23] Adam Holewinski, Juan-Carlos Idrobo and Suljo Linic. High-performance Ag–Co alloy catalysts for electrochemical oxygen reduction. Nature Chemistry. 2014;6(9):828–834. DOI: 10.1038/nchem.2032

[24] Lizhi Yuan, Luhua Jiang, Jing Liu, Zhangxun Xia, Suli Wang and Gongquan Sun. Facile synthesis of silver nanoparticles supported on three dimensional graphene oxide/carbon black composite and its application for oxygen reduction reaction. Electrochimica Acta. 2014;135:168–174. DOI: 10.1016/j.electacta.2014.04.137

[25] Shin-Ae Park, Hankwon Lim and Yong-Tae Kim. Enhanced oxygen reduction reaction activity due to electronic effects between Ag and Mn3O4 in alkaline media. ACS Catalysis. 2015;5(6):3995–4002. DOI: 10.1021/acscatal.5b00495

[26] Junya Ohyama, Yui Okata, Noriyuki Watabe, Makoto Katagiri, Ayaka Nakamura, Hidekazu Arikawa, Ken-ichi Shimizu, Tatsuya Takeguchi, Wataru Ueda and Atsushi Satsuma. Oxygen reduction reaction over silver particles with various morphologies and surface chemical states. Journal of Power Sources. 2014;245(1):998–1004. DOI: doi:10.1016/j.jpowsour.2013.07.034

[27] Raju P. Gupta. Lattice relaxation at a metal surface. Physical Review B. 1981;23(12): 6265–6270. DOI: http://dx.doi.org/10.1103/PhysRevB.23.6265

[28] Weiyin Li and Fuyi Chen. A density functional theory study of structural, electronic, optical and magnetic properties of small Ag–Cu nanoalloys. Journal of Nanoparticle Research. 2013;15(7):1809. DOI: 10.1007/s11051-013-1809-9

[29] F. Baletto, C. Mottet, and R. Ferrando. Growth simulations of silver shells on copper and palladium nanoclusters. Physical Review B. 2002;66(15):155420. DOI: http://dx.doi.org/10.1103/PhysRevB.66.155420

[30] Roy L. Johnston. Evolving better nanoparticles: Genetic algorithms for optimising cluster geometries. Dalton Transactions. 2003;(22):4193–4207. DOI: 10.1039/B305686D

[31] B. Delley. An all-electron numerical method for solving the local density functional for polyatomic molecules. The Journal of Chemical Physics. 1990;92(1):508. DOI: 10.1063/1.458452

[32] B. Delley. From molecules to solids with the DMol3 approach. The Journal of Chemical Physics. (2000);113(18): 7756. DOI: 10.1063/1.1316015

[33] J. Oviedo and R. E. Palmer. Amorphous structures of Cu, Ag, and Au nanoclusters from first principles calculations. The Journal of Chemical Physics. 2002;117(21):9548. DOI: 10.1063/1.1524154

[34] Jonathan P. K. Doye and David J. Wales. Global minima for transition metal clusters described by Sutton–Chen potentials. New Journal of Chemistry. 1998,;22(7):733-744. DOI: 10.1039/A709249K

[35] Wenqiang Ma, Fuyi Chen, Nan Zhang and Xiaoqiang Wu. Oxygen reduction reaction on Cu-doped Ag cluster for fuel-cell cathode. Journal of Molecular Modeling. 2014; 20:2454. DOI: 10.1007/s00894-014-2454-7

[36] Kihyun Shin, Da Hye Kim, Sang Chul Yeo and Hyuck Mo Lee. Structural stability of AgCu bimetallic nanoparticles and their application as a catalyst: A DFT study. Catalysis Today. 2012;185(1):94–98. DOI: 10.1016/j.cattod.2011.09.022

[37] Kihyun Shin, Da Hye Kim and Hyuck Mo Lee. Catalytic Characteristics of AgCu Bimetallic Nanoparticles in the Oxygen Reduction Reaction. ChemSusChem. 2013;6(6): 1044–1049. DOI: 10.1002/cssc.201201001

[38] N. Zhang, F. Y. Chen and X.Q. Wu. Global optimization and oxygen dissociation on polyicosahedral Ag32Cu6 core-shell cluster for alkaline fuel cells. Scientific Reports (2015) 5, Article number: 11984. DOI: 10.1038/srep11984

[39] Rutger a. Van Santen and Matthew Neurock. Concepts in Theoretical Heterogeneous Catalytic Reactivity. Catalysis Reviews: Science and Engineering. 1995 ;37(4):557–698. DOI: 10.1080/01614949508006451

[40] Waldemar Mróz, Bogusław Budner, Wojciech Tokarz, Piotr Piela and Michael L. Korwin-Pawlowski. Ultra-low-loading pulsed-laser-deposited platinum catalyst films for

polymer electrolyte membrane fuel cells. Journal of Power Sources. 2015;273(1):885–893. DOI: 10.1016/j.jpowsour.2014.09.173

[41] Sadesh Kumar Natarajan and Jean Hamelin. High-performance anode for polymer electrolyte membrane fuel cells by multiple-layer Pt sputter deposition. Journal of Power Sources. 2010;195(22):7574–7577. DOI: 10.1016/j.jpowsour.2010.06.027

[42] Yachao Jin and Fuyi Chen. Facile preparation of Ag-Cu bifunctional electrocatalysts for zinc-air batteries. Electrochimica Acta. 2015;158:437–445. DOI: 10.1016/j.electacta.2015.01.151

[43] M. J. Montenegro and M. Döbeli. Pulsed laser deposition of La0.6Ca0.4CoO3(LCCO) films. A promising metal-oxide catalyst for air based batteries. Physical Chemistry Chemical Physics. 2002;4(12):2799–2805. DOI: 10.1039/B200120A

[44] Shaomin Zhu, Zhu Chen, Bing Li, Drew Higgins, Haijiang Wang, Hui Li and Zhongwei Chen Nitrogen-doped carbon nanotubes as air cathode catalysts in zinc-air battery. Electrochimica Acta. 2011;56(14):5080–5084. DOI: 10.1016/j.electacta.2011.03.082

[45] Yimin Lei, Fuyi Chen, Yachao Jin and Zongwen Liu. Ag-Cu nanoalloyed film as a high-performance cathode electrocatalytic material for zinc-air battery. Nanoscale Research Letters. 2015;10:197. DOI: 10.1186/s11671-015-0900-9

[46] Xiaoqiang Wu, Fuyi Chen, Yachao Jin, Nan Zhang and Roy L. Johnston. Silver–copper nanoalloy catalyst layer for bifunctional air electrodes in alkaline media ACS Application of Materrial Interfaces. 2015;7(32):17782–17791. DOI: 10.1021/acsami.5b04061

[47] Jia-Jun Han, Ning Li and Tian-Yun Zhang. Ag/C nanoparticles as an cathode catalyst for a zinc-air battery with a flowing alkaline electrolyte. Journal of Power Sources. 2009;193(2):885–889. DOI: 10.1016/j.jpowsour.2009.02.082

[48] Ying Wanga,Yu Liuc and Xiaojian Lu. Silver-molybdate electrocatalysts for oxygen reduction reaction in alkaline media. Electrochemistry Communications. 2012;20:171–174. DOI: 10.1016/j.elecom.2012.05.004

[49] Yanguang Li, Ming Gong and Yongye Liang. Advanced zinc-air batteries based on high-performance hybrid electrocatalysts. Nature Communications. 2013;4:1805. DOI: 10.1038/ncomms2812

[50] Yachao Jin, Fuyi Chen, Yimin Lei and Xiaoqiang Wu. A Silver-copper alloy as an oxygen reduction electrocatalyst for an advanced zinc-air battery. Chemical Catalysis Catalytical Chemitry. 2015;7(15):2377–2383. DOI: 10.1002/cctc.201500228

[51] Sang Woo Han, Yunsoo Kim and Kwan Kim. Dodecanethiol-derivatized Au/Ag bimetallic nanoparticles: TEM, UV/VIS, XPS, and FTIR analysis. Journal of Colloid and Interface Science. 1998;208(1):272–278. DOI: 10.1006/jcis.1998.5812

[52] Xianfeng You, Feng Chen, Jinlong Zhang and Masakazu Anpo. A novel deposition precipitation method for preparation of Ag-loaded titanium dioxide. Catalysis Letters. 2005;102(3-4):247–250. DOI: 10.1007/s10562-005-5863-5

[53] C. C. Chusuei, M. A. Brookshier and D. W. Goodman. Correlation of relative X-ray photoelectron spectroscopy shake-up intensity with CuO particle size. Langmuir. 1999;15(8):2806–2808. DOI: 10.1021/la9815446

[54] Po-Chieh Li, Chi-Chang Hu, Tai-Chou Lee, Wen-Sheng Chang and Tsin Hai Wang. Synthesis and characterization of carbon black/ manganese oxide air cathodes for zinc-air batteries. Journal of Power Sources. 2014;269:88–97. DOI: 10.1016/j.jpowsour.2014.06.108

[55] Ting-Hao Yang, Shanmuganathan Venkatesan, Chien-Hung Lien, Jen-Lin Chang and Jyh-Myng Zen. Nafion/lead oxide–manganese oxide combined catalyst for use as a highly efficient alkaline air electrode in zinc–air battery. Electrochimica Acta. 2011;56(17):6205–6210. DOI: 10.1016/j.electacta.2011.03.094

[56] Jia Wei Desmond Ng, Maureen Tang and Thomas F. Jaramillo. A carbon-free, precious-metal-free, high-performance O_2 electrode for regenerative fuel cells and metal–air batteries. Energy & Environmental Science. 2014;7(6):2017–2024. DOI: 10.1039/C3EE44059A

[57] Dong Un Lee, Ja-Yeon Choi, Kun Feng, Hey Woong Park, and Zhongwei Chen. Advanced Extremely Durable 3D Bifunctional Air Electrodes for Rechargeable Zinc-Air Batteries. Advanced Energy Materials. 2014;4(6):1301389. DOI: 10.1002/aenm.201301389

[58] Guojun Du, Xiaogang Liu, Yun Zong, T.S. Andy Hor, Aishui Yu and Zhaolin Liu. Co3O4 nanoparticle-modified MnO_2 nanotube bifunctional oxygen cathode catalysts for rechargeable zinc–air batteries. Nanoscale. 2013;5(11):4657–4661. DOI: 10.1039/C3NR00300K

Section 5

DNA-Based Biocatalysts

Chapter 16

Native and Synthetic G-quartet-based DNAzyme Systems – Artificial Enzymes for Biotechnological Applications

Stefan Loic

Additional information is available at the end of the chapter

http://dx.doi.org/10.5772/62165

Abstract

Catalysis of chemical reactions is crucial for both chemical industry and research. However, scientists are not the first ones to use catalysts in their laboratory. In fact, they are also essential for nature which designs plenty of biocatalysts, playing a pivotal role in living systems. For a long time, it was thought that only enzymes had this property. However, since the beginning of the 1980s, it is known that ribonucleic acids (also termed RNA) can acquire this ability, making them compulsory for key reactions (*e.g.*, for the translation of messenger RNA in the ribosome). Based on that, chemists designed several synthetic DNA catalysts (termed DNAzymes) for a large variety of reactions and applications. Among the DNA structures used, G-quadruplexes are guanine-rich noncanonical DNA structures (*i.e.*, differing from duplex DNA) composed of native G-quartets and particularly interesting for their ability to catalyze reactions of peroxidation. This peroxidase-mimicking system found plenty of applications detailed in this chapter. Moreover, optimizations of experimental conditions are also discussed and highlight the versatility and easy-to-use characteristics of G-quadruplexes DNA. Also, synthetic G-quartets, mainly TASQ (for *template-assembled synthetic G-quartets*), developed by chemists showed their ability to mimic G-quadruplexes, thanks to the presence of a G-quartet. Thus, synthetic G-quartets proved their capability to catalyze peroxidase-mimicking reactions, and these new exciting nature-mimicking catalytic systems are presented in detail in this chapter.

Keywords: DNAzyme, G-quadruplex, hemin, G-quartet, TASQ

1. Introduction

Catalysts are essential in the fields of both synthetic applications and biological functions. Indeed, while they make thousands of chemical reactions possible for industry and research, they play a pivotal role in living systems. All along the years, nature develops strategies to catalyze biochemical reactions. The most representative catalysts are undoubtedly the enzymes, but they are not the only ones. Interestingly, ribonucleic acids (also known as RNA) are also able to play this role and to catalyze key reactions, like in the active site of the ribosome during the translation of messenger RNA (mRNA). Besides, it is strongly supposed that the origins of the prebiotic life were based on the use of RNA as both the carrier of the genetic information, and a catalyst: RNA was a self-sufficing molecule. This theory was termed "RNA world."[1–6]

Inspired by the role of RNA as a catalyst, chemists developed new catalytic systems based on deoxyribonucleic acids, also termed DNA.[7, 8] DNA offers more advantages, like its better stability compared to the RNA equivalent.[9] These aspects are developed in section 2 of this chapter.

Among all the DNA structures used as catalysts (*e.g.*, canonical duplex structures or noncanonical triplexes, etc.), author would like to highlight the reader on the G-quadruplex structures.[10] These noncanonical edifices, composed of a stacking of native G-quartets, are introduced in section 3. This presentation is followed by the story of the discovery of G-quadruplexes as catalysts, and then, by a rationalization of how this chemical mechanism works, and how chemists can modulate experimental conditions to obtain the efficiency desired. To complete the presentation, the large range of applications of these noncanonical structures is commented on and shows how versatile and effective quadruplex DNA are.

Based on the observation that the catalytic activity of G-quadruplex is mainly due to the presence of native G-quartets,[11] several groups designed new molecules able to form a synthetic G-quartet. The most representative examples are TASQ (for *template-assembled synthetic G-quartet*), [12–16] composed of four guanines grafted on a template, and able to self-assemble into an intramolecular G-quartet. In section 4 of this chapter, the concept of TASQ is first clarified and then their catalytic activity is specified. Finally, the very first applications proposed in the literature are described, and pave the way to the use of synthetic molecules to mimic natural enzymes, like peroxidases.

2. From enzyme to DNAzyme

2.1. From enzyme to ribozyme

Long regarded as the only biomolecules able to catalyze chemical reactions, proteins are not the exclusive edifices playing this pivotal role in biological systems. Indeed, at the beginning of the 1980s, it was discovered that RNA (for *ribonucleic acids*) share also this property and are involved in numerous biological process.[1, 2] Some of the most representative examples are certainly the RNase P catalyzing transfer RNA (or tRNA) maturation [17], the riboregulators (also termed riboswitches) incorporated in certain messenger RNA (or mRNA), and controlling transcription or translation *in cellulo*.[18, 19] Active site of the ribosomes, composed of ribosomal RNA (or rRNA) and catalyzing the protein synthesis in the cytoplasm, also has to be highlighted.[20] This ability of RNA to catalyze an enzymatic reaction was termed "ribozyme," obtained by contracting the words "ribonucleic acids" and "enzyme." This fantastic biological and chemical breakthrough was honored in 1989 by the Nobel Prize in Chemistry, attributed to the two pioneers of this field: Sidney Altman and Thomas R. Cech.[3, 4] The potential of these ribozymes quickly drew the scientific community's attention which developed the very first artificial ones in 1990.[21] The RNA sequences were first designed to catalyze single-strand RNA cleavage and then for RNA ligation, porphyrin metalation, or more classic organic chemical reactions like Diels–Alder and Michael reactions.[22]

2.2. From ribozyme to DNAzyme

Interestingly, not any natural catalytic deoxyribonucleic acids (or DNA) has been found as yet in nature. However, their higher chemical stability compared to RNA and proteins makes DNA catalysts of choice for chemists who want to develop innovative applications in a larger range of experimental conditions. Indeed, the DNA stability against heat treatment and hydrolysis is evaluated at 1.000 and 100.000 times higher than for RNA and proteins, respectively. [9] Furthermore, obtaining specific designed DNA sequences is increasingly easy, thanks to the automated DNA synthesizers for both academic laboratories and the industry. In parallel, several companies are specialized in the custom DNA synthesis, and offer the opportunity for everybody to work now with DNA catalysts. Other advantages of using DNA as a catalyst can be highlighted here, like the possibility to functionalize it (*e.g.*, fluorescent probes or specific other moieties), to graft it on a solid support (*e.g.*, polymers, gold surface), and on top of that, the virtually unlimited number of catalysts that may be obtained by modulating both the number and the nature of the nucleotides.[7, 23, 24] Thus, chemists can design the specific sequence meeting their requests in term of application and efficiency.

It was not until 1994 that the first instance of an artificial catalytic DNA, termed "deoxyribozyme" (by analogy with the ribozymes) or most widely named "DNAzyme," was published. [25] Since then, plenty of applications has been developed, from the most original to the most complex ones, using different kinds of DNA structures. As a matter of fact, DNA is a highly versatile molecule that can self-assemble into several tridimensional organizations, depending of the sequences and conditions.[7, 23, 24] The most familiar form is undoubtedly the double-helix (also termed duplex) DNA form, used as a DNAzyme for enantioselective Diels–Alder and Friedel–Craft reactions.[7, 22] Nevertheless, other noncanonical DNA structures were also studied for their ability to catalyze chemical reactions. The two most representative structures are the triplex and the G-quadruplex DNA forms, with a clear predominance for the latter ones, which constitute the next section of this book chapter.[23, 24]

3. Native G-quartet-based DNAzymes: G-quadruplexes

3.1. Structure of G-quadruplexes

G-quadruplexes (Figure 1), formed from G-rich DNA strands, are composed of a stacking of several native G-quartets. Each quartet results from the self-assembly of four guanines in a same plan, self-stabilized by eight hydrogen bonds, from the Hoogsteen and Watson–Crick faces of the guanine moieties. The additional π-stacking interactions between G-quartets, and the bonding of cationic ions (*e.g.*, K^+, Na^+), increase the global stability of the tridimensional edifice.[10, 26, 27] In a structural point of view, the G-quadruplexes can mainly differ from the total numbers of constitutive G-quartets (from two to several thousands)[28–30] or strands (from one to four),[31–35] from the orientation of the strands (leading to several conformations named antiparallel, parallel, and hybrid), and finally from the length, DNA bases composition, and position of the loop(s) (which can be edgewise, diagonal, or chain-reverse).[36–41] All of these parameters are linked to the global stability of the edifice, like the number of G-quartets (as it is discussed later in this chapter), and are interdependent.

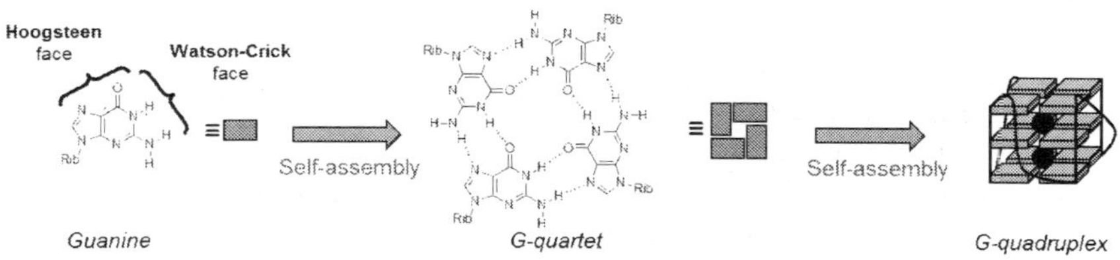

Figure 1. Schematic representation of the self-assembly of guanines *via* the Watson–Crick and Hoogsteen faces to form G-quartets, and G-quadruplexes.

These noncanonical structures are well known in a biological context, because they are strongly suspected to play important roles in key cellular events, like chromosomal instability, or regulation of gene expression. These aspects are far from the scope of this chapter, and author incites curious readers to have a look to some reviews cited hereafter.[26, 27, 42–44]

3.2. The seeds of the G-quartet ability to catalyze peroxidase-like reactions

In 1996, Y. Li and D. Sen developed and published the fourth known DNAzyme system,[45] able to catalyze the incorporation of metals (*i.e.*, Cu(II) and Zn(II)) into a specific porphyrin, named mesoporphyrin IX, or MPIX. To select the best DNA catalyst for their system, they used the *in vitro* SELEX (for *systematic evolution of ligands by exponential enrichment*) method that highlighted one sequence, termed PS5.ST1, from an initial pool of DNA sequences.[45, 46] Interestingly, three main observations were crucial:

a. the sequence of the strand was guanine-rich,

b. the presence of alkaline cations was required (with a catalytic activity 300 times higher with K^+ than with Na^+), and

c. the addition of another porphyrin derivative, the *N*-methyl mesoporphyrin IX (or NMM), well known for its interaction with G-quadruplexes and unable to be metallated due to the steric hindrance of the methyl group,[47–49] inhibited the incorporation of the metals into the MPIX. Altogether, these data suggested for the very first time that G-quadruplexes could adopt catalytic properties.

The subsequent step has been taken in 1998 when the same group showed that hemin, a Fe(III) cofactor playing a pivotal role in many enzymes (*e.g.*, in catalases, monooxygenases, and peroxidases), is activated by the presence of a G-quadruplex. More precisely, authors proved that the activity of the *horseradish peroxidase* (also termed HRP), composed of an hemin surrounded by a protein environment, can be mimicked using the same hemin but in the presence of G-quadruplexes.[50] Since then, many experiments using different sequences and morphologies of G-quadruplexes unambiguously confirmed the mandatory role of the G-quadruplex structure, for which the accessible G-quartet (*i.e.*, the external one), able to interact with hemin, constitutes the key step of the reaction.[24, 51–55]

3.3. Mechanism and factors influencing the catalytic activity

After almost 20 years of research on the G-quadruplex peroxidase-mimicking systems (Figure 2), the precise mechanism of the catalytic cycle is not fully understood. Nevertheless, in 2012, L. Stefan *et al.* proposed a first mechanism in nine main steps focused on the iron–porphyrin complex (*i.e.*, the hemin).[56] It was built on the basis of plenty of work published for the hemoprotein systems (primarily peroxidases and catalases).[57, 58] The mechanism is not scrutinized here, and readers are urged to see the references for additional information. However, the main factors influencing the catalysis are described hereafter and might be used as a roadmap.

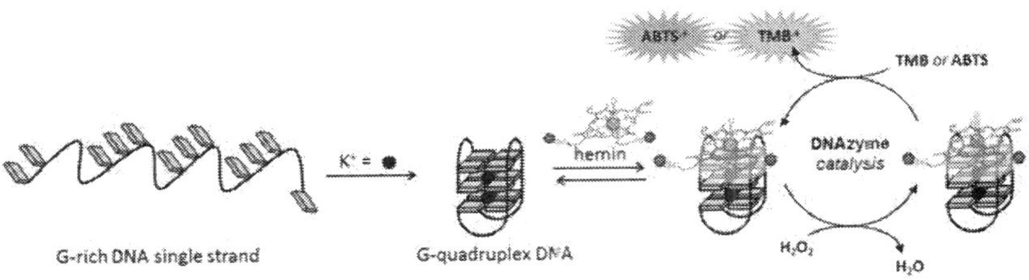

Figure 2. Schematic representation of the DNAzyme activity promoted by a G-quadruplex.

As explained before, the morphology of the G-quadruplexes used (*i.e.*, number of G-quartets, strands, type and length of loops, etc.) influences the catalytic efficiency of the reaction. The activation of the hemin rests on the presence of a hydrophobic binding site, playing the role of the "binding pocket" in enzymes. In this DNAzyme context, accessible G-quartets favor the interaction with hemin and, besides, protect the porphyrin from the oxidative degradation due to the oxygen peroxide. The existence of an axial ligand giving electronic density to the iron atom is another major key point.[11] Like in the native HRP in which an histidine has this function, it is assumed that in the G-quadruplex, one of the guanines of the external G-quartet is devoted to this, by flipping out of the plan (exactly like what was observed in another study with platinum complex binding to a G-quadruplex).[59, 60] Moreover, based on molecular modeling, it was also supposed that this effect could be due, in special cases, to a cytosine from a nearby loop, intercalated between the accessible G-quartet and the hemin, and creating π-

stacking interactions.[55] This insertion of the loop is clearly not a *sine qua non* condition, because several G-quadruplexes without loops and/or cytosine are effective as biotechnological catalysts (*e.g.*, d(($G_3T_n)_3G_3$) (n = 1–4),[54] d(($TG_4)_4$),[56] d(($T_4G_4)_4$),[23]or d(($T_4G_6T_4)_4$)).[61] Notwithstanding, their activities are less high than G-quadruplexes with loops, for which the catalytic efficiency can be sorted as follows: antiparallel ones > hybrid forms > parallel ones. [62] The presence of a polar environment and several H-bond donors and acceptors on the distal face of the hemin also constitute a positive point.[11] All these aforementioned criteria are directly linked to the inherent design of the DNA sequence,[63] but other experimental parameters can be used to modulate the pseudo-enzymatic activity of the G-quadruplex. Among all the conditions (non-exhaustive list), chemists can easily inflect the pH, the nature of the buffer, the nature and concentration of salts (*e.g.*, K^+ and Na^+ are important for the G-quadruplex structuration), the presence and nature of a surfactant (*e.g.*, Triton X-100, Tween 20, Brij 56), the temperature, and, last but not least, the adjunction of an "additive."[9, 50, 64–66] Indeed, some additional compounds can be used to amplify the response of the catalysis. In particular, the use of adenosine triphosphate (or ATP), and its derivatives, was studied by the groups of D.-M. Kong and D. Monchaud.[56, 67] Interestingly, the role of this small molecule is tricky, and Monchaud's team tried to decipher its actual role. In fact, ATP was proposed to favor several equilibria in the hemin oxidation/reduction process, as it was described in detail in 2012.[56] The four main effects of ATP are:

a. to facilitate the fixation of H_2O_2 on the iron atom of hemin,

b. to fuel the two one-electron transfer steps of the catalytic cycle,

c. to protect the reaction partners against oxidation, and

d. to modulate the pH of the reaction mixture, like what is observed in cells.[68]

3.4. Applications of G-quadruplex-based DNAzymes

The *horseradish peroxidase,* mimicked by G-quadruplex-based DNAzyme systems, is a very well-known protein in a biochemical context in which, grafted to a biomolecule (mainly an antibody), it is used to oxidize a colorimetric or fluorogenic substrate in ELISA (for *enzyme-linked immunosorbent assay*). This high-throughput assay is, for example, used for the clinical detection of anti-HIV antibodies from biological samples.[69, 70] Moreover, the *horseradish peroxidase* is also used for the purification of waste waters,[71] to develop biosensors, or as a reagent for organic synthesis. This popular enzyme, extracted from the root of the horseradish, is able to oxidize many substrates (*e.g.*, aromatics, phenols, indoles, amines) in the presence of hydrogen peroxide. Its properties were discovered by L. A. Planche in 1810, who noticed that a tincture of guaiac resin became blue when a piece of fresh horseradish root was placed in it. The chemical reaction occurred here is now expected to be due to the peroxidase-catalyzed oxidation of 2,5-di-(4-hydroxy-3-methoxyphenyl)-3,4-diphenylfuran to the corresponding bis-methylenequinone (also known as "guaiacum blue").[72]

Due to all the advantages described before concerning the use of DNA instead of protein to catalyze reactions, chemists have used for more than 15 years DNAzyme to develop and create plenty of new applications. Indeed, the far better modularity of the G-quadruplex-based peroxidase-mimicking systems constitutes a new powerful and invaluable tool for scientists, mainly for *in vitro* detection of a large range of molecules.

Trying to be as exhaustive as possible, a list of the principal applications found in the literature is detailed hereinafter.

a. Detection of monovalent cations like K^+[73–76] or Ag^+.[77–79] These corresponding methods are mainly based on the properties of monovalent cations to fold G-rich DNA strands to G-quadruplex. In other words, if the ion is present in the solution, the G-quadruplex is formed, and then it can interact with hemin, and subsequently, with the addition of H_2O_2, it is able to catalytically oxidize a colorimetric, such as ABTS (for 2,2′-azino-bis(3-ethylbenzothiazoline-6-sulfonic acid)[80, 81] or TMB (3,3′,5,5′-tetramethyl-benzidine)[82–84] or a fluorimetric probe.[85, 86]

b. Detection of bivalent cations like Cu^{2+},[87–91] with a very low detection limit of 1 nM using the DNAzyme system published by F. Wang *et al.* [92] Also detection of Hg^{2+}[92–98] and Pb^{2+}[96, 99–108] are well represented, in particular because they cause huge damage to human body through food, water, and air. Recently, a dual label-free sensor was developed by H. Li *et al.*, and it can be used to measure the concentration of these ions to the limit of 3.9 nM and 4.8 nM, for lead(II) and mercury(II), respectively.[89]

c. Detection of organic molecules. Among them, C. Yang *et al.* focused their research on the ochratoxin A, a toxin that can contaminate a large number of food commodities (*e.g.*, cereals, spices, coffee, milk).[109] The International Agency for Research on Cancer has classified this compound as a possible human carcinogen, and the European Commission has set the maximum level at 5 nM (equal to 2 μg.kg^{-1}) for grape-based beverages, like wine. In their publication, the authors proposed a powerful DNAzyme system with a detection limit of 2.5 nM, making it suitable for this analysis, respectful of the European standards. Moreover, the detection of glucose in urine by naked eyes can be performed using a sensor composed of both 23-mer G-quadruplex and glucose oxidase.[110] Thus, the DNAzyme system is coupled to a classic enzymatic reaction, and offers the opportunity to detect glucose from a concentration of 1 μM. Cocaine is also the focus of several studies, and DNAzymes have a high sensitivity up to 5 μM,[111] or 2.5 μM,[92] depending of the methodology used.

d. Detection of proteins, like nucleolines,[112] thrombin,[113–120] lysozymes,[113, 116, 121], and the vascular endothelial growth factor (VEGF), which plays a major role in the cancer development.[122] Also, Z. Ye *et al.* designed a DNAzyme system able to evaluate the concentration of the estrogen receptor alpha (ERα), a transcription growth factor involved in the hormone-dependent breath cancer development.[123] In the same way, a DNA-zyme-based immunosensor for the highly sensitive detection of the spore wall proteins of *Nosema bombycis*, from biological samples (*i.e.*, silkworm blood), is a promising bio-technological tool for the diagnosis of pébrine disease, a silkworm infection with significant economic impacts.[124] Furthermore, a 96-well plate assay is reported by Y. He *et al.* for the detection of leptin, a protein playing a key role as a "satiety hormone" and influencing basal metabolism, hematopoiesis, reproduction, and angiogenesis.[125] This method has a sensitivity up to 2 pg.mL^{-1}, 50 times better than the results obtained with ELISA (with a limit of 100 pg.mL^{-1}). Likewise, L. Stefan *et al.* reported the very first steps of a general strategy to make G-quadruplex DNA catalysts easily immobilizable, for the development of high-throughput ELISA-type assay.[126] In this study, a 96-well plate coated with streptavidin was functionalized using a cyclododecapeptide termed RAFT (for *regioselectivity addressable functionalized template*),[127] equipped on one side with a G-quadruplex (intra- or intermolecular), and with a biotin on the other side. The first step was the optimization of experimental conditions to access the best detection limit. In this case, it was found that for a catalysis limited to 2 hours (a fixed time decided by the authors to have a catalytic response in an acceptable amount of time), reactions have to be carried out in a cacodylic buffer (10 mM, with 10 mM KCl and 90 mM NaCl) at pH 4.4, with 1 μM hemin, 400 μM TMB, and finally 2 mM H_2O_2 to trigger the process. Using this protocol compatible with biochemical applications, the authors used only 2 pmol of the catalysis per well to detect streptavidin from a commercially available pre-coated microplate (300 pmol/well). This default of RAFT-quadruplex was chosen both to avoid unspecific associations and to limit the consumption of the catalyst. To propose scientists different ways to work with these kinds of DNAzyme systems, two protocols were developed. The more user-friendly one is a three-step procedure: first, a 200 μL solution of RAFT-quadruplex + hemin solubilized in the *ad hoc* buffer described before was poured in the wells; after an incubation time of half an hour, a washing step (using a cacodylic buffer at pH 7.2) was performed to remove all unbound materials; and finally, a solution, including the luminescent probe TMB, was put inside all the wells. Reactions started when hydrogen peroxide solutions were put inside. The variation of absorbance at 370 nm was monitored using a 96-plate UV–Vis reader with one measure every 2 minutes. This work highlighted the fact that surface-immobilized DNAzymes are interesting alternatives to develop practically convenient, simple, and rapid biophysical assays. Nevertheless, this work constitutes another brick in the wall of the development of effective DNAzyme-based assays.[126]

e. Detection of nucleic acids is also the target of several G-quadruplex DNAzyme process. Indeed, they are applied for the detection of single-strand DNA,[128] or other DNA analytes,[62] including the smart "DNA machine" developed by I. Willner's team in which its sensitivity is equal to 10^{-14} M of the target sequence (a 29-mer corresponding to a domain of single-strand DNA hepatitis B viral gene).[129] The detection of genetically modified organisms can also be done, as proved by B. Qiu *et al.* with both cauliflower mosaic virus 35S promoter, and lectin gene.[130] This approach was improved in 2014 to reach a detection limit of 5 nM.[131] Interestingly, a nonclassic fluorescence probe, the 2′,7′-dichlorodihydrofluorescein diacetate (or H_2DCFDA), was used instead of the usual ABTS, TMB, or luminol.

f. Moreover, DNAzyme is a powerful biotechnological tool to detect micro RNA (also termed mRNA),[132–134] particularly when it is coupled to a rolling circle amplification which allows a sensitivity of 0.3 fM,[135] or as "DNAzyme Ferris wheel" like nanostructures, as proposed by W. Zhou *et al.* in 2015,[136] with a detection limit of 5×10^{-13} M for miR-141 (a biological marker of the human prostate cancer) by naked eyes. Furthermore, scientists created DNAzyme sensors to detect small nucleotides like adenosine triphosphate (or ATP),[137–140] cyclic diguanilate c-di-GMP,[141] and also 8-OHdG (for 8-hydroxy-2′-deoxyguanosine) from the urine, that is associated with various cancers, diabetes, and neurological diseases.[142]

g. The modification of one nucleotide inside a DNA sequence can also be verified using G-quadruplex-based DNAzyme. Actually, the method published by M. Deng *et al.* is able to detect the substitution of one DNA base in a 39-mer strand.[143] In parallel, I. Willner's team proposed an alternative method using functionalized magnetic particles,[144] whereas in 2015, D. Verga *et al.* described a smart methodology to discriminate by naked eyes a single nucleotide variation.[145] Enzymatic activities were also investigated in the presence of DNAzyme-based sensors. It was the case for the methyltransferase,[146] cholesterol oxidase,[147] glucose oxidase,[148] or telomerase.[149–151] This last example is of great interest because telomerase, overexpressed in a tumoral context, constitutes a high-potential biological marker for the diagnosis of cancers.[152, 153] The first DNAzyme method to detect it from cellular lysates was proposed by R. Freeman *et al.*, and is based on the concept that only cellular extracts containing telomerase (*i.e.*, the cancer cells) can elongate a primer.[154] Thus, the longer G-rich DNA strand, composed of several specific $(TTAGGG)_n$ repeats, is able to form a G-quadruplex that can act as the catalyst of a DNAzyme process. Consequently, an "ON" signal, due to the catalytic oxidation of the colorless ABTS to the corresponding green oxidized product, is the proof of the telomerase activity, that is, that the lysate was from a cancer cell. Conversely, healthy cells contain nonactive telomerase, and the corresponding lysate has consequently no effect on the primer. Too short to form a G-quadruplex, the DNAzyme catalysis cannot occur, leading to no change of the media color/absorbance: it is the "OFF" signal. Inspired by this work, the D. Monchaud's team decided to use additives (*i.e.*, template-assembled synthetic G-quartet) to improve the detection limit of G-quadruplex composed of a repeat of the $(TTAGGG)_n$ motif.[155] This research will be discussed later in this chapter. Another work from L.-J. Wang is based on the same concept as the Freeman's one, but using a 93-mer nucleotide telomerase substrate primer which leads to a detection limit as low as 0.1 aM ($= 10^{-18}$) able to detect telomerase activity from HeLa cells.[156]

h. Detection of antibodies to develop ELISA-type immunoassays[157, 158] or for immunohistochemistry assays are also extremely promising. G-quadruplex DNAzyme enables the visualization of the prostate-specific antigen (PSA), a high-potential tumor marker, directly in solid tissue sections.[159] In parallel, an assembly of antibody/gold nanoparticle/DNAzyme system was develop by M. Shi *et al.* in 2014[160] and is used to quantify one of the most important carbohydrate tumor marker, the antigen CA19-9, from human serum samples. The detection limit obtained was 0.016 $U.mL^{-1}$, which is the lowest one reported in the literature. Using a close approach with a G-rich DNA sequence grafted on the gold nanoparticles surface, a new DNAzyme biosensor was proposed as a direct antigen–antibody detection assay.[161]

i. Interestingly, the DNAzyme methodology was also used to detect bigger living systems, like the bacteria *Escherichia coli* O157:H7[162] or *Alicyclobacillus acidoterrestris*.[163] In the first example, authors used graphene oxide/thionine/gold nanoparticles coated SiO_2 nanocomposites to immobilize DNA, while for the second one, a more classic approach was used, in which G-quadruplex–hemin complexes oxidize the colorless guaiacol, produced by the bacteria, to tetraguaiacol, which is amber.

j. Developed to target G-quadruplexes *in vivo*, *in cellulo*, or *in vitro* for biological applications, G-quadruplex ligands are molecules able to interact with G-quadruplexes and to stabilize them.[164–166] This ability can be evaluated, thanks to the DNAzyme process, because of a competition between the ligands researchers want to try, and the hemin. In other words, a good ligand takes the place of the hemin; hemin is subsequently not activated and, consequently, leads to a decrease of the signal intensity (measured by UV–Vis or by fluorescence).[61, 167–169]

k. To finish this laundry list, it is essential to mention other ingenious applications for the G-quadruplex DNAzymes, like the development of logic gates[170–173] as the INHIBIT one published by T. Li *et al.*,[174] or as the AND one proposed by J. Chen *et al.*, described as a keypad lock security system.[175]

To summarize, all these cases, which represent the range from the more applied to the more conceptual scientific applications of the same DNAzyme catalysis, illustrate how using DNA instead of enzyme to catalyze a reaction puts out a new avenue in terms of polyvalence. It is believed that this list will increase more and more in the next years. But the precise understanding of the mechanism constitutes also an exciting challenge. On the one hand, this progress should offer scientists the possibility to fine-tune the experimental conditions (*e.g.*, sequence of the G-quadruplex DNA strand(s), length of the loop(s), addition of a boosting agent, etc.). On the other hand, a better comprehension should help to enlighten chemists about the mechanistic aspect of the oxidation states of the hemin, in both biological and DNAzyme systems.

4. Synthetic G-quartet-based DNAzymes: template-assembled synthetic G-quartets (or TASQ)

4.1. Concept and structure of TASQ

Based on the idea that G-quadruplexes, mimicking the natural *horseradish peroxidase*, lead to an increase of the range of applications, mainly due to the higher stability of the DNA compared to the protein and permitting a use in a bigger range of experimental conditions (temperature, buffer, ion strength, etc.),[7, 23, 176] few research teams decided to develop G-quadruplex-mimicking systems.

Because the pivotal step of the catalytic cycle is based on the activation of hemin by interaction with one of the external G-quartet,[11] the group of Dr. D. Monchaud decided to synthesize the very first example of a water-soluble molecule, composed of four guanine residues, and able to form, intramolecularly, a synthetic G-quartet.[14]

Historically, the very first observation of synthetic G-quartets was made by I. Bang in 1910, who was able to form gels from a concentrate solution of guanosine monophosphate. However, the hypothesis of the self-assembly of guanine derivatives to G-quartet arrangements was only published more than half a century later by M. Gellert, M. Lipsett, and D. Davies in 1962. The dried fibers obtained could be analyzed by X-ray diffraction and helped the authors to propose the initial supramolecular structure shown in Figure 1.[177] Interestingly, a recent study proved that the length of the fibers can reach from 8 nm to 30 nm, corresponding to from 24 to 87 stacked G-quartets, respectively.[178]

During the last decades, plenty of examples using synthetic G-quartets as a supramolecular motif were developed and well summarized in several reviews.[10, 179, 180] The applications of these systems concern pH-sensitive hydrogel probes,[181] synthetic transmembrane Na^+ transporter,[182] and other ionic channels,[183–185] enantioselective systems controlled by the cation used,[186] combinatorial chemistry,[187] and also molecular electronics and liquid crystals.[179] Among all these examples, in which the elementary brick is made of one or two guanines, only few stem from a four-guanine-based compound. This assessment is inquisitive and nonintuitive because G-quartets are composed of four guanines. It was why the team of J. Davis developed in 2000 and 2003 1,3-alternate calix[4]arene derivatives functionalized by four guanines.[188, 189] This smart system in which two guanines are on one side, while the two others are on the other side of the calixarene template, was used as both cation (inside the G-quartet) and anion (thanks to the H-bonds between protons of the amide groups and the anion) receptors.[188] The formation of the G-quartet is intermolecular between two guanines from one molecule, and two guanines from another one. It took the scientific community until 2008 to propose the first intramolecular synthetic G-quartet molecules, termed TASQ.

The name "TASQ," for *template-assembled synthetic G-quartet*, was introduced by the research group of J. C. Sherman to describe molecules built around a template and functionalized by four guanines, able to interact each other to self-assemble into an intramolecular G-quartet.[12] This concept probably derived from a modification of the TASP (for *template-assembled synthetic peptide*) development,[190] in which four peptide sequences were used instead of DNA bases. The aim of these models was the understanding of protein interactions, thanks to their spatial proximity when grafted to the same scaffold.

Thus, the first TASQ were synthesized from a highly lipophilic calixarene moiety substituted by four 2′,3′-*O*-isopropylideneguanosine.[12] The intramolecular formation of the G-quartet was demonstrated by ^{1}H NMR, NOESY, COSY, and HMQC and was definitively proved by X-ray diffraction in 2012.[191] Notwithstanding, no application of these systems was published, and their hydrophobic properties were probably the reason for that. Interestingly, they rectified this point using first phosphate groups to functionalize the calixarene on the opposite side of the guanosines[192] and then, subsequently, with the phosphate group intercalating between the guanine moieties and the scaffold (due to the use of 3′-monophosphate guanosines).[193]

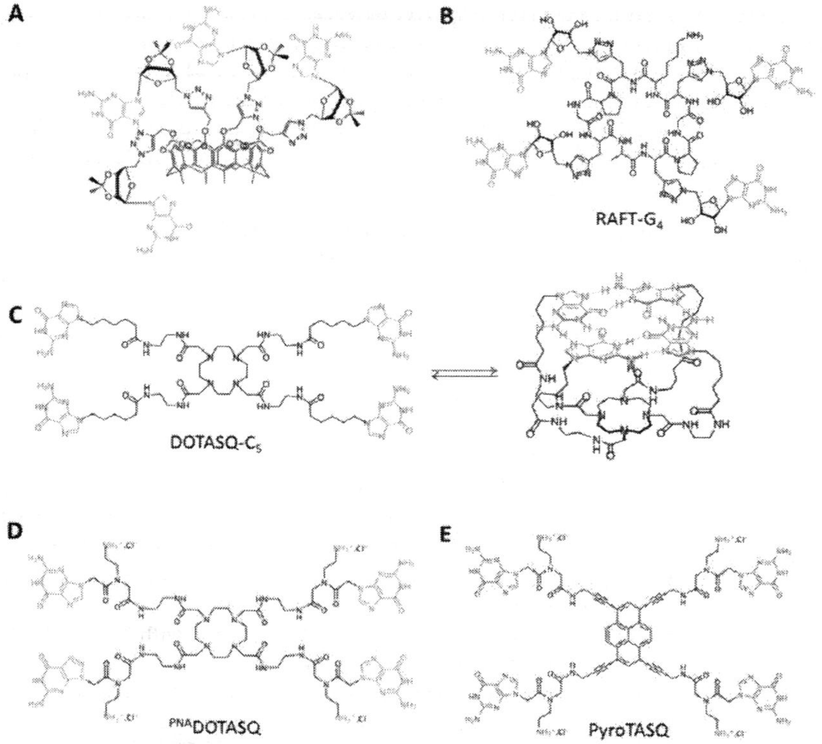

Figure 3. Chemical structures of the first TASQ developed by J. C. Sherman (**A**), [12] RAFT-G_4 (**B**), [13] DOTASQ-C_5 in equilibrium between its open (left) and closed conformations (right) (**C**), [14] ^{PNA}DOTASQ (**D**), [15] and PyroTASQ (E). [16]

In parallel, the group of E. Defrancq was focused on the functionalization of a cyclodecapeptide termed RAFT (for *regioselectivity addressable functionalized template*) by DNA strands.[127, 194–196] This research led the team to synthesize a TASQ from a RAFT equipped with four guanosines.[13] Thanks to a collaboration with D. Monchaud's team, this molecule was studied as a DNAzyme (see section 4.2) and showed that the G-quartet intramolecular formation (confirmed by circular dichroism, and NMR studies) was able to catalyze the hemin oxidation/reduction cyclic reaction.[197]

However, the very first example of a water-soluble TASQ was proposed by L. Stefan *et al.* in 2011.[14] The key point was the choice of the *ad hoc* template, able to drive the G-quadruplex formation, and being fully soluble in water. It was why the cyclen macrocycle was chosen, for its high solubility, its ability of metal chelation, and its C_4-type symmetry, identical to the one of a G-quartet.[198, 199] Even if this last criteria was not a *sine qua non* condition, it seemed to be favorable, like in the calixarene-templated TASQ developed by J. C. Sherman's group. Thus, DOTASQ were born from a DOTA (for 1,4,7,11-tetraazacyclododecane-*N,N′,N″,N‴*-tetraacetic acid) moiety in which the four "arms" were functionalized by four alkylguanine groups. The term DOTASQ is a portmanteau word created from both acronyms DOTA (*i.e.*, the template) and TASQ (*i.e.*, the supramolecular property of the molecule). For the sake of comparison, four DOTASQ were synthesized: DOTASQ-C_1 and DOTASQ-C_5 (C_1 and C_5 suffixes indicating the length of the alkyl chain), and the terbium(III) equivalents Tb.DOTASQ-C_1 and Tb.DOTASQ-C_5, respectively.[14]

Furthermore, another TASQ was developed with peptide nucleic acid guanine (also termed PNA guanines) arms and was called ^{PNA}DOTASQ.[15]

The peroxidase-mimicking catalytic activities of these compounds were evaluated and are described in the next sections. As a remark, other TASQ were developed by D. Monchaud's team like PyroTASQ (pyrene as a template),[16] NaphtoTASQ (naphthalene as a template) [200], and also PorphySQ and ^{PNA}PorphySQ (porphyrin moieties as templates).[201, 202]

4.2. Catalytic activities of synthetic G-quartets

The formation of a synthetic G-quartet was expected to have the same properties as a native one from a G-quadruplex. Indeed, this synthetic G-quartet is able to mimic the external (also termed accessible) G-quartet of the biological edifice.[12–16, 201] In this section, the main experimental data and results are highlighted to prove the feasibility of this new strategy of nature-mimicking catalysts. To begin with, the very first example of this approach, published by D. Monchaud's team, is presented and then the improvement with ^{PNA}DOTASQ is explained. A collaboration between this group and the E. Defrancq's one led to the study of the RAFT-G_4, introduced in the previous section (see section 4.1), and detailed here.

In parallel, the work of H. O. Sintim's group, focused on intermolecular G-quartets (*i.e.*, not TASQ) using c-di-GMP (for cyclic diguanylic acid) as a catalyst, will be presented.[141, 203, 204]

DOTASQs as a pre-catalyst: As explained before, the main step of the catalytic reaction is the activation of the hemin by interaction with the G-quartet. To verify this *sine qua non* condition, titrations of 1 μM hemin with DOTASQ were performed in cacodylic buffer.[205] The characteristic UV–Vis band of hemin from 350 nm to 400 nm was observed and, interestingly, the signal increased in all the cases with DOTASQ-C_1, DOTASQ-C_5, and 22AG, the telomeric G-quadruplex sequence d[$AG_3(T_2AG_3)_3$] used as a reference. Moreover, for the three systems, a stoichiometry of about 1:1 was found, and the dissociation constants were calculated using the following equation:

$$K_d = \frac{[\text{hemin}][\text{cat.}]}{[\text{hemin} + \text{cat}]} \qquad (1)$$

Thus, the subsequent K_d were found: 170 nM for DOTASQ-C_1, 135 nM for DOTASQ-C_5, and 235 nM for the native G-quadruplex 22AG. These results were duplicated with other concentrations of hemin, and similar values were obtained. They show that both native and synthetic G-quartets are able to interact with hemin, forming the key step of the peroxidation reaction. Inspired by the optimal experimental conditions published by P. Travascio *et al.*, and developed for G-quadruplexes,[50] primitive experiments were carried out with 1 μM hemin, 2 mM ABTS, 600 μM H_2O_2, and from 0 μM to 50 μM DOTASQ, in Caco.KTD buffer. This buffer designated a cacodylic acid buffer composed of 10 mM lithium cacodylate (Caco), 10 mM KCl (letter K), 90 mM LiCl, with addition of 0.1 % (v/v) DMSO (letter D) favoring the solubilization of hemin, and 0.05 % (w/v) Triton X-100 (letter T), a nonionic surfactant promoting disaggregation of hemin.

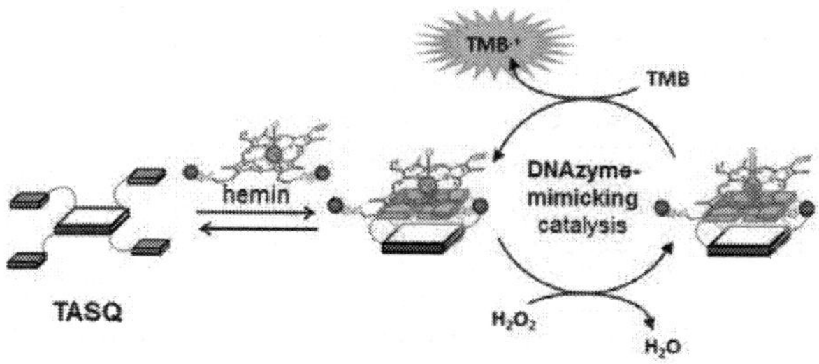

Figure 4. Schematic representation of the peroxidase-like activity promoted by the use of a TASQ. [205, 206]

The catalytic activity was therefore evaluated by UV–Vis absorbance, in 1-mL-quartz cuvettes, measuring the formation of $ABTS^{\cdot+}$ (also proposed as $ABTS^{\cdot-}$ in the scientific literature), the oxidized product of ABTS which absorbs at 420 nm.[80, 81] All the results were compared to a control experiment, strictly composed of all the same reagents, except the lack of the pre-catalyst (*i.e.*, the hemin is alone with ABTS and hydrogen peroxide, and not catalyzed by G-quadruplexes or TASQs). After less than 1 hour, all the absorbance signals were constant and proved the feasibility of the TASQ-catalyzed peroxidation reaction concept.

However, to confirm that the catalysis of hemin was due to the formation of the intramolecular G-quartet, a DOTASQ-C_5 derivative, termed Prot.DOTASQ-C_5, was synthesized. In fact, Prot.DOTASQ-C_5 has "protected guanines" with 6-*O*-benzyl groups in the four guanine moieties. Thus, the formation of G-quartet is impossible because of the rupture of the H-bonds involving the carboxylic group in position 6. The catalytic activity of this compound is null, confirming that the formation of an intramolecular G-quartet inside the TASQ is mandatory to activate hemin and to catalyze the reaction.[205]

For the sake of comparison of the efficiency of the DOTASQ, apparent rate constants k_{cat} were calculated, dividing the initial rate (V_0) by the concentration of the catalyst ([cat.]), using Eq (2).

$$k_{cat} = \frac{V_0}{[\text{cat.}]} = \frac{\Delta Abs_i}{\varepsilon_{ABTS^{\cdot+}}} \times \frac{1}{[\text{cat.}]} \quad (2)$$

Thus, k_{cat} of 0.36 h^{-1} and 0.29 h^{-1} were obtained for DOTASQ-C_5 and DOTASQ-C_1, respectively. For structural reasons, DOTASQ-C_5 is able to form its intramolecular G-quartet easier than DOTASQ-C_1. This better stability of the G-quartet favors the π-stacking of hemin and, consequently, the catalytic activity.

To compare TASQ with G-quadruplexes, a constant of k_{cat} = 9.71 h^{-1} was got for 22AG, which is 25 times more efficient than DOTASQ-C_5. This difference seems to be curious because the K_d of DOTASQ and 22AG with hemin were close. A similar assessment was reported in the literature but with opposite results, with PS2.M and the RNA equivalent rPS2.M.[51] Indeed, their catalytic activities were similar, although the K_d values were 27 nM and 900 nM, respectively.

It can be hypothesized that, even if the key step was defined as the hemin/G-quartet interaction, the environment around the G-quartet plays also a pivotal role for the catalytic efficiency. The presence of loops, the accessibility of the G-quartet to hemin, and other factors described before (see section 3.3) are critical for all the steps of the catalysis.[11, 56]

Finally, to eliminate all the doubts, another substrate was used instead of ABTS, and TMB was chosen. Its oxidation is a two-step reaction producing first a charge transfer complex, with absorbance at 370 nm and 652 nm, and a final diimine product absorbing at 450 nm.[82–84] Using different experimental conditions than before, the catalytic activity of 50 μM DOTASQ-C_5 was evaluated in the presence of 1 μM hemin, 500 μM TMB, and 1.5 mM H_2O_2 in Caco.KTD buffer, and were compared with the same reaction with 2 μM 22AG. Like with ABTS, experiments with TMB approved the ability of TASQ to catalyze peroxidase-like catalysis. Obtained apparent catalytic constants were 0.02 h^{-1} for DOTASQ-C_5 against 0.27 h^{-1} for 22AG.[205] Altogether, these results constitute the proof of concept of the use of native and synthetic G-quartets as a universal platform for the peroxidation reactions. However, efficiency had to be improved, and better results were obtained with the second generation of TASQ: the ^{PNA}DOTASQ.

^{PNA}DOTASQ as a pre-catalyst: The main structural modifications made from DOTASQ to ^{PNA}DOTASQ were the substitution of the original alkyl-arms by PNA guanine moieties. Thus, the new properties worn by this new TASQ are numerous and not detailed here (this molecule was also developed as a smart G-quadruplex ligand),[15] but the main point which must be highlighted here is the introduction of a total of four cationic charges, thanks to the presence of four pendant primary amine side chains. These positive charges (at physiological or acidic pH) are of the utmost importance to increase the interaction with hemin. Indeed, hemin is an Fe(III)–porphyrin holding two anionic charges due to two carboxylic groups. Electrostatic interactions were expected to facilitate the hemin approach and association with the G-quartet.

To verify this hypothesis, R. Haudecoeur *et al.* performed the catalytic reaction with TMB, using the same protocol described before, from 0 μM to 50 μM ^{PNA}DOTASQ. For the sake of comparison, experiments with the same range of concentration were carried out with DOTASQ-C_5 (*i.e.*, the most efficient DOTASQ) and also with 22AG at 2 μM as a reference.[206]

Interestingly, the results showed the far better catalytic ability of ^{PNA}DOTASQ compared to the DOTASQ. Indeed, the initial rates were evaluated at 1.25 μM.min^{-1} and 0.05 μM.min^{-1} at

50 μM, respectively, corresponding to an average increase of efficiency of 2600 %. At this stage, ^{PNA}DOTASQ is the more active TASQ catalyst for peroxidase-like reactions, and is closer to the activity of G-quadruplex, with a factor of 11 between it and 22AG (*i.e.*, in the experiments, 22 μM ^{PNA}DOTASQ offers the same catalytic response than 2 μM of the natural G-quadruplex). ^{PNA}DOTASQ confirms the role of intramolecular G-quartets to activate hemin and to perform DNAzyme-like catalysis, and also that improvement of the efficiency is just at the very beginning of the development. The modification of the arms is a positive point, but the role of the template, directly linked to the ability of synthetic G-quartet formation, is by consequence indirectly linked to the catalytic properties.

RAFT-G_4 as a pre-catalyst: To evaluate the role of the template on the DNAzyme-like activity, the TASQ developed by E. Defrancq's team was tested.[197] Termed RAFT-G_4, the molecule is composed of a cyclodecapeptide and four guanosine arms in which one triazole per arm is intercalated between the ribose and the peptide scaffold (playing a role in the stabilization of the intramolecular G-quartet).[13] As a first step, UV–Vis titrations of hemin by RAFT-G_4 were performed to check the interaction, and compared with DOTASQ-C_5. Interestingly, very similar signals shifted from the signal of hemin alone were obtained in the presence of both TASQ for the two peaks at 363 nm and 394 nm, corresponding to the aggregated and disaggregated forms of the hemin (*i.e.*, catalytically active), respectively.[50]

Afterwards, catalytic experiments were carried out with 1 μM hemin, 2 mM ABTS, and 600 μM H_2O_2 in Caco.KTD buffer, with concentrations of DOTASQ-C_5 and RAFT-G_4 from 10 mM to 100 mM. Positively, the results revealed the ability of RAFT to be a catalyst, even if the efficiency was lower than for the DOTASQ-C_5. To optimize the catalysis, the use of 10 mM ATP in Caco.KTD buffer at pH 4.8 (protocol based on previous studies with G-quadruplexes) was performed and showed the positive effect of these modifications of the experimental conditions. More precisely, this optimization led to an improvement of DOTASQ and RAFT-G_4 efficiency by factors 1.8 and 5.1, respectively.[197]

These series of experiments triggered two main conclusions. First, the ability of a TASQ to be a good catalyst is dependent on the template, the nature of the guanine arms, and the stability of the G-quartet, and it seems to be for now difficult to rationalize their impacts. The multiplication of examples of TASQ in the literature will be an invaluable chance to understand the role of each structural part of the molecules. Author wagers that new molecules like PyroTASQ[16] or NaphtoTASQ[200] will help to decipher a little more the reasons of these differences.

Another nice synthetic G-quartet system used as DNAzyme-like catalyst was developed by H. O. Sintim's team.[141] In this case, the molecule is not a TASQ forming an intramolecular G-quartet, but the cyclic diguanylic acid (also termed c-di-GMP), composed of two guanine residues. The authors proposed that c-di-GMP are able to form a discrete G-quadruplex formed by two intermolecular G-quartets (*i.e.*, from four c-di-GMP molecules) at the micromolar level, but with the *sine qua non* presence of the intercalating proflavin molecule. In this case, it can be considered that proflavin assumes the role of a non-covalently linked template, unlike in TASQ. Experiments carried out with 0.5 μM hemin, 30 μM proflavin, 2 mM ABTS, and 2 mM H_2O_2 in Tris-HCl buffer (50 mM, pH 7.9) revealed that a catalytic activity can be distinguished for concentrations of c-di-GMP higher than 2 μM. Several control experiments were performed and clearly highlighted the fact that the formation of intermolecular G-quartets is required to observe catalytic activities.

All these data firmly confirm the role of synthetic G-quartets, both intramolecular and intermolecular, as efficient and promising nature-mimicking systems. In spite of the increasing excitation around this field, it is nevertheless honest to confess that the use of TASQ as catalysts for peroxidase-like experiments is, for the moment, far from the success of the DNAzyme, mainly in term of applications.[7, 23, 24] However, the very first ones were developed and are presented hereinafter.

4.3. Applications of synthetic G-quartets

One of the main topics of natural G-quartets is the biological role of G-quadruplex structures, found in many sequences in the human genome (estimated at about 350.000 sequences).[207, 208] This alternative higher-order DNA structure (*i.e.*, a noncanonical double-helix one)[209] is strongly suspected to play important roles in key cellular events, like chromosomal instability or regulation of gene expression.[26, 27, 42–44] Notably, the extremities of chromosomes, termed telomeres, are composed of a G-overhang strand with several $(TTAGGG)_n$ repeats from an average of 100–200 bases.[26, 27, 210] Interestingly, this single strand (stabilized by a protein complex named shelterin) is the substrate for a key enzyme, the telomerase.[211–213] Although this field is of particular interest and extremely exciting, it is far from the scope of this chapter, and author recommends some excellent reviews to go further insight. In few words, telomerase is able to synthesize TTAGGG repeats, leading to the elongation of the telomeres. It is however only active in a vast majority of cancer cells (85 % of the tested cell lines) but inactive in somatic healthy cells.[214] Telomerase is thus considered as a cancer marker with a strong therapeutic potential.[152, 153] Consequently, direct or indirect evaluation of the activity of this enzyme from cell lysates could be of the utmost importance for the tumor diagnosis.

Based on the DNAzyme technology, R. Freeman *et al.* developed an assay to detect telomerase activity in an indirect way. If it is active, a primer is elongated to long $(TTAGGG)_n$ strands able to form G-quadruplexes.[154] However, long single-strand sequences can form multimers, [215] composed of several G-quadruplex structures, with a "stacking"[216–218] or a "beads-on-a-string"[219, 220] global structure. In the first case, several G-quadruplexes from the same strand are stacked together, while in the second case, G-quadruplexes are independent, without any interactions between them.

A study from L. Stefan *et al.* [155] showed that mainly for a 46-base long strand ($d[AG_3(T_2AG_3)_7]$ composed of two G-quadruplexes separated only by three nucleobases), two different kinds of G-quartets able to activate hemin can be defined. Indeed, the first accessible G-quartets are the external ones, and are identical to the ones found in a single G-quadruplex. However, the second hemin interaction site is at the interface between two G-quadruplexes. The teams of L. Petraccone and H. Sugiyama demonstrated that this "pocket" (composed of two G-quartets) is more hydrophobic than the classic external G-quartet and favors therefore the interactions with hydrophobic organic molecules, like hemin.[221, 222] Interestingly, this internal binding site clearly has a positive effect on the hemin binding and on its catalytic activity. The efficiency of the DNAzyme response is due to the hydrophobic properties of this internal site that favors hemin fixation and its protection against degradation. Moreover, once stacked to an internal G-quartet, the distal face of the Fe(III)–porphyrin is "through contact with" the second G-quartet that can play a key role during the oxidation/reduction process of the iron during the catalysis (*e.g.*, mimicking the action of histidine like in the natural proteins, favoring H_2O_2 deprotonation).[11] This hydrophobic site, composed of two native G-quartets able to "sandwich" the hemin, is thus more active than only one G-quartet. The increase of activity leads surely to a better detection limit.

With this in mind, it was decided by D. Monchaud's group to create an artificial high-activity hemin binding site to improve the detection of telomeric G-quadruplex sequences, thanks to the use of TASQ. For the detection of 22AG, the addition of TASQ was expected to modify the characteristics of one or two of the external sites, to one or two pseudo-internal sites. Indeed, thanks to a like-likes-like process, TASQ interact with G-quadruplexes *via* a synthetic G-quartet/native G-quartet recognition, identical to a classic G-quartet/G-quartet interaction. Thus, the "binding pocket" created between the external native G-quartet of the G-quadruplex and the intramolecular synthetic G-quartet of a TASQ is an artificial high-activity hemin binding site.[155]

To verify it, a DNAzyme experiment was carried out in Caco.KTD in a 96-well plate with 1 μM hemin, 2 mM ABTS, and 600 μM H_2O_2, in the presence or absence of 50 μM DOTASQ-C_5, with different concentrations of 22AG (from 65 nM to 8 μM). Results highlighted the fact that the catalysis is more efficient in the presence of TASQ, and offer the possibility to decrease the detection limit from 4 μM to 500 nM. This improvement confirms the concept of the pseudo-internal high-activity hemin binding site that can be considered as an equivalent of the "binding pocket" of natural enzymes. In term of initial rates, whereas the optimal concentration of DOTASQ-C_5 can increase it by factor 2.7 at 40 equivalents, ^{PNA}DOTASQ was able to double it only at one equivalent.[155]

To conclude, TASQ can be used as "boosters" of the catalytic activity of DNAzyme, and permit to improve the detection limit by speeding up the rate of oxidation of the substrate (*e.g.*, ABTS or TMB). This effect could be very interesting to detect smaller concentrations of G-quadruplexes, in particular of telomeric G-quadruplexes, to determine with a better signal-to-noise ratio telomerase concentrations from cell lysates.

Another application detailed here is the development of a DNAzyme-mimicking system to detect the bacterial signaling molecule c-di-GMP, published by H. O. Sintim *et al.*[141] The detection of this molecule, able to form biofilms in several clinical relevant bacterial pathogens, is crucial to limit hospital infections. Interestingly, the target molecule is also, by itself, the catalyst of the peroxidation reaction (see section 4.2), because of its ability to self-assemble to form G-quartets. This method was validated with *E. coli* overexpressing a diguanylate cyclase WspRD70E from crude bacterial lysates.

The last but not the least application is the use of TASQ to create fully synthetic process able to mimic nature. Indeed, from a fully natural process with the *horseradish peroxidase*, DNA strands and hemin are still natural products in DNAzyme. However, in TASQ-based catalysis, only hemin is natural, because organic synthetic molecules (*i.e.*, TASQ) replace DNA. To pursue the evolution, another extremely soluble Fe(III)–porphyrin, the 5,10,15,20-tetrakis(4-sulfonatophenyl)porphyrin (also termed FeTPPS), was used instead of hemin. This molecule was well known for its excellent water solubility, its stability, its resistance to highly oxidative conditions, and known to perform H_2O_2-mediated oxidations for more than two decades.[223–225] After a first titration of the FeTPPS with DOTASQ-C_5, the catalytic experiments were carried out in Caco.K (without Triton X-100 and DMSO, because of the excellent solubility of this porphyrin compared to hemin). The same protocol was used in the presence of ABTS, and results showed an initial rate of 0.36 h^{-1} for DOTASQ-C_5, while no activity was detected with 22AG. This point must be due to electrostatic repulsions between the four negative charges of the sulfonate derivative and the negatively charged phosphate groups of the DNA strand. In sum, the use of FeTPPS with a TASQ was the last step to create a fully synthetic process mimicking a well-known natural process of peroxidation.[205]

5. Conclusion

As demonstrated all along this chapter, native and synthetic G-quartets are powerful catalysts for peroxidase-like process.

On the one hand, the DNAzyme field is fed by the numerous examples of G-quadruplexes used as a native catalytic platform, that found dozens of applications, from the detection of products in biological samples, to the evaluation of heavy metal concentrations, proteins activity, or to develop logic gates for new DNA-based nanotechnologies.[7, 23, 24] G-quadruplex DNA are extremely versatile structures that can be folded from plenty of sequences, in several media, and their functionalization to add probes, functional groups, or to graft them on a solid support are important and invaluable advantages. The design of the G-quadruplex structures, that is, of the catalyst, is far easier than for enzymes, because small modifications of a protein commonly lead to a modification of the active site and then, consequently, to a partial or total loss of activity. Indeed, all the G-quadruplexes are virtually able to catalyze peroxidase-mimicking reactions, because the key part of these noncanonical structures is one of the external G-quartet which is, by definition, the basic unit of G-quadruplexes.[10, 180] As a result, these easy-to-use DNAzyme systems are ready to be applied in chemistry and biology laboratories, mainly for their adaptability and stability, but also in the medical field in which DNAzyme can be considered as a cheaper alternative to the natural *horseradish peroxidase*, mainly to tag relevant biomolecules, like antibodies in ELISA protocols.

On the other hand, the only use of the minimal catalytically active part of the G-quadruplexes, the G-quartet, was presented in this chapter, mainly thanks to the use of TASQ. These template-assembled synthetic G-quartets, able to form an intramolecular G-quartet, proved that a small synthetic molecule can selectively interact with hemin to catalyze peroxidase-mimicking reactions. Interestingly, all the TASQ highlighted here (DOTASQ,[14, 205] PNADOTASQ, [15, 206] or RAFT-G_4)[13, 197]) offer different activities, and the best edifice is definitely PNADOTASQ, which is closer to the efficiency of G-quadruplex-based DNAzymes, even if a decrease by a factor of 10, approximately, is observed. However, the use of synthetic molecules instead of natural structures (*i.e.*, enzymes or DNA) is an undeniable advantage, in particular because these TASQ are easily synthesized in four straightforward steps from commercially available products, with good yields. Their modifications and improvements are only limited by imagination, and the new molecules designed by D. Monchaud's team clearly demonstrate this point.[16, 200] Interestingly, a fully synthetic system based only on nonnatural components was described in this chapter for its ability to reproduce the natural enzymatic process. In other words, TASQ permit to mimic a natural catalysis originally made by the *horseradish peroxidase* with only synthetic molecules made by chemists.[55] It can be postulated that the mechanism behind this reaction is probably very close to the natural one, even if more data are needed to confirm this proposition.

To conclude, this chapter showed the role of nature-mimicking catalytic systems, using noncanonical DNA G-quadruplex structures as native G-quartets, or synthetic G-quartets with TASQ. Even if the efficiency of these systems is for now not as high as the natural *horseradish peroxidase*, it is weighted against plenty of advantages in terms of applications, experimental conditions, versatility, and chemical modifications. Step by step, the scientific community puts new bricks in the wall and paves the way to more efficient nature-mimicking catalytic systems, closer and closer to what nature is able to do.

Author details

Stefan Loic*

Address all correspondence to: dr.stefanloic@gmail.com

Polytechnic School of Milan, Italy

References

[1] Kruger K., Grabowski P. J., Zaug A. J., Sands J., Gottschling D. E., Cech T. R. Self-splicing RNA: Autoexcision and autocyclization of the ribosomal RNA intervening sequence of tetrahymena. Cell. 1982;31(1):147–157. DOI: 10.1016/0092-8674(82)90414-7

[2] Guerrier-Takada C., Gardiner K., Marsh T., Pace N., Altman S. The RNA moiety of ribonuclease P is the catalytic subunit of the enzyme. Cell. 1983;35(3):849–857. DOI: 10.1016/0092-8674(83)90117-4

[3] Altman S. Enzymatic cleavage of RNA by RNA (Nobel Lecture). Angewandte Chemie International Edition in English. 1990;29(7):749–758. DOI: 10.1002/anie.199007491

[4] Cech T. R. Self-splicing and enzymatic activity of an intervening sequence RNA from tetrahymena (Nobel Lecture). Angewandte Chemie International Edition in English. 1990;29(7):759–768. DOI: 10.1002/anie.199007591

[5] Talini G., Branciamore S., Gallori E. Ribozymes: Flexible molecular devices at work. Biochimie. 2011;93(11):1998–2005. DOI 10.1016/j.biochi.2011.06.026

[6] Robertson M. P., Joyce G. F. The origins of the RNA world. Cold Spring Harbor Perspectives in Biology. 2012;4(4):1–22. DOI: 10.1101/cshperspect.a003608

[7] Silverman S. K. DNA as a versatile chemical component for catalysis, encoding, and stereocontrol. Angewandte Chemie International Edition. 2010;49(40):7180–7201. DOI: 10.1002/anie.200906345

[8] Krishnan Y., Simmel F. C. Nucleic acid based molecular devices. Angewandte Chemie International Edition. 2011;50(14):3124–3156. DOI: 10.1002/anie.200907223

[9] Kong D.-M. Factors influencing the performance of G-quadruplex DNAzyme-based sensors. Methods. 2013;64(3):199–204. DOI: 10.1016/j.ymeth.2013.07.013

[10] Davis J. T., Spada G. P. Supramolecular architectures generated by self-assembly of guanosine derivatives. Chemical Society Reviews. 2007;36(2):296–313. DOI: 10.1039/B600282J

[11] Sen D., Poon L. C. H. RNA and DNA complexes with hemin [Fe(III) heme] are efficient peroxidases and peroxygenases: How do they do it and what does it mean? Critical Reviews in Biochemistry and Molecular Biology. 2011;46(6):478–492. DOI: 10.3109/10409238.2011.618220

[12] Nikan M., Sherman J. C. Template-assembled synthetic G-quartets (TASQs). Angewandte Chemie International Edition. 2008;47(26):4900–4902. DOI: 10.1002/anie.200704199

[13] Murat P., Gennaro B., Garcia J., Spinelli N., Dumy P., Defrancq E. The use of a peptidic scaffold for the formation of stable guanine tetrads: Control of a H-bonded pattern in water. Chemistry - A European Journal. 2011;17(21):5791–5795. DOI: 10.1002/chem.201003556

[14] Stefan L., Guedin A., Amrane S., Smith N., Denat F., Mergny J.-L., Monchaud D. DOTASQ as a prototype of nature-inspired G-quadruplex ligand. Chemical Communications. 2011;47(17):4992–4994. DOI: 10.1039/C0CC04960C

[15] Haudecoeur R., Stefan L., Denat F., Monchaud D. A model of smart G-quadruplex ligand. Journal of the American Chemical Society. 2013;135(2):550–553. DOI: 10.1021/ja310056y

[16] Laguerre A., Stefan L., Larrouy M., Genest D., Novotna J., Pirrotta M., Monchaud D. A twice-as-smart synthetic G-quartet: PyroTASQ is both a smart quadruplex ligand

and a smart fluorescent probe. Journal of the American Chemical Society. 2014;136(35):12406–12414. DOI: 10.1021/ja506331x

[17] Mondragon A. Structural studies of RNase P. Annual Review of Biophysics. 2013;42:537–557. DOI: 10.1146/annurev-biophys-083012-130406

[18] Breaker D. R. Riboswitches and the RNA world. Cold Spring Harbor Perspectives in Biology. 2012;4(2):1–15. DOI: 10.1101/cshperspect.a003566

[19] Garst A. D., Edwards A. L., Batey R. T. Riboswitches: Structures and mechanisms. Cold Spring Harbor Perspectives in Biology. 2011;3(6):1–13. DOI: 10.1101/cshperspect.a003533

[20] Moore P. B., Steitz T. A. The roles of RNA in the synthesis of protein. Cold Spring Harbor Perspectives in Biology. 2010;3(11):1–17. DOI: 10.1101/cshperspect.a003780

[21] Robertson D. L., Joyce G. F. Selection in vitro of an RNA enzyme that specifically cleaves single-stranded DNA. Nature. 1990;344(6265):467–468. DOI: 10.1038/344467a0

[22] Silverman S. K. Nucleic acid enzymes (ribozymes and deoxyribozymes): In Vitro selection and application. In: Begley T. P., editor. Wiley Encyclopedia of Chemical Biology. Wiley & Sons; Hoboken, New Jersey, USA. 2008. p. 1–17. DOI: 10.1002/9780470048672.wecb406

[23] Kosman J., Juskowiak B. Peroxidase-mimicking DNAzymes for biosensing applications: A review. Analytica Chimica Acta. 2011;707(1–2):7–17. DOI: 10.1016/j.aca.2011.08.050

[24] Willner I., Shlyahovsky B., Zayats M., Willner B. DNAzymes for sensing, nanobiotechnology and logic gate applications. Chemical Society Reviews. 2008;37(6):1153–1165. DOI: 10.1039/B718428J

[25] Breaker R. R., Joyce G. F. A DNA enzyme that cleaves RNA. Chemistry & Biology. 1994;1(4):223–229. DOI: 10.1016/1074-5521(94)90014-0

[26] Yang D., Okamoto K. Structural insights into G-quadruplexes: Towards new anticancer drugs. Future Medicinal Chemistry. 2010;2(4):619–646. DOI: 10.4155/fmc.09.172

[27] Xu Y. Chemistry in human telomere biology: Structure, function and targeting of telomere DNA/RNA. Chemical Society Reviews. 2011;40(5):2719–2740. DOI: 10.1039/C0CS00134A

[28] Davis J. T. G-quartets 40 years later: From 5′-GMP to molecular biology and supramolecular chemistry. Angewandte Chemie International Edition. 2004;43(6):668–698. DOI: 10.1002/anie.200300589

[29] Borovok N., Molotsky T., Ghabboun J., Porath D., Kotlyar A. Efficient procedure of preparation and properties of long uniform G4–DNA nanowires. Analytical Biochemistry. 2008;374(1):71–78. DOI: 10.1016/j.ab.2007.10.017

[30] Lyonnais S., Pietrement O., Chepeliansky A., Gueron S., Lacroix L., Le Cam E., Mergny J.-L. Functionalization of DNA G-wires for patterning and nanofabrication. Nucleic Acids Symposium Series (Oxford). 2008;52:689–690. DOI: 10.1093/nass/nrn348

[31] Crungelj M., Hud N. V., Plavec J. The solution structure of d(G4T4G3)2: A bimolecular G-quadruplex with a novel fold. Journal of Molecular Biology. 2002;320(5):911–924. DOI: 10.1016/S0022-2836(02)00569-7

[32] Mergny J.-L., De Cian A., Ghelab A., Sacca B., Lacroix L. Kinetics of tetramolecular quadruplexes. Nucleic Acids Research. 2005;33(1):81–94. DOI: 10.1093/nar/gki148

[33] Guo X., Liu S., Yu Z. Bimolecular quadruplexes and their transitions to higher-order molecular structures detected by ESI-FTICR-MS. Journal of the American Society for Mass Spectrometry. 2007;18(8):1467–1476. DOI: 10.1016/j.jasms.2007.05.003

[34] Zhou J., Bourdoncle A., Rosu F., Gabelica V., Mergny J.-L. Tri-G-quadruplex: Controlled assembly of a G-quadruplex structure from three G-rich strands. Angewandte Chemie International Edition. 2012;51(44):11002–11005. DOI: 10.1002/anie.201205390

[35] Tran P. L. T., De Cian A., Gros J., Moriyama R., Mergny J.-L. Tetramolecular quadruplex stability and assembly. In: Chaires J. B., Graves D., editors. Quadruplex Nucleic Acids. 1st ed. Berlin, Heidelberg: Springer-Verlag; 2012. p. 243–273. DOI: 10.1007/128_2012_334

[36] Mergny J.-L., Phan A.-P., Lacroix L. Following G-quartet formation by UV-spectroscopy. FEBS Letters. 1998;435(1):74–78. DOI: 10.1016/S0014-5793(98)01043-6

[37] Hazel P., Huppert J., Balasubramanian S., Neidle S. Loop-length-dependent folding of G-quadruplexes. Journal of the American Chemical Society. 2004;126(50):16405–16415. DOI: 10.1021/ja045154j

[38] Phan A. T., Kuryavyi V., Patel D. J. DNA architecture: From G to Z. Current Opinion in Structural Biology. 2006;16(3):288–298. DOI: 10.1016/j.sbi.2006.05.011

[39] Burge S., Parkinson G. N., Hazel P., Todd A. K., Neidle S. Quadruplex DNA: Sequence, topology and structure. Nucleic Acids Research. 2006;34(19):5402–5415. DOI: 10.1093/nar/gkl655

[40] Dai J., Carver M., Yang D. Polymorphism of human telomeric quadruplex structures. Biochimie. 2008;90(8):1172–1183. DOI: 10.1016/j.biochi.2008.02.026

[41] Doluca O., Withers J. M., Filichev V. V. Molecular engineering of guanine-rich sequences: Z-DNA, DNA triplexes, and G-quadruplexes. Chemical Reviews. 2013;113(5):3044–3083. DOI: 10.1021/cr300225q

[42] Brooks T. A., Kendrick S., Hurley L. Making sense of G-quadruplex and i-motif functions in oncogene promoters. FEBS Journal. 2010;277(17):3459–3469. DOI: 10.1111/j.1742-4658.2010.07759.x

[43] Neidle S. Human telomeric G-quadruplex: The current status of telomeric G-quadruplexes as therapeutic targets in human cancer. FEBS Journal. 2010;277(5):1118–1125. DOI: 10.1111/j.1742-4658.2009.07463.x

[44] Balasubramanian S., Hurley L. H., Neidle S. Targeting G-quadruplexes in gene promoters: A novel anticancer strategy? Nature Reviews Drug Discovery. 2011;10(4): 261–275. DOI: 10.1038/nrd3428

[45] Li Y., Sen D. A catalytic DNA for porphyrin metallation. Nature Structural Biology. 1996;3(9):743–747. DOI: 10.1038/nsb0996-743

[46] Li Y., Geyer R., Sen D. Recognition of anionic porphyrins by DNA aptamers. Biochemistry. 1996;35(21):6911–6922. DOI: 10.1021/bi960038h

[47] Nicoludis J. M., Barrett S. P., Mergny J.-L., Yatsunyk L. A. Interaction of human telomeric DNA with N-methyl mesoporphyrin IX. Nucleic Acids Research. 2012;40(12): 5432–5447. DOI: 10.1093/nar/gks152

[48] Paramasivan S., Rujan I, Bolton P. H. Circular dichroism of quadruplex DNAs: Applications to structure, cation effects and ligand binding. Methods. 2007;43(4):324–331. DOI: 10.1016/j.ymeth.2007.02.009

[49] Ragazzon P., Chaires J. B. Use of competition dialysis in the discovery of G-quadruplex selective ligands. Methods. 2007;43(4):313–323. DOI: 10.1016/j.ymeth.2007.08.003

[50] Travascio P., Li Y., Sen D. DNA-enhanced peroxidase activity of a DNA aptamer-hemin complex. Chemistry & Biology. 1998;5(9):505–517. DOI: 10.1016/S1074-5521(98)90006-0

[51] Travascio P., Bennet A. J., Wang D. Y., Sen D. A ribozyme and a catalytic DNA with peroxidase activity: Active sites versus cofactor-binding sites. Chemistry & Biology. 1999;6(11):779–787. DOI: 10.1016/S1074-5521(99)80125-2

[52] He Y., Tian Y., Mao C. Human telomeric DNA sequences have a peroxidaseapoenzyme activity. Molecular BioSystems. 2009;5(3):238–240. DOI: 10.1039/B822031J

[53] Kong D.-M., Cai L.-L, Guo J.-H., Wu J., Shen H.-X. Characterization of the G-quadruplex structure of a catalytic DNA with peroxidase activity. Biopolymers. 2009;91(5): 331–339. DOI: 10.1002/bip.21135

[54] Kong D.-M., Yang W., Wu J., Li C.-X., Shen H.-X. Structure–function study of peroxidase-like G-quadruplex-hemin complexes. Analyst. 2010;135(2):321–326. DOI: 10.1039/B920293E

[55] Poon L. C.-H., Methot S. P., Morabi-Pazooki W., Pio F., Bennet A. J., Sen D. Guanine-rich RNAs and DNAs that bind heme robustly catalyze oxygen transfer reactions. Journal of the American Chemical Society. 2011;133(6):1877–1884. DOI: 10.1021/ja108571a

[56] Stefan L., Denat F., Monchaud D. Insights into how nucleotide supplements enhance the peroxidase-mimicking DNAzyme activity of the G-quadruplex/hemin system. Nucleic Acids Research. 2002;40(17):8759–8772. DOI: 10.1093/nar/gks581

[57] Jantschko W., Furtmuller P. G., Zederbauer M., Lanz M., Jakopitsch C., Obinger C. Direct conversion of ferrous myeloperoxidase to compound II by hydrogen peroxide: An anaerobic stopped-flow study. Biochemical and Biophysical Research Communications. 2003;312(2):292–298. DOI: 10.1016/j.bbrc.2003.10.117

[58] Jones P., Dunford H. B. The mechanism of compound I formation revisited. Journal of Inorganic Biochemistry. 2005;99(12):2292–2298. DOI: 10.1016/j.jinorgbio.2005.08.009

[59] Ourliac-Garnier I., Elizondo-Riojas M.-A., Redon S., Farrell N. P., Bombard S. Cross-links of quadruplex structures from human telomeric DNA by dinuclear platinum complexes show the flexibility of both structures. Biochemistry. 2005;44(31):10620–10634. DOI: 10.1021/bi050144w

[60] Bertrand H., Bombard S., Monchaud D., Teulade-Fichou M.-P. A platinum–quinacridine hybrid as a G-quadruplex ligand. JBIC Journal of Biological Inorganic Chemistry. 2007;12(7):1003–1014. DOI: 10.1007/s00775-007-0273-3

[61] Cheng X., Liu X., Bing T., Cao Z., Shangguan D. General peroxidase activity of G-quadruplex–hemin complexes and its application in ligand screening. Biochemistry. 2009;48(33):7817–7823. DOI: 10.1021/bi9006786

[62] Nakayama S., Sintim O. H. Colorimetric split G-quadruplex probes for nucleic acid sensing: Improving reconstituted DNAzyme's catalytic efficiency via probe remodeling. Journal of the American Chemical Society. 2009;131(29):10320–10333. DOI: 10.1021/ja902951b

[63] Nakayama S., Wang J., Sintim H. O. DNA-based peroxidation catalyst—what is the exact role of topology on catalysis and is there a special binding site for catalysis? Chemistry - A European Journal. 2011;17(20):5691–5698. DOI: 10.1002/chem.201002349

[64] Yang X., Fang C., Mei H., Chang T., Cao Z., Shangguan D. Characterization of G-quadruplex/hemin peroxidase: Substrate specificity and inactivation kinetics. Chemistry - A European Journal. 2011;17(51):14475–14484. DOI: 10.1002/chem.201101941

[65] Nakayama S., Sintim H. O. Investigating the interactions between cations, peroxidation substrates and G-quadruplex topology in DNAzyme peroxidation reactions using statistical testing. Analytica Chimica Acta. 2012;747:1–6. DOI: 10.1016/j.aca.2012.08.008

[66] Kosman J., Juskowiak B. Optimization study of the catalytic activity of DNAzymes based on telomeric G-quadruplexes. Central European Journal of Chemistry. 2012;10(2):368–372. DOI: 10.2478/s11532-011-0152-1

[67] Kong D.-M., Xu J., Shen H.-X. Positive effects of ATP on G-quadruplex-hemin DNAzyme-mediated reactions. Analytical Chemistry. 2010;82(14):6148-6153. DOI: 10.1021/ac100940v

[68] Humez S., Monet M., van Coppenolle F., Delcourt P., Prevarskaya N. The role of intracellular pH in cell growth arrest induced by ATP. American Journal of Physiology - Cell Physiology. 2004;287(6):C1733–C1746. DOI: 10.1152/ajpcell.00578.2003

[69] Azevedo A. M., Martins V. C., Prazeres D. M. F., Vojinovic V., Cabral J. M. S., Fonseca L. P. Horseradish peroxidase: A valuable tool in biotechnology. Biotechnology Annual Review. 2003;9:199–247. DOI: 10.1016/S1387-2656(03)09003-3

[70] Zakharova G. S., Oporov I. V., Tishkov V. I. Horseradish peroxidase: Modulation of properties by chemical modification of protein and heme. Biochemistry (Moscow). 2011;76(13):1391–1401. DOI: 10.1134/S0006297911130037

[71] Gungor A. A. Use of different plant derived peroxidases for the removal of phenol from water. Asian Journal of Chemistry. 2011;23(8):3710–3712.

[72] Veitch N. C. Horseradish peroxidase: A modern view of a classic enzyme. Phytochemistry. 2004;65(3):249–259. DOI: 10.1016/j.phytochem.2003.10.022

[73] Li T., Wang E., Dong S. G-quadruplex-based DNAzyme as a sensing platform for ultrasensitive colorimetric potassium detection. Chemical Communications. 2009;(5): 580–582. DOI: 10.1039/B815814B

[74] Yang X. Li T., Li B., Wang E. Potassium-sensitive G-quadruplex DNA for sensitive visible potassium detection. Analyst. 2010;135(1):71–75. DOI: 10.1039/B913036E

[75] Sun H., Li X., Li Y., Fan L., Kraatz H.-B. A novel colorimetric potassium sensor based on the substitution of lead from G-quadruplex. Analyst. 2013;138(3):856–862. DOI: 10.1039/C2AN36564B

[76] Wang G., Chen L., Zhu Y., He X., Xu G., Zhang X. Development of an electrochemical sensor based on the catalysis of ferrocene actuated hemin/G-quadruplex enzyme for the detection of potassium ions. Biosensors and Bioelectronics. 2014;61:410–416. DOI: 10.1016/j.bios.2014.05.052

[77] Kong D.-M., Cai L.-L., Shen H.-X. Quantitative detection of Ag+ and cysteine using G-quadruplex–hemin DNAzymes. Analyst. 2010;135(6):1253–1258. DOI: 10.1039/B925168E

[78] Zhou X.-H., Kong D.-M., Shen H.-X. G-quadruplex–hemin DNAzyme-amplified colorimetric detection of Ag+ ion. Analytica Chimica Acta. 2010;678(1):124–127. DOI: 10.1016/j.aca.2010.08.025

[79] Zhou X.-H., Kong D.-M., Shen H.-X. Ag+ and cysteine quantitation based on G-quadruplex–hemin DNAzymes disruption by Ag+. Analytical Chemistry. 2010;82(3):789–793. DOI: 10.1021/ac902421u

[80] Childs R. E., Bardsley W. G. The steady-state kinetics of peroxidase with 2,2′-azino-di-(3-ethyl-benzthiazoline-6-sulphonic acid) as chromogen. Biochemical Journal. 1975;145(1):93–103.

[81] Kadnikova E. N., Kostic N. M. Oxidation of ABTS by hydrogen peroxide catalyzed by horseradish peroxidase encapsulated into sol–gel glass: Effects of glass matrix on reactivity. Journal of Molecular Catalysis B: Enzymatic. 2002;18(1–3):39–48. DOI: 10.1016/S1381-1177(02)00057-7

[82] Josephy P. D., Eling T., Mason R. P. The horseradish peroxidase-catalyzed oxidation of 3,5,3′,5′-tetramethylbenzidine. Free radical and charge-transfer complex intermediates. The Journal of Biological Chemistry. 1982;257(7):3669–3675.

[83] Marquez L. A., Dunford H. B. Mechanism of the oxidation of 3,5,3′,5′-tetramethylbenzidine by myeloperoxidase determined by transient- and steady-state kinetics. Biochemistry. 1997;36(31):9349–9355. DOI: 10.1021/bi970595j

[84] Li B., Du Y., Li T., Dong S. Investigation of 3,3′,5,5′-tetramethylbenzidine as colorimetric substrate for a peroxidatic DNAzyme. Analytica Chimica Acta. 2009;651(2): 234–240. DOI: 10.1016/j.aca.2009.09.009

[85] Nakayama S., Sintim O. H. Biomolecule detection with peroxidase-mimicking DNAzymes; expanding. Molecular BioSystems. 2009;6(1):95–97. DOI: 10.1039/b916228c

[86] Cai Y., Li N., Kong D.-M., Shen H.-X. Fluorogenic substrate screening for G-quadruplex DNAzyme-based sensors. Biosensors and Bioelectronics. 2013;49:312–317. DOI: 10.1016/j.bios.2013.05.034

[87] Yin B.-C., Ye B.-C., Tan W., Wang H., Xie C.-C. An allosteric dual-DNAzyme unimolecular probe for colorimetric detection of copper(II). Journal of the American Chemical Society. 2009;131(41):14624–14625. DOI: 10.1021/ja9062426

[88] Zhang L., Zhu J., Ai J., Zhou Z., Jia X., Wang E. Label-free G-quadruplex-specific fluorescent probe for sensitive detection of copper(II) ion. Biosensors and Bioelectronics. 2013;39(1):268–273. DOI: 10.1016/j.bios.2012.07.058

[89] Li H., Huang X.-X., Cai Y., Xiao H.-J., Zhang Q.-F., Kong D.-M. Label-free detection of Cu2+ and Hg2+ ions using reconstructed Cu2+-specific DNAzyme and G-quadruplex DNAzyme. PLoS ONE. 2013;8(10):1–9. DOI: 10.1371/journal.pone.0073012

[90] Ge C., Luo Q., Wang D., Zhao S., Liang X., Yu L., Xing X., Zeng L. Colorimetric detection of copper(II) ion using click chemistry and hemin/G-quadruplex horseradish peroxidase-mimicking DNAzyme. Analytical Chemistry. 2014;86(13):6387–6392. DOI: 10.1021/ac501739a

[91] Xu M., Gao Z., Wei Q., Chen G., Tang D. Hemin/G-quadruplex-based DNAzyme concatemers for in situ amplified impedimetric sensing of copper(II) ion coupling with DNAzyme-catalyzed precipitation strategy. Biosensors and Bioelectronics. 2015;74:1–7. DOI: 10.1016/j.bios.2015.05.056

[92] Wang F., Orbach R., Willner I. Detection of metal ions (Cu2+, Hg2+) and cocaine by using ligation DNAzyme machinery. Chemistry - A European Journal. 2012;18(50): 16030–16036. DOI: 10.1002/chem.201201479

[93] Li T., Dong S., Wang E. Label-free colorimetric detection of aqueous mercury ion (Hg2+) using Hg2+-modulated G-quadruplex-based DNAzymes. Analytical Chemistry. 2009;81(6):2144–2149. DOI: 10.1021/ac900188y

[94] Kong D.-M., Wang N., Guo X.-X., Shen H.-X. "Turn-on" detection of Hg2+ ion using a peroxidase-like split G-quadruplex–hemin DNAzyme. Analyst. 2010;135(3):545–549. DOI: 10.1039/B924014D

[95] Jia S.-M., Liu X.-F., Li P., Kong D.-M., Shen H.-X. G-quadruplex DNAzyme-based Hg2+ and cysteine sensors utilizing Hg2+-mediated oligonucleotide switching. Biosensors and Bioelectronics. 2011;27(1):148–152. DOI: 10.1016/j.bios.2011.06.032

[96] Lin Y.-W., Liu C.-W., Chang H.-T. Fluorescence detection of mercury(II) and lead(II) ions using aptamer/reporter conjugates. Talanta. 2011;84(2):324–329. DOI: 10.1016/j.talanta.2011.01.016

[97] Zhang Z., Yin J., Wu Z., Yu R. Electrocatalytic assay of mercury(II) ions using a bifunctional oligonucleotide signal probe. Analytica Chimica Acta. 2013;762:47–53. DOI: 10.1016/j.aca.2012.12.008

[98] Yuan Y., Gao M., Liu G., Chai Y., Wei S., Yuan R. Sensitive pseudobienzyme electrocatalytic DNA biosensor for mercury(II) ion by using the autonomously assembled hemin/G-quadruplex DNAzyme nanowires for signal amplification. Analytica Chimica Acta. 2014;811:23–28. DOI: 10.1016/j.aca.2013.11.051

[99] Elbaz J., Shlyahovsky B., Willner I. A DNAzyme cascade for the amplified detection of Pb2+ ions or L-histidine. Chemical Communications. 2008;(13):1569–1571. DOI: 10.1039/B716774A

[100] Lan T., Furuya K., Lu Y. A highly selective lead sensor based on a classic lead DNAzyme. Chemical Communications. 2010;46(22):3896–3898. DOI: 10.1039/B926910J

[101] Li C.-L., Liu K.-T., Lin Y.-W., Chang H.-T. Fluorescence detection of lead(II) ions through their induced catalytic activity of DNAzymes. Analytical Chemistry. 2011;83(1):225–230. DOI: 10.1021/ac1028787

[102] Li T., Dong S., Wang E. A lead(II)-driven DNA molecular device for turn-on fluorescence detection of lead(II) ion with high selectivity and sensitivity. Journal of the American Chemical Society. 2010;132(38):13156–13157. DOI: 10.1021/ja105849m

[103] Li T., Wang E., Dong S. Lead(II)-induced allosteric G-quadruplex DNAzyme as a colorimetric and chemiluminescence sensor for highly sensitive and selective Pb2+ detection. Analytical Chemistry. 2010;82(4):1515–1520. DOI: 10.1021/ac902638v

[104] Yang X., Xu J., Tang X., Liu H., Tian D. A novel electrochemical DNAzyme sensor for the amplified detection of Pb2+ ions. Chemical Communications. 2010;46(18):3107–3109. DOI: 10.1039/C002137G

[105] Lin Z., Li X., Kraatz H.-B. Impedimetric immobilized DNA-based sensor for simultaneous detection of Pb2+, Ag+, and Hg2+. Analytical Chemistry. 2011;83(17):6896–6901. DOI: 10.1021/ac2014096

[106] Pelossof G., Tel-Vered R., Willner I. Amplified surface plasmon resonance and electrochemical detection of Pb2+ ions using the Pb2+-dependent DNAzyme and hemin/G-quadruplex as a label. Analytical Chemistry. 2012;84(8):3703–3709. DOI: 10.1021/ac3002269

[107] He H.-Z., Leung K.-H., Yang H., Chan D. S.-H., Leung C.-H., Zhou J., Bourdoncle A., Mergny J.-L., Ma D.-L. Label-free detection of sub-nanomolar lead(II) ions in aqueous solution using a metal-based luminescent switch-on probe. Biosensors and Bioelectronics. 2013;41:871–874. DOI: 10.1016/j.bios.2012.08.060

[108] Zhang B., Chen J., Liu B., Tang D. Amplified electrochemical sensing of lead ion based on DNA-mediated self-assembly-catalyzed polymerization. Biosensors and Bioelectronics. 2015;69:230–234. DOI: 10.1016/j.bios.2015.02.041

[109] Yang C., Lates V., Prieto-Simon B., Marty J.-L., Yang X. Aptamer-DNAzyme hairpins for biosensing of Ochratoxin A. Biosensors and Bioelectronics. 2012;32(1):208–212. DOI: 10.1016/j.bios.2011.12.011

[110] Bo H., Wang C., Gao Q., Qi H., Zhang C. Selective, colorimetric assay of glucose in urine using G-quadruplex-based DNAzymes and 10-acetyl-3,7-dihydroxy phenoxazine. Talanta. 2013;108:131–135. DOI: 10.1016/j.talanta.2013.03.001

[111] Nie J., Zhang D.-W., Tie C., Zhou Y.-L., Zhang X.-X. A label-free DNA hairpin biosensor for colorimetric detection of target with suitable functional DNA partners. Biosensors and Bioelectronics. 2013;49:236–242. DOI: 10.1016/j.bios.2013.05.020

[112] Li T., Shi L., Wang E., Dong S. Multifunctional G-quadruplex aptamers and their application to protein detection. Chemistry - A European Journal. 2009;15(4):1036–1042. DOI: 10.1002/chem.200801282

[113] Li D., Shlyahovsky B., Elbaz J., Willner I. Amplified analysis of low-molecular-weight substrates or proteins by the self-assembly of DNAzyme–aptamer conjugates. Journal of the American Chemical Society. 2007;129(18):5804–5805. DOI: 10.1021/ja070180d

[114] Li T., Wang E., Dong S. G-quadruplex-based DNAzyme for facile colorimetric detection. Chemical Communication. 2008;(31):3654–3656. DOI: 10.1039/B805565C

[115] Zhu J., Li T., Hu J., Wang E. A novel dot-blot DNAzyme-linked aptamer assay for protein detection. Analytical and Bioanalytical Chemistry. 2010;397(7):2923–2927. DOI: 10.1007/s00216-010-3802-9

[116] Zhang Y., Li B., Jin Y. Label-free fluorescent detection of thrombin using G-quadruplex-based DNAzyme as sensing platform. Analyst. 2011;136(16):3268–3273. DOI: 10.1039/C1AN00002K

[117] Yi H., Xu W., Yuan Y., Bai L., Wu Y., Chai Y., Yuan R. A pseudo triple-enzyme cascade amplified aptasensor for thrombin detection based on hemin/G-quadruplex as signal label. Biosensors and Bioelectronics. 2014;54:415–420. DOI: 10.1016/j.bios.2013.11.036

[118] Xie S., Chai Y., Yuan Y., Bai L., Yuan R. A novel electrochemical aptasensor for highly sensitive detection of thrombin based on the autonomous assembly of hemin/G-quadruplex horseradish peroxidase-mimicking DNAzyme nanowires. Analytica Chimica Acta. 2014;832:51–57. DOI: 10.1016/j.aca.2014.04.065

[119] Zheng Y., Chai Y., Yuan Y., Yuan R. A pseudo triple-enzyme electrochemical aptasensor based on the amplification of Pt–Pd nanowires and hemin/G-quadruplex. Analytica Chimica Acta. 2014;834:45–50. DOI: 10.1016/j.aca.2014.04.060

[120] Wu H., Zhang K., Liu Y., Wang H., Wu J., Zhu F., Zou P. Binding-induced and label-free colorimetric method for protein detection based on autonomous assembly of hemin/G-quadruplex DNAzyme amplification strategy. Biosensors and Bioelectronics. 2015;64:572–578. DOI: 10.1016/j.bios.2014.09.096

[121] Zhang J., Chai Y., Yuan R., Yuan Y., Bai L., Xie S., Jiang L. A novel electrochemical aptasensor for thrombin detection based on the hybridization chain reaction with hemin/G-quadruplex DNAzyme-signal amplification. Analyst. 2013;138(16):4558–4564. DOI: 10.1039/C3AN00396E

[122] Freeman R., Girsh J., Jou A. F.-J., Ho J.-A. A., Hug T., Dernedde J., Willner I. Optical aptasensors for the analysis of the vascular endothelial growth factor (VEGF). Analytical Chemistry. 2012;84(14):6192–6198. DOI: 10.1021/ac3011473

[123] Ye Z., Zhang W., Liu L., Chen G., Shen Z., Zhou N. Fabrication of a colorimetric biosensing platform for the detection of protein–DNA interaction. Biosensors and Bioelectronics. 2013;46:108–112. DOI: 10.1016/j.bios.2013.02.028

[124] Wang Q., Song Y., Chai Y., Pan G., Li T., Yuan Y., Yuan R. Electrochemical immunosensor for detecting the spore wall protein of Nosema bombycis based on the amplification of hemin/G-quadruplex DNAzyme concatemers functionalized Pt@Pd nanowires. Biosensors and Bioelectronics. 2014;60:118–123. DOI: 10.1016/j.bios.2014.03.075

[125] He Y., Wang X., Zhang Y., Gao F., Li Y., Chen H., Wang L. An ultrasensitive chemiluminescent immunosensor for the detection of human leptin using hemin/G-quadruplex DNAzymes-assembled signal amplifier. Talanta. 2013;116:816–821. DOI: 10.1016/j.talanta.2013.07.074

[126] Stefan L., Lavergne T., Spinelli N., Defrancq E., Monchaud D. Surface-immobilized DNAzyme-type biocatalysis. Nanoscale. 2014;6(5):2693–2701. DOI: 10.1039/C3NR05954E

[127] Murat P., Bonnet R., Van der Heyden A., Spinelli N., Labbe P., Monchaud D., Teulade-Fichou M.-P., Dumy P., Defrabcq E. Template-assembled synthetic G-quadruplex (TASQ): A useful system for investigating the interactions of ligands with constrained quadruplex topologies. Chemistry - A European Journal. 2010;16(20): 6106–6114. DOI: 10.1002/chem.200903456

[128] Weizmann Y., Beissenhirtz M. K., Cheglakov Z., Nowarski R., Kotler M., Willner I. A virus spotlighted by an autonomous DNA machine. Angewandte Chemie International Edition. 2006;45(44):7384–7388. DOI: 10.1002/anie.200602754

[129] Weizmann Y., Cheglakov Z., Willner I. A Fok I/DNA machine that duplicates its analyte gene sequence. Journal of the American Chemical Society. 2008;130(51):17224–17225. DOI: 10.1021/ja806222e

[130] Qiu B., Zheng Z.-Z., Lu Y.-J., Lin Z.-Y., Wong K.-Y., Chen G.-N. G-quadruplex DNAzyme as the turn on switch for fluorimetric detection of genetically modified organisms. Chemical Communications. 2011;47(5):1437–1439. DOI: 10.1039/C0CC02933E

[131] Jiang X., Zhang H., Wu J., Yang X., Shao J., Lu Y., Lin Z., Chen G. G-quadruplex DNA biosensor for sensitive visible detection of genetically modified food. Talanta. 2014;128:445–449. DOI: 10.1016/j.talanta.2014.05.002

[132] Meng X., Zhou Y., Liang Q., Qu X., Yang Q., Yin H., Ai S. Electrochemical determination of microRNA-21 based on bio bar code and hemin/G-quadruplet DNAenzyme. Analyst. 2013;138(12):3409–3415. DOI: 10.1039/C3AN36788F

[133] Tian T., Xiao H., Zhang Z., Long Y., Peng S., Wang S., Zhou X., Liu S., Zhou X. Sensitive and convenient detection of microRNAs based on cascade amplification by catalytic DNAzymes. Chemistry - A European Journal. 2013;19(1):92–95. DOI: 10.1002/chem.201203344

[134] Wang X.-P., Yin B.-C., Wang P., Ye B.-C. Highly sensitive detection of microRNAs based on isothermal exponential amplification-assisted generation of catalytic G-quadruplex DNAzyme. Biosensors and Bioelectronics. 2013;42:131–135. DOI: 10.1016/j.bios.2012.10.097

[135] Zhang P., Wu X., Yuan R., Chai Y. An "off–on" electrochemiluminescent biosensor based on DNAzyme-assisted target recycling and rolling circle amplifications for ultrasensitive detection of microRNA. Analytical Chemistry. 2015;87(6):3202–3207. DOI: 10.1021/ac504455z

[136] Zhou W., Liang W., Li X., Chai Y., Yuan R., Xiang Y. MicroRNA-triggered, cascaded and catalytic self-assembly of functional "DNAzyme ferris wheel" nanostructures for

highly sensitive colorimetric detection of cancer cells. Nanoscale. 2015;7(19):9055–9061. DOI: 10.1039/C5NR01405K

[137] He H.-Z., Ma V. P.-Y., Leung K.-H., Chan D. S.-H., Yang H., Cheng Z., Leung C.-H., Ma D.-L. A label-free G-quadruplex-based switch-on fluorescence assay for the selective detection of ATP. Analyst. 2012;137(7):1538–1540. DOI: 10.1039/C2AN15999F

[138] Liu F., Zhang J., Chen R., Chen L., Deng L. Highly effective colorimetric and visual detection of ATP by a DNAzyme–aptamer sensor. Chemistry & Biodiversity. 2011;8(2):311–316. DOI: 10.1002/cbdv.201000130

[139] Bao T., Shu H., Wen W., Zhang X., Wang S. A sensitive electrochemical aptasensor for ATP detection based on exonuclease III-assisted signal amplification strategy. Analytica Chimica Acta. 2015;862:64–69. DOI: 10.1016/j.aca.2014.12.049

[140] Liu Y., Lei J., Huang Y., Ju H. "Off-on" electrochemiluminescence system for sensitive detection of ATP via target-induced structure switching. Analytical Chemistry. 2014;86(17):8735–8741. DOI: 10.1021/ac501913c

[141] Nakayama S., Roelofs K., Lee V. T., Sintim O. H. A C-di-GMP–proflavine–hemin supramolecular complex has peroxidase activity—implication for a simple colorimetric detection. Molecular BioSystems. 2012;8(3):726–729. DOI: 10.1039/C2MB05430B

[142] Liu H., Wang Y.-S., Wang J.-C., Xue J.-H., Zhou B., Zhao H., Liu S.-D., Tang X., Chen S.-H., Li M.-H., Cao J.-X. A colorimetric aptasensor for the highly sensitive detection of 8-hydroxy-2′-deoxyguanosine based on G-quadruplex–hemin DNAzyme. Analytical Biochemistry. 2014;458:4–10. DOI: 10.1016/j.ab.2014.04.031

[143] Deng M., Zhang D., Zhou Y., Zhou X. Highly effective colorimetric and visual detection of nucleic acids using an asymmetrically split peroxidase DNAzyme. Journal of the American Chemical Society. 2008;130(39):13095–13102. DOI: 10.1021/ja803507d

[144] Willner I., Cheglakov Z., Weizmann Y., Sharon E. Analysis of DNA and single-base mutations using magnetic particles for purification, amplification and DNAzyme detection. Analyst. 2008;133(7):923–927. DOI: 10.1039/B802015A

[145] Verga D., Welter M., Steck A.-L., Marx A. DNA polymerase-catalyzed incorporation of nucleotides modified with a G-quadruplex-derived DNAzyme. Chemical Communications. 2015;51(34):7379–7381. DOI: 10.1039/C5CC01387A

[146] Li W., Liu Z., Lin H., Nie Z., Chen J., Xu X., Yao S. Label-free colorimetric assay for methyltransferase activity based on a novel methylation-responsive DNAzyme strategy. Analytical Chemistry. 2010;82(5):1935–1941. DOI: 10.1021/ac902670c

[147] Li R., Xiong C., Xiao Z., Ling L. Colorimetric detection of cholesterol with G-quadruplex-based DNAzymes and $ABTS2^{-}$. Analytica Chimica Acta. 2012;724:80–85. DOI: 10.1016/j.aca.2012.02.015

[148] Golub E., Niarov A., Freeman R., Zatsepin M., Willner I. Photoelectrochemical biosensors without external irradiation: Probing enzyme activities and DNA sensing us-

ing hemin/G-quadruplex-stimulated chemiluminescence resonance energy transfer (CRET) generation of photocurrents. The Journal of Physical Chemistry C. 2012;116(25):13827–13834. DOI: 10.1021/jp303741x

[149] Pavlov V., Xiao Y., Gill R., Dishon A., Kotler M., Willner I. Amplified chemiluminescence surface detection of DNA and telomerase activity using catalytic nucleic acid labels. Analytical Chemistry. 2004;76(7):2152–2156. DOI: 10.1021/ac0352191

[150] Xiao Y., Pavlov V., Gill R., Bourenko T., Willner I. Lighting up biochemiluminescence by the surface self-assembly of DNA–hemin complexes. ChemBioChem. 2004;5(3): 374–379. DOI: 10.1002/cbic.200300794

[151] Li Y., Li X., Ji X., Li X. Formation of G-quadruplex-hemin DNAzyme based on human telomere elongation and its application in telomerase activity detection. Biosensors and Bioelectronics. 2011;26(10):4095–4098. DOI: 10.1016/j.bios.2011.03.041

[152] Hartley C. B. Telomerase and cancer therapeutics. Nature Reviews Cancer. 2008;8(3): 167–179. DOI: 10.1038/nrc2275

[153] Corey D. R. Telomeres and telomerase: From discovery to clinical trials. Chemistry & Biology. 2009;16(12):1219–1223. DOI: 10.1016/j.chembiol.2009.12.001

[154] Freeman R., Sharon E., Teller C., Henning A., Tzfati Y., Willner I. DNAzyme-like activity of hemin–telomeric G-quadruplexes for the optical analysis of telomerase and its inhibitors. ChemBioChem. 2010;11(17):2362–2367. DOI: 10.1002/cbic.201000512

[155] Stefan L., Denat F., Monchaud D. Deciphering the DNAzyme activity of multimeric quadruplexes: Insights into their actual role in the telomerase activity evaluation assay. Journal of the American Chemical Society. 2011;133(50):20405–20415. DOI: 10.1021/ja208145d

[156] Wang L.-J., Zhang Y., Zhang C.-Y. Ultrasensitive detection of telomerase activity at the single-cell level. Analytical Chemistry. 2013;85(23):11509–11517. DOI: 10.1021/ac402747r

[157] Zhou J., Lai W., Zhuang J., Tang J., Tang D. Nanogold-functionalized DNAzyme concatemers with redox-active intercalators for quadruple signal amplification of electrochemical immunoassay. ACS Applied Materials & Interfaces. 2013;5(7):2773–2781. DOI: 10.1021/am400652g

[158] Zong C., Wu J., Xu J., Ju H., Yan F. Multilayer hemin/G-quadruplex wrapped gold nanoparticles as tag for ultrasensitive multiplex immunoassay by chemiluminescence imaging. Biosensors and Bioelectronics. 2013;43:372–378. DOI: 10.1016/j.bios.2012.12.051

[159] Thirstrup D., Baird G. S. Histochemical application of a peroxidase DNAzyme with a covalently attached hemin cofactor. Analytical Chemistry. 2010;82(6):2498–2504. DOI: 10.1021/ac902887j

[160] Shi M., Zhao S., Huang Y., Zhao L., Liu Y.-M. Signal amplification in capillary electrophoresis based chemiluminescent immunoassays by using an antibody–gold nanoparticle–DNAzyme assembly. Talanta. 2014;124:14–20. DOI: 10.1016/j.talanta.2014.02.032

[161] Noorsharmimi O., Qiuting L., Gee J. T., Yee S. C., Rahmah N., Jorn G., Theam S. L. Development of an antigen-DNAzyme based probe for a direct antibody-antigen assay using the intrinsic DNAzyme activity of a daunomycin aptamer. Sensors. 2014;14(1):346–355. DOI: 10.3390/s140100346

[162] Li Y., Deng J., Fang L., Yu K., Huang H., Jiang L., Liang W., Zheng J. A novel electrochemical DNA biosensor based on HRP-mimicking hemin/G-quadruplex wrapped GOx nanocomposites as tag for detection of Escherichia coli O157:H7. Biosensors and Bioelectronics. 2015;63:1–6. DOI: 10.1016/j.bios.2014.07.012

[163] Liu T., Zhang X., Zhu W., Liu W., Zhang D., Wang J. A G-quadruplex DNAzyme-based colorimetric method for facile detection of Alicyclobacillus acidoterrestris. Analyst. 2014;139(17):4315–4321. DOI: 10.1039/C4AN00643G

[164] Monchaud D., Teulade-Fichou M.-P. A hitchhiker's guide to G-quadruplex ligands. Organic and Biomolecular Chemistry. 2008;6(4):627–636. DOI: 10.1039/B714772B

[165] Georgiades S. N., Abd Karim N. H., Suntharalingam K., Vilar R. Interaction of metal complexes with G-quadruplex DNA. Angewandte Chemie International Edition. 2010;49(24):4020–4034. DOI: 10.1002/anie.200906363

[166] Vy Thi Le T., Han S., Chae J. Park H.-J. G-quadruplex binding ligands: From naturally occurring to rationally designed molecules. Current Pharmaceutical Design. 2012;18(14):1948–1972. DOI: 10.2174/138161212799958431

[167] Kong D.-M., Wu J., Ma Y.-E., Shen H.-X. A new method for the study of G-quadruplex ligands. Analyst. 2008;133(9):1158–1160. DOI: 10.1039/B804627A

[168] Fu L., Li B., Zhang Y. Label-free fluorescence method for screening G-quadruplex ligands. Analytical Biochemistry. 2012;421(1):198–202. DOI: 10.1016/j.ab.2011.10.021

[169] Qiao Y., Deng J., Jin Y., Chen G., Wang L. Identifying G-quadruplex-binding ligands using DNA-functionalized gold nanoparticles. Analyst. 2012;137(7):1663–1668. DOI: 10.1039/C2AN16051J

[170] Miyoshi D., Inoue M., Sugimoto N. DNA logic gates based on structural polymorphism of telomere DNA molecules responding to chemical input signals. Angewandte Chemie International Edition. 2006;45(46):7716–7719. DOI: 10.1002/anie.200602404

[171] Shlyahovsky B., Li Y., Lioubashevski O., Elbaz J., Willner I. Logic gates and antisense DNA devices operating on a translator nucleic acid scaffold. ACS Nano. 2009;3(7):1831–1843. DOI: 10.1021/nn900085x

[172] Zhou M., Kuralav F., Windmiller J. R., Wang J. DNAzyme logic-controlled biofuel cells for self-powered biosensors. Chemical Communications. 2012;48(32):3815–3817. DOI: 10.1039/C2CC30464C

[173] Zhu J., Li T., Zhang L., Dong S., Wang E. G-quadruplex DNAzyme based molecular catalytic beacon for label-free colorimetric logic gates. Biomaterials. 2011;32(30):7318–7324. DOI: 10.1016/j.biomaterials.2011.06.040

[174] Li T., Wang E., Dong S. Potassium−lead-switched G-quadruplexes: A new class of DNA logic gates. Journal of the American Chemical Society. 2009;131(42):15082–15083. DOI: 10.1021/ja9051075

[175] Chen J., Zhou S., Wen J. Concatenated logic circuits based on a three-way DNA junction: A keypad-lock security system with visible readout and an automatic reset function. Angewandte Chemie International Edition. 2015;54(2):446–450. DOI: 10.1002/anie.201408334

[176] Willner I., Shlyahovsky B., Zayats M., Willner B. DNAzymes for sensing, nanobiotechnology and logic gate applications. Chemical Society Reviews. 2008;37(6):1077–1280. DOI: 10.1039/b718428j

[177] Gellert M., Lipsett M. N., Davies D. R. Helix formation by guanylic acid. Proceedings of the National Academy of Sciences of the United States of America. 1962;48(12): 2013–2018.

[178] Wong A., Ida R., Spindler L., Wu G. Disodium guanosine 5′-monophosphate self-associates into nanoscale cylinders at pH 8: A combined diffusion NMR spectroscopy and dynamic light scattering study. Journal of the American Chemical Society. 2005;127(19):6990–6998. DOI: 10.1021/ja042794d

[179] Sivakova S., Rowan S. J. Nucleobases as supramolecular motifs. Chemical Society Reviews. 2005;34(1):9–21. DOI: 10.1039/B304608G

[180] Lena S., Masiero S., Pieraccini S., Spada G. P. Guanosine hydrogen-bonded scaffolds: A new way to control the bottom-up realisation of well-defined nanoarchitectures. Chemistry - A European Journal. 2009;15(32):7792–7806. DOI: 10.1002/chem.200802506

[181] Ghoussoub A., Lehn J.-M. Dynamic sol–gel interconversion by reversible cation binding and release in G-quartet-based supramolecular polymers. Chemical Communications. 2005;5(46):5763–5765. DOI: 10.1039/B512527H

[182] Kaucher M. S., Harrell W. A., Davis J. T. A unimolecular G-quadruplex that functions as a synthetic transmembrane Na+ transporter. Journal of the American Chemical Society. 2006;128(1):38–39. DOI: 10.1021/ja056888e

[183] Arnal-Herault C., Pasc A., Michau M., Cot D., Petit E., Barboiu M. Functional G-quartet macroscopic membrane films. Angewandte Chemie International Edition. 2007;46(44):8409–8413. DOI: 10.1002/anie.200702605

[184] Ma L., Melegari M., Colombini M., Davis J. T. Large and stable transmembrane pores from guanosine–bile acid conjugates. Journal of the American Chemical Society. 2008;130(10):2938–2939. DOI: 10.1021/ja7110702

[185] Ma L., Harrell W. A., Davis J. T. Stabilizing guanosine-sterol ion channels with a carbamate to urea modification in the linker. Organic Letters. 2009;11(7):1599–1602. DOI: 10.1021/ol9002477

[186] Shi X., Fettinger J. C., Davis J. T. Homochiral G-quadruplexes with Ba2+ but not with K+: The cation programs enantiomeric self-recognition. Journal of the American Chemical Society. 2001;123(27):6738–6739. DOI: 10.1021/ja004330v

[187] Sreenivasachary N., Lehn J.-M. Gelation-driven component selection in the generation of constitutional dynamic hydrogels based on guanine-quartet formation. Proceedings of the National Academy of Sciences of the United States of America. 2005;102(17):5938–5943. DOI: 10.1073/pnas.0501663102

[188] Kotch F. W., Sidorov V., Lam Y.-F., Kayser K. J., Li H., Kaucher M. S., Davis J. T. Water-mediated association provides an ion pair receptor. Journal of the American Chemical Society. 2003;125(49):15140–15150. DOI: 10.1021/ja0364165

[189] Sidorov V., Kotch F. W., El-Kouedi M., Davis J. T. Toward artificial ion channels: Self-assembled nanotubes from calix[4]arene–guanosine conjugates. Chemical Communications. 2000;(23):2369–2370. DOI: 10.1039/B007332F

[190] Huttunen-Hennelly H. E. K., Sherman J. C. The design, synthesis, and characterization of the first cavitand-based de novo hetero-template-assembled synthetic proteins (Hetero-TASPs). Organic & Biomolecular Chemistry. 2007;5(22):3637–3650. DOI: 10.1039/B711869D

[191] Nikan M., Patrick B. O., Sherman J. C. Crystal structure of a template-assembled synthetic G-quadruplex. ChemBioChem. 2012;13(10):1413–1415. DOI: 10.1002/cbic.201200262

[192] Nikan M., Bare G. A. L., Sherman J. C. Synthesis of a water-soluble triazole-linked cavitand–guanosine conjugate. Tetrahedron Letters. 2011;52(15):1791–1793. DOI: 10.1016/j.tetlet.2011.02.017

[193] Bare G. A. L., Liu B., Sherman J. C.. Synthesis of a single G-quartet platform in water. Journal of the American Chemical Society. 2013;135(32):11985–11989. DOI: 10.1021/ja405100z

[194] Murat P., Cressend D., Spinelli N., Van der Heyden A., Labbe P., Dumy P., Defrancq E. A novel conformationally constrained parallel G quadruplex. ChemBioChem. 2008;9(16):2588–2591. DOI: 10.1002/cbic.200800457

[195] Bonnet R., Murat P., Spinelli N., Defrancq E. Click–click chemistry on a peptidic scaffold for easy access to tetrameric DNA structures. Chemical Communications. 2012;48:5992–5994. DOI: 10.1039/C2CC32010J

[196] Bonnet R., Lavergne T., Gennaro B., Spinelli N., Defrancq E. Construction of anti-parallel G-quadruplexes through sequential templated click. Chemical Communications. 2015;51(23):4850–4853. DOI: 10.1039/C4CC09744K

[197] Stefan L., Duret D., Spinelli N., Defrancq E., Monchaud D. Closer to nature: An ATP-driven bioinspired catalytic oxidation process. Chemical Communications. 2013;49(15):1500–1502. DOI: 10.1039/C2CC38317A

[198] Hermann P., Kotek J., Kubicek V., Lukes I. Gadolinium(III) complexes as MRI contrast agents: Ligand design and properties of the complexes. Dalton Transactions. 2008;23:3027–3047. DOI: 10.1039/B719704G

[199] Mewis R. E., Archibald S. J. Biomedical applications of macrocyclic ligand complexes. Coordination Chemistry Reviews. 2010;254(15–16):1686–1712. DOI: 10.1016/j.ccr.2010.02.025

[200] Laguerre A., Hukezalie K., Winckler P., Katranji F., Chanteloup G., Pirrotta M., Perrier-Cornet J.-M., Wong J. M. Y., Monchaud D. Visualization of RNA-quadruplexes in live cells. Journal of the American Chemical Society. 2015;137(26):8521–8525. DOI: 10.1021/jacs.5b03413

[201] Xu H.-J., Stefan L., Haudecoeur R., Vuong S., Richard P., Denat F., Barbe J.-M., Gros C. P., Monchaud D. Porphyrin-templated synthetic G-quartet (PorphySQ): A second prototype of G-quartet-based G-quadruplex ligand. Organic & Biomolecular Chemistry. 2012;10(27):5212–5218. DOI: 10.1039/C2OB25601K

[202] Laguerre A., Desbois N., Stefan L., Richard P., Gors C. P., Monchaud D. Porphyrin-based design of bioinspired multitarget quadruplex ligands. ChemMedChem. 2014;9(9):2035–2039. DOI: 10.1002/cmdc.201300526

[203] Nakayama S., Kelsey I., Wang J., Roelofs K., Stefane B., Luo Y., Lee V. T., Sintim H. O. Thiazole orange-induced c-di-GMP quadruplex formation facilitates a simple fluorescent detection of this ubiquitous biofilm regulating molecule. Journal of the American Chemical Society. 2011;133(13):4856–4864. DOI: 10.1021/ja1091062

[204] Nakayama S., Kelsey I., Wang J., Sintim H. O. c-di-GMP can form remarkably stable G-quadruplexes at physiological conditions in the presence of some planar intercalators. Chemical Communications. 2011;47(16):4766–4768. DOI: 10.1039/C0CC05432A

[205] Stefan L., Xu H.-J., Gros C. P., Denat F., Monchaud D. Harnessing nature's insights: Synthetic small molecules with peroxidase-mimicking DNAzyme properties. Chemistry - A European Journal. 2011;17(39):10857–10862. DOI: 10.1002/chem.201101337

[206] Haudecoeur R., Stefan L., Monchaud D. Multitasking water-soluble synthetic G-quartets: From preferential RNA–quadruplex interaction to biocatalytic activity. Chemistry - A European Journal. 2013;19(38):12739–12747. DOI: 10.1002/chem.201300791

[207] Huppert J. L., Balasubramanian S. Prevalence of quadruplexes in the human genome. Nucleic Acids Research. 2005;33(9):2908–2916. DOI: 10.1093/nar/gki609

[208] Todd A. K., Johnston M., Neidle S. Highly prevalent putative quadruplex sequence motifs in human DNA. Nucleic Acids Research. 2005;33(9):2901–2907. DOI: 10.1093/nar/gki553

[209] Ghosh A., Bansal M. A glossary of DNA structures from A to Z. Acta Crystallographica Section D: Biological Crystallography. 2003;59(4):620–626. DOI: 10.1107/S0907444903003251

[210] De Cian A., Lacroix L., Douarre C., Temime-Smaali N., Trentesaux C., Riou J.-F., Mergny J.-L. Targeting telomeres and telomerase. Biochimie. 2008;90(1):131–155. DOI: 10.1016/j.biochi.2007.07.011

[211] Szostak J. W. DNA ends: Just the beginning (Nobel Lecture). Angewandte Chemie International Edition. 2010;49(41):7386–7404. DOI: 10.1002/anie.201000635

[212] Blackburn E. H. Telomeres and telomerase: The means to the end (Nobel Lecture). Angewandte Chemie International Edition. 2010;49(41):7405–7421. DOI: 10.1002/anie.201002387

[213] Greider C. W. Telomerase discovery: The excitement of putting together pieces of the puzzle (Nobel Lecture). Angewandte Chemie International Edition. 2010;49(41):7422–7439. DOI: 10.1002/anie.201002408

[214] Kim N. W., Piatyszek M. A., Prowse K. R., Harley C. B., West M. D., Ho P. L., et al. Specific association of human telomerase activity with immortal cells and cancer. Science. 1994;266(5193):2011–2015. DOI: 10.1126/science.7605428

[215] Phan A. T. Human telomeric G-quadruplex: Structures of DNA and RNA sequences. FEBS Journal. 2010;277(5):1107–1117. DOI: 10.1111/j.1742-4658.2009.07464.x

[216] Haider S., Parkinson G. N., Neidle S. Molecular dynamics and principal components analysis of human telomeric quadruplex multimers. Biophysical Journal. 2008;95(1): 296–311. DOI: 10.1529/biophysj.107.120501

[217] Petraccone L., Garbett N. C., Chaires J. B., Trent J. O. An integrated molecular dynamics (MD) and experimental study of higher order human telomeric quadruplexes. Biopolymers. 2010;93(6):533–548. DOI: 10.1002/bip.21392

[218] Xu Y., Ishizuka T., Kurabayashi K., Komiyama M. Consecutive formation of G-quadruplexes in human telomeric-overhang DNA: A protective capping structure for telomere ends. Angewandte Chemie International Edition. 2009;48(42):7833–7836. DOI: 10.1002/anie.200903858

[219] Martadinata H., Heddi B., Lim K. W., Phan A. T. Structure of long human telomeric RNA (TERRA): G-quadruplexes formed by four and eight UUAGGG repeats are stable building blocks. Biochemistry. 2011;50(29):6455–6461. DOI: 10.1021/bi200569f

[220] Yu H.-Q., Miyoshi D., Sugimoto N. Characterization of structure and stability of long telomeric DNA G-quadruplexes. Journal of the American Chemical Society. 2006;128(48):15461–15468. DOI: 10.1021/ja064536h

[221] Shinohara K.-I., Sannohe Y., Kaieda S., Tanaka K.-I., Osuga H., Tahara H., et al. A chiral wedge molecule inhibits telomerase activity. Journal of the American Chemical Society. 2010;132(11):3778–3782. DOI: 10.1021/ja908897j

[222] Cummaro A., Fotticchia I., Franceschin M., Giancola C., Petraccone L. Binding properties of human telomeric quadruplex multimers: A new route for drug design. Biochimie. 2011;93(9):1392–1400. DOI 10.1016/j.biochi.2011.04.005

[223] Labat G., Seris J.-L., Meunier B. Oxidative degradation of aromatic pollutants by chemical models of ligninase based on porphyrin complexes. Angewandte Chemie International Edition in English. 1990;29(12):1471–1473. DOI: 10.1002/anie.199014711

[224] Lente G., Espenson J. H. A kinetic study of the early steps in the oxidation of chlorophenols by hydrogen peroxide catalyzed by a water-soluble iron(III) porphyrin. New Journal of Chemistry. 2004;28(7):847–852. DOI: 10.1039/B400482E

[225] Lente G., Espenson J. H. Unusual kinetic role of a water-soluble iron(III) porphyrin catalyst in the oxidation of 2,4,6-trichlorophenol by hydrogen peroxide. International Journal of Chemical Kinetics. 2004;36(8):449–455. DOI: 10.1002/kin.20018